# Routledge Handbook of Trends and Issues in Tourism Sustainability, Planning and Development, Management, and Technology

The Handbook offers a comprehensive overview of theoretical and practical perspectives for tracking and interpreting trends and issues in tourism sustainability, planning and development, management, and technology.

Tourism is a dynamic and unpredictable industry and understanding its trends and issues is critical for the successful and sustainable development of the private and public sector. As such, this Handbook proposes clear definitions and provides a systematic classification scheme for such analysis. It reviews trends and issues in four thematic areas of tourism – sustainability; planning and development; management; and technology – and includes contributions from 83 leading tourism scholars from across the globe. The Handbook provides insights on the differences between domestic, outbound, and inbound markets and acknowledges that the supply sub-sectors of tourism are diverse, highlighting variations by geographic regions.

The book emphasises the necessity of prioritising sustainability and the achievement of the UN's Sustainable Development Goals. Students and professionals interested in tourism, hospitality, and sustainability will find a wealth of multidisciplinary knowledge in this Handbook.

**Alastair M. Morrison** is Research Professor at the University of Greenwich in London, UK and was Associate Dean and Distinguished Professor Emeritus at Purdue University, USA specialising in the area of tourism and hospitality marketing in the School of Hospitality and Tourism Management.

**Dimitrios Buhalis** is Director of the eTourism Lab and Deputy Director of the International Centre for Tourism and Hospitality Research at Bournemouth University Business School in England.

# Routledge Handbook of Trends and Issues in Tourism Sustainability, Planning and Development, Management, and Technology

**Edited by Alastair M. Morrison and Dimitrios Buhalis**

Routledge
Taylor & Francis Group

LONDON AND NEW YORK

Designed cover image: © Getty Images

First published 2024
by Routledge
4 Park Square, Milton Park, Abingdon, Oxon OX14 4RN

and by Routledge
605 Third Avenue, New York, NY 10158

*Routledge is an imprint of the Taylor & Francis Group, an informa business*

*British Library Cataloguing-in-Publication Data*
A catalogue record for this book is available from the British Library

ISBN: 978-1-032-27197-2 (hbk)
ISBN: 978-1-032-27198-9 (pbk)
ISBN: 978-1-003-29176-3 (ebk)

DOI: 10.4324/9781003291763

Typeset in Times New Roman
by SPi Technologies India Pvt Ltd (Straive)

To Sheng Hua, Andy, and Alick
To Maria and Stella

# Contents

# Figures

# Tables

# Editors

**Alastair M. Morrison** is Research Professor at the University of Greenwich in London, UK and was Associate Dean and Distinguished Professor Emeritus at Purdue University, USA, specialising in the area of Tourism and Hospitality Marketing in the School of Hospitality and Tourism Management. Professor Morrison is ranked in the top 2% of scientists in the world based on the science-wide author database developed by Elsevier and Stanford University in 2019. He has published several books and around 350 academic articles and conference proceedings, as well as over 50 research monographs related to marketing and tourism. He is Co-Editor-in-Chief of the *International Journal of Tourism Cities* and Fellow of the International Academy for the Study of Tourism. Professor Morrison has served as President of the International Tourism Studies Association, Chairman of the Travel & Tourism Research Association (TTRA) – Canada Chapter, Board Member of the CenStates TTRA Chapter, Vice President of the International Society of Travel and Tourism Educators, and Chairman of the Association of Travel Marketing Executives.

**Dimitrios Buhalis** is Director of the eTourism Lab and Deputy Director of the International Centre for Tourism and Hospitality Research at Bournemouth University Business School in England. He is a Strategic Management and Marketing expert specialising in Information Communication Technology applications in the Tourism, Travel, Hospitality and Leisure industries. He is the Editor-in-Chief of Tourism Review and the Encyclopedia of Tourism Management and Marketing. Professor Buhalis has written and co-edited more than 25 books and 300 scientific articles and is recognised as a Highly Cited Researcher by Clarivate™ with more than 61,000 citations and an h-index of 105 on Google Scholar. He is ranked in the top 2% of scientists in the world based on the science-wide author database developed by Elsevier and Stanford University since 2019. Professor Buhalis is a past President of the International Federation for Information Technologies in Travel and Tourism (IFITT) and Vice President of the International Academy for the Study of Tourism.

# Contributors

**Olayinka Afolabi** is an Information Technology specialist and a smart tourism destination researcher at the Eastern Mediterranean University, Cyprus. Olayinka has professional degrees in Management Information Systems and Information Technology, and a PhD in Tourism Management with publications on research on smart tourism destination and mobile application location-based services privacy concerns and risk. Olayinka is currently a data engineer, with an interest in big data analysis and cloud computing.

**Vera Antunes** is a PhD student in Communication Sciences at the University of Beira Interior, Covilhã, Portugal. Her thesis is titled "The Importance of Strategic Communication in the Projection of Portugal as a Thermal Tourism Destination" and is funded by the FCT. She is a researcher at LabCom and a member of different associations in communications, including SOPCOM, ECREA, and IAMCR. Her research focuses on thermal tourism and strategic communication. She has published her results in prestigious scientific journals and presented at national and international conferences.

**Wolfgang Georg Arlt** is the founder and CEO of the Hamburg-based China Outbound Tourism Research Institute and one of the leading experts on the biggest global outbound tourism source market – China. He has been connected to China and tourism development for more than 40 years. His first visit to Mainland China took place in 1978 and he worked as a tour operator until 2001, when he became a full-time professor for International Tourism Management. Prof. Dr. Arlt has acted as organiser, keynote speaker, or chair for conferences around the world and has been quoted in all major global media. He is also the author or editor of several books about Chinese outbound tourism and a Fellow of the Royal Geographical Society and the Royal Asiatic Society.

**Maria Criselda Badilla** is Associate Professor at the University of the Philippines, Diliman. She is currently Director for Academic Affairs of the Asian Institute of Tourism. She is a marketing specialist for the preparation of Tourism Master Plans for local government units in the Philippines. Her research interests are tourism marketing, destination branding, image formation, e-tourism, and education tourism.

**Ahmad Bahar** is a lecturer and researcher in marine tourism and marine ecotourism at the Faculty of Marine Science and Fisheries, Hasanuddin University in Indonesia. He is an editor for marine tourism at the *Sapa Laut* journal at Haluoleo University and an editor for the *Lutjanus* journal at the Pangkep Agricultural Polytechnic. He teaches Master's and Doctoral programmes at Hasanuddin University, including a Master's in Tourism Management. He also gives public lectures and keynotes at the Makassar Tourism Polytechnic. He wrote the book *Marine Tourism Village on Hoga Island Wakatobi*. Together with the WWF-Indonesia team, he compiled the book *Observing and Interacting with Marine Animals and Diving-Snorkeling-Recreational Fishing-Jetski-Parasailing*.

**Adela Balderas-Cejudo** is Lecturer at ESIC University and the Basque Culinary Center, Associate Professor and Researcher at Deusto Business School, University of Deusto, Spain and Research Fellow at the Oxford Institute of Population Ageing, University of Oxford, UK. She has an Executive MBA, a Master's in Marketing, and a Master's in Coaching and is currently a member of the Humanism in Management and Economics research team. She is Visiting Professor at national and international universities (Ecole hôtelière de Lausanne, Switzerland; Cornell and George Washington University, USA; Regensburg, Germany; Xiamen, China). Her research has mainly focused on senior tourism and leadership.

**Josune Baniandrés** is a professor in the area of People Management at Deusto Business School, University of Deusto, Spain. She holds the positions of Vice Dean for Faculty and Director of the Department of Management at this institution. She is currently a member of the Humanism in Management and Economics research team. Her research has mainly focused on organisational entrepreneurship, servant leadership, and gender mainstreaming in business management. She is currently researching human resource management and impostor syndrome.

**Almudena Barrientos-Báez** is a doctoral assistant at the Complutense University of Madrid, Spain. She holds a doctorate with International Mention Cum Laude in Education. She was Director of the Qualifying Master's Degree in Teacher Training at the European University of Madrid. She has contributed to the Master's in Management of Protocol, Production, Organization, and Design of Events – Communication Area – University of Camilo José Cela; the Master's in Management of Tourist Accommodation at the University of Girona; and the Degree in Tourism (EUTI-ULL) and Teaching at the University of Valencia. Her research work is linked directly and also transversely to Neuromarketing, Neurocommunication, Gender, Use of Social Networks, Communication, PPRR, and ICT. She is part of the Concilium Research Group of the Complutense University where Neuromarketing is one of the main study topics.

**Mobina Beheshti** is a faculty member at the Final International University, Department of Software Engineering, in Cyprus. She commenced her academic career in Information Technology at Eastern Mediterranean University (EMU), Cyprus. She continued her master's studies in the fields of Information Systems and Information and Communication Technology in Education at EMU. Mobina earned her PhD in Computer Education and Instructional Technology at Near East University, Cyprus. Mobina specialises in Educational Science and Instructional Technology applications. She has published a book and more than ten scientific articles indexed in educational fields (e.g. SSCI, Scopus, Google Scholar).

**Vanessa S. Bernauer** is a doctoral researcher and lecturer at the Chair of Human Resource Management at Helmut Schmidt University, University of the Federal Armed Forces Hamburg in Germany. She conducts research on equality, diversity, inclusion, and identity in organisations with a focus on service work in the luxury segment, alternative forms of organising in digitalised workplaces, and sexual orientation. Her work has been published in leading international journals such as *Human Resource Management Journal* and *Equality, Diversity and Inclusion* (EDI). As a guest editor she co-edits

Special Issues with (inter)national colleagues, e.g. in EDI. She has conceptualised and conducted courses and workshops for the Academy of Management.

**Marta Buenechea-Elberdin** is a professor in the area of People Management at Deusto Business School, University of Deusto, Spain. She is currently researching on resilient leadership, human resource management trends, and impostor syndrome. Her research has mainly focused on the impact of intellectual capital on innovation in companies and how this relationship can be modified by contingency variables. She has published articles in indexed journals recognised in her field.

**Neil Carr** is a professor in the Department of Tourism, University of Otago in New Zealand. His work is grounded in notions of welfare, wellbeing, and rights. He has explored these within the contexts of families, animals, and sex, utilising the lenses of leisure and tourism to do so. The brains behind all of this, his canine pals, are only stymied by their lack of opposable thumbs, which therefore gives them the ideal excuse to laze around for most of the day rather than being stuck to a keyboard.

**Carl Francis Castro** is Chairperson of the School of International Tourism and Hospitality Management at La Consolacion College Manila, Philippines. He has written and co-authored a reference book for Senior High in the Philippines, three book chapters in international publications, and five scholarly articles published in local journals. As a registered environmental planner and an advocate of sustainable development, Castro helps local government units in their comprehensive land use plans and local climate change action plans.

**Duygu Çelebi** is a full-time lecturer in the Department of Gastronomy and Culinary Arts at Yasar University, Turkey. She holds Bachelor's and Master's degrees in Tourism Management and a PhD in Business Administration from Yasar University. Her research interests include tourism marketing, destination marketing, social media marketing, special interest tourism, gastronomy, and also social entrepreneurship in the gastronomy field.

**Muhammet Necati Çelik** is Research Assistant and PhD Candidate in the Department of Tourism Management at Alanya Alaaddin Keykubat University, Turkey. He completed his Master's thesis on the sustainability of accommodation enterprises in 2019 and he is in the last year of his doctoral degree. He is developing several scientific publications and collaborations on sustainable tourism, destination management, stakeholder management, hospitality management, and overtourism issues.

**Evrim Çeltek** has been working as Associate Professor at the Recreation Management Department, Tokat Gaziosmanpaşa University, located in Tokat, Turkey. She obtained her Master of Science in tourism business administration from Sakarya University (Turkey) and she obtained her PhD in tourism and hotel management from Anadolu University (Turkey). Her research focused on tourism marketing and smart technologies. She has published several articles, books, and book chapters about mobile marketing, e-commerce, mobile commerce, advergame, digital marketing, augmented reality, virtual reality, gamification, and electronic customer relationship management.

**Suk Ha Grace Chan** is Assistant Professor at the City University of Macao, Macao SAR, China. Her expertise is in tourism marketing and event management. Over the last few decades, she has worked across many tourism sectors and has accumulated valuable industry experience. She published SSCI papers related to hospitality human resources management and sustainability. Her research direction focuses on tourism marketing and human resources management.

**Jihon Choe** is a former Master's student with the White Lodging-J.W. Marriott, Jr. School of Hospitality & Tourism Management at Purdue University in West Lafayette, Indiana.

**Prokopis Christou** is Assistant Professor in Tourism at the Cyprus University of Technology. He is the author of the books *The History and Evolution of Tourism* and *Philosophies of Hospitality and Tourism: Giving and Receiving*. He specialises in qualitative research, as well as various socio-philosophical experiential issues related to tourism and hospitality, such as the notion of wellbeing. His work has been published in prestigious academic journals, such as *Annals of Tourism Research* and *Tourism Management*.

**Jonathon Day** is Associate Professor in Purdue's School of Hospitality and Tourism Management in the USA. He is committed to ensuring tourism is a force for good in the world. In addition to over 50 academic articles and chapters, he is the author of *Introduction to Sustainable Tourism and Responsible Travel* and the co-author of *The Tourism System*, 8th edition. Dr Day's research interests focus on sustainable tourism, responsible travel, and strategic destination governance within the tourism system. He is the lead researcher and director of the Sustainable Tourism and Responsible Travel Lab and the chair of the Travel Care Code Initiative.

**Kaitano Dube** is one of the leading tourism geographers and an Ecotourism Management Lecturer at Vaal University of Technology in South Africa. He is a National Research Foundation Rated Researcher. He serves as Faculty of Human Science Research, Innovation and Commercialisation Professor. He is Associate Editor for *Cogent Social Sciences* and *Frontiers in Sustainable Tourism*, and is also Editorial Board Member for *Tourism Geographies*. Dube has published highly cited work in the areas of tourism, aviation climate change, sustainability, and COVID-19.

**Cristina Estevão** is Assistant Professor at the University of Beira Interior, in the city of Covilhã, Portugal. She holds a PhD in Management and completed post-doctoral studies in Tourism. She is a researcher at the NECE – Research Center in Business Sciences and her research areas are Management, Competitiveness, Marketing, and Strategy. She has published two books, book chapters, as well as several scientific articles at the national and international level.

**Ekaterina Glebova** is pursuing her academic interests in research at the intersection of sports and technological transformation at the University Paris Saclay in France. She has over ten years of international experience in marketing, consultancy, and business development. She has published numerous book chapters and articles in peer-reviewed journals (e.g. *Journal of Sport Management and Marketing, Frontiers in Psychology, Physical Culture and Sport: Studies and Research*). Ekaterina holds a few visiting faculty positions, including at the Hungarian University of Sports Science and the EDHEC Business School in Nice.

**Ronald E. Goldsmith** joined the marketing faculty at Florida State University in 1981 and retired as Professor Emeritus in 2016. He received his PhD in Marketing from the University of Alabama at Tuscaloosa in the USA. His research interests include the areas of consumer behaviour, marketing research, services marketing, and strategic marketing.

He won a Teaching Incentive Program award for outstanding teaching in 1995. Goldsmith published approximately 300 journal articles and conference papers as well as the book *Consumer Psychology for Marketing*. He was for many years the North American Editor of the *Service Industries Journal*.

**Gisela Gonçalves** (PhD in Communication Sciences) is a professor at the University of Beira Interior, in Covilhã, Portugal. She is the coordinator of the Master's in Strategic Communication, Advertising and Public Relations, and a researcher at LabCom, where she manages the publisher LabCom Books. She is the author of numerous publications on strategic communications and public relation ethics, and has published her results in prestigious scientific journals and books. She co-edited the *Routledge Handbook of Nonprofit Communication* (2022). She is the vice president of SOPCOM, the Portuguese Association of Communication Sciences.

**Norhazliza Halim** is Senior Lecturer at the Faculty of Built Environment and Surveying, University of Technology Malaysia (UTM) and graduated with a PhD in Tourism Management from the University of Tasmania, Australia. Norhazliza is also the Head of the Tourism Planning Research Group and UTM Living Lab Urban Planning. She is a research expert in Sustainable Tourism Development, Ecotourism, Rural and Community Tourism, Smart Tourism, and Tourism Demand with more than 15 years of experience in teaching, research, and consultancy. She also has ten years of experience as a tourism consultant in Special Area Development, Policy Review, Strategic Planning, and Regional Development Planning, and has published around 50 academic articles and conference proceedings.

**Yue Yvonne He** is based at the Faculty of International Tourism and Management at City University of Macao, Macao SAR, China. She majored in International Tourism and Hospitality Management. She published many papers during her Master's degree and her research spans many fields, such as Marketing, Customer Satisfaction, Buying Behaviour, Brand Image, Event Management, and Corporate Social Responsibility. Yvonne has a strong interest in the perspective of the female market and female buying behaviour which hasn't been well discussed in previous hospitality research. Her research direction also focuses on the female perspective and extends to servicescape, female buying behaviour, and female perception.

**Sotiris Hji-Avgoustis** is Professor at the Ball State University's Miller College of Business in the USA. He joined the university in 2013 after a 22-year academic career at Indiana University, Indianapolis, where his leadership responsibilities included chairing the Department of Tourism, Conventions, and Event Management (nine years). At Ball State University, he chaired the Department of Family and Consumer Sciences (four years) and the AACSB-accredited Department of Management (three years). His research interests are in city tourism with an emphasis on cultural tourism, sports tourism, meetings, incentives, conferences, and expositions (MICE), and social entrepreneurship in Black-majority urban neighbourhoods.

**Hanh My Thi Huynh** is Lecturer at the Faculty of Business Administration, the University of Danang – University of Economics in Vietnam. She obtained a PhD at Grenoble Alpes University, France. Her research interest focuses on Human Resource Management, Organisational Behaviour, and Cross-cultural Management. She has published in scientific journals, including *Current Issues in Tourism* and *Tourism Management Perspectives*.

**Chutong Jiang** is a former undergraduate student at the White Lodging-J.W. Marriott, Jr. School of Hospitality & Tourism Management at Purdue University. In spring of 2019 he was an undergraduate research student at the Visitor Harassment Research Unit at Purdue in West Lafayette, Indiana.

**Daisy Kanagasapapathy** is a tourism expert currently attached to the Royal Melbourne Institute of Technology ( RMIT) Vietnam. She completed her PhD in Tourism at Bournemouth University, UK. Her PhD examined the tourist flow experience at Maritime Greenwich, where she collaborated with English Heritage, Historic England, Greenwich Tourism, and UNESCO. She has more than ten years of industry experience serving the National Tourist Board. She has organised high-profile events for academic, professional, and governmental audiences involving the Prime Minister's Department. She has also worked closely with major industry players such as the UN World Tourism Organization, the Pacific Asia Travel Association, the International Congress and Convention Association, and the International Association of Professional Congress Organisers. Her research focuses on heritage experience, destination marketing, sustainability, and sports tourism.

**Yeung (William) Kong** is Research Director in the office of the Hon Kenneth Leung, The Legislative Council of the Hong Kong SAR, China. His research interests are policy analysis and implementation, regional development, and tourism and public policy.

**Marine L'Hostis** is a lecturer at the tourism faculty of the University of Angers (ESTHUA) in France and is a member of the research unit "Espaces et Sociétés". In 2020, she defended her geography PhD on the spread of Chinese tourism in France and the influence of spatial representations and spatial capital. In addition to the spatial dimension of Chinese international tourism, she studies its social dimension and the processes through which Chinese society has started developing tourist practices. Her interests also include the diplomatic and political implications of Chinese outbound tourism.

**Yen E. Lam-González** is Research Associate at TiDES – the University Institute for Tourism and Sustainable Economic Development and the UNESCO Chair of Tourism and Sustainable Economic Development at the University of Las Palmas de Gran Canaria in Spain. She has published around 80 books, book chapters, academic articles, and conference proceedings. She has participated in more than 20 cooperative, research, and innovation projects funded by the World Bank, the United Nations, and the European Union in the economy, environment, and tourism fields.

**Huu Nghia Le** is Research Lecturer at the School of Tourism, UEH College of Business, UEH University, in Vietnam. He specialises in the area of tourism geography and travel marketing and management. He has wide experience in the tourism industry. His research interest has focused on sustainable tourism development, destination branding, and tourism for inclusive growth.

**George W. Leeson** is Professorial Fellow at the Oxford Institute of Population Ageing, and Senior Research Fellow & Dean of Degrees, Kellogg College, University of Oxford. His first degree was in Mathematics, followed by a Master's in Applied Statistics, from Oxford. His doctoral work was in Demography. He directs the Institute's research networks in Latin America (LARNA) and in Central and Eastern Europe (EAST) and also the Centre for Migration and Ageing Populations (MAP Centre). Dr Leeson has directed the Danish Longitudinal Future Study, which elucidates the attitudes and aspirations of future generations of older people in Denmark, and he is Principal Investigator with Professor Sarah Harper on the Global Ageing Study, a survey of 44,000 men and women aged 40–80 in 24 countries.

**Carmelo J. León** is Director of ECOMAS – Research Group of Economy, Environment, Sustainability and Tourism at the University of Las Palmas de Gran Canaria in Spain. He is the Director of the UNESCO Chair in Tourism and Sustainable Economic Development. In the past he held the position of Director at TiDES. He has been the lead researcher of more than 20 national undertakings of European and international research, cooperation, and innovation projects in the economy, environment, and tourism fields. He has spoken at numerous international conferences and written more than 200 scientific articles, books, and book chapters. He has been cited more than 1,300 times (Clarivate™) and holds an h-index of 34 (Google Scholar).

**Javier de León** works at the University of Las Palmas de Gran Canaria in Spain. He is Co-Director of the UNESCO Chair in Tourism and Sustainable Economic Development and the Institute of Tourism and Sustainable Economic Development (TiDES). He held the position of Cooperation Director at the Vice-rectorate of Cooperation and Internationalisation. He has participated in more than 20 national, European, and international projects in the economy, environment, and tourism fields. He has written over 100 scientific articles, books, and book chapters. He holds an h-index of 18 on Google Scholar.

**Jing Li** is Associate Professor at Jinan University. Her research interests are the relationship between tourists' ICT (non)usage and their positive psychology, environmentally responsible tourist behaviours, and value co-creation in tourism development. She studies tourist experience from the lens of positive psychology, particularly focusing on rewarding outcomes from mindful ICT (non)usage. Her perspectives on tourists' social media involvement can be presented as social media playing important roles in decision-making while minimisation of unnecessary engagement en route is beneficial.

**Tomasz Napierała** works at the Faculty of Geographical Sciences, University of Lodz Poland. He cooperates with the CiTUR Centre for Tourism Research, Development and Innovation in Portugal. His scientific interests are focused on spatial aspects of competition in the lodging industry, mainly the spatial volatility of prices and its impact on competitiveness. He is experienced in leading international projects granted by the Erasmus+ Programme and European Economic Area and Norwegian grants. Since 2020 Tomasz has proudly held the position of Secretary General of the Polish Geographical Society.

**Thi Hong Hai Nguyen** is Associate Professor at the School of Management & Marketing, Greenwich Business School, University of Greenwich, in London. She holds a PhD in Tourism Management from the School of Hotel and Tourism Management at the Hong Kong Polytechnic University. Her current research focuses on heritage tourism, tourist behaviour, special events and festivals, and destination management.

**Annmarie Nicely** is Associate Professor at Purdue University and Lead Researcher at the Visitor Harassment Research Unit at Purdue in Indiana. Her area of expertise is behavioural issues in hospitality business and tourism communities. She is one of the leading scholars in visitor harassment.

**Neha Nimble** is an intersectional feminist and a social science researcher. She is working as Senior Research Manager at Ashoka University's Centre for Social Impact and Philanthropy. She has previously worked with the Tata Institute of Social Sciences, Mumbai where she also undertook her PhD on Gendered and Intersectional Constructions within Tourism Livelihoods. Overall, her areas of interest and expertise include gender, intersectionality, livelihoods, community-based tourism, human trafficking, and social exclusion. Other than being invited to speak at numerous international conferences, panels, and seminars, she has published a number of related academic papers in national and international publications.

**Nabila Norizan** is currently a PhD candidate in Urban and Regional Planning at the University of Technology Malaysia. She is also a researcher for the Tourism Planning Research Group, specialising in tourism management, tourism planning and policy, and sustainable urban planning research. Nabila is also a player in the hospitality industry with her last role in revenue management; in her work, she applies collaborative design, design thinking, and big/open data to invite new perspectives within tourism innovation.

**Adesola Osinaike** is Senior Lecturer and Acting Director for Tourism, Hospitality and Events at Canterbury Christ Church University, UK, where she oversees undergraduate and postgraduate courses. Her research interests are Hospitality Organisational Strategies and Behaviour, Revenue Management, Performance Measurements, Organisational Culture, Sustainable Business Practices, and Management of Small and Medium-sized Enterprises. She has presented her research at international conferences and contributed to book chapters and journals. She acts as a reviewer for articles for international journal publishers.

**Lemonia (Lenia) Papadopoulou-Kelidou** is Lecturer at CITY College University of York Europe Campus in Thessaloniki Greece. She is a UNWTO Online Academy instructor for the course "Negotiations in Tourism". She holds a PhD in Negotiations in Tourism from the University of the Aegean, Greece; an MSc in Economics and Finance from the University of Southampton, UK; a BA in Economics from the University of East Anglia, UK; and a Master's Diploma in Negotiations from the Economic University of Athens, Greece. Her research interests are focused on negotiations in hospitality, destinations, yachting, and aviation. She is an experienced Negotiator in International Relations and a Consultant and Trainer in Business and Tourism Negotiations.

**Claire Papaix** is a transport economist specialised in urban mobility and wellbeing appraisal. Her teaching activities and key research interests in Montpellier Business School include sustainable transitions, transport and health, movement, and space. Previously, she was Programme Leader and Senior Lecturer in Transport and Business Logistics at the University of Greenwich, supervising PhD studies on transport and wellbeing analysis, before undertaking a research sabbatical to acquire knowledge in geo-spatial analysis. She has experience in small business creation and events management having co-founded a start-up in the sectors of sport and social business. She is also a certified yoga teacher.

**Andreas Papatheodorou** is Professor in Industrial and Spatial Economics with Emphasis on Tourism at the Department of Tourism Economics and Management, University of the Aegean, Greece, where he also directs the MSc Programme in Strategic Management of Tourism Destinations and Hospitality Enterprises. An Oxford University MPhil

(Economics) and DPhil (Economic Geography) holder, he has published extensively in the areas of air transport and tourism and been involved in a large number of research and consulting projects. He is Editor-in-Chief of the *Journal of Air Transport Studies* and an Associate Editor of *Annals of Tourism Research*. Andreas is President of the Hellenic Aviation Society and participates in the Panel of Experts of the United Nations World Tourism Organization.

**Jinah Park** is Research Assistant Professor at the Hospitality and Tourism Research Centre, School of Hotel and Tourism Management, The Hong Kong Polytechnic University, Hong Kong SAR, China. Her research interests are tourist behaviour, mobility, and destination marketing. She has a solid publication record in top-tier tourism and hospitality academic journals such as *Annals of Tourism Research*, *Tourism Management*, the *Journal of Destination Marketing & Management*, the *Journal of Hospitality & Tourism Research*, and the *Journal of China Tourism Research*.

**Eduardo Parra-López** is Professor of Business Organization in the Digital Economy and Tourism at the University of La Laguna, Santa Cruz de Tenerife, in Spain. He is Director of the Chair of Sustainable Territories, Socio-economic Development, and Tourism at La Laguna University. He is Co-editor of the *Journal of Destination Marketing & Management* and Editor of *Spanish Abstracts for Tourism Review*. His experience in projects and institutions has focused mainly on tourism. He has been a Visiting Professor at: the University of Strathclyde (Scotland); IMI International Centre for Tourism (Switzerland, Lucerne); Universidad del Valle (Guatemala); and Universidad Católica La Paz (Bolivia). He also contributes as Visiting Professor at the Universities of Vigo, Málaga, Oviedo, Almería, Lleida, and Valencia in Spain for their Master's programmes in Tourism.

**Ian Patterson** PhD is Visiting Professor at the Silk Road International University of Tourism and Cultural Heritage in Samarkand, Uzbekistan. Previously he was Associate Professor in the School of Business (Tourism) at the University of Queensland (2001–2015) and Griffith University (1991–2000) in Australia. His research interests include senior tourism, tourism behaviour, health and wellness tourism, sustainable tourism, and qualitative analysis. He has produced over 100 publications in textbooks, book chapters, and peer-reviewed journal papers. He was Co-Editor, *Annals of Leisure Research*, 2004–2011 and received Life Membership of the Australian and New Zealand Association for Leisure Studies in 2015. He was awarded the Tourism Medal of Uzbekistan in 2022.

**Adam Pawlicz** is a scientist with 20 years of research experience in the hospitality and e-tourism area. He is associate professor at Szczecin University, Poland and a visiting professor at Klaipeda University, Lithuania and Vidzeme University, Latvia. He has co-authored over 100 publications and is a member of the scientific committee of various international journals such as *SWU "Neofit Rilski"*, Bulgaria. His current research focuses on the impact of sharing economy and online travel agencies in the hospitality market.

**Marko Perić** is Full Professor and Vice Dean for International Affairs at the University of Rijeka, Faculty of Tourism and Hospitality Management (Croatia). His research and teaching activities have concentrated in the area of strategic management in tourism and hospitality, project management, and sports management. His work has been published in prestigious international journals such as the *International Journal of Contemporary Hospitality Management*, *The Service Industries Journal*, *Small Business Economics: An*

*Entrepreneurship Journal, Tourism Management Perspectives*, the *Journal of Sport & Tourism*, and *Event Management*. He is Co-Editor-in-Chief of the *Tourism and Hospitality Management* journal.

**Ige Pırnar** currently works as Head of Department of Business Administration, Yasar University. She studied for her MBA at Bilkent University (1989) and her PhD at Ankara University in Business Administration (1998). She won a scholarships for a Certificate on "Tourism Education in Universities", UNWTO, New Delhi, 1990, and a teaching assistantship scholarship, Bilkent University, 1987–1989. Pırnar has eight books in Turkish and one in English on the topics of *International Business: Key Concepts, International Tourism Management, Convention and Meetings Management", Total Quality Management in Tourism, Direct Marketing, Public Relations in Tourism, International Services Marketing, Food & Beverage Management*, and *Quality Management in Services*; she has also produced many articles and conference proceedings. Her areas of expertise are tourism management, international tourism, global marketing, international management, and entrepreneurship.

**Sima Rahimizhian** studied for a BSc in Information Technology at Eastern Mediterranean University (EMU) in North Cyprus. She continued her Master's and doctoral studies in tourism management at EMU. She is currently a senior instructor at EMU's School of Computing and Technology. Sima is a researcher and expert, with publications and citations in fields such as e-tourism, information technology in management, innovation, and information system management.

**Maria Rigou** obtained her MSc in International Tourism and Hospitality Management from Cyprus University of Technology. She has attended and delivered various management and workplace seminars. For several years, she has been working in the hospitality industry in managerial positions. Her research interests involve employee/workplace relations and employee wellbeing. Her research output has been presented in prestigious academic conferences.

**Farzad Safaeimanesh** is Faculty Member and Coordinator of the School of Tourism & Culinary Arts at Final International University in Kyrenia, North Cyprus. He is Co-Editor of the *European Journal of Tourism & Hospitality Management*. His research interests include Information and Communication Technologies in Tourism, Service Quality, Consumer Behaviour, Choice Modelling, and Quantitative & Qualitative Research Methodologies.

**María de los Ángeles Pérez Sánchez** is a PhD candidate at the School of Management, Tourism Department, Zhejiang University Hangzhou, Zhejiang Province, China. She is a lecturer at the Tourism School of Zhejiang University City College, Hangzhou and the Zhejiang International University. Her research focuses on aspects of hotel branding, social psychology of leisure, tourism, and new technologies and consumer behaviour.

**Alexis Saveriades** is Assistant Professor of Tourism Policy, Planning and Development at the Cyprus University of Technology. He heads the Academy of Professions in Hospitality and Tourism at the Cyprus University of Technology, and serves as a member of the University's Senate. Alexis enjoys extensive teaching experience at the undergraduate and postgraduate level. He is actively engaged in policy formulation and strategic planning at the regional and national level, as well as in the management of tourism destinations, destination governance, and meta-governance. He has authored a book entitled *Tourism Policy*.

**Muruvvet Deniz Sezer** is a research assistant and Ph.D. student in the Business Administration Department in Yasar University, Turkey. She received her BSc and MSc degrees from Eskisehir Osmangazi and Dokuz Eylul University, Department of Industrial Engineering. She worked as a visiting researcher at the University of Groningen, the Netherlands from October 2022 to March 2023. She has published works in various esteemed journals. She is involved in the research projects "Towards Net-zero Emissions-hydrogen Solutions in Airports, University of Groningen, Netherlands" and "British Council–Newton Fund Research Environment Links Turkey/U.K.: Circular and Industry 4.0 Driven Sustainable Solutions for Reducing Food Waste in Supply Chains in Turkey".

**Shweta Singh** is a recent graduate of the White Lodging-J.W. Marriott, Jr. School of Hospitality & Tourism Management at Purdue University in West Lafayette, Indiana. Her areas of expertise include marketing research, consumer behaviour, and tourism marketing. Between 2017 and 2018 she was a graduate research assistant with the Visitor Harassment Research Unit at Purdue.

**Haiyan Song** is Mr and Mrs Chan Chak Fu Professor in International Tourism and Deputy Dean Research in the School of Hotel and Tourism Management at The Hong Kong Polytechnic University, Hong Kong SAR, China. His research interests are in the areas of tourism demand analysis, service quality management, tourism supply chain management, and wine economics. He has published widely in such journals as *Annals of Tourism Research*, *Tourism Management*, and the *Journal of Travel Research*. Professor Song is Editor-in Chief of the *Journal of China Tourism Research* and Associate Editor of *Annals of Tourism Research*.

**Diep Ngoc Su** is Lecturer at the Faculty of Tourism, The University of Danang – University of Economics in Vietnam. She holds a PhD degree from Swinburne University of Technology, Australia. Her research interests are tourist behaviour, travel behaviour, and sustainable consumption behaviour. She has over 25 publications in high-ranked journals such as *Current Issues in Tourism*, the *International Journal of Contemporary Hospitality Management*, the *Journal of Hospitality Marketing and Management*, *Tourism Management Perspectives*, and the *Journal of Cleaner Production and Transportation Research Part A: Policy and Practices*.

**Chaitanya Suárez-Rojas** is Research Associate at the University Institute for Tourism and Sustainable Economic Development, University of Las Palmas de Gran Canaria in Spain. She is also Research Visitor at the Centre for Environmental and Resource Economics, Umeå University in Sweden. She has participated in several research, cooperation, and innovation projects and at international conferences in the economy, environment, and tourism fields. She has published around 20 scientific articles, book chapters, and conference proceedings.

**Nellie (Magdalena Petronella) Swart** is Associate Professor in Tourism at the University of South Africa, and a Certified Meeting Professional. She has authored and co-authored scientific articles, books, book chapters, and conference proceedings, and serves on the editorial board of the *International Journal of Tourism Cities*. The capacitating of tourism communities and women through community projects is a priority. Nellie is the Chair of the Tourism Educators South Africa and supports the South African government and tourism stakeholders on various committees. In 2022 she was recognised as one of the 100 Most Powerful People in Africa Hospitality and is a G100 City Chair.

**Binglin Martin Tang** is a PhD student at City University of Macao, Macao SAR, China. His major study is in International Tourism Management and he focuses on smart tourism, cultural heritage tourism, and big data analysis. He has published papers in the *International Journal of Social Science Research*, analysing the hospitality and tourism sectors.

**Kailasam Thirumaran** works at James Cook University in Singapore. He received his PhD from the National University of Singapore. He has worked in the tourism industry for over ten years, in Singapore and the United States. At James Cook University Singapore, he specialises in business and tourism management subjects. He coined the term "affinity tourism" which refers to the propensity of guests to partake in the "familiar" and "similar" cultural experiences of their hosts. His research focuses on service excellence, cultural, entrepreneurship, and luxury tourism. He welcomes collaboration with scholars and PhD candidates in luxury and heritage tourism.

**Thinaranjeney Thirumoorthi** is Senior Lecturer in the Department of Management, Faculty of Business and Economics, University of Malaya. She started her career as a lecturer in 2016 and currently teaching Principles of Marketing, Strategic Marketing, Global Marketing, and Hospitality Management for undergraduate and postgraduate programmes, as well as supervising PhD and Master's students. Her main research areas are backpacking tourism, medical tourism, gastronomic tourism, scuba diving tourism, and community-based tourism.

**Lorna Thomas** is Principal Lecturer and the Director of Partnerships at the Christ Church Business School, Canterbury Christ Church University, in the UK. She has been an academic and researcher for over 25 years. She has been a Course Director for many of those years, mostly for undergraduate and postgraduate tourism courses. Her research interests include areas such as service delivery, queue management, cultural heritage, ethics, and sustainability. Lorna has been involved with collaborative partners, which includes setting up franchise and validated courses internationally and in the UK.

**Jens Thraenhart** is CEO of Barbados Tourism Marketing Inc. Jens was appointed Chief Executive Officer of Barbados Tourism Marketing Inc. (BTMI), the official national tourism organisation of the island nation of Barbados in the Caribbean. BTMI is responsible for promoting Barbados as a sustainable tourism destination and the ultimate Caribbean vacation island by implementing marketing strategies that help the responsible development of tourism. Jens completed his doctorate at the Hong Kong Polytechnic University.

**Rodoula H. Tsiotsou** is Professor of Services Marketing and Director of the Marketing Laboratory MARLAB at the Department of Business Administration, University of Macedonia, Greece. She received her PhD from Florida State University, in the USA. She is Associate Editor for the *Journal of Services Marketing* and the *International Journal of Consumer Studies*. She has produced three books and more than 100 scientific publications in a variety of international scientific journals and conference proceedings. She is ranked in the Top 2% of Scientists in the World based on the Science-wide Author Database developed by Elsevier and Stanford University.

**Paul Tully** is a PhD candidate at the University of Otago in New Zealand. He holds a bachelor's in Tourism Management from the University of Central Lancashire, UK, and an MSc in Sociology and Social Research from Newcastle University, UK. His research interests revolve around the critical study of leisure with a current focus on human–animal relationships. Recent works include projects on donkeys, farm animals, and cats in human leisure environments.

**Cinà van Zyl** is Professor in Tourism Management at the University of South Africa. She previously chaired the Department of Applied Management. Her research interest is in the leisure and business tourism market, as well as the festival and sport events sector. Cinà is a member of the Editorial Board for the *Journal of Transport and Supply Chain Management* as well as Theme Editor of the *International Journal of Tourism Cities*. She is author and co-author of journal publications, has contributed to several books and read papers at national and international conferences.

**Alan Yen** is Associate Professor and Molinaro Fellow at the Ball State University's Miller College of Business, in Indiana, USA. His primary teaching areas are lodging and customer service. His research interests include hotel marketing and human resource management issues in the hospitality industry. He currently serves as the co-editor of the *International Journal of Hospitality & Tourism Administration*.

**Xu Zhao** is a transport engineer at Beijing Transport Institute, China. He was awarded his PhD at the University of Greenwich, UK. During his PhD studies, his research mainly focused on sustainable commuting behaviour and wellbeing. He used qualitative and quantitative research methods, such as thematic analysis, structural equation modelling, and the logit model to analyse travel behaviour and transport mode choice. Currently, Xu is focusing on intelligent transportation systems, Mobility as a Service (MaaS), and carbon trading in sustainable mobility. He has also worked widely with policymakers and business groups domestically and internationally.

**Lina Zhong** is Professor at Beijing International Studies University in Beijing, China, and is Executive Dean of the Institute for Big Data Research in Culture and Tourism. She has published around 15 books and 80 academic articles in the area of tourism and hospitality. Professor Zhong is a reviewer of the *International Journal of Tourism Research*, the *Asia Pacific Journal of Tourism Research*, *Industrial Management and Data Systems*, and the *Journal of Hospitality Marketing and Management*.

**Lixian Zhou**, a Master's student in Tourism Management at Jinan University, is interested in studying tourists' perception of Information Communication Technology applications and destination marketing strategies based on social media. Her methodological priority is designing experiments to manifest causal relationships.

**Yufang Zhou** is an expert in the area of transportation carbon trading at Beijing Transport Institute, China. She is also the director's assistant of the Beijing Key Laboratory of Transport Energy Conservation and Emission Reduction. She gained her Master's degree at Columbia University in the USA. She led her team to work on carbon neutral, carbon peak, and carbon trading areas. She worked on developing Beijing Mobility-as-a-Service and carbon trading, and helped to achieve the low-carbon transformation of urban transport.

**Dan Zhu** is a doctoral student at The Ohio State University. During spring of 2018 she was a graduate research aide with the Visitor Harassment Research Unit at Purdue in West Lafayette, Indiana.

**Mengyao Zhu** is a graduate student at Beijing International Studies University in China and a research assistant at the Institute for Big Data Research in Culture and Tourism. Her research interests are smart tourism, destination management, and cultural tourism. She has published several academic articles related to technology applications to tourism and big data research in tourism and hospitality.

# Introduction

*Alastair M. Morrison and Dimitrios Buhalis*

Editors: Alastair M. Morrison and Dimitrios Buhalis

Routledge Handbook of Trends and Issues in Global Tourism, publication in 2023

- The main themes and objectives of the Handbook are to:
  - 1. Highlight the importance of tracking and interpreting global trends and issues in tourism.
  - 2. Propose clear definitions and a systematic classification scheme for trends and issues.
  - 3. Identify and describe the major trends and issues affecting tourism.
  - 4. Review the interactions among trends and issues, and the results of their combined impacts.
  - 5. Delineate theoretical and practical approaches for tracking and interpreting trends and issues.
  - 6. Provide a comprehensive academic and practitioner reference source on trends and issues in global tourism.

*Figure I.1* Editors, themes, and objectives of book.

Photo courtesy of Unsplash.com, Katerina Kerdi.

## Rationale for the Handbook

The main aim of the *Routledge Handbook of Trends and Issues in Global Tourism* is to provide a comprehensive and considered text on trends and issues within tourism worldwide. The following quote expresses the difficulties that exist in finding systematic and credible information on trends and issues:

> Pinpointing trends and issues in international tourism is challenging in itself as there are no books or comprehensive documentation sources for these data. The proverbial "searching for a needle in a haystack" is a good metaphor for anyone trying to gather this information in order to look ahead. Also, the changes impacting tourism are so dynamic and often unpredictable that searching for trends and issues can be an endless and certainly is an ongoing task.
>
> (Morrison, 2021, p. 219)

This situation was exacerbated by the COVID-19 pandemic that began in early 2020 and its effects continued through to 2022 and 2023. In fact, a concoction of trends and issues were already in play when the global public health crisis arrived; however, scholars were tending to treat these separately without duly considering their interrelationships and combined effects. Thus, the main reason for writing the *Routledge Handbook on Trends and Issues in Global Tourism* was to raise this theme from a peripheral and ad hoc topic to a mainstream subject in tourism and hospitality education and research, and in so doing address a gap in the market of related books.

## Handbook volumes

There are two volumes of the Handbook: the *Routledge Handbook on Trends and Issues in Global Tourism* and the *Routledge Handbook of Trends and Issues in Tourism Sustainability, Planning and Development, Management, and Technology*. The first volume covers the major trends and issues in global tourism; the second reviews the trends and issues in four thematic areas of tourism. There are 78 chapters in total contributed by 147 authors from more than 30 countries.

## Themes and objectives

The main themes and objectives of the Handbook are to:

1. Highlight the importance of tracking and interpreting global trends and issues in tourism;
2. Propose clear definitions and a systematic classification scheme for trends and issues;
3. Identify and describe the major trends and issues affecting tourism;
4. Review the interactions among trends and issues, and the results of their combined impacts;
5. Delineate theoretical and practical approaches for tracking and interpreting trends and issues;
6. Provide a comprehensive academic and practitioner reference source on trends and issues in global tourism.

**Structure**

There is no single book on the market that comprehensively deals with global trends and issues in tourism, and that is the main differentiator for the *Routledge Handbook of Trends and Issues in Global Tourism*. Notwithstanding this, the book uses a systematic classification of trends and issues that is not found elsewhere, while having a set of contributing authors from among the leading tourism scholars in the world and thereby solidifying its credibility and authority.

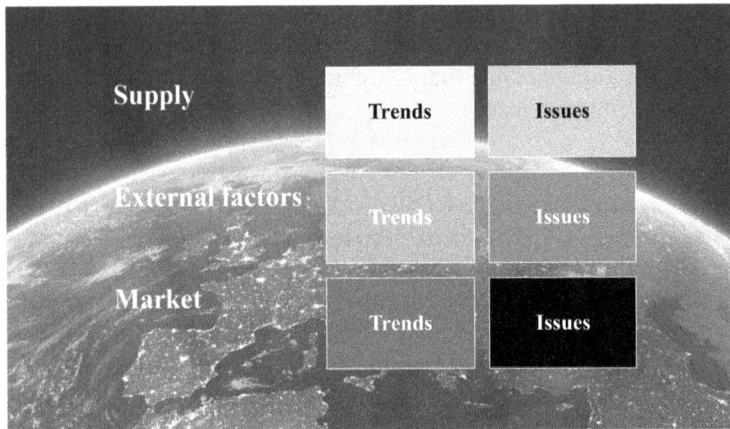

*Figure I.2* Classification scheme for trends and issues.

Photo courtesy of Microsoft 365.

This innovative Handbook structure is based on the proposed classification scheme. This avoids the "laundry list" approaches by acknowledging the fundamental sources of trends and issues. The Handbook accepts that not all tourism markets are the same and that they need separate attention with respect to trends and issues. Therefore, the contents reflect major trip-purpose market segments including leisure/pleasure, business/MICE, and VFR (visiting friends and relatives), while also recognising differences among domestic, outbound, and inbound markets. It is acknowledged that the supply sub-sectors of tourism are diverse and tend to be dissimilarly influenced by specific trends and issues, and these differences are reflected in the text as are variations by geographic regions.

The basic structure of the contents of the two volumes is as follows:

**Routledge handbook of trends and issues in global tourism**

Part I: Supply-side trends:
- Transportation;
- Attractions, culture, and heritage tourism;
- Technology;
- Policies and issues;
- Destination management.

Part II: External factor trends.

Part III: Market-led trends.

**Routledge handbook of trends and issues in tourism sustainability, planning and development, management, and technology**

- Sustainability;
- Planning and development;
- Management;
- Technology.

*Areas of emphasis*

The contributing authors were asked to focus on certain emphasis areas. These included:

- Major trends and issues in the topic field up to 2021–2022, including the work of key contributors.
- Stakeholders and their interest and involvement with the topic field.
- Predictions for 2032 to 2042 to 2052.
- Opportunities and challenges with these trends and/or issues: Who wins and who loses?
- Do they contribute to the Sustainable Development goals and how so?
- What are the catalysts?
- Stakeholder, business, and policy implications: Who needs to do what, and when?

**Reference**

Morrison, A. M. (2021). Reflections on trends and issues in global tourism. In P. U. C. Dieke, B. King, and R. Sharpley (eds.). *Tourism in development: Reflective essays*, pp. 218–232. Wallingford. UK: CABI Publishing.

# Part I

# Sustainability

**Wolfgang Georg Arlt** writes on *Beyond sustainability: The meaningful tourism paradigm* in Chapter 1. This chapter introduces a new paradigm for the future development of global travel and tourism: Meaningful tourism. Based on the methods and insights of positive psychology and the spin-offs of positive tourism and positive sustainability as well as the advanced versions of experience economy, meaningful tourism claims to find answers to a number of shortcomings in the discussion about sustainable tourism, responsible tourism, regenerative tourism, green-growth tourism, de-growth tourism, steady-state tourism, circular economy tourism, and other such approaches. Firstly, meaningful tourism starts not from the question of what to reduce, give up, forego, or at least feel ashamed about in our tourism practices. It rather concentrates on forms of tourism which ensure benefits and satisfaction for all stakeholders involved. Secondly, meaningful tourism does not consider the different stakeholders to be in competition with each other, with the need to balance their interests or to treat the interests of one of them as more important than those of the others. Stakeholders are neither competitors nor enemies, their interests do not have to be contradictory.

Chapter 2, authored by **Jonathon Day**, is on *Emerging themes in sustainable tourism*. Sustainable tourism differs from other 'tourisms'. It is not a market segment or a product type, like adventure or business travel; rather it is a deliberate approach to implementing tourism in a way that contributes to socio-environmental sustainability. Sustainability, much like quality, is an approach to conducting tourism – an operating system for tourism. Although sustainable tourism has been one of the dominant themes in tourism research over the past several decades, there remains much to learn about the intersection of sustainability and tourism. This chapter examines the trends and issues facing sustainable tourism in the coming years.

**Ekaterina Glebova** and **Marko Perić** describe the *Tourism Great Reset: The inclusive, sustainable, and innovative reality* in Chapter 3. The Great Reset is the name of the 50th annual meeting of the World Economic Forum (WEF), where participants discussed sustainably rebuilding society and the economy following the COVID-19 pandemic. It is based on three main pillars: (1) creating conditions for a stakeholder economy; (2) building 'resilient, equitable, and sustainable' infrastructure projects, in the spirit of a green economy; (3) 'harness the innovations of the Fourth Industrial Revolution"'. This vision is being gradually disseminated and applied to all fields of life and industries; tourism is not an exception. In this chapter, the authors discourse on the intersection of today's tourism industry and the Great Reset concept in a holistic manner, intending to give a 'helicopter view' on the researched phenomena through a prism of qualitative analysis. Consequently,

DOI: 10.4324/9781003291763-1

they identify and outline current tourism trends and future perspectives through the prism of ongoing socio-economic metamorphoses and the Great Reset values. It involves reflections about tourism's new normal based on a literature review and experts' commentary analysis, followed by real examples and cases. To this end, it allows us to understand, explain, describe, and visualise the process and direction of modern transformations in the tourism industry, identifying the directions for future research and development.

*A dialogue for tourism, climate change and philanthropy* is the topic of Chapter 4 by **Kaitano Dube** and **Cinà van Zyl**. Aiming to open up dialogue on climate change, tourism, and philanthropy, the authors seek to explore the role of tourism, climate change, and philanthropy by addressing some challenges that host communities face in view of the climate change backlash. A systematic review addresses the gap in the literature. Tourism, climate change, and philanthropy are a growing area that will likely expand as climate change awareness increases amongst tourism stakeholders. Most significant efforts in tourism, climate change, and philanthropy aim to address SDG 13 on climate change action and SDG 12 on sustainable and responsible consumption. To date, action against climate change has been implemented through the adoption of market-based measures that allow tourism role players to participate in the carbon market and the industry players who adopt green technology. Future research is needed on tourism philanthropy to aim at fostering climate resilience amongst host communities as an integral part of responsible tourism.

**Ian Patterson** and **Adela Balderas-Cejudo** discuss *Baby Boomers and sustainable tourism: The need for a new research agenda* in Chapter 5. Sustainable tourism has become an increasingly popular field of research because it provides a balance between the three dimensions of tourism development – the environmental, economic, and socio-cultural. Sustainable tourists are generally opposed to traditional mass tourism or, as it is termed today, overtourism. The aim of this chapter is to present a systematic review of the current literature to determine whether older tourists are undertaking new travel trends post-COVID-19 that specifically promote sustainable tourism practices. One of the older age segments termed Baby Boomers (born between 1946 and 1964) have a real and enduring love for the outdoors and soft adventure tourism and are strong supporters of sustainable tourism. Sustainable tourism involves practices that minimise the negative impact of visitors on the natural environment, while respecting local cultures. This chapter also predicts some of the future trends of sustainable tourism such as the need to escape the stress of city life, increasing preference for green hotels, and a greater demand for experiential tourism.

Chapter 6 poses the question *Sustainable tourist: How big is your footprint?* and is authored by **Daisy Kanagasapapathy**. This chapter explores tourists' sustainable consumption behaviour by trying to understand how tourists adopt a more sustainable tourism lifestyle. Specifically, the chapter provides an overview of Malaysian tourism and unsustainable consumption behaviours, addressing the following: How sustainable are you when you travel?

**Chaitanya Suárez-Rojas, Carmelo J. León, Javier de León**, and **Yen E. Lam-González** present *Whale-watching tourism: Future sustainability trends* in Chapter 7. In whale-watching tourism, ensuring responsible human–cetacean interactions has raised critical academic debate over recent decades. This chapter reviews the empirical evidence with the aim of progressing towards new trends that yield managerial responses for reconciling whale-watching tourism with sustainability principles. A co-word analysis of the last 30 years of scientific literature has been conducted to explore the evolution of the leading topics, relate these to some industry milestones, and identify the research and managerial gaps. The evidence urges a new socio-ecological relationship approach in whale watching for sustainability, to be achieved by 2030 and beyond, by reorienting management practices to a more integrative approach based on scientific breakthroughs and collaborative stakeholder networks.

*Fair pricing in tourism: From profitability towards sustainability* is reviewed by **Tomasz Napierała** and **Adam Pawlicz** in Chapter 8. Fair pricing used to refer to the problem of the buyer's judgement of a seller's price decision. Nowadays, when sustainability becomes a vital and fundamental consideration of tourism development, fair pricing should be discussed in a much wider context. The ethical context of fair pricing in tourism should be emphasised, as fair pricing is recognised as one of the transparent and ethical practices of tourism enterprises. Fair pricing is a significant element of corporate social responsibility and supports other goals like beneficial partnerships of tourism enterprises with local communities and suppliers, providing equal opportunities and remuneration for the employees of tourism enterprises, as well as increasing the sustainable treatment of scarce natural resources. Fair pricing relates not only to the judgement of tourists (price versus value), but also to the opinion of the employees of tourism enterprises (price related to salaries), suppliers (price of finished product related to prices of raw materials and services), local communities (price versus income), and other stakeholders dependent on the tourism industry in different ways. Finally, fair pricing must solve the issue of so-called 'cheap nature'. This means that fair pricing relates to the valuation of ecosystem services, as tourism consumes environmental assets and significantly influences the cultural landscape. Fair pricing might contribute to the protection of bio- and geodiversity, the climate, and the cultural landscape. The aim of the chapter is to: (1) define fair pricing and discuss the evolutionary changes of the definition, (2) highlight the emerging importance of fair pricing in tourism, (3) review theoretical concepts, empirical results, and practical applications of fair pricing, and (4) discuss the contribution of fair pricing to achieving sustainable development goals for tourism.

Chapter 9 by **Ahmad Bahar** is on *Tourism sustainability is a big problem in the development of marine tourism in Indonesia*. In the last two decades, marine tourism has developed very rapidly in Indonesia. The number of tourist visits continues to increase in almost all such destinations. This is because the objects of marine tourism attraction are widespread in Indonesia and varied. For example, for manta ray diving, according to 2015 Manta Watch data, although there are more manta ray spots in the Maldives, the chances of encountering these protected biota are three times higher in Komodo National Park, West Manggarai Regency. To achieve the target of 20 million foreign tourists, the government has opened ten new marine tourism destinations. The growth of marine tourism causes damage to destinations, such as significant damage to coral reef ecosystems, the disappearance of mola fish (*Mola mola*) and whale sharks (*Rhincodon typus*) from destinations due to the booming tourist visits. The results of Bahar's research (2015) showed a decrease in hard coral cover from 58% (good category) to 20% (bad category) within 12 years in several dive spots located within the tourism zone in Wakatobi National Park. The contribution of the tourism sector to gross domestic product is only 10.4% with the contribution from the marine tourism sector only around 10%. Another problem is that some districts, such as Wakatobi Regency, which initially made the marine tourism sector a leading one for regional development, are now turning to the fisheries sector as their leading sector. One of the reasons is the low regional income from the marine tourism sector.

# 1 Beyond sustainability

## The meaningful tourism paradigm

*Wolfgang Georg Arlt*

### Introduction: The need for a new paradigm

In May 2022, Julian Kirchherr published a remarkable academic paper entitled "Bullshit in the Sustainability and Transitions Literature: A Provocation". He strongly criticises the trend of what he calls very frankly "scholarly bullshit" being published in large numbers using buzzwords like circular economy and sustainability to get high levels of citations regardless of the mediocre quality of the papers. He singles out the category of "activist rants" as the worst of such contributions, as "these articles frequently remain experiential instead of turning theoretical and/or empirical; they attempt to build their legitimacy through general, feel-good claims instead of substantive arguments" (Kirchherr, 2022).

In practical application, economic development and cost-saving measures have been criticised for not giving more attention than local residents wellbeing and nature preservation (Sorensen & Grindsted, 2021), with better quality experiences of the tourists or the situation of the employees seldom given attention in practical implications and academic discussions.

In 2015, 17 Sustainable Development Goals (SDGs) were defined by the United Nations. As a result, a whole industry developed around the discussion, measurement and evaluation of these SDGs. Unfortunately, as Biermann et al. (2022) and the Sustainable Development Goals Report 2022 (UN, 2022a) have shown, practically, for all goals, the situation has either not improved or even become worse. Government leaders and industry bosses can be found hiding behind SDG flags in their offices, SDG buttons on their lapels and SDG logos on their glossy pamphlets with little more than words to show.

The term "sustainability" itself has not only been misused but has also become ubiquitous in general discussion, often with the simple meaning of something existing for a longer period of time (Crane, 2013). The term also has very different connotations in different languages. In English, sustainability is an active process of keeping something alive. Sustainability here is a situation, which is favourable for the action of sustaining, with no explicit reference to time. The German term used for sustainability – *Nachhaltigkeit* – however, is based on the idea of a carefully crafted perfect system which once in place does not need any outside action anymore, and originated in the 18th century. In French, the temporal dimension is stressed in *durabilité* and concentrates on the fact that something can exist for a long time and is not easy to destroy. The Chinese term *Ke Chi Xu Xing* (可持续性) has a complex meaning as it combines "possibility", "existence" and "not stopping over a long period of time" with a multi-dimensional term. Sometimes *Naiyong* (耐用) is used instead, which concentrates on durability.

DOI: 10.4324/9781003291763-2

Discussions on sustainability are heavily biased towards a concentration on environmental questions. For example, the UNEP (2022) report with the title *Transforming Tourism in the Pan-European Region for a Resilient and Sustainable Post-COVID World* solely concentrates on the environment despite the broader approach proclaimed in the title, similar to other recent publications (Conely, 2022, European Travel Commission, 2022).

The SARS-CoV-2 pandemic has accelerated the search for a new approach to global tourism and provided the opportunity to rethink tourism development. As Taleb Rifai, the former Secretary General of UNWTO optimistically stated in 2011: "Tourism will not bounce back but will leap forward into a new world, a new normal: A better and more sustainable world" (Rifai, 2021).

The UNEP report, similar to many other sources, shows that demand for mainstream sustainable tourism is gaining momentum and that most travellers want to travel more sustainably in the future. However, many respondents have complained that they did not perceive any changes in the reuse and recycling of materials, waste management or increased energy efficiency in the tourism sector (UNEP, 2022).

The need for a paradigm shift in global tourism, going beyond environmental questions, existed long before the SARS-CoV-2 pandemic. The world has developed economically at an ever-growing speed over the last 30 years without a parallel growth of political institutions managing globalisation – the climate catastrophe, the rise of despotism and the concentration of wealth in ever fewer hands can be seen as the result.

Global tourism spending has even outpaced global GDP growth (UNWTO, 2021), yet likewise no regulatory body has emerged to define and enforce limits to the acquisition of public goods, such as beaches and city centres, to look after the carrying capacity of nature and host communities to support the quality of the experience of guests and hosts or to fight the elephant in the room of the tourism industry: seasonality and, connected to seasonality, the dominance of low pay and seasonal jobs.

The very success of international tourism, with five times the number of trips in 2019 compared to 1980 (COTRI Analytics, 2022), has made a mockery of the idea of hospitality and has run in many places a juggernaut over local nature, local culture, authenticity, diversity and serendipity, negatively impacting the satisfaction levels of all stakeholders involved. Gigantic cruise ships, all-inclusive resorts, overcrowded beaches and Disneyfied tourist cities are all examples of forms of tourism organisation which in effect prevent meaningful guest–host encounters and lead to price wars between providers of increasingly identical service offers.

Already in 2018 and 2019, a debate under the headline of "overtourism" had developed as a result of the growing resistance of host communities, signalling the need to change the structure of global tourism. Tourism started to change its image from being a provider of joy, jobs and peace to an image of being a force of destruction and pollution and a reason to develop "flight shame" (Chiambaretto et al., 2021). At the same time, the growing number of non-Western tourists often felt treated as second-class customers (Arlt, 2020).

During the pandemic, a plethora of discussions about the necessity of new tourism, adherence to SDGs, and the need for tourists to finally start to behave in a more sustainable and responsible way evolved (Bhuiyan et al., 2021, Persson-Fischer & Liu, 2021). The solutions discussed were varied but concentrated very often on reducing the number of trips and visitors through regulations or increased prices.

Higgins-Desbiolles et al. (2019) offered the idea of a community-centred tourism framework that "begins with the redefinition of tourism in order to place the rights of local communities above the rights of tourists for holidays and the rights of tourism corporates to make a profit" (p. 1936). Why locals should have higher levels of rights and how to deal with the many forms of travelling which are not for holidays remain, however, unanswered questions. UNESCO World Heritage Sites, for instance, are designated to support their preservation and the ability of everybody to visit them, and not to be considered the property of the current local community with which they may or may not be historically connected.

Pearson (2022), in the case of Sedona in Arizona, even mused about ceasing, after 20 years, to visit the place anymore so as to reduce by one visitor the large number of visitors who are disturbing nature and the solemn atmosphere; and Fahey (2022) decided for environmental reasons never again to use an aircraft despite being a travel journalist.

However, with international tourism restarting in spring 2022, most discussions and pledges for a better new normal were quickly forgotten and below-cost air tickets and overcrowded city centres appeared again with breath-taking speed. Return tickets from, for example, Hamburg to Palma de Mallorca were again on offer for less than €50 (Opodo, 2022). Citizens of Barcelona complained that Las Ramblas was fuller than ever with tourists (AFP, 2022). Even the cruise ships started spewing out 5,000 short-time visitors onto hapless cities and islands (Heritage Tribune, 2022).

The necessity to rethink tourism is also based on the insight that tourism goes beyond recreation. Only a minority of the people affluent enough to afford tourism, about one-fifth of humankind with respect to international tourism and about two-thirds with respect to domestic tourism (Arlt, 2022b), travel to rest their limbs and muscles. Instead, travel is done for increasingly diverse and often mixed purposes including, for example, digital nomads, combining online work and travel, visiting friends and relatives (VFR), visiting the ever-growing group of friends and family living in different places, as well as business and meetings, incentives, conferences, and exhibitions (MICE), which cannot be completely substituted by Zoom calls. Religion, culture, education, health, special interests, second homes and many more need to be added, creating a whole universe of unforced mobility motivations. The majority of people who can afford tourism, especially international tourism, spend their working time sitting in front of a computer or in meetings rather than sweating in a coal mine, a factory or on a field.

The wish to refresh the brain and the wish for self-actualisation by new experiences and new inspiration as well as gaining social capital had gained importance already before the COVID-19 pandemic. Binkhorst and Den Dekker (2009) stated that tourists are driven by a growing desire to engage in spontaneous, self-expressed and self-determined tourism experiences.

The sudden realisation of the fragility of life and the possibility of a shattering of the perceived stability of personal circumstances regardless of personal wealth has resulted in increased criticism of shallow consumerism and 3B (Beach, Beer, Boredom), 3S (Sea, Sun, Sex, or in the Chinese variation: Sightseeing, Shopping, Selfies) tourism, further increasing the demand for meaningful tourism beyond simple recreation and sightseeing. Meaningful tourism also takes into account the changes in the age structure of travellers, with international (leisure) travellers increasingly belonging to the age cohort of 50+ years. In Europe for example, EU residents aged 55+ accounted for 41% of all tourism nights for private purposes spent in 2019 (EUROSTAT, 2021). It also reflects the changes in the set-up of families travelling, which increasingly no longer follow the Mama, Papa plus two shared biological children pattern and the fact that about one out of ten travellers has some physical or mental disabilities. Meaningful tourism further acknowledges the fact that about one-third of all travellers in international tourism, hospitality and aviation no longer predominantly belong to North American or European cultural backgrounds (UNWTO, 2020). Therefore, it needs to be recognised that what is considered as meaningful is informed not only by personal tastes or previous experiences but also by the cultural background of the stakeholder (Wen et al., 2021).

The post-pandemic situation has also highlighted the increased difficulties of finding and retaining well-qualified personnel. Low levels of payment, unsatisfactory working conditions, seasonal employment and limited career chances had already resulted in problems finding sufficient numbers of staff before the pandemic. However, during the pandemic, many employees were forced to look for other employment and found in many cases work with similar levels of payment but much less demanding in terms of working hours and better social standing (Kwok, 2021). In debates about sustainability, employees are seldom part of the discussion. Meaningful tourism acknowledges that employees of service providers in tourism, hospitality and transportation are stakeholders on the same level as other stakeholders.

## Theoretical foundations of meaningful tourism

### *Positive psychology, positive sustainability and positive tourism*

To provide a positive solution for post-pandemic tourism development, the paradigm of meaningful tourism has been developed by me. It is based on the need of moving forward from the concepts of sustainable tourism and responsible tourism as discussed above. Important elements informing the paradigm are the concepts of positive psychology, positive tourism, positive sustainability and the experience economy as part of the transformation economy. Quality, benefits and satisfaction for all stakeholders are the key elements of meaningful tourism.

Sustainable and responsible tourism in most cases provides negative proposals for what to stop doing and for what to pay more for, without providing many benefits in return except a good conscience. The endless discussions of "if only everybody would behave more civilised, if only earning money would not be the main purpose of tourism business" are all failing to describe what the stakeholders would get in return for behavioural changes.

A major starting point for the meaningful tourism paradigm is the positive psychology approach. Positive psychology focuses on the thriving and flourishing of humans and the factors contributing to success, virtue and happiness (Seligman & Csikszentmihalyi, 2014). Wellbeing theory as part of positive psychology, according to Seligman (2012), identifies five dimensions that lead to a person's happiness and fulfilment, which include positive emotions, engagement, relationships, meaning and accomplishment, known as PERMA, which has been applied in different topics of tourism research.

The first dimension, positive emotions, refers to hedonic accounts of happiness that encompass different emotions ranging from excitement to calmness. Engagement is associated with the concept of flow, suggesting a mental state of full immersion. Relationships is based on the fact that humans as social beings pursue relationships with other humans as a major source of positive emotions. Meaning addresses the five dimensions of connectedness, coherence, purpose, resonance and significance, especially in transformative experiences. Finally, accomplishment relates to the experience of success increasing self-esteem

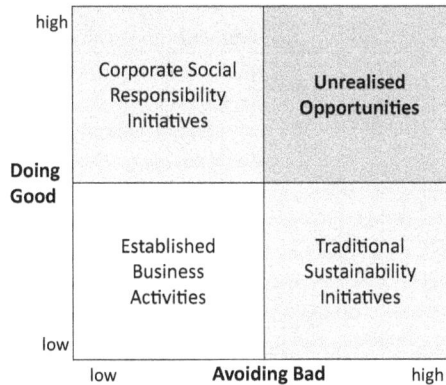

```
high
            Corporate Social           Unrealised
            Responsibility            Opportunities
            Initiatives

Doing
Good
            Established              Traditional
            Business                 Sustainability
            Activities               Initiatives

low
      low              Avoiding Bad          high
```

*Figure 1.1* Innovation opportunities in positive sustainability.
Author's design based on Lichtenthaler (2021).

(Neuhofer & Celuch, 2020). For the meaningful tourism paradigm, PERMA is used to identify the experience of quality, benefits and satisfaction.

The positive sustainability approach should not be confused with the trendy "NetPositive" term, which concentrates on putting more back into the environment than you take out, again concentrating mostly on the environment. Recently two authors (Kühnen et al., 2022, Lichtenthaler, 2021) have used the term "positive sustainability", though not related to tourism or other service industries. Lichtenthaler even created a new term "positainability" to describe the step of not only avoiding bad actions but promoting good actions. He points out that companies miss opportunities by leaving the upper-right corner of a portfolio empty, which would signal high levels of avoiding bad and doing good activities (Figure 1.1). This idea can be transformed for the approach of meaningful tourism by checking for each activity if it is creating good and avoiding bad for all stakeholders involved, instead of only trying to avoid bad results for one or only a few of all stakeholders.

In 2017, a collection of articles was published under the title of "positive tourism" (Filep et al., 2017). In the foreword, Philip L. Pearce points out that, since the beginning of the 2000s, as positive psychology developed to augment the study of individual problems, deficits and difficulties by concentrating on how to achieve greater happiness and subjective wellbeing, tourism psychologists like himself and Mihaly Csikszentmihaly have been instrumental in using these approaches to understand if and how tourism can enhance the life experience of those involved in it and travel. The pandemic and the untimely death of Philip L. Pearce have slowed the work in this field, but it is a major part of the foundation of the meaningful tourism paradigm.

In the literature discussing tourists' wellbeing, affective pleasure and inner feeling of self-growth in the process of satisfying various sensory needs and achieving travel goals (Filep & Laing, 2019; Laing & Frost 2017; Lengieza et al., 2019; Rahmani et al., 2018; Smith & Diekmann, 2017), the terms "hedonic" and "eudaimonic" are used. Tourist eudaimonia is defined as reflecting tourists' inner feelings of self-growth such as realising individual potential and self-fulfillment (Figure 1.2), while tourist hedonia is used to describe tourists' affective pleasure from satisfaction of sensory needs such as enjoying delicious foods or a beautiful scenery (Su et al., 2021). According to Csikszentmihaly and Coffey (2017),

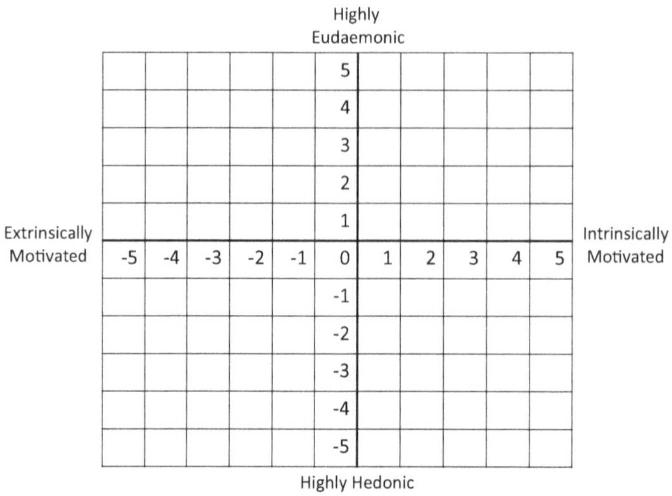

*Figure 1.2* Eudaemonic measurement grid.

a traveller can switch between the different forms of motivation and either hedonic or eudaemonic activities within one trip.

### Experience economy and transformative experiences

Pine and Gilmore provided in 1999 the foundation for understanding the paradigm shift from a service economy to an experience economy, wherein pleasurable, enjoyable and memorable experiences become dominant also in tourism. The notion of "today meaningful tourism experiences" incorporates the values of the three generations of the experience economy: staged experiences, co-creative experiences and transformative experiences. In the first step, designed, pleasurable and enjoyable experiences were discussed (Pine & Gilmore, 1999); in the second step personalised and extended interactions with other tourists and tourist stakeholders were added to the observations (Mehmetoglu & Engen, 2011); and in the current third step life-changing transformations are analysed (Kirillova et al., 2017; Soulard et al., 2019).

Pine and Gilmore's approach concentrated on four realms of experiences – education, entertainment, aesthetics and escapism – that manifest across two dimensions: customer participation and connection. This approach emphasises the role of tourism service providers in delivering staged experiences to tourists. However, the approach insufficiently reflects the self-creation, co-creation and shaping of experiences by the tourists themselves, as physical and virtual networking are increasingly responsible for driving meaningful interaction between different stakeholders (Chirakranont & Sakdiyakorn, 2022, Xu et al., 2016). Whereas in the so-called experience economy 2.0, tourists are seen to generate their own unique experience, in parallel with the shift from the experience economy towards the transformation economy (Pine & Gilmore, 2011; Kirillova et al., 2017), there are concepts of a meaningful tourism experience in the experience economy 3.0 as one that promotes tourists with an experience existential authenticity (Pung et al., 2020).

| Staged experience | Co-creative experience | Transformative experience |
|---|---|---|
| A commercially supplied holistic set of pleasurable and memorable experiences curated by tourism providers via four realms, namely, education, entertainment, aesthetics and escapism | A co-construction of tourism experiences and values through personalised and extended interactions between the tourists and tourism providers, as well as other tourism stakeholders | Tourism experiences that leads to tourists' transformation such as changes in one's self-understanding, revision of one's belief system, and alterations in one's behaviour and lifestyle |

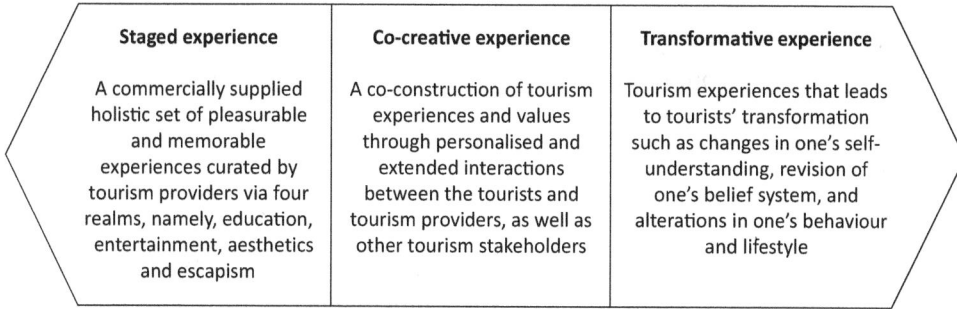

*Figure 1.3* Staged, co-creative and transformative experiences.

For the meaningful tourism paradigm, the different forms of experience provide important tools towards the analysis and understanding of the necessary adaptations of services, strategies and policies for all stakeholders concerned so as to provide the perceived quality, benefits and satisfaction according to the specific needs of each shareholder, including but not limited to travellers. However, unlike some participants in the academic debate about experience economy 3.0, meaningful tourism does not regard eudaemonic activities as superior to hedonic activities.

**Meaningful tourism**

From the very beginning of modern tourism in the middle of the 19th century, "tourists" has always been used as a term to distinguish between free-willed travellers and the sheep-like tourist. The root of the stigmatisation can be seen in the resentment felt towards the lower social classes as they catch up with those above them. According to D'Eramo (2021), the stages of the travellers growing contempt for the tourist correspond to the spread of leisure travel, from the aristocracy to the bourgeoisie in the 19th century, and from the bourgeoisie to the proletariat in the 20th century. For the 21st century, one may add, the spread from Western travellers to non-Western travellers is the next step.

The term "meaningful travel" has been used by different organisations in the form of volunteering vacations or as part of business-to-business digital training and resource tools (Bouskill & Corbeil, 2021). Airbnb declared in its annual report for 2021 that people will travel more meaningfully in 2021, defining it as travel that creates meaningful memories, time with loved ones, discovery and learning (Airbnb, 2021).

Several operators offer tours that "allow travellers to visit destinations with the knowledge that in doing so, they are supporting local people and businesses in honest work, they are deeply immersing themselves in other cultures, and they are experiencing a destination meaningfully" (González & Hubbard, 2022) or that help to get a "better and more meaningful connection to the people and land of this great country" (Mccreesh, 2022). Likewise, a website called Conscioustravel.com promotes tourism which "creates meaningful and sustainable livelihoods for those people and enterprises on which it depends" (Pollock, 2022).

With the exception of the last example, all the companies carefully avoid the term "tourism". However, as Eduardo Santander, the CEO of the European Travel Commission points out in the online Meaningful Tourism Training programme (Arlt, 2022a), the better new normal needs to cover all those going abroad, not just the elite few who see themselves as travellers and look down on the tourists.

The careful distinction between hating the deplorable tourist hordes while revelling in one's travelling is however not working anymore in the times of a climate catastrophe. "Some people now talk proudly of being 'flight-free' as they might about being vegan. An anti-travel movement is gaining momentum; in some circles, tales of far-off places have gone from badges of enlightenment to something like guilty secrets" (Robbins, 2021).

### The way to the meaningful tourism paradigm

With the growth of domestic and especially international tourism after World War II, concern about the economic, social and especially ecological consequences of mass tourism began to be discussed when Kaspar (1973) called for what he termed "environmental ecology" as a new dimension of the tourism debate. A decade later, Krippendorf (1984) argued in his publication *Ferienmenschen* (*Holiday People*) about a new form of tourism that (re) creates a harmony between nature and tourists.

The notion of sustainability was established by the Brundtland report, *Our Common Future* (World Commission on Environment and Development [WCED], 1987). This report triggered the development which led to the 1992 UN Conference on the Environment and Development in Rio de Janeiro and the UN World Summit on Sustainable Development 2002 in Johannesburg.

It also initiated the term "sustainability", whereby "sustainable development" was defined as "development that meets the needs of the present without compromising the ability of future generations to meet their own needs" (WCED, 1987).

Of the five main criteria for sustainable tourism, which were introduced by Inskeep (1991), namely economic, environmental and social responsibility, as well as the responsibility for visitor satisfaction and global justice and equity, only the first three aspects were taken up in the following extensive discussions, as the United Nations organisations, including the World Tourism Organization (UNWTO), supported a concept of sustainable tourism resting only on the three environmental, socio-cultural and economic pillars.

One result of the growing criticism of the negative effects of tourism was an enlarged concept of responsible tourism which was developed by Goodwin (2011) in his publication *Taking Responsibility for Tourism*. Responsible tourism recognises that consumers, suppliers and governments all have responsibilities to address the impacts of mainstream tourism. Sustainability is still the goal, but a goal which can only be achieved by people taking responsibility.

The last few years have seen new concepts entering the discourse, especially regenerative tourism and tourism as part of a circular economy (Sorensen & Grindsted, 2021), but none of these cover the interests of all stakeholders involved or are able to give practical advice as to how stakeholders could be convinced to see an advantage in such radical changes beyond the lofty feeling of "gratitude in our hearts, for the blessing of being able to travel once more" (Tollman, 2022). Furthermore, in most of the discussions, tourism implicitly concentrates on the discussion of the leisure tourism of holidays and rich Western travellers who are asked to reduce the number of trips after decades of travelling instead of newly entering the social strata affluent enough to visit other countries.

### The meaningful tourism paradigm

Half a century has passed since the publication of the Brundtland report and the start of the discussions about sustainable tourism. The period of the pandemic provided a radical example of how the world would look without international tourism and an opportunity to move forward in the development of new concepts.

Meaningful tourism as a new paradigm is a result of this experience, taking onboard all original five dimensions of sustainability, giving the aspects of global justice and equity more importance and introducing a more balanced and holistic set of criteria and key performance indicators (KPIs). The equal rights of guests and hosts are recognised but also the interests of employees working in tourism and hospitality service providers and of the companies themselves at a time when the shortage of staff and the negative economic effects of the pandemic are core issues. Not least are also to be seen the interests and responsibilities of governments on different levels and of the environment and by default future generations.

The meaningful tourism paradigm also reflects the changed demand. According to UNEP (2022), the percentage of travellers stating that the pandemic had made them want to travel in a more sustainable way in the future than before has risen globally to 61%, with more than 80% answering affirmative in Vietnam, India, Colombia and China. The lowest levels with an increase of less than 40% are reported from source markets which had already before the pandemic higher levels of awareness, namely Israel, Germany and the Netherlands

As the rebound of tourism in 2022 showed, it helps little to give schoolmasterly commands to tourists to put the interests of the host communities and the environment before their own interests. It also does not help to speculate on what percentage increase in prices customers would agree to pay for a greener tourism product, or by how much the wages in the industry have to be increased to bring employees back, if customers are not offered more than just a better conscience for their extra money and the employees are not offered more than money to increase their job satisfaction.

In most of the large number of studies and strategy papers for post-pandemic tourism development, two elephants in the room continue to be ignored: the one-size-fits-all approach of offering standard products by many service providers and many destinations, which results in pronounced levels of seasonality; and connected to that the dominance of low-pay, low-recognition, no-career and seasonal-only jobs resulting in the systemic difficulty of finding enough, let alone qualified, staff in all branches of the tourism and hospitality industry.

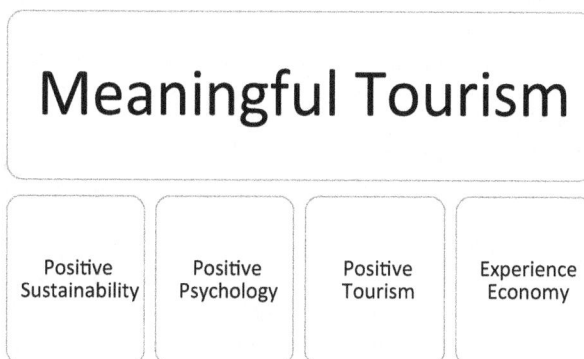

*Figure 1.4* The meaningful tourism paradigm.
Author's design.

Tourism source markets and the demands and interests of market segments are more divided than ever, 3S trips are replaced more and more by the search for experience and immersion, and looking at buildings becomes replaced by meeting interesting people. As has been pointed out above, the need for bodily relaxation is not the key purpose of leisure tourism anymore, and the dominance of younger customers originating in Western countries will very likely come to an end within this decade.

Offering similar products to huge numbers of customers is mostly in the interest of tour operators, for whom the term "niche" remains a dirty word. However, tour operators and especially mass-market package tours are losing importance within the tourism industry thanks to increasingly experienced and Internet-enabled individual travellers. In a recent survey, for instance, only 7% of Chinese identified travelling in package tour groups as their preference (Parulis-Cook & Wang, 2022). Airlines, hotels and restaurants do not mind much if travellers fly to and stay in a destination for a multitude of reasons as long as they turn into their customers. Using all parts of a country and all seasons of a year for touristic offers is possible almost anywhere and anytime if it is based on the identification of the right global market segment and the right product adaptation, which in many cases will include the participation of the local population.

The meaningful tourism paradigm is taking all these developments into consideration. It is based on a return to quality, benefits and satisfaction for all stakeholders involved, namely the guests, the host communities, the employees of service providers, the companies, governments and a livable environment for future generations, with quality and satisfaction measured by the stakeholders themselves.

In a nutshell, the positive effects and benefits for all stakeholders involved include:

- For guests/visitors to enjoy tourism services creating satisfaction based on the benefits of products which are more precisely adapted to their specific demands, going beyond relaxation and sightseeing towards new experiences.
- For host communities to benefit from instead of being encumbered by encounters with visitors.
- For the staff in companies providing tourism and hospitality services to benefit from better year-round working conditions, recognition as hosts instead of servants, full-time careers and meaningful work.
- For service providing companies to have a sustainable perspective, year-round business, motivated staff, higher margins and lower marketing costs thanks to online and offline recommendations by satisfied guests.
- For governments to obtain employment opportunities for their citizens, enjoy increased tax income and generate a more evenly spatial and temporal distribution of tourism and friendly international relations.
- For the environment and future generations, the chance of mitigated environmental damage, based on the feeling of embeddedness and belonging by all stakeholders.

Guests who are provided with exactly what they want, and even a bit of what they did not know they wanted, will turn into product ambassadors, offering free recommendation marketing instead of expensive and decreasingly efficient offline and social media marketing. Host communities will see the advantages of receiving visitors interested in interacting with them, and employees will value better pay and year-round jobs with the possibility of feeling like hosts again instead of servants. Companies will be able to ask for higher prices against perceived better quality, will be able to use their resources year-round and will have less problems in retaining and training their staff. Governments will receive more taxes and will be able to use tourism as a regional development tool. With a feeling of belonging in the sense of a kinship economy, guests and hosts alike can be expected to treat the natural environment with more care.

Experience shows that putting up signs of *"Verboten!"* and flygskam (flight-shaming) campaigns will not change the behaviour of the majority of tourists. Doing something good for the environment alone will not convince guests to pay substantially more money for the same service. Travelling has, unlike elitist views, made a comeback after the pandemic, as a human right, not a privilege, so pricing and/or taxing the bottom half out of the market is not an option either. Distinguishable benefits, which are aligned to the needs of all stakeholders involved, are necessary to transfer the meaningful tourism paradigm into practical action and to change the attitude and perception of all stakeholders involved.

## Outlook

The concept of positive sustainability as well as the paradigm of meaningful tourism will need much further research and debate. Both promise to be important tools in moving towards a positive approach to the future.

In the industry, a number of positive developments are recognisable. Already in 2016, the destination management organisation (DMO) for Copenhagen declared "the end of tourism" and the advent of "localhood" as a long-term vision that supports the inclusive co-creation of future destinations where human relations are the focal point. Locals and visitors would no longer just coexist but interact around shared experiences of localhood (Wonderful Copenhagen, 2016). The decision of Vancouver Island Tourism Board to morph into a "social enterprise created to ensure that travel is a force for good", called 4VI in 2022 (Arlt, 2022b), is another example, as is the insight of the president of the Greek Tourism Confederation SETE that "Happy residents bring happy tourists" (Pantziou, 2022). In the hospitality industry, calls to use the advantages of digitalisation to enable human employees freed from repetitive work to restart host interactions with guests are gaining attention (Moxness, 2021). However, even the 4VI and SETE leaders still speak about the need to "balance" the interests of guests and hosts, as if they were enemies, instead of supporting ways to align their interests as well as those of the other stakeholders in a benefits-for-all situation.

In the Northern summer of 2022, revenge tourism blossomed with travellers willing to pay increased prices and to endure under-staffed airports and hotels. The change will not be brought about by the industry or governments, despite the positive examples mentioned above, but by the demands of tourists, the host communities, the employees and the NGOs which care about the environment.

The summer of 2022 also brought home the fact that the tourism development of the coming decades will take place in an unstoppable climate crisis with stronger heatwaves, more burning forests and dried-up rivers. "The negative trend in climate will continue at least until the 2060s, independent of our success in climate mitigation", as Petteri Taalas, Secretary-General of the World Meteorological Organization of the UN stated at a press conference in July 2022 (UN, 2022b).

Given this constantly deteriorating frame for tourism development, meaningful tourism will become more and more important. Meaningful tourism will solve the problem of host communities who feel they have been taken hostage by the tourism masses by gaining economically and socially from visitors really interested in their region. Meaningful tourism will decrease the problem of environmental degradation by making visitors more careful through feeling part of the community. Meaningful tourism will help companies and governments to earn money to sustain the service providers, region and employees to enjoy good working conditions and a sense of providing hospitality in their daily work.

## References

AFP. (2022). Mass tourism returns to Barcelona – So does debate. Retrieved from https://www.france24.com/en/live-news/20220513-mass-tourism-returns-to-barcelona-so-does-debate

Airbnb (ed.) (2021). From Isolation to Connection—Travel in 2021. Travel report 2021. San Francisco: Airbnb.

Arlt, W. G. (ed.) (2020). *Welcoming the New Chinese Outbound Tourists: Guest relationships with Chinese visitors in the 2020s*. Hamburg: COTRI China Outbound Tourism Research Institute

Arlt, W.G. (2022a). *Meaningful Tourism Training*. Hamburg: MTC.

Arlt, W.G. (2022b). Why the world needs Meaningful Travel. *PhoCusWright*. Retrieved from https://www.phocuswire.com/why-the-world-needs-meaningful-tourism

Bhuiyan, M.A., Crovella, T., Paiano, A & Alves, H. (2021). A Review of Research on Tourism Industry, Economic Crisis and Mitigation Process of the Loss: Analysis on Pre, During and Post Pandemic Situation. *Sustainability* 13(18): 10314.

Biermann, F. et al. (2022). Scientific evidence on the political impact of the Sustainable Development Goals. *Nature Sustainability* 5: 795–800.

Binkhorst, E. & Den Dekker, T. (2009). Agenda for co-creation tourism experience research. *Journal of Hospitality Marketing & Management* 18(2–3): 311–327.

Bouskill, D., & Corbeil, D. (2021). Planet D. Retrieved from https://theplanetd.com/

Chiambaretto, P., Mayenc, E., Chappert, H., Engsig, J., Fernandez, A.S., & Le Roy, F. (2021). Where does flygskam come from? The role of citizens' lack of knowledge of the environmental impact of air transport in explaining the development of flight shame. *Journal of Air Transport Management* 93: 102049.

Chirakranont, R., Sakdiyakorn, M. (2022). Conceptualizing meaningful tourism experiences: Case study of a small craft beer brewery in Thailand. *Journal of Destination Marketing & Management* 23: 100691.

Conely, C. (2022). 5 Ways to Promote Sustainable Travel: a Guide for the Small DMO. Retrieved from https://crowdriff.com/resources/home/5-ways-promote-sustainable-travel-small-dmo

COTRI Analytics. (2022). *Global Tourism Database*. Hamburg.

Crane, H. (2013). How can we secure sustainable jobs for the most disadvantaged? *The Guardian*. Retrieved from https://www.theguardian.com/public-leaders-network/2013/oct/11/sustainable-jobs-disadvantaged

Csikszentmihaly, M., & Coffey, J. (2017). Why do we travel? A positive psychological model for travel motivation. In: Filep, S., Laing, S. J. & Csikszentmihalyi, M. (eds.). *Positive Tourism*. Abingdon: Routledge, pp. 122–132.

D'Eramo, M. (2021). *The World in a Selfie - An Inquiry into the Tourist Age*. New York: Verso.

European Travel Commission. (2022). Sustainable Travel in an Era of Disruption: Impact of COVID-19 on Sustainable Tourism Attitudes. Retrieved from https://etc-corporate.org/reports/sustainable-travel-in-an-era-of-disruption-impact-of-COVID-19-on-sustainable-tourism-attitudes/

EUROSTAT. (2021). Tourism trends and ageing. Retrieved from https://ec.europa.eu/eurostat/statistics-explained/index.php?title=Tourism_trends_and_ageing&oldid=526328

Fahey, D. (2022). Why this travel writer decided to never fly again. *Lonely Planet*. Retrieved from https://www.lonelyplanet.com/amp/articles/slow-travel-without-flying

Filep, S., Laing, S. J. & Csikszentmihalyi, M. (eds.) (2017). *Positive Tourism*. Abingdon: Routledge.

Filep, S., & Laing, S. J. (2019). Trends and directions in tourism and positive psychology. *Journal of Travel Research* 58(3): 343–354.

Fletcher, R, Mas, I.M., Blanco-Romero, A., & Blazquez-Salom, M. (2019). Tourism and degrowth: An emerging agenda for research and praxis. *Journal of Sustainable Tourism* 27(12): 1745–1763.

Goodwin, H. (2011). *Taking Responsibility for Tourism*. Oxford: Goodfellow Publishers.

González, K., & Hubbard, R. (2022). Meaningful travel experiences. Retrieved from https://innovations oftheworld.com/meaningful-travel-experiences/

Heritage Tribune. (2022). Palma to limit cruise ships to protect itself from Overtourism. Retrieved from https://heritagetribune.eu/spain/palma-to-limit-cruise-ships-to-protect-itself-from-overtourism/

Higgins-Desbiolles, F., Carnicelli, S., Krolikowski, C., Wijesinghe, G., & Boluk, K. (2019). Degrowing tourism: Rethinking tourism. *Journal of Sustainable Tourism* 27(12): 1926–1944.

Inskeep, E. (1991). *Tourism Planning: an Integrated and Sustainable Development Approach*. New York: John Wiley & Sons.

Kaspar, C. (1973). Fremdenverkehrsökologie – eine neue Dimension der Fremdenverkehrslehre. (Engl: Tourism ecology – A new dimension of tourism science). In: Walter, E. (Ed.), *Festschrift zur Vollendung des 65. Lebensjahres von P. Bernecker*. Haupt: Wien, 139–143.

Kirchherr, J. (2022). Bullshit in the Sustainability and Transitions Literature: A Provocation. *Circular Economy and Sustainability* 3(1): 1–6. Open access, published May 20, 2022.

Kirillova, K., Lehto, X., & Cai, L. (2017). Tourism and existential transformation: An empirical investigation. *Journal of Travel Research* 56(5): 638–650.

Krippendorf, J. (1984). *Die Ferienmenschen. Für ein neues Verständnis von Freizeit und Reisen*. Zürich: Orell Füssli.

Kühnen, M., Silva, S., & Hahn, R. (2022). From negative to positive sustainability performance measurement and assessment? A qualitative inquiry drawing on framing effects theory. *Business Strategy and the Environment* 31(5): 1985–2001. https://doi.org/10.1002/bse.2994

Kwok, L. (2021). What if Labor Shortage is a Long-Term Threat to the Hospitality and Tourism Industry? Retrieved from https://www.hospitalitynet.org/opinion/4106680.html

Laing, J.H., & Frost, W. (2017). Journeys of well-being: Women's travel narratives of transformation and self-discovery in Italy. *Tourism Management* 62: 110–119.

Lengieza, M. L., Hunt, C. A., & Swim, J. K. (2019). Measuring eudaimonic travel experiences. *Annals of Tourism Research* 74: 195–197.

Lichtenthaler, U. (2021). Why being sustainable is not enough: embracing a net positive impact. *Journal of Business Strategy* 44(1): 13–20, article publication date: 21 December 2021.

Mccreesh, M. (2022). Where adventure travel in Taiwan begins. Retrieved from https://origin wild.com.tw/

Mehmetoglu, M., & Engen, M. (2011). Pine and Gilmore's concept of experience economy and its dimensions: An empirical examination in tourism. *Journal of Quality Assurance in Hospitality & Tourism* 12(4): 237–255.

Moxness, P. (2021). *Spin the Bottle Service*. Atlanta: How2Conquer Publishers.

Neuhofer, B., Celuch, K. (2020). Experience design and the dimensions of transformative festival experiences. *International Journal of Contemporary Hospitality Management* 32(9): 2881–2901.

Opodo. (2022). Flight offers Hamburg-Mallorca. Retrieved from http://www.opodo.de

Pantziou, E. (2022). Greece Aiming to Find Balance Between Tourism Activity and Sustainability. Retrieved from https://news.gtp.gr/2022/04/13/greece-aiming-to-find-balance-between-tourism-activity-and-sustainability/

Parulis-Cook, S., Wang, M.F. (2022). *Chinese Traveler Sentiment Report: Spring 2022*. Beijing: Dragon Trail. Retrieved from https://dragontrail.com/resources/blog/china-traveler-sentiment-survey-spring-2022

Pearson, St. (2022). Overtourism Has Reached a Dangerous Tipping Point—Am I Part of the Problem? Retrieved from https://www.outsideonline.com/adventure-travel/essays/sedona-overtourism-last-tourist

Persson-Fischer, U., Liu, S. (2021). The Impact of a Global Crisis on Areas and Topics of Tourism Research. *Sustainability* 13(2): 906.

Pine, B.J., Gilmore, J.H. (1999). *The experience economy: Work is theatre & every business a stage.* Boston: Harvard Business School Press.

Pine, B.J., Gilmore, J.H. (2011). *The experience economy*. Boston: Harvard Business Press.

Pollock, A. (2022). A higher purpose. Retrieved from http://www.conscious.travel/

Pung, J.M., Gnoth, J., & Del Chappa, G. (2020). Tourist transformation. Towards a conceptual model. *Annals of Tourism Research* 81: 102885.

Rahmani, K., Gnoth, J., & Mather, K. (2018). Hedonic and eudaimonic well-being: A psycholinguistic view. *Tourism Management* 69: 155–166.

Rifai, T. (2021). Interview with Mediterranean Observer. Retrieved from https://mediterranean. observer/crises-opportunities-and-sustainable-tourism-dr-taleb-rifai/

Robbins, T. (2021). Why we travel — and why we shouldn't stop. *Financial Times*, May 21, 2021. Retrieved from https://www.ft.com/content/fd66019e-d960-4187-968f-ca2206081f30

Seligman, M.E. (2012). *Flourish: A visionary new understanding of happiness and well-being.* New York: Simon and Schuster.

Seligman, M.E. and Csikszentmihalyi, M. (2014). *Positive psychology: An introduction, Flow and the foundations of positive psychology.* Dordrecht: Springer, 279–298.

Smith, M. K., & Diekmann, A. (2017). Tourism and wellbeing. *Annals of Tourism Research* 66: 1–13.

Sorensen, F., Grindsted, T. (2021). Sustainability approaches and nature tourism development. Annals of Tourism Research 91 (2021).

Soulard, J., McGehee, N. G., & Stern, M. (2019). Transformative tourism organizations and glocalization. Annals of Tourism Research, 76, 91–104.

Su, L.J., Tang, B.J., & Nawijn, J. (2021). How tourism activity shapes travel experience sharing: Tourist well-being and social context. *Annals of Tourism Research* 91.

Tollman, G. (2022). Travel at a crossroads – There's no turning back. Retrieved from https://www. travelpulse.com/news/impacting-travel/venice-postpones-tourist-entry-fees-until-2023.html

UN. (2022a). Sustainable Development Goals Report 2022. Retrieved from: https://unstats.un.org/ sdgs/report/2022/The-Sustainable-Development-Goals-Report-2022.pdf

UN. (2022b). WMO warns of frequent heatwaves in decades ahead. Retrieved from https://news. un.org/en/story/2022/07/1122822

UNEP. (2022). *Transforming tourism in the Pan-European region for a resilient and sustainable post-COVID world.* Paris: UNEP.

UNWTO. (2020). UNWTO World Tourism Barometer. Vol. 18, Issue 3, June 2020. Retrieved from https://www.e-unwto.org/doi/epdf/10.18111/wtobarometereng.2020.18.1.3

UNWTO. (2021): International tourism growth continues to outpace the economy. Retrieved from https://www.unwto.org/international-tourism-growth-continues-to-outpace-the-economy

WCED. (1987). Report of the World Commission on Environment and Development: Our Common Future. In United Nations General Assembly document A/42/427. New York: UN.

Wen, J., Kozak, M., Yang, S. and Liu, F. (2021). COVID-19: potential effects on Chinese citizens' lifestyle and travel. *Tourism Review* 76(1): 74–87.

Wonderful Copenhagen. (2016). The End of Tourism. Retrieved from https://localhood.wonderful copenhagen.dk/wonderful-copenhagen-strategy-2020.pdf

Xu, F., Tian, F., Buhalis, D., Weber, J., & Zhang, H. (2016). Tourists as mobile gamers. Gamification for tourism marketing. *Journal of Travel & Tourism Marketing* 33(8): 1124–1142.

# 2 Emerging themes in sustainable tourism

*Jonathon Day*

## Introduction

Humanity faces concurrent existential crises, each requiring immediate action. Environmental issues, including climate change and biodiversity loss, and socio-economic issues including gender equality and reducing inequities, reducing hunger and poverty, will require focused attention in the coming decades. These challenges have been categorised as sustainability issues and the United Nations Sustainable Development Goals (SDGs) are a broadly accepted set of goals and actions designed to address these issues.

The environmental and social crises highlighted by the SDGs are impacting the tourism system. Additionally, while tourism enjoyed steady growth from the beginning of the jet age until the COVID-19 pandemic, the events of the early 2020s raise questions about the possible scenarios facing society as we consider the future of tourism. Assumptions of liberal values and increasing mobilities allowing unfettered growth in tourism seem worth re-examining. Tourism's contributions to these problems, as part of the problem and the solution, are being questioned.

Sustainable tourism differs from other "tourisms". Sustainable tourism is not a market segment or a product type, like adventure or business travel; rather it is a deliberate approach to implementing tourism in a way that contributes to socio-environmental sustainability. Sustainability, much like quality, is an approach to conducting tourism – an operating system for the tourism system.

Although sustainable tourism has been one of the dominant themes in tourism research over the past several decades (Moyle, Moyle, Ruhanen, Weaver, & Hadinejad, 2020), there remains much to learn about the intersection of sustainability and tourism. This chapter examines the trends and issues facing sustainable tourism in the coming years.

## Sustainability and tourism

*Our Common Future* (WCED, 1987), also known as the Brundtland Report, was an influential catalyst in stimulating discussion and action around sustainable development. The appeal of sustainable development quickly led to the growth in sustainable tourism research and policy (Ruhanen, Moyle, & Moyle, 2019; Ruhanen, Weiler, Moyle, & McLennan, 2015). Since the Brundtland report was released, increasing attention has been placed on two interrelated issues: the sustainability of tourism itself and the ability of tourism to contribute to sustainable development. The UNWTO (2015) is optimistic that tourism can help achieve these important goals – particularly when tourism is conducted sustainably. But there is value in recognising the difference between sustainable tourism and tourism for

DOI: 10.4324/9781003291763-3

sustainable development. Sustainable tourism is a set of practices designed to ensure tourism maximises its positive impacts and minimises its negative impacts to ensure it maintains its social licence to operate. Sustainable tourism practices contribute directly to SDG 12 (responsible consumption and production). Sustainable development through tourism is the achievement of sustainability goals using tourism-related activities. As such, tourism has the potential to contribute to each of the SDGs. Nevertheless, as Boluk, Cavaliere, and Higgins-Desbiolles (2019) note, addressing the ways in which tourism may contribute to the SDGs requires critical analysis. While tourism may contribute to achieving the goals, this is far from a certainty. Indeed, tourism has the potential to exacerbate social and environmental issues and hinder the achievement of SDGs.

## The nature of sustainable tourism

To appreciate the challenges with implementing sustainability principles and practices in the tourism system, it is important to recognise the nature of sustainability and the tourism system. Sustainable tourism is defined by UNEP-UNWTO (2005) as "tourism that takes full account of its current and future economic, social, and environmental impacts, addressing the needs of visitors, the industry, the environment, and host communities" (p. 12). This definition incorporates three important elements:

- A focus on tactical and strategic time frames;
- Balancing the so-called "triple bottom line" of economic, social/cultural, and environmental issues;
- Concern for multiple stakeholders.

Although this definition is appealing, even a cursory critical assessment of these concepts reveals they are vague and require a significant number of specific actions, some of which may be difficult to implement. Unsurprisingly, despite the appeal of sustainable tourism to academics, policymakers, tourism operators, and travellers, several authors express frustration that sustainable tourism has not been more widely adopted (Dodds & Butler, 2010; Maxim, 2014; Ruhanen et al., 2015).

Tourism is a complex adaptive system (Farrell & Twining-Ward, 2004; Morrison, Lehto, & Day, 2018), and implementing sustainable tourism represents a wicked problem (Day, 2021). As such, implementation requires recognition of the complex, complicated, and dynamic nature of sustainable tourism. In addressing the adoption of sustainable tourism practices, several factors must be considered. First, implementing sustainable tourism is often discussed holistically, as though sustainable tourism can be achieved with a single action, rather than a raft of activities each of which contributes to sustainability. This approach oversimplifies the work and conceals the challenges. Hunter (1997) advocates for considering sustainable tourism as an "overarching paradigm which incorporates a range of approaches" (p. 850). As such, sustainable tourism is not a single activity but a portfolio of activities that must be undertaken. In the following section, "Deconstructing Sustainable Tourism", the range of activities required to achieve sustainability is discussed.

Second, in addition to requiring a range of activities, implementing sustainable tourism programmes is complex and requires many independent actors working toward similar goals. The fragmented nature of the industry, the predominance of small and medium-sized enterprises, and limitations of governing bodies add to this challenge within tourism. Several articles have examined the need for alignment of action, if not collaboration or coordination (Dredge, 2006; Maxim, 2016). Additionally, sustainable tourism requires action by actors, such as policymakers, outside the tourism industry who are unaware of the specific issues associated with tourism.

Third, one characteristic of systems is that they stack. Tourism is embedded in the larger socio-economic system and ecosystem (Day, 2016). Destination systems, transportation systems, enterprises, and even individuals are all examples of component systems within the larger tourism system. Sustainability and sustainable tourism, like many social issues, require analysis across system levels. For example, at a macro-level, as noted in the policy example, sustainable tourism is dependent on actions in the broader society not specific to tourism per se. Sustainable tourism also requires action at the tourism system and destination levels. For example, at a destination community level it would require community-level policies and programmes to support sustainability in the destination as well as programmes directly related to sustainable tourism operations. Additionally, it would require individual tourism businesses within the destination to adopt their own sustainability initiatives. Sustainable tourism also requires action at the meso-level. Tourism enterprises must behave sustainably for destinations to be sustainable. A growing body of literature addresses corporate social responsibility, including environmental and social issues. Sustainable tourism also requires individual (micro-level) actions. Individual travellers to the destination would need to make sustainable and responsible decisions during their visit. Individuals within businesses must implement change programmes to support adoption of sustainable tourism programmes. Each level of analysis has unique issues and challenges but to ensure sustainability within the tourism system, each level must contribute to sustainability goals.

Fourth and finally, as with all wicked problems (Day, 2021), the solution to sustainability in each destination and each sector of the system will be unique. While general principles may guide action there are not right or wrong answers, each specific response will be determined by a range of situational factors, from local stakeholders, the environment, culture, and politics, to name but a few.

**Deconstructing sustainable tourism**

Deconstructing sustainable tourism into its components is helpful for understanding the phenomenon (Table 2.1). Considerable effort has been spent on identifying the components of sustainable tourism programmes and the metrics by which sustainable tourism may be measured. Early research identified the proliferation of sustainable tourism certifications and the diversity of approaches. Nevertheless, the key elements of sustainable tourism programmes are increasingly harmonised.

A catalyst for this alignment is the Global Sustainable Tourism Council (GSTC), which established criteria using the International Social and Environmental Accreditation and Labelling (ISEAL) Code of Good Practices (Bricker & Schultz, 2011). The GSTC has determined criteria for destinations and businesses. Each set of these criteria – destinations, hotels, and tour operators – include criteria and indicators organised into four categories: Sustainable Management, Socio-Economic Sustainability, Cultural Sustainability, and Environmental Sustainability (GSTC, 2019). There are 38 destination criteria and 26 hotel/tour operator criteria, and the elements of sustainability include a broad range of activities including not only environmental, but also socio-economic and cultural issues. Destinations or organisations seeking to meet threshold levels of sustainability actions must address multiple issues simultaneously. Environmental issues include energy conservation, water management, waste management, and environmental conservation. Socio-economic issues

*Table 2.1* Socio-economic and environmental issues addressed in sustainable tourism criteria

Socio-economic issues:
- Equitable distribution of economic costs and benefits of tourism;
- Decent work and career opportunities;
- Supporting entrepreneurs and fair trade;
- Preventing exploitation and discrimination;
- Ensuring property and user rights;
- Safety and security issues including access to health services and crime prevention;
- Protection of cultural assets (tangible and intangible);
- Cultural interpretation;
- Visitor management in destinations;
- Stakeholder participation in destination community development.

Environmental issues:
- Climate change mitigation and adaptation;
- Energy conservation;
- Transition to renewable energy sources including transportation;
- Water stewardship and quality;
- Waste and pollution management and reduction (including solid waste, food, light, noise);
- Protection of sensitive environments;
- Protection of biodiversity (including wildlife interaction).

This list was developed based on analysis of the GSTC Destination Criteria Version 2.0

include workforce development, support for local businesses and the provision of safe, decent work opportunities, and protection of vulnerable or marginalised people from exploitation. Cultural and heritage issues include the protection of tangible and intangible cultural heritage assets and effective management of heritage locations. Each of these activities must take place with deliberate management. To meet the requirements of each criterion, additional activity including planning, programmes, and performance management systems must be undertaken (Day & Romanchek, 2020). Furthermore, destinations require a policy and legislative framework in place to support these activities.

## Emerging trends and key issues

Recognising the systemic nature and complexity of sustainable tourism provides insight into the many changes observed in sustainable tourism. It also helps explain the various emerging trends that will impact sustainable tourism. Some of these changes are the result of larger socio-economic issues impacting the tourism system; others are unique to the system. As Gibson (1999) has observed, "the future is here. It is just unevenly distributed". Many of the issues and trends sustainable tourism will face in the coming decades are already upon us (Table 2.2).

### *Climate crisis*

The urgency of the climate crisis will ensure this is a dominant theme in the coming years. The climate crisis represents a range of challenges for the tourism system that will undoubtedly attract considerable attention. Scott and Gössling (2022) warn that "the last three decades of research have failed to prepare the industry for the net-zero transition and climate disruption that will transform tourism over the next three decades" (p. 1). The transformation

*Table 2.2* Challenges facing sustainability in tourism in the coming decades

---

**Climate change and tourism**: Our response to the climate crisis will be the primary issue facing tourism in the coming decades. How tourism mitigates and adapts in the face of the direct and indirect challenges of climate change will be the defining theme of the coming decades.

**New approaches to tourism production**: Meeting our sustainability challenges will require new ways of producing tourism products and experiences.

**Evolving stakeholders and changing expectations of tourism**: There are growing concerns that tourism generates benefits that distribute across the tourism system rather than just benefiting some:

- Locals-first tourism development and community oriented tourism: Destination communities more meaningfully involved in planning and development to ensure tourism contributes to their quality of life.
- Diversity, equity, and inclusion in the tourism system: There is growing discussion – and action – to ensure the benefits of tourism extend to diverse groups.
- Changing consumers: Researchers and practitioners must expand their understanding of the role individuals play in sustainability in the tourism system.

**Operationalising sustainable tourism**: Implementing sustainable tourism is a complicated, complex challenge. There is a growing need for a greater understanding of implementing these programmes that will continue in the coming years:

- Implementing change in complex, adaptive systems: Implementation of sustainability in complex systems like destinations requires a greater understanding of long-term collaborative actions.
- Diffusion of innovation: Understanding the drivers of innovation adaptation across the tourism system will be critical to more comprehensive sustainable tourism programmes.
- Meaningful metrics: Effective sustainable tourism performance management will require better data, particularly at the sub-national/regional level.

**Broadening perspectives**: Sustainability must be embedded in all tourism activities. It is not a type of tourism but an approach to conducting tourism activities:

- Sustainable urban tourism: Cities are a driver of tourism growth and so it is critical that greater attention is paid to the implementation of sustainable tourism in urban centres.
- Sustainable space tourism: Space tourism is emerging as a new form of tourism and now is the time to tackle issues of sustainable tourism in space.

**Regenerating sustainable tourism**: Ensuring that sustainable tourism practices resonate with stakeholders over the long term is important. Sustainable tourism's ability to incorporate new paradigms or priorities will be necessary to achieve long-term goals.

---

of the tourism system will require system actors to adopt mitigation activities designed to reduce the generation of greenhouse gases (GHG) and adaptive strategies necessary to respond to the many changes climate change will bring. As climate science improves, the direct impacts of climate change are becoming increasingly clear. Less clear are the indirect societal impacts that will occur as a result of it (Day et al., 2021).

Weaver (2011) has questioned whether sustainable tourism can survive climate change, implying that the primacy of the issue detracts from other sustainability issues. Appreciation of the intersectionality of climate change with other environmental and social issues is growing, and it is increasingly clear that framing these as either/or issues is unnecessary and unproductive. Responses to climate change – mitigation and efforts and adaptations to a changing tourism system – intersect with a range of sustainable tourism activities. For example, climate change mitigation efforts are closely related to environmental performance, including energy conservation, the transition to renewable forms of energy, and water and waste management. Even social issues, such as the protection of vulnerable populations and the distribution of benefits from tourism, intersect with climate change.

### Circularity and new approaches to tourism production

The challenges created by unsustainable levels of consumption are forcing industries, including tourism, to consider new production processes. Achieving sustainable production and consumption (SDG 12) will require new approaches to production, and the tourism system must adopt new business practices and adapt to changes in the system. Proponents of the circular economy put forward practices that eliminate waste from the production cycle. The circular economy is based on three principles, driven by design: eliminate waste and pollution, circulate products and materials at their highest value, and regenerate nature (EllenMacArthurFoundation, 2017). Research into the implementation of circularity as a concept in the production of tourism and hospitality is still nascent (Rodríguez, Florido, & Jacob, 2020), but the principles have been applied across the tourism system in destination communities and enterprises.

Stimulated by the climate crisis, many of the current forms of production associated with tourism can be expected to change. For example, transportation, a key element of the tourism system, is undergoing significant changes, and the implications for tourism will be significant. The transition from cars with combustion engines to electric vehicles (EVs) will bring systemic changes. Drive travel is the foundation of domestic travel in many countries, and the current tourism system has developed in conjunction with petroleum-based automobiles. Already hotels and restaurants are exploring providing new charging services and new EV-touring products, like travel apps and tourism trails, are being produced. The transition from petroleum-based fuels in other transportation sectors, such as aviation and cruising, is progressing, albeit more slowly.

### Evolving stakeholders and changing expectations

Tourism development has been criticised for its business focus, with tourism research primarily adopting a neoliberal industry-based orientation (Cave & Dredge, 2020; Ruhanen et al., 2015). Researchers' focus on supporting tourism business development and satisfying consumer needs has relegated other stakeholders to the sidelines – despite the fact that stakeholder engagement is a defining element of sustainability. Indeed, stakeholder engagement is a key criterion of sustainable tourism. Nevertheless, until recently, most destinations have not addressed resident sentiments regarding tourism, let alone engaged meaningfully in discussions about the type of tourism growth desired in each destination. If sustainable tourism seeks to maximise benefits for a range of stakeholders, including destination community residents and other participants in tourism endeavours, then additional work is required to meet this goal. The recognition of the importance of engaging with a broader range of stakeholders, the importance of diversity, equity, and inclusion (DEI), and the distribution of benefits (and costs) of tourism are generating new themes in sustainable tourism.

These issues in tourism are part of broader societal-level discussions. There are concerns that capitalism and neoliberal policies lack guide rails to ensure positive social outcomes are a part of the current discourse. The rise of stakeholder capitalism, embodied in the Business Roundtable's recent statement on the purpose of the firm (BRT, 2019), is evidence of broad awareness of the issue. Increasing awareness of structural racism and gender discrimination as well as other DEI issues in recent years provides greater context for the changes taking place in tourism.

### Overtourism, locals-first tourism development, and community-oriented tourism

The growth in tourism numbers has exposed deficiencies in visitor management and inequities in the distribution of the positive and negative impacts of tourism on destination communities. One of the defining issues of the pre-pandemic period was the growth of overtourism, in which destination communities from across the globe strained under increasing numbers of visitors. The outcry of overtourism placed a spotlight on the burden of tourism to many destination stakeholders. The issues around overtourism highlight the increasingly fragile state of the social licence for tourism in destination communities, and unsurprisingly local-first, resident-centred tourism growth is an emerging trend that is expected to continue (Girma, 2022).

Lessons from overtourism can be applied more broadly across the tourism system. From a system perspective, feedback loops in tourism can be slow, resulting in issues festering discontent. As a result, when tipping points are reached, it is challenging to get ahead of what seems like a sudden emergency. Closer monitoring of issues within the system and proactive responses to emerging challenges are required to avoid the type of tourism backlash seen with overtourism.

### Diversity, equity, and inclusion in the tourism system

Intersecting with the shift toward resident-focused tourism is the issue of the equitable distribution of tourism's benefits. The dominant rhetoric justifying tourism highlights its ability to generate economic benefits, including jobs and increased tax revenues for the local community. Researchers have noted that the distribution of benefits is often concentrated with a few entities and that these organisations may be separate from the destination community in which they operate. Ensuring that tourism's benefits are distributed equitably is an emerging issue. Sharpley (2000) notes tourism often reinforces rather than diminishes socio-economic disparities, while Jamal and Camargo (2014) highlight key issues of justice required for sustainable development.

One aspect associated with the distribution of benefits from tourism is its ability to generate economic opportunities for individuals seeking jobs and career opportunities. Although tourism has long been thought to have high turnover rates, the post-pandemic "great resignation" in the USA has impacted the hospitality industry more than any other sector of the economy. The industry's ability to attract and retain talented people is a sustainable tourism issue. Tourism may be promoted as creating jobs, but its ability to contribute to SDG 8, which includes creating decent work, should not be taken for granted. Greater focus on workforce development research will be necessary (Baum, 2018; Mooney, Robinson, Solnet, & Baum, 2022).

Another emerging set of related issues deals with DEI. Again, these issues have been incorporated in established criteria for sustainable tourism but have received only scant attention from researchers. Diversity issues are interwoven throughout the tourism system, and awareness of the range of activities that need to be addressed is growing. Aspects of DEI have received some attention. At the enterprise (meso-) level, companies – in particular lodging companies – have been progressive in implementing DEI programmes. Unfortunately, these programmes are far from universal. Additionally, some programmes support minority and women-owned businesses. Progress in the inclusion of Black business owners (Benjamin & Dillette, 2021) will support sustainability efforts. While many tourism enterprises and lodging brands have embraced DEI initiatives within their workforces, breaking down systematic and structural racism and intolerance represents a significant challenge.

Diversity issues in tourism extend beyond workforce issues. Who you are impacts your ability to engage with the tourism system. Freedom of mobility is impacted by race, sexuality, and gender. In the USA, explorations of racism's impacts on tourism have included the impact of racial violence on Black travel (Duffy, Pinckney, Benjamin, & Mowatt, 2019) and the Black travel experience (Dillette, Benjamin, & Carpenter, 2019). Research on the transformation of racialised tourism is in its infancy. Recognising racial bias in dominant narratives of history and culture is critical to providing balanced narratives and interpretation (Benjamin, Kline, Alderman, & Hoggard, 2016). Ensuring that the tourism system is more equitable and inclusive requires considerable research and action.

### *Changing consumers*

The role of the individual and their relationship to sustainability will be increasingly pertinent. Despite some insights showing systemic factors have the greatest impacts on sustainable tourism, significant areas for improvement exist related to travellers and sustainability.

Increasing the sustainability of tourism is dependent on changing the behaviour of individuals, travellers, and other stakeholders. The need for greater understanding of the process of changing behaviours and designing choices that lead to positive outcomes is an important stream for future research. Applications of behavioural science techniques and behavioural economics to encourage sustainability-related behaviours are emerging (Souza-Neto, Marques, Mayer, & Lohmann, 2022).

Another emerging area of study is the intersection of emotional and mental health issues and concerns for sustainability-related issues and tourism. Mkono and Hughes (2020) highlight a wide range of travel-related behaviours, including flying, accommodation choice, and the impact of travel on local communities, to name a few, as triggers for eco-guilt or eco-shame. Indeed, preliminary evidence shows that growing feelings of eco-guilt and eco-shame are impacting attitudes toward tourism and travel behaviours (Bahja, Alvarez, & Fyall, 2022; Mkono & Hughes, 2020).

The impacts of individual attitudes and behaviours on tourism system changes are also important. The growth of social movements like flight shaming, driven by individual concerns for the environment and amplified by social media, is leading to changing markets and policy changes. Consumer enthusiasm for new, environmentally friendly products such as EVs will have transformative impacts on the tourism system.

### Operationalising sustainable tourism

Dodds and Butler (2010) recognise that the challenge of sustainable tourism is in its implementation. While considerable research has addressed what should be included in comprehensive sustainable tourism programmes, less attention has been paid to the implementation and ongoing management of these programmes. At least three broad issues, and streams of research, must be addressed in more detail in the coming years: the implementation of change, diffusion of innovation, and metrics for sustainability programme performance management.

*Implementing change in complex adaptive systems*

Unsurprisingly, the implementation of sustainable tourism programmes presents a range of challenges. At the macro-level, implementing sustainable tourism is complicated by a lack of clearly defined roles, responsibilities, and authority. For example, although destination management organisations (DMOs) are becoming more involved in destination stewardship, they have limited ability or resources to implement change. In many cases, implementing sustainable tourism programmes in destination systems requires collaboration, cooperation, and alignment of purpose. There is substantial opportunity to explore the ways organisations in the destination system can contribute to destination-wide sustainability improvements.

At the enterprise level, the introduction of sustainability to business programmes is a change management activity. Of course, change management is one of the dominant themes in organisational behaviour and business literature.

*Diffusion of innovation and catalysts for the adoption of sustainability activities*

Given this perceived lack of progress in sustainable tourism, it is surprising that greater attention has not been paid to the diffusion of sustainable tourism practices in the system. Although focusing on achievement levels of the complete portfolio is useful, it is important to consider that each element has its own unique pattern of adoption. Some actions, such as energy-conservation programmes or towel-reuse programmes in hotels, are ubiquitous, while others have been adopted more slowly. With a few exceptions (Dabphet, Scott, & Ruhanen, 2012; Smerecnik & Andersen, 2011), little attention has been paid to the adoption of sustainability-related technologies, work practices, or policies.

Similarly, little attention has been paid to other processes or actions likely to enhance the saturation of sustainability practices in the tourism system. Research on the impacts of sustainable supply chain management is still in its infancy, as is research on the effectiveness of DMO-led tourism programmes designed to support adoption of sustainable tourism practices.

*Meaningful metrics*

Developing meaningful metrics for managing sustainable tourism will be critical for systematic improvements in performance. Again, progress is uneven across the range of activities required. At an enterprise level, there is growing attention to the performance management of sustainability-related activities for corporate social responsibility (CSR) reporting as well as environmental, social, and governance (ESG) reporting. At a destination level, particularly at sub-national and regional levels, it is more challenging to find effective tourism-related metrics. Further complicating the issue is a lack of clarity regarding who should collect the data.

## Broadening the perspective

Although sustainable tourism is a set of principles that can be applied to any tourism organisation, much discussion around sustainable tourism has focused on tourism in natural and rural environments (Ruhanen et al., 2015). It is critical for sustainable tourism principles to be adopted by cities, regions, nations, and beyond.

*Sustainable tourism in the city*

Urbanisation has been a defining trend of the past 50 years, and the top 300 cities account for more than 45% of tourism arrivals (WTTC, 2018). The growth of urban tourism raises a range of sustainability issues, but as Maxim (2014) notes, sustainable tourism has received scant attention in the context of urban tourism.

Urban tourism systems are among the most complex systems. Although the principles of sustainable tourism are universal, urban tourism faces unique challenges. Some issues have become acute, with overtourism becoming a major social issue in some cities. Perhaps more significantly, cities will be at the forefront of changes associated with climate change. Cities will need to deal with issues as diverse as water scarcity, new infrastructure needs, climate-induced population changes, and the need to build resilience in the face of extreme weather, and each will impact tourism within the city.

*Sustainable space tourism*

In the coming years, sustainable tourism principles must extend beyond the Earth as travel to space proliferates. The growth of space tourism is expected to accelerate in the coming decades. Even as a relatively small sector of the overall tourism system, the emerging field raises significant issues for sustainability and tourism. The expenditure of fossil fuels to achieve orbital and sub-orbital travel, and its implications for climate change, raises ethical issues for individual travellers and the suppliers, and other issues are emerging. For example, the preservation of heritage sites on the Moon has been raised as an issue. Given the current manifest destiny approach to space exploration, questions of the preservation of natural spaces beyond the Earth are likely to emerge. In their review of 109 articles on space tourism, Zhang and Wang (2020) identify several themes that intersect with sustainability including natural environment issues, economics and finance, general social-anthropology, politics, policy and legal issues, and general management.

**Regenerating sustainable tourism**

While regenerative practices align with circularity, they can be seen as a distinct emerging theme. The growing interest in regenerative tourism can be seen as a confluence of research streams focused on achieving better outcomes from tourism and broad dissatisfaction with the perceived low ambitions of the sustainability movement.

Some of the regenerative movement's appeal is that it overcomes growing frustration with the implementation of sustainability. JWT Innovation Group notes that "sustainability as we know it is dead. Doing less harm is no longer enough. The future of sustainability lies in regeneration: seeking to restore and replenish what we have lost, to build economies and communities that thrive, and allow the planet to thrive too" (Stafford, Tilley, & Britton, 2018, p. 2). Elkington (2020), the thought leader credited with introducing the concept of the triple-bottom-line, has noted that doing less harm is no longer enough and proposes regenerative approaches to sustainability.

Critiques of sustainable tourism identify similar frustrations with the application of sustainability to tourism. While sustainable and regenerative tourism seek to maximise the benefits of tourism, critics note that sustainable tourism is often framed in terms of the neoliberal growth paradigm (Ateljevic, 2020; Bellato, Frantzeskaki, & Nygaard, 2020). Bellato et al. (2020) propose regenerative tourism as a transformative approach through which "tourism living systems facilitate encounters, create connections and develop recip-rocal and mutually beneficial relationships through travel experiences, uniquely reflecting tourism places" (p. 17). Regenerative tourism seeks to ensure that tourism systems thrive and flourish. Regenerative approaches to tourism intersect with increasing interest in community-based and resident-focused tourism. They also align with increasing interest in non-traditional business models including social enterprises and community collectives.

The embrace of regenerative approaches to tourism development has invigorated sus-tainable tourism. Although it is reasonable to accept differences between sustainable tour-ism and regenerative tourism, both share many of the same principles, and perhaps more importantly, regenerative activities contribute to sustainability.

### What does it mean? Interpretation and implementation

As tourism, and society in general, grapple with the challenges we face, sustainability-related issues will remain critically important to the tourism system. Tourism is a dynamic, adaptive, ever-changing system, and all aspects of it will be changed by issues related to sustainability. Specifically, we must recognise that marginalising sustainable tourism in this context marginalises the topic universally.

Although tourism has been recognised as a complex adaptive system for some time, scant evidence of the implications of this insight has had negative impacts, with much of the research in the field remaining fragmented and siloed.

Sustainability requires greater systems thinking. It must be considered holistically, but also as the sum of its parts. These issues have been presented separately, but it is critical to recognise their intersectionality. Our approaches to sustainable tourism must also be adap-tive and flexible, reflecting the changes we face.

Greater understanding of the relationships between levels in the system is also needed. For instance, while significant research exists on enterprise-level CSR, little is available on sustainability across supply chains or the diffusion of sustainability practices in destination communities.

The work of ensuring the sustainability of tourism is increasingly urgent. Tourism oper-ates with an implicit social licence based on the general assumption that it has a net positive contribution to the destination communities. However, stakeholders are increasingly chal-lenging this assumption. Communities are responding to overtourism with legislation reducing visitors, and climate advocates are flight-shaming travellers. It is critical that tour-ism organisations address the concerns of these stakeholders to ensure the sustainability of the tourism system. Without improving the sustainability of tourism, assuming the social licence for it seems increasingly unfounded.

### Contributions of this discussion

This chapter has addressed several critical issues associated with sustainability and tour-ism. The seeds of future trends have already been planted. The challenges are outlined here and will likely remain with us for the coming decades. Environmental issues including cli-mate change and biodiversity loss will greatly affect society and the tourism system. Simi-larly, issues of diversity, equity, and inclusion and the distribution of tourism's benefits and costs will remain as destination communities and the system itself grapple with issues asso-ciated with growing tourism on a finite planet. It is worth noting that many of the chal-lenges facing tourism originate beyond the tourism system and represent broader socio-economic and environmental challenges. This insight highlights the importance of systems thinking when addressing sustainability issues in tourism.

The issues raised in this chapter highlight several important issues and trends in sustainability, recognising the primacy of climate change as a sustainability issue and the intersectionality of climate change with other environmental and social challenges. The chapter has also noted the tourism system is dynamic, and new approaches to production will impact the system. The chapter has addressed growing concerns about the distribution of tourism's positive and negative impacts within destination communities. The promises of tourism have been unevenly distributed, resulting in growing stakeholder engagement in tourism decisions that impact the quality of life and highlighting the need for greater attention to DEI issues. The chapter has examined the knowledge gap in our understanding of emerging consumer attitudes and behaviours. It has highlighted the critical importance of moving sustainability from an appealing concept to the standard operating procedure for enterprises and destinations. Finally, the chapter has addressed the critical need for sustainability to embrace new paradigms to meet society's expectations. The regenerative movement builds on the foundations of sustainability, providing new energy and inspiration as we face the challenges of the 21st century.

## References

Ateljevic, I. (2020). Transforming the (tourism) world for good and (re)generating the potential 'new normal'. *Tourism Geographies*, *22*(3), 467–475. doi:10.1080/14616688.2020.1759134

Bahja, F., Alvarez, S., & Fyall, A. (2022). A critique of (ECO)guilt research in tourism. *Annals of Tourism Research*, *92*, 103268. doi:10.1016/j.annals.2021.103268

Baum, T. (2018). Sustainable human resource management as a driver in tourism policy and planning: A serious sin of omission? *Journal of Sustainable Tourism*, *26*(6), 873–889. doi:10.1080/09669582.2017.1423318

Bellato, L., Frantzeskaki, N., & Nygaard, C. A. (2020). Regenerative tourism: A conceptual framework leveraging theory and practice. *Tourism Geographies*, *22*(3), 503–513. doi:10.1080/14616688.2022.2044376

Benjamin, S., & Dillette, A. K. (2021). Black travel movement: Systemic racism informing tourism. *Annals of Tourism Research*, *88*, 103169. doi:10.1016/j.annals.2021.103169

Benjamin, S., Kline, C., Alderman, D., & Hoggard, W. (2016). Heritage site visitation and attitudes toward african american heritage preservation: An investigation of North Carolina Residents. *Journal of Travel Research*, *55*(7), 919–933. doi:10.1177/0047287515605931

Boluk, K., Cavaliere, C. T., & Higgins-Desbiolles, F. (2019). A critical framework for interrogating the United Nations Sustainable Development Goals 2030 Agenda in tourism. *Journal of Sustainable Tourism*, *27*(7), 847–864. doi:10.1080/09669582.2019.1619748

Bricker, K., & Schultz, J. (2011). Sustainable tourism in the USA: A comparative look at global sustainable tourism criteria. *Tourism Recreation Research*, *36*(3), 215–229.

BRT. (2019). Our commitment: Statement on the purpose of a corporation. Retrieved from https://opportunity.businessroundtable.org/ourcommitment/

Cave, J., & Dredge, D. (2020). Regenerative tourism needs diverse economic practices. *Tourism Geographies*, *22*(3), 503–513. doi:10.1080/14616688.2020.1768434

Dabphet, S., Scott, N., & Ruhanen, L. (2012). Applying diffusion theory to destination stakeholder understanding of sustainable tourism development: a case from Thailand. *Journal of Sustainable Tourism*, *20*(8), 1107–1124. doi:10.1080/09669582.2012.673618

Day, J. (2016). *An Introduction to Sustainable Tourism and Responsible Travel* (Beta ed.). West Lafayette, IN: Placemark Solutions.

Day, J. (2021). Sustainable Tourism in City is a Wicked Problem. In A. Morrison & J. A. Coca-Stefaniak (Eds.), *Routledge Handbook of Tourism Cities*. New York: Routledge.

Day, J., Chin, N., Sydnor, S., Widhalm, M., Shah, K., & Dorworth, L. (2021). Implications of climate change on tourism and outdoor recreation: An Indiana, USA, case study. *Climatic Change, 169*, 1–21.

Day, J., & Romanchek, J. L. (2020). *Sustainable Tourism for Destinations: Insights from the GSTC Destination Criteria 2.0 for Sustainable Tourism*. Retrieved from.

Dillette, A. K., Benjamin, S., & Carpenter, C. (2019). Tweeting the black travel experience: Social media counternarrative stories as innovative insight on #TravelingWhileBlack. *Journal of Travel Research, 58*(8), 1357–1372. doi:10.1177/0047287518802087

Dodds, R., & Butler, R. (2010). Barriers to implementing sustainable tourism policy in mass tourism destinations. *Tourismos: An International Multidisciplinary Journal of Tourism, 5*(1), 35–53.

Dredge, D. (2006). Policy Networks and the local organization of tourism. *Tourism Management, 27*(2), 269–280.

Duffy, L. N., Pinckney, H. P., Benjamin, S., & Mowatt, R. (2019). A critical discourse analysis of racial violence in South Carolina, U.S.A.: implications for traveling while Black. *Current Issues in Tourism, 22*(19), 2430–2446. doi:10.1080/13683500.2018.1494143

Elkington, J. (2020). *Green Swans: The Coming Boom in Regenerative Capitalism*. New York: Fast Company Press.

EllenMacArthurFoundation. (2017). What is a circular economy. Retrieved from https://www.ellenma carthurfoundation.org/circular-economy/concept

Farrell, B., & Twining-Ward, L. (2004). Reconceptualizing tourism. *Annals of Tourism Research, 31*(2), 274–295. doi:10.1016/j.annals.2003.12.002

Gibson, W. (1999, 11/30/1999) *The Science of Science Fiction/Interviewer: B. Gladstone*. Talk of the Nation, National Public Radio, NPR.

Girma, L. (2022). Communities Move Beyond Spectator Role for Travels Future. In *Megatrends Defining Travel in 2022*. New York: Skift.

GSTC. (2019). *GSTC Destination Criteria with Performance Indicators and SDGs*. Retrieved from Washington, DC.

Hunter, C. (1997). Sustainable tourism as an adaptive paradigm. *Annals of Tourism Research, 24*(4), 850–867.

Jamal, T., & Camargo, B. A. (2014). Sustainable tourism, justice and an ethic of care: toward the Just Destination. *Journal of Sustainable Tourism, 22*(1), 11–30. doi:10.1080/09669582.2013.786084

Maxim, C. (2014). Drivers of success in implementing sustainable tourism policies in urban areas. *Tourism Planning & Development, 12*(1), 1–11. doi:10.1080/21568316.2014.960599

Maxim, C. (2016). Sustainable tourism implementation in urban areas: A case study of London. *Journal of Sustainable Tourism, 24*(7), 971–989. doi:10.1080/09669582.2015.1115511

Mkono, M., & Hughes, K. (2020). Eco-guilt and eco-shame in tourism consumption contexts: Understanding the triggers and responses. *Journal of Sustainable Tourism, 28*(8), 1223–1244. doi:10.1080/09669582.2020.1730388

Mooney, S., Robinson, R., Solnet, D., & Baum, T. (2022). Rethinking tourism's definition, scope and future of sustainable work and employment: editorial for the Journal of Sustainable Tourism special issue on "locating workforce at the heart of sustainable tourism discourse". *Journal of Sustainable Tourism, ahead-of-print*(ahead-of-print), 1–19. doi:10.1080/09669582.2022.2078338

Morrison, A., Lehto, X., & Day, J. (2018). *The Tourism System* (8th ed.). Dubuque, Iowa: Kendall Hunt.

Moyle, B., Moyle, C.-L., Ruhanen, L., Weaver, D., & Hadinejad, A. (2020). Are we really progressing sustainable tourism research? A bibliometric analysis. *Journal of Sustainable Tourism, 29*(1), 106–122. doi:10.1080/09669582.2020.1817048

Rodríguez, C., Florido, C., & Jacob, M. (2020). Circular economy contributions to the tourism sector: A critical literature review. *Sustainability (Basel, Switzerland), 12*(11), 4338. doi:10.3390/su12114338

Ruhanen, L., Moyle, C.-L., & Moyle, B. (2019). New directions in sustainable tourism research. *Tourism Review, 74*(2), 138–149. doi:10.1108/tr-12-2017-0196

Ruhanen, L., Weiler, B., Moyle, B. D., & McLennan, C.-L. J. (2015). Trends and patterns in sustainable tourism research: A 25-year bibliometric analysis. *Journal of Sustainable Tourism, 23*(4), 517–535. doi:10.1080/09669582.2014.978790

Scott, D., & Gössling, S. (2022). A review of research into tourism and climate change - Launching the annals of tourism research curated collection on tourism and climate change. *Annals of Tourism Research, 95*. doi:10.1016/j.annals.2022.103409

Sharpley, R. (2000). Tourism and sustainable development: exploring the theoretical divide. *Journal of Sustainable Tourism, 8*(1), 1–19. doi:10.1080/09669580008667346

Smerecnik, K., & Andersen, P. (2011). The diffusion of environmental sustainability innovations in North American hotels and ski resorts. *Journal of Sustainable Tourism, 19*(2), 171–196. doi:10.1080/09669582.2010.517316

Souza-Neto, V., Marques, O., Mayer, V. F., & Lohmann, G. (2022). Lowering the harm of tourist activities: A systematic literature review on nudges. *Journal of Sustainable Tourism, ahead-of-print*(ahead-of-print), 1–22. doi:10.1080/09669582.2022.2036170

Stafford, M., Tilley, S., & Britton, E. (2018). *The New Sustainability: Regeneration*. Retrieved from New York: Wunderman Thompson: https://intelligence.wundermanthompson.com/trend-reports/the-new-sustainability-regeneration/

UNEP-UNWTO. (2005). *Making Tourism More Sustainable - A Guide For Policy makers*. Retrieved from Madrid: http://www.unep.fr/shared/publications/pdf/dtix0592xpa-tourismpolicyen.pdf

UNWTO. (2015). *Tourism and the Sustainable Development Goals*. Retrieved from Madrid, Spain.

WCED. (1987). *Our Common Future*. Oxford: Oxford University Press.

Weaver, D. (2011). Can sustainable tourism survive climate change? *Journal of Sustainable Tourism, 19*(1), 5–15. doi:10.1080/09669582.2010.536242

WTTC. (2018). *Travel and Tourism: City Travel and Tourism Impact 2018*. Retrieved from London, UK.

Zhang, Y., & Wang, L. (2020). Progress in space tourism studies: A systematic literature review. *Tourism Recreation Research, ahead-of-print*(ahead-of-print), 1–12. doi:10.1080/02508281.2020.1857522

# 3 Tourism great reset

## The inclusive, sustainable, and innovative reality

*Ekaterina Glebova and Marko Perić*

### Introduction

In June 2020, more than 3,000 of the world's most high profile progressive elites gathered in-person and online at the 50th annual meeting of the World Economic Forum (WEF) in Davos, Switzerland. The purpose of this meeting announced as "The Great Reset" was to discuss sustainably rebuilding society and the economy following the multidimensional crisis of the pandemic that disrupted and reshaped all industries. The Great Reset is based on three main pillars: (1) creating conditions for a stakeholder economy; (2) building resilient, equitable, and sustainable infrastructure projects, in the spirit of a green economy; and (3) harnessing the innovations of the Fourth Industrial Revolution (Schwab and Malleret, 2020).

These ideas have been further captured and formed the basis of the campaign "The Great Reset" (https://greatreset.com/), when 700 activists from the advertising and communications industry united and launched a creative industry movement intended to embed the constructive environmental behaviours noticed during the lockdown in 2020. They have been led by a network of "purpose disruptors", industry insiders with an idea of how to change advertising to tackle climate change. They claim that advertising is adding an extra 28% to the annual carbon footprint of every single person in the UK (Purpose Disruption, 2021), and they engage the advertising industry to adopt the concept of "advertised emissions".

The Great Reset vision is being gradually disseminated and applied to all fields of life and industry; tourism is not an exception (Buhalis, 2022). Tourism is a massive field (Butcher, 2020) nowadays, affordable and accessible for all. The growth and development (quantitative and qualitative) of the tourism industry in recent decades can be explained by many evident factors as the growth in population, greater discretionary income, more leisure, improved mobility, digitalisation, and urbanisation (Butler and Wall, 1985; Buhalis, 2022). However, on closer inspection, we may find many difficulties and challenges hidden behind the positive aspects. These became particularly visible during the pandemic. Following the notable growth in 2019, 2020 was "the worst year in tourism history", according to the World Tourism Organization (2021). Travel was banned, social distancing prevented closer encounters between people, and tourism was put on hold. Despite state interventions, travel companies went bankrupt and many people lost their jobs (Nhamo, Dube, & Chikodzi, 2020). However, this crisis presented opportunities for an economic and social reset (Everingham & Chassagne, 2020; Hall, Scott, & Gössling, 2020). As a complex phenomenon reflecting all the social, economic, technological, and political changes and metamorphoses of the COVID-19 pandemic and its aftermath (Fletcher et al., 2021), tourism might be at the centre of this transformation.

DOI: 10.4324/9781003291763-4

According to the theoretical founders of the Great Reset, the authors of the epony-mous book Schwab & Malleret (2020), the reset is classified into macro-, micro-, and individual levels (Figure 3.1). The macro-reset is subclassified into economic, societal, geopolitical, environmental, and technological kinds (dimensions) of the reset. The mac-ro-reset flows in the framework of the three prevailing secular forces: (1) interdependence, (2) velocity, and (3) complexity. At the micro-level, we find micro-trends (the acceleration of digitisation, resilient supply chains, governments and business, stakeholder capitalism,

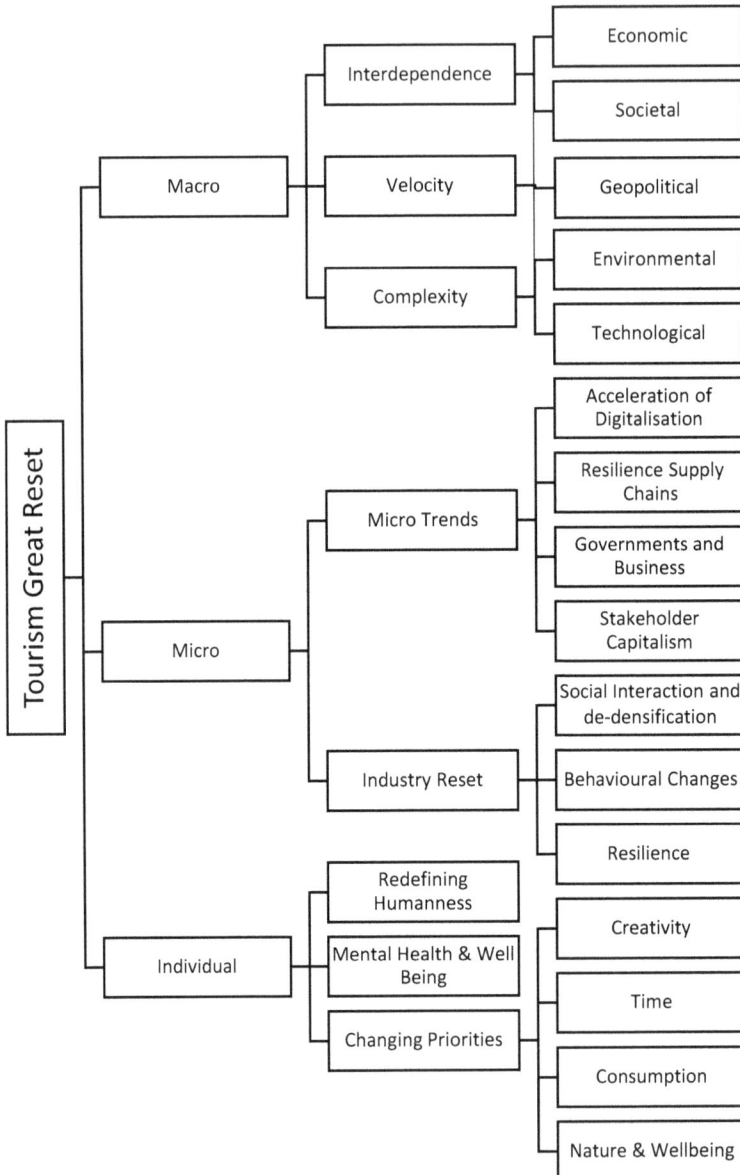

*Figure 3.1* Tourism Great Reset.

*Source*: Schwab and Malleret (2020).

and environmental, social, and governance) and industry resets (social interaction and de-densification, behavioural changes, resilience). An individual reset embraces redefining humanness, mental health and wellbeing, and changing priorities (creativity, time, consumption, nature, and wellbeing).

In the following sections we will describe the phenomenon of the Tourism Great Reset in greater detail, at the macro-, micro-, and individual levels (Figure 3.1). This will involve plenty of interrelated constructs and introduce the key stakeholders, industry trends, relevant issues, and forecasting.

### The stakeholders and their interest and involvement

The tourism industry is usually responsible for the flow of millions of travellers all over the world. They are moving in all directions, for different purposes, by various means. This involves (directly and indirectly) many stakeholders and builds a tourism ecosystem, including social relations between government, business, and civil society (Presenza & Cipollina, 2010). Technological innovations and changing consumer habits are revolutionising the industry, evolving the roles of different stakeholders, and shaking the whole tourism ecosystem. The growth of social, economic, environmental, and geopolitical issues in tourism, directly affects stakeholder networks. Furthermore, the pandemic 2020 crisis directed all hospitality stakeholders at all levels to rethink, reimagine, and reshape their business and social practices, which would reform hospitality into the next new normal and state of uncertainty (Sigala, 2021).

The sustainable development and green economy constructs are far from being new and innovative now, and they are entirely framed by global initiatives linked to ecology, wellbeing, and tackling the threats of climate change (Meadows et al., 1972; Reddy & Wilkes, 2013). The green economy and Sustainable Development Goals (SDGs) are considered a potential approach to reducing poverty and achieving low-carbon sustainable growth (UNEP, 2011; UNEP et al., 2012; United Nations, 2015).

The WEF and multiple UN organisations (e.g., UNESCO, UNEP, UNDP) take an active, even leading, part in building and disseminating the green economy and SDGs in the tourism field and beyond. However, this cannot be possible without an effective collaboration with various private and public institutions, international organisations, and governments. One of the key constructs of the Great Reset is stakeholder capitalism (Schwab, 2021). In contrast to shareholder capitalism, theoretically, it seeks to create and establish long-term business values and strategies beneficial for all the stakeholders, without a focus on shareholders. Thus, from the perspective of stakeholder capitalism, an ideal business model in tourism should be focused on the interests of all the stakeholders in the field: tourists, tourism industry employees, and employers, local communities (Sunley, 1999), companies, service providers, public and private organisations, municipalities and governments, equally to the interests of shareholders (Perić, Vitezić, & Đurkin, 2017). This means that the overall competitive landscape has been changed in a way that sustainability has become the driving force of the firm and its decision-making. The wider community, including the environment, has been acknowledged as a true stakeholder and become the priority for decision-makers. The reflection of this reversal can be seen in the emergence of so-called sustainability business models (SBMs) or business models for sustainability (BMfS) (Stubbs & Cocklin, 2008). Accepting sustainability as a business strategy seems to be beneficial, since it can create new opportunities and generate additional revenues even while profit is needed for the firm's survival, though it is not the firm's only mission (Perić et al., 2017). However, the new normal has forced businesses to quest for sustainability and resilience even further. We suggest that organisations need resilient business models (RBMs) that will incorporate sustainability and resilience into their strategy and, in parallel, consider a wide range of stakeholder interests. An RBM is therefore holistic, hybrid, flexible, and capable of comprehending and dealing with external forces negatively affecting the tourism industry. Besides capturing economic values, RBMs create value for customers as well as for a wider community. This requires a systemic approach, great flexibility of the internal system, and a long-term focus on the community and environment perspective.

Does this make the role of shareholders and entrepreneurship a less attractive activity, increasing responsibilities and decreasing potential income? We believe it does not. It reshapes the vision and point of view on the business itself, transforming values. According to Schwab (2021), it is dedicated to the global economy that works for progress, people, and the planet, contributing to SDGs and considering the needs of all tourism ecosystem stakeholders, and society at large. We conclude that this means that economic and societal profits must be co-created simultaneously.

Another challenge refers to employment. The future of employment in tourism and hospitality may seem vague (Baum et al., 2019), as total automation, digitalisation, robotisation (Ivanov & Webster, 2020), sanitary and environmental norms and standards become established. This is coupled with the current economic crisis and the consequent bankruptcy of many tourism-related companies, so the question about the future of employment in tourism remains open. This is a huge issue because the industry is traditionally based on the concurrent interplay between a provider/host and a tourist. On the one hand, machines (or robots) might increasingly replace humans in providing services and forecasting activities (Fan, Gao, & Han, 2022). On the other hand, there remains a need for human empathy, judgement, and evaluation, especially in areas where the desired result is difficult to translate into something machine-friendly. Tourism companies will tend to hire people who can make responsible decisions (which requires ethical judgement), engage and encourage customers and employees (which requires emotional intelligence), and identify new opportunities (which requires creativity). Such employees, who can design and/or implement innovative business solutions through technology, represent talents that become crucial for the companies. No doubt the need for a human-to-human encounter will continue to be present. This would be particularly evident in the high-tier tourism and hospitality market. Universal digitalisation and the presence of intelligent systems in companies will have a synergistic effect on the competitiveness of these companies. This will also contribute to the growth and development of the economy as a whole. Another change is expected in the designing and managing of organisations. We believe the management and organisation structures as we know them have come to an end. A specific combination of people and technology requires new solutions. The human factor will remain present in tourism and hospitality, and where there are people, there is a need for management and organisation – only the role of people and managers, the demands placed on them as well as their competencies, will change and differ from the traditional ones we are used to.

### The Tourism Great Reset: Issues and trends, present and future

The tourism macro-reset is occurring at the global scale, with progress across all five macro-categories (Schwab & Malleret, 2020; Figure 3.1), and which captures all fields of modern human life. At the micro-level the tourism reset seems to be a long, stressful, and complex series of constant modifications and adaptation to a new normal. In this sense, the Fourth Industrial Revolution has shifted only from the Internet and the client-server business models to ubiquitous mobility, bridging the digital and physical environment. This transition and connection between real and virtual environments are further supported by advanced robotics and artificial intelligence (AI), which enable automation and optimisation in completely new ways. Indeed, the notion of physical and geographical distance is shrinking in a globalised world, not only metaphorically but also literally. No innovation can remain a secret for long, resulting in great opportunities to bring industry and society as a whole to a new level.

Tourism depends on mobility, a fundamental human need. Currently, the mobility and transportation network is at the initial stage of massive transformation, as innovative tools and solutions enable transportation to be reshaped so as to fit SDGs. This raises many ethical questions and managers seek out ways to develop a transport system that is smart, inclusive, and eco-friendly, sometimes sacrificing customer-centricity and consumer comfort. Environmental issues and concerns are replacing the client, and experts are looking for a solution to find new ways to make modern mobility systems more sustainable. For instance, the transportation of tourists is a major generator of all human-induced carbon dioxide ($CO_2$) emissions that directly participate in global warming (Grofelnik, Perić, and Wise, 2020). This is particularly true for the global aviation industry. All airline companies actively demonstrate their concern and action through their official announcements and advertising: Emirates, Wizz Air, Air France, among others. Furthermore, the airline industry has pledged to be carbon neutral by 2050 (World Economic Forum Agenda, 2021) by focusing on four key pillars: (1) green fuel, (2) offsets, (3) direct eye capture, and (4) electric and hydrogen. Even the Fédération Internationale de l'Automobile (FIA), the governing body of motor sport, is driving towards new technology and sustainable energy and has decided to introduce some changes. The next generation of top category World Rally Championship (WRC) cars, one of the flagships of motorsport that attracts a global audience, featured a plug-in hybrid unit in 2022.

Despite these commendable initiatives and efforts, this may not be enough. Other innovative solutions are emerging on a regular basis. The Virgin Hyperloop One system is a disruptive innovation in the field of transportation. This system uses capsules that accelerate electrically through a low-pressure tube and, thanks to aerodynamics, floats above the track using magnetic levitation. It is estimated that the maximum speed for a passenger capsule or light cargo will be 670 miles per hour or 1,080 kilometres per hour, which is two to three times faster than levitating trains, and even 10 to 15 times faster than a traditional railway. The whole system is completely autonomous and closed, safe, and clean. It can draw power from any energy source available on the route. If the sun or wind can be used as the energy source, then the whole system is 100% free of direct $CO_2$ emissions. It eliminates human driver errors and weather hazards, and is more efficient in comparison with similar transportation systems. Virgin Hyperloop One is 40% faster than air transport, and per kilometre is about five times cheaper. Compared to road truck transport, the Virgin Hyperloop One is about 50% more expensive per kilometre travelled, but therefore about six times faster. The speed of transport will also reduce the required storage space and associated costs by about 25%. Managers of airlines and road and rail companies might be concerned and should prepare adequate strategies to remain competitive. Technological development moves transportation even beyond our standards and expectations (technological characteristics, ergonomics). Thus, predictions for mass space tourism are an example. We understand space tourism as human space travel for recreational purposes

(Cohen & Spector, 2019). It seeks to allow tourists to experience being astronauts and travel in space for leisure or business. There are many uncertainties and open questions, notably the legal regulation of space tourism (Ryzhenko & Halahan, 2020; Toivonen, 2021), and environmental, ethical, and security issues.

Through all human history, people have been building physical infrastructure; however, with the Great Reset, the focus has moved to digital infrastructure needs. As indicated earlier, 21st-century tourists are hard to perceive without the usage of digital technologies which are key to smart tourism and mobility. However, the building, development, and maintenance of innovative and connected digital infrastructure are time-consuming and complex and the cost remains high. It also embraces various issues, like privacy, data protection, and cyber security. Contrary to popular opinion, virtual tourism pre-dates the computer age. For example, from 1923 to 1934 in the USSR, Soviet virtual tourism was supposed to teach the population how to think about the past, the present, and the future. Hirsch (2003) explains how and why the Ethnographic Department of the Russian Museum can be seen as a venue for virtual tourism, where all visitors immersively experienced an idealised narrative about the socialist transformation of the Soviet Union. This programme was managed by ethnographers and political activists. Thus, virtual tourism is capable of providing travel in time and space, in existing and non-existing destinations, in imaginable and unimaginable ways, without the physical displacement of the tourist: it is virtual and provided by a consuming video content, most probably involving immersive technologies (Beck & Egger, 2018; Voronkova, 2018). Since it does not require any physical mobility and real-life social interaction, scholars noticed the potential of virtual tourism in the recovery of the industry during the COVID-19 pandemic (Lu et al., 2022). Virtual tourism is gradually evolving with the development and mass diffusion of technological products, notably drones (Mirk & Hlavacs, 2014), cameras, media production, and marketing, and immersive technologies (virtual, augmented, mixed reality, immersive environments, Metaverse) (Voronkova, 2018; Glebova, 2020). There is another reason why the virtual might become a generally accepted way of experiencing tourism (see Figure 3.2).

The Museum of the Future in Dubai attracts attention around the world not only due to its truly unique design and futuristic interior and exterior features, but also because it provides exclusive cutting-edge immersive experiences to visitors, following the philosophy that "to reimagine the future, we must be open to new possibilities. Join us on a journey to 2071. Pioneer new worlds and ways of living and return to shape the world for the better". For example, they propose to "inhabit the skies" by virtually travelling to a space station located 600 kilometres above the Earth or "discover new worlds" by exploring the space station and learning about the community of people living and working there. Another immersive experience option is "witness the wonders of nature" in the framework of the Heal Institute; and a digital Amazon section offers immersion in a mixed reality re-creation of the Amazon rainforest alongside the interplay of hundreds of species.

Adding to Lu et al. (2022), tourism is characterised by four major challenges, some of which have already been mentioned earlier in this chapter. First, there is a focus on tourists' experiences. Second, providers have to ensure all safety and security preconditions regarding travel, including the prevention of terrorist attacks, the safe practising of tourism activities, as well as the implementation of a whole set of new measures related to health issues concerning the COVID-19 pandemic. Third, nature, as a key resource for developing tourism, must be preserved from the excessive use and negative influences of tourism activities. Fourth, there is a rise in technology-driven business models. Considering these challenges, it follows that if tourism providers want to respect the sustainable carrying capacity of a

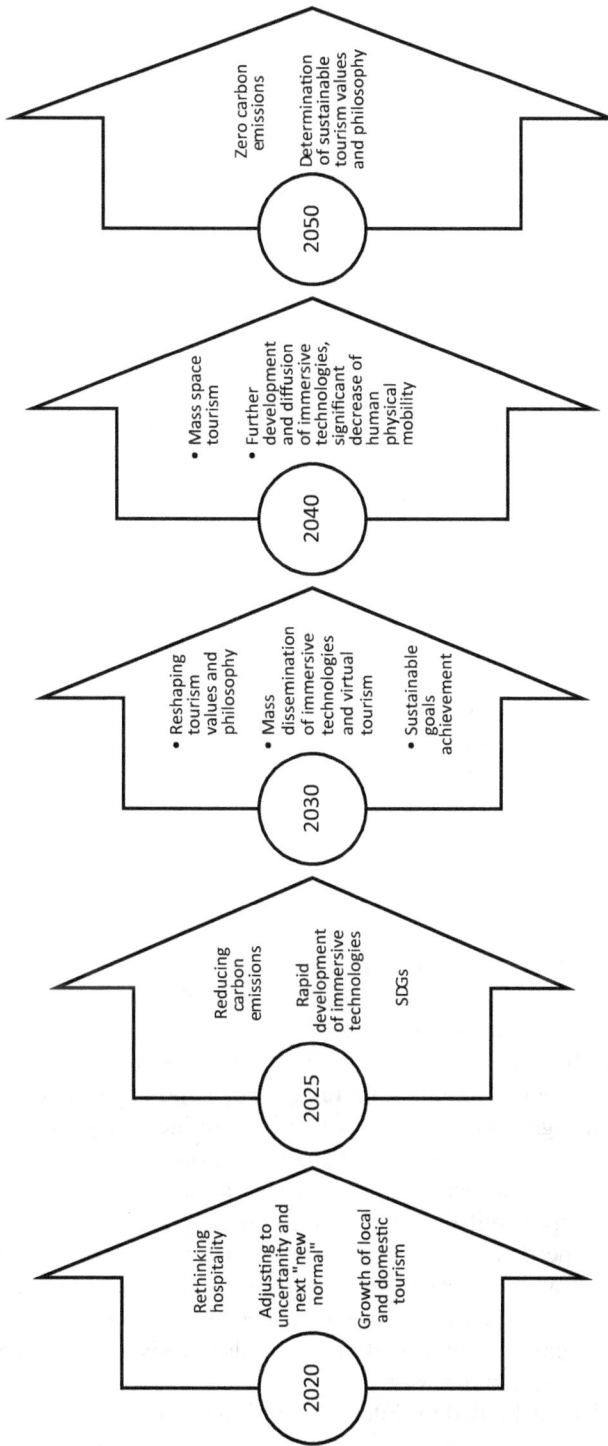

*Figure 3.2* Tourism Great Reset timeline: Forecast attempt.

destination and maintain social distances between tourists, they should limit the supply. If the supply is limited, the price will usually go up, resulting in real experiences not being available to everyone. Only tourists who are wealthy authentic seekers will be able to afford such experiences. In parallel, others will ask for cheaper alternatives to replace real experiences. For this reason, tourism providers will start to offer virtual experiences in specialised hubs that will be equipped with state-of-the-art devices (e.g., visiting a museum, diving experiences). Therefore, virtual tourism is a pathway to inclusion because it is accessible to all potential consumers, regardless of demographics and social, economic, and health status.

Individual reset is linked to changes in consumption, behaviour, employment, and lifestyle in general. International and long-distance tourism tend to be replaced by local and domestic alternatives. This is not only beneficial for economies, but also stimulates infrastructure (physical and virtual), destination image development, and the social, historical, and cultural authenticity of hosting communities. Various campaigns are dedicated to raising tourists' awareness about environmental and social issues and promoting sustainable tourism practices amongst them through different channels. Outer journeys should be a trigger for the transformation of tourists' inner consciousness and changes in daily routines of consumption. In the foreseeable future, ideologically, it seems to be a pointer to a truly meaningful customer experience and journey, shifting tourism values and philosophy through the paradigm of social, economic, and environmental points. Stevens (2021) highlights eight paradigm shifts that will shape how tourism evolves in the post-COVID period. There is a need for hybridity and transformations that will offer novel and co-created solutions that do not fit the traditional perception of tourist offers. Providers need to be flexible when working with other stakeholders in a tourism eco-system, welcome new ideas, and appreciate individuals' talents even more than traditional skills or qualifications. The culmination would be a need to define new metrics of success that will encompass value to the provider and value to society. As discussed by Sheldon (2020), we also believe that the future will favour transformed tourists.

## Conclusion

Tourism is facing an iterative series of social, economic, ethical, political, and health system crises, and the best solution for these problems and challenges would be a collective and coherent response by all actors in the modern tourism field. The Tourism Great Reset requires a well-thought-out and devised new social algorithm that will use new technologies and digitalisation to ensure a fairer, more equitable, and environmentally sustainable society globally. Total digitalisation and automation are inevitably followed by industry transformation: how tourism is consumed, managed, and conceptualised. The uptake and wide diffusion of digital technologies will provide information access for tourists and exciting virtual experience opportunities. Big data collection and analysis will allow the personalisation of consumer needs and deliver enhanced customer experiences (Stylos et al.). Such experiences will have a deeper meaning for tourists whose daily behaviour will be changed. This can be considered a factor of the further development of sustainable tourism and, consequently, the achievement of sustainable goals. This is how tourism has a chances to be inclusive, sustainable, and innovative.

In this chapter, we have exhibited the intersection of tourism industry and the Great Reset concept in a holistic manner. We have identified and outlined current tourism trends and future perspectives through the prism of ongoing socio-economic metamorphoses and the

Great Reset values. This involved reflections about tourism's new normal based on a literature review and expert analysis, followed by real examples and cases. Studying the researched phenomena through a prism of qualitative analysis has allowed us to understand, explain, describe, and visualise the process and direction of modern transformations in the tourism industry. This contributes to tourism theory and practice by identifying the prospective directions for future research and development. How to achieve more inclusive, sustainable, and resilient tourism business models as well as examining tourists (transformed) behaviours and experiences will remain the priority of future studies. The overall transformation will not happen overnight and all stakeholders should be involved in this process. Tourism and hospitality practitioners could use this chapter's argumentation when evaluating new business opportunities and developing innovative tourism offerings. Policymakers and educational institutions should not wait for this change to happen by itself. They need to introduce novel content in educational curricula (at all levels) to alter the ways of thinking and behaving of current and future business leaders as well as tourists. The pursuit of an economic, societal, geopolitical, environmental, and technological reset should be a major challenge for academics and practitioners in the future.

## References

Baum, T., Solnet, D., Robinson, R., & Mooney, S. K. (2019). Tourism employment paradoxes, 1946-2095: A perspective article. *Tourism Review*, 75(1), 252–255. DOI: 10.1108/TR-05-2019-0188.

Beck, J., & Egger, R. (2018). Emotionalise Me: Self-Reporting and Arousal Measurements in Virtual Tourism Environments. In Stangl, B. & Pesonen, J. (eds), *Information and Communication Technologies in Tourism 2018* (pp. 3–15). Cham: Springer.

Buhalis. (2022). *Encyclopedia of Tourism Management and Marketing*. Edward Elgar Publishing. DOI: 10.4337/9781800377486.

Butcher, J. (2020). Constructing mass tourism. *International Journal of Cultural Studies*, 23, 136787792091192. DOI:10.1177/1367877920911923.

Butler, R., & Wall, G. (1985). Introduction: Themes in research on the evolution of tourism. *Annals of Tourism Research*, 12, 287–296. DOI:10.1016/0160-7383(85)90001-5.

Cohen, E., & Spector, S. (Eds.). (2019). *Space Tourism: The Elusive Dream*. Bingley, UK: Emerald Group Publishing.

Everingham, P., & Chassagne, N. (2020). Post COVID-19 ecological and social reset: moving away from capitalist growth models towards tourism as Buen Vivir. *Tourism Geographies*, 22(3), 555–566. DOI: 10.1080/14616688.2020.1762119.

Fan, H., Gao, W., & Han, B. (2022). How does (im)balanced acceptance of robots between customers and frontline employees affect hotels' service quality?. *Computers in Human Behavior*, 133, 107287. DOI: 10.1016/j.chb.2022.107287.

Fletcher, R., Blanco-Romero, A., Blázquez-Salom, M., Cañada, E., Murray Mas, I., & Sekulova, F. (2021). Pathways to post-capitalist tourism. *Tourism Geographies*. DOI: 10.1080/14616688.2021.1965202.

Glebova, E. (2020). Définir la réalité étendue dans les sports : Limitations, facteurs et opportunités. In Desbordes, M., & Hautbois, C. (Eds.), *Management du sports 3.0* (pp. 271–293). Paris: Economica.

Grofelnik, H., Perić, M., & Wise, N. (2020). Applying carbon footprint method possibilities to the sustainable development of sports tourism. *WIT Transactions on Ecology and the Environment*, 248(Sustainable tourism IX), 153–163. DOI: 10.2495/ST200131.

Hall, C. M., Scott, D., & Gössling, S. (2020). Pandemics, transformations and tourism be careful what you wish for. *Tourism Geographies*, 22(3), 577–598. DOI: 10.1080/14616688.2020.1759131.

Hirsch, F. (2003). Getting to know "The peoples of the USSR": Ethnographic exhibits as soviet virtual tourism, 1923–1934. *Slavic Review*, 62(4), 683–709.

Ivanov, S., & Webster, C. (2020). Robots in tourism: A research agenda for tourism economics. *Tourism Economics*, 26(7), 1065–1085. DOI: 10.1177/1354816619879583.

Lu, J., Xiao, X., Xu, Z., Wang, C., Zhang, M., & Zhou, Y. (2022). The potential of virtual tourism in the recovery of tourism industry during the COVID-19 pandemic. *Current Issues in Tourism*, 25(3), 441–457. DOI: 10.1080/13683500.2021.1959526.

Meadows, D. H., Meadows, D. L., Randers, J., & Behrens III, W. W. (1972). *The Limits to Growth: A Report for the Club of Rome's Project on the Predicament of Mankind*. New York: Universe Books.

Mirk, D., & Hlavacs, H. (2014, July). Using Drones for Virtual Tourism. In *International Conference on Intelligent Technologies for Interactive Entertainment* (pp. 144–147). Cham: Springer.

Nhamo, G., Dube, K., & Chikodzi, D. (2020). *Counting the Cost of COVID-19 on the Global Tourism Industry*. Cham: Springer.

Perić, M., Vitezić, V., & Đurkin, J. (2017). Business model concept: An integrative framework proposal. *Managing Global Transitions*, 15(3), 255–274. DOI: 10.26493/1854-6935.15.255-274.

Presenza, A., & Cipollina, M. (2010). Analyzing tourism stakeholders network. *Tourism Review*, 65, 17–30. DOI: 10.1108/16605371011093845.

Purpose Disruption. (2021). Advertised Emissions. The carbon emissions generated by UK advertising: the bigger picture. Report.

Reddy, M. V., & Wilkes, K. (2013). Tourism and Sustainability: Transition to a Green Economy. In Reddy & Wilkes (Eds.), *Tourism, Climate Change and Sustainability* (pp. 3–24). Routledge, chapter 1.

Ryzhenko, I., & Halahan, O. (2020). International legal regulation of space tourism. *Advanced Space Law*, 5, 83–90. DOI: 10.29202/asl/2020/5/8.

Schwab, K. (2021). *Stakeholder Capitalism: A Global Economy that Works for Progress, People, and Planet*. New York: John Wiley & Sons.

Schwab, K., & Malleret, T. (July 9, 2020). *COVID-19: The Great Reset. Agentur Schweiz*. Forum Publishing. ISBN 978-2-940631-12-4. http://reparti.free.fr/schwab2020.pdf [last access 16/05/2022].

Sheldon, P. J. (2020). Designing tourism experiences for inner transformation. *Annals of Tourism Research*, 83, 102935. DOI: 10.1016/j.annals.2020.102935.

Sigala, M. (2021) The Great Reset: Hospitality Redefined. In Wilks, J., Pendergast, D., Leggat, P. A., & Morgan, D. (Eds.), *Tourist Health, Safety and Wellbeing in the New Normal*. Singapore: Springer. DOI: 10.1007/978-981-16-5415-2_20.

Stevens, T. (2021). Sports tourism: playing the new game. *Tourism and Hospitality Management*, 27(3), 717–722. DOI: 10.20867/thm.27.3.10.

Stubbs, W., & Cocklin, C. (2008). Conceptualizing a "Sustainability business model". *Organization & Environment*, 21(2), 103–127. DOI: 10.1177/1086026608318042.

Stylos, N., Zwiegelaar, J. and Buhalis, D. (2021). Big data empowered agility for dynamic, volatile, and time-sensitive service industries: the case of tourism sector. *International Journal of Contemporary Hospitality Management*, 33(3), 1015–1036. https://doi.org/10.1108/IJCHM-07-2020-0644

Sunley, P. (1999). Space for stakeholding? Stakeholder capitalism and economic geography. *Environment and Planning A: Economy and Space*, 31(12), 2189–2205. DOI: 10.1068/a312189.

Toivonen, A. (2021). *Sustainable Space Tourism: An Introduction*. Bristol, UK; Blue Ridge Summit, PA: Channel View Publications.

UNEP, FAO, IMO, UNDP, IUCN, World Fish Center, GRIDArendal. (2012). Green economy in a blue world. www.unep.org/greeneconomy and www.unep.org/, regional seas, ISBN: 978-82-7701-097-7

United Nations. (2015). Transforming our world: the 2030 agenda for sustainable DEVELOPMENT. sustainabledevelopment.un.org, A/RES/70/1

United Nations Environment Programme (UNEP). (2011). Towards a green economy: Pathways to sustainable development and poverty eradication. https://sdgs.un.org/publications/unep-2011-towards-green-economy-pathways-sustainable-development-and-poverty

Voronkova, L. P. (2018, December). Virtual Tourism: on the Way to the Digital Economy. In *IOP Conference Series: Materials Science and Engineering* (Vol. 463, No. 4, p. 042096). IOP Publishing.

World Economic Forum Agenda. (2021). 4 ways airlines are planning to become carbon neutral. https://www.weforum.org/agenda/2021/11/airlines-industry-climate-neutral-travel/ [accessed 28/01/2022].

World Tourism Organization. (2021). *International Tourism Highlights*, 2020 Edition. Madrid: UNWTO. DOI: 10.18111/9789284422456.

World Tourism Organization (UNWTO). (2021). 2020: Worst year in tourism history with 1 billion fewer international arrivals. https://www.unwto.org/news/2020-worst-year-in-tourism-history-with-1-billion-fewer-international-arrivals#:~:text=2020%3A%20Worst%20Year%20in%20Tourism%20History%20with%201%20Billion%20Fewer%20International%20Arrivals,-All%20Regions&text=Global%20tourism%20suffered%20its%20worst,World%20Tourism%20Organization%20(UNWTO) [accessed 25/01/2022].

# 4 A Dialogue for tourism, climate change, and philanthropy

*Kaitano Dube and Cinà van Zyl*

## Introduction

Climate change is an existential challenge facing the world with global environmental, political, and social consequences. Hence, climate change threatens the existence of many economic sectors across the world, including tourism. In order to protect lives, livelihoods, and economies, urgent action must be taken from an adaption and mitigation perspective, particularly in areas with fragile ecosystems and economies. Given the threat posed by climate change, many academics have described it as an emergency (Davidson et al., 2020). In acknowledging the threat, the United Nations (UN) described the 2021 AR6 report as "Code Red for Humanity" to demonstrate the magnitude of the threat posed by climate change on humanity and the environment (IPCC, 2021; UN, 2021). Despite being a significant contributor to climate change, the tourism industry is also particularly vulnerable to its adverse impacts (Dube, 2022).

The tourism value chain is replete with activities and technologies that release carbon emissions into the atmosphere, contributing to climate change. Transport services and accommodation establishments are the most significant contributors to the sector's global carbon emissions (Chen et al., 2018; Dube, 2021a). There have been growing calls for the sector to do more to offset and drastically reduce its carbon footprint in pursuit of sustainability from a Sustainable Development Goal (SDG) perspective, which is an ethical, justice, and climate change perspective (Stanković et al., 2021; Fan et al., 2022). In that regard, tourists are increasingly demonstrating their willingness to offset the carbon footprint that accumulates during travel (Berger et al., 2022) as the sustainability ethos in the tourism sector gathers momentum. This position has been buttressed by the pandemic that increased consciousness around good environmental practice (Tasci et al., 2021). Doing more to protect the environment is a moral and business imperative inspired by the desire to improve the environment that tourism relies on for its sustenance. Stanković et al. (2021) argue that the issue of ecotourism must equally have an enforceable legal footing to force sustainability amongst tourists in particular. With philanthropy being a voluntary promotion of human welfare, philanthro-tourism can be defined as the process of travelling outside the country of residence for the purpose of using resources and skills to put something back into the communities visited in a responsible, sustainable, and ethical way (van Zyl, 2021a). With the many dimensions of the climate change issue, there have been other attempts to address the environmental and carbon footprint of tourism that has triggered the emergence of environmental philanthropism in the tourism sector. This has manifested as some form of climate philanthropy, such as funding for climate change action, either for mitigation or adaptation activities (Liedong

DOI: 10.4324/9781003291763-5

et al., 2022). The role of philanthropists in the climate change space is well recognised and is visible in the formation and funding of operations at the United Nations Framework Convention on Climate Change (UNFCCC) and their central role in the convening of the annual Conference of Parties (COP) and other gatherings (Morena, 2017; Pill, 2021; Michelson, 2021). The increasing pressure for the tourism sector to be more sustainable (Santos et al., 2022) has significant implications for various economic sectors. This chapter examines the place, contextual trends, and possible future of climate philanthropism in travel, tourism, and hospitality.

**Major trends and issues surrounding climate change and philanthropism**

Climate extremes have been witnessed to be on the increase over the past decade (2010–2020) and the emergence of the COVID-19 pandemic has reinvigorated the demand for the travel and tourism sector to act more responsibly and act toward achieving sustainability (Almeida & Silva, 2020; Nhamo et al., 2020). The issue of environmental philanthropy is not new to the tourism sector, and there has been a fair amount of coverage of how tourism can assist in environmental protection from a business perspective in the context of responsible and sustainable tourism. The tourism industry has often treated the issue of environmental philanthropism as part of its corporate philanthropy (Hallak et al., 2012) through various Corporate Social Investment (CSI) initiatives.

Ashley and Haysom (2006) argue that philanthropy in tourism cases is exhibited in the context of pro-poor tourism business approaches. Lengieza et al. (2021) observed that most tourists who resort to environmental philanthropism are motivated during their visits to address environmental issues. To this end, climate philanthropy can only be defined within the context in which travel, tourism, and hospitality have tackled environmental and social challenges to achieve sustainability. It has to be understood that tourism climate philanthropy concerns the actions that a business takes to address climate challenges, ranging from mitigation to climate adaptation or resilience that host communities extend to bolster efforts at such actions. Such actions are taken as part of a business's CSI efforts or as carbon-offsetting projects by tourists and travel, hospitality, and tourism businesses.

"Tourism climate philanthropy" is a fairly new term since urgency over climate change is a relatively new development and only slightly older than a decade, given that the second major conference on tourism and climate change only took place in Davos, Switzerland, in October 2007 (UNWTO, UNEP, 2008). The issue of tourism, climate change, and philanthropy has not been dealt with separately. Those who have sought to deal with the issue have not made it a standalone issue. Nonetheless, academics have been proposing that all tourism stakeholders should embrace climate change philanthropy to deal with the challenges of climate change on society. Korstanje et al. (2019) argue that part of tourism climate philanthropy deals with encouraging tourists to donate towards clean energy investment, cleaning up physical remains, and funding scientific research in renewable technology such as solar. If such technology is directed at equipping rural and marginalised communities, it can assist in addressing climate change. Often impoverished communities depend on fossil fuels to power their energy needs, contributing to carbon emissions. Investment through philanthropy in such communities has a significant impact on addressing climate change and providing much-needed technology to transform lives in a multifaceted way.

Sinclair-Maragh (2016) notes that tourism and climate change philanthropy in tourism and hospitality must address the SDGs. This can be done by facilitating a process where tourists and travellers for tourism enterprises are given an opportunity to mitigate their environmental footprint, for example by allowing them to offset their carbon footprint by buying carbon credits. Cobbla (2015) argues that tourists have to proactively mitigate climate change. In many instances, this is unlikely to happen if tourism businesses do not take an active and leading role in making this a reality by deliberately providing platforms where tourists can be engaged in climate change efforts, particularly in and around the areas where tourism business operates. This means refining how we define sustainable and responsible tourism. The definition has to deliberately encompass climate change philanthropy as a critical aspect.

Several academics believe that tourism climate philanthropy has to anchor its efforts on climate change mitigation, either through donations or as part of CSI by tourists and tourism enterprises (Ladkoo, 2019; Mabibibi et al., 2021; Pill, 2022; Liedong et al., 2022). The absence of any mention of how the tourism industry can assist in climate change adaptation is of particular concern. Given that a number of tourist resorts and destinations are already battling the irreversible impacts of climate change, there is a strong case for obtaining urgent philanthropic funds to address adaptation needs, particularly in developing countries and Small Island States, where the climate change cost is already nearing or surpassing the GDP of those countries. Without such funding, some resorts and destinations currently categorised as last-chance destinations for sustainability are at significant risk from the physical and biophysical effects posed by extreme weather events and the social and political upheavals that emanate from these adverse impacts. More importantly, evidence shows that climate change can drastically reduce working hours and hamper the productivity of tourism operations owing to extreme heat (Sibitane et al., 2022), flooding, and other severe weather events such as tropical cyclones and fires, which often reduce tourist resorts to disaster zones.

Nonetheless, this should not discount the need for philanthropic injections into climate change mitigation because tourism remains one of the biggest carbon emitters. As such, tourism poses a direct and indirect threat to climate change, inadvertently threatening the sector's own sustainability and viability. The global focus on mitigation and the lack of focus on climate change adaptation have caused a big outcry from the developing world, battling the harsh backlash of climate change. The sector has long been blamed for participating in greenwashing and paying lip service to the issue of climate change, particularly in business-led climate change initiatives.

### Tourism stakeholders' interest in climate change philanthropic projects

There has been some documentation on how the tourism industry is assisting communities and funding environmental projects to better adapt to climate change in Southern Africa. As part of its SDG localisation, the Grootbos Nature Reserve has been making frantic efforts to use some of its tourism products to restore the plant fynbos, which is severely threatened by fire (Dube & Nhamo, 2021) (Figure 4.1). The investment by the company and tourists was geared to assisting with the restoration of part of the Cape Floral Kingdom World Heritage Site in the Gansbaai area, benefiting communities by restoring biodiversity and creating employment. Through the Grootbos Foundation, the tourism company (Grootbos) assists in ecosystem service restoration. Funds from the Foundation, financed by tourists and revenue from the Grootbos Nature Reserve, are being used to clear alien vegetation, helping to protect the fynbos species from the impacts of climate change, and controlling fire in the Hermanus area. Through their attempts and those of other organisations such as the Dyer Foundation, conservation projects have also led to greater

*Figure 4.1* Fynbos conservation and rehabilitation at Grootbos Nature Reserve, South Africa.

efforts in the research and understanding of the transitions occurring within the fynbos region. Most importantly, the area has been kept attractive to tourists as it has been left largely undisturbed.

The Grootbos bee project aims to protect bees under threat of extinction from global changes and invest in food security so that the host community can overcome the challenges imposed by climate change impacts. Other safari companies that have invested in climate change philanthropy in Southern Africa include &Beyond, which has numerous projects addressing mitigation and adaptation efforts (Dube et al. 2020). Some iconic projects include funding community projects in breeding endangered animal species and paying for the protection and conservation of Mnemba Island in Mozambique. The safari company also partnered with several universities to research, manage, and protect the natural heritage worldwide. In other protected areas, such as the South African National Parks, the philanthropists can play a crucial role in the protection of climate-change-endangered species, such as the work that is being conducted in the Richtersveld National Park to protect the semi-desert plants and desert plants that are facing extinction from prolonged droughts, worsened by climate change (Figure 4.2). The established Botanical Garden is an important research hub, and tourists can play a central role in funding such conservation areas across the world.

Reid et al. (2017) note that those hotels that adopted tourism sustainability practices in the Asia Pacific region were making philanthropic donations to community outreach projects to offset climate change vulnerabilities amongst host communities. According to

*Figure 4.2* Desert botanical gardens in Richtersveld National Park.

Peña-Miranda et al. (2021), these hotels adopted philanthropic projects to mitigate climate change at a minimal level as part of their Council for Scientific and Industrial Research (CSIR) projects. Hashmi (2015) reports that as part of their sustainability strategy, the Lancaster London Hotel has heavily invested in green technology in response to the competitive nature of the industry. This includes investing in community projects for climate change mitigation, such as installing electric charging stations.

In the travel sector, particularly the aviation industry, most philanthropic efforts have been directed at ensuring carbon emissions are kept in check, with most work aimed at carbon offsets (Guix et al., 2022; Ritchie et al., 2021; Dube, 2021). Tourists, aircraft manufacturers, and the airline industry fund these offset projects. The main projects are designed to ensure that investment is made in some of the most disadvantaged communities so as to foster environmental care. These philanthropic activities mostly aim at funding carbon-sink projects to compensate for the carbon-emission pollution generated by aviation tourism. Through the projects, tourists and aviation companies purchase credits equivalent to their emissions as compensation by means of voluntary mechanisms.

**The future of tourism and climate change philanthropy**

The current drive and demand from tourists who urge that more effort should be put into climate change action in line with the demands that are espoused in SDG 13 on climate change action and the Tourism Glasgow Declaration suggest that tourism climate philanthropy is likely to see an upsurge in the foreseeable future. The fact that SDG achievement is predicated on the world managing to contain the adverse impacts of climate change and its successes in ensuring environmental protection make climate change action an urgent matter for the tourism industry (Scott & Gössling, 2021; Scott, 2021). Given that the industry is a consumptive and non-consumptive user of the environment, the increasingly environmentally conscious consumer will demand that tourism enterprises do more to protect the fragile ecosystem, including climate action, which is the foundation of the tourism industry. There is evidence that witnessing some of the extreme weather events inflicted by climate change has forced some tourists to be in favour of pro-environmental protection which is likely to influence pro-environmental protection action and funding (Salim et al., 2022). The recent increasing focus on building climate change resilience and climate change adaptation will likely dominate the list of tourist demands at destinations and across the entire tourism value chain.

In developing countries, responsible tourism behaviour is likely to be shaped by the desire to assist host communities to better deal with the impacts of climate change in terms of building resilience. Given the need in some destinations to protect fragile ecosystems, such as coral reefs, heritage sites, and other protected areas, such as national parks whose biodiversity is threatened by climate change, tourism companies and tourists will probably acquire a keen interest in investing to save those resources, in a way which addresses the dictates of the SDGs (Dimopoulos et al., 2021). The SDG framework has been the focal point of sustainable tourism, with several academics recommending its usage in the sector to address the economic, social, and environmental challenges associated with tourism development (Rasoolimanesh et al., 2020a, 2020b; Liburd et al., 2020).

Given the preceding observations, philanthropism is likely to be influenced primarily by the need of tourism stakeholders to address the SDGs relating to affordable and clean energy (SDG7) and industry innovation and infrastructure (SDG9). This will come from intensified efforts in carbon offsets and investments in clean energy production, primarily as the industry's market-based measures respond to climate change. Evidence shows that critical aviation hubs such as airports are investing in carbon markets to deal with tourism aviation's carbon emissions, especially under the Carbon Offsetting and Reduction Scheme for International Aviation (CORSIA) (Pérez-Morón & Marrugo-Salas, 2021). Dube (2021) notes that some airports in various parts of the world, as part of their market-based measures, opt to support rural communities with energy-efficient technologies for cooking, so as to assist communities to diversify from carbon-intensive technologies. Such initiatives are fairly common in the sector as are others such as building solar projects for poor communities as part of their carbon offset strategy.

Equally important to the tourism sector will be philanthropic efforts aimed at ensuring the protection of oceans (SDG14, Life below Water), SDG 15 (Life on Land), and SDG 11, specifically 11.4, on strengthening efforts to protect and safeguard the world's cultural and natural heritage. Coastal areas and oceans are critical hubs for coastal and marine tourism, and the drive and investment in those areas will be largely due to the need to protect the tourism product in those areas (van Zyl, 2021).

With increasing evidence, several heritage sites are potential candidates for last-chance tourism due to climate change and other anthropogenic impacts (Vousdoukas et al., 2022; Trojanowska, 2022). The COVID-19 pandemic greatly affected several heritage agencies' capacity to manage and conserve their heritage sites, particularly those in developing countries, which will worsen the challenges imposed by climate change. The inevitable tendency is to turn to philanthropists for funding climate change solutions.

Evidence also reveals that protected areas such as national parks, which have been battling the adverse impacts of climate change, have witnessed a greater challenge of poaching after the pandemic wiped out the income and revenue from tourism and governments. These were the main support system for conservation funding (Smith et al., 2021). To protect the ever-declining biodiversity (Nhamo et al., 2021), protected areas will have to tap into other forms of funding, and philanthropists could play a role in financing some of the conservation operations. In this space, we are likely to see some significant leveraging by protected area agencies so that they tap into the carbon markets. The philanthropists' role in this regard would be to support enterprises which could assist in developing the carbon market in protected areas and support the host communities in such areas to adapt to and mitigate climate change. This could come in the form of offsets. Supporting communities to deal with the adverse impacts of climate change will reduce poaching pressure, because poverty will probably intensify in developing countries, forcing many to poach for survival.

Climate philanthropists in protected areas can also assist in financing nature-based solution projects as part of a responsible tourism approach, which could help to improve land degraded by fires, droughts, and soil erosion, which have been worsened by climate change. Organisations and tourists alike can assist in this regard to ensure sustainability. In recent decades, as an appreciation of the role of tourism in aiding the climate disaster increases, more elite philanthropism will probably emerge from tourists. Climate philanthropism will be an essential hallmark of many organisations' marketing strategies and campaigns in the future.

## Predictions for 2032 onwards

### *More tourism degrowth and regenerative tourism*

A new era of environmentalism is likely to influence tourists in demanding more action be taken for responsible consumption (SDG12), climate change action (SDG13), and more biodiversity conservation for marine and terrestrial tourism conservation areas (SDG 14 and SDG15). The current debate on tourism degrowth and regenerative tourism is likely to shape tourism policy and practice going forward, given the increased understanding of the threats to tourism that are directly and indirectly linked to climate change. Indeed, this point is aptly expressed by Hall et al. (2020), who propose that tourism degrowth may be a panacea in addressing the challenges posed by tourism growth. Climate change is directly linked to tourism growth, overtourism, and, by proxy, capitalism, a key driver. Cave and Dredge (2020) state that regenerative tourism is equally concerned with ensuring that tourism addresses its failures in the climate change space.

### *Rise in climate change philanthro-tourism*

There is, therefore, anticipation that the tourism industry will make greater philanthropic efforts to address some of its environmental evils by offsetting its carbon emissions in community projects that address various community needs through partnering with communities under the SDG 17 partnership. Communities will increasingly anticipate that tourism enterprises will give back to those communities which have suffered from mass tourism. This expectation is extended to all tourism role players, such as aircraft manufacturers, airlines, airport operators, safari companies, and tourists who for too long have enjoyed the fruits of tourism at the expense of host communities which often lose their ecosystem services due to tourism development. There will be a call for new creative ways of dealing with all transport means in tourism.

*Host community's importance*

Climate change philanthropism will be the focal point of debate as the world marches towards NetZero and just transitions, which would demand greater participation from tourism stakeholders in mitigating and building climate resilience for vulnerable communities. To this end, the tourism industry must formulate policies promoting climate philanthropy. It also demands making greater investments in host communities for overcoming the challenges brought about by climate change and other environmental challenges linked directly and indirectly to tourism enterprises. We should get locals involved to drive the green agenda for the new environmentalism.

*Increase in meaningful travel, tourism, and hospitality*

Now that sustainable travel is here to stay, travellers will want their next holiday to be more meaningful. Future travellers will want to travel better, to consciously support a destination that has suffered from climate change effects. Industry experts in the lead need to plan multi-month and multi-destination family sabbaticals to long-haul destinations. The family could take their own tutor along to keep children up to date with the school curriculum while they are away, building responsible future travellers. The sabbatical would involve a

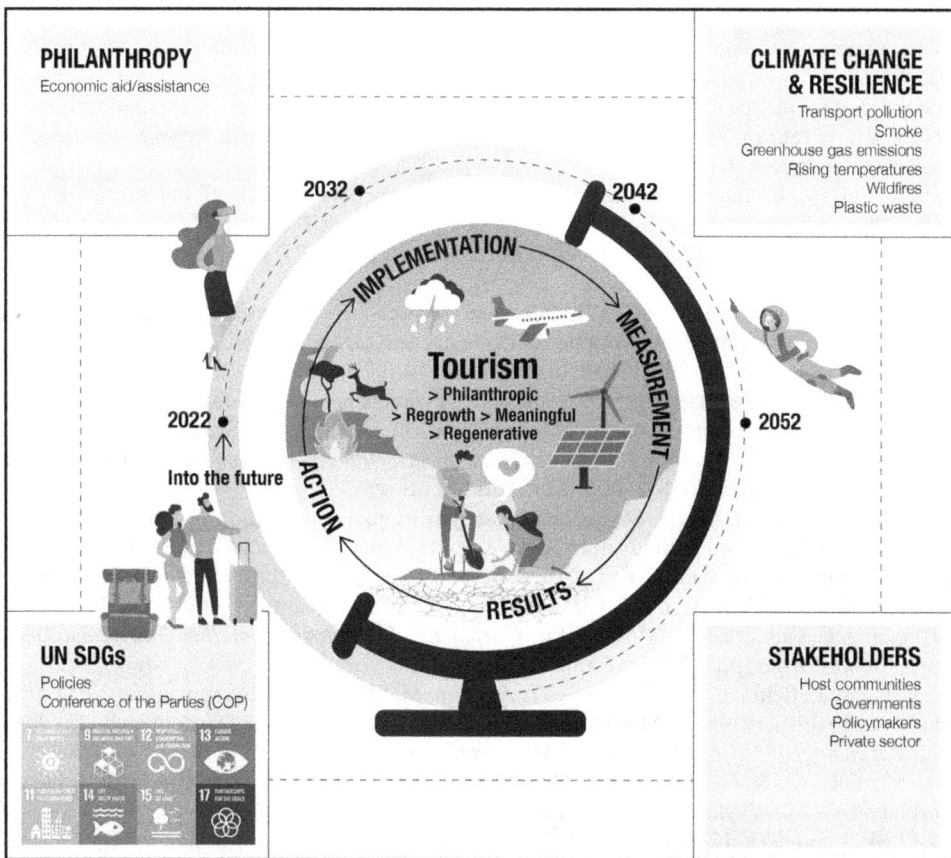

*Figure 4.3* Visual abstract: A dialogue for tourism, climate change, and philanthropy.

key element of giving back, where the family volunteers to help with conservation activities and wildlife/nature monitoring (van Zyl, 2021).

## References

Almeida, F., & Silva, O. (2020). The impact of COVID-19 on tourism sustainability: Evidence from Portugal. *Advances in Hospitality and Tourism Research (AHTR)*, *8*(2), 440–446.

Ashley, C., & Haysom, G. (2006). From philanthropy to a different way of doing business: Strategies and challenges in integrating pro-poor approaches into tourism business. *Development Southern Africa*, *23*(2), 265–280.

Berger, S., Kilchenmann, A., Lenz, O., & Schlöder, F. (2022). Willingness-to-pay for carbon dioxide offsets: Field evidence on revealed preferences in the aviation industry. *Global Environmental Change*, *73*, 102470.

Cave, J., & Dredge, D. (2020). Regenerative tourism needs diverse economic practices. *Tourism Geographies*, *22*(3), 503–513.

Chen, J., Zhao, A., Zhao, Q., Song, M., Baležentis, T., & Streimikiene, D. (2018). Estimation and factor decomposition of carbon emissions in China's tourism sector. *Problemy Ekorozwoju*, *13*(2), 91–101.

Cobbla, J. (2015). Using tourism to mitigate against climate change: The case of the Caribbean. Destination competitiveness, the environment and sustainability: Challenges and cases. *CABI Series in Tourism Management Research*, *2*, 73.

Davidson, K., Briggs, J., Nolan, E., Bush, J., Håkansson, I., & Moloney, S. (2020). The making of a climate emergency response: Examining the attributes of climate emergency plans. *Urban Climate*, *33*, 100666.

Dimopoulos, D., Queiros, D. & van Zyl, C. (2021). Perspectives on the impact of external risks on the future of dive tourism at a high latitude reef complex in the Indian Ocean Region. *Journal of the Indian Ocean Region*, *17*(2).

Dube, K. (2021a). Climate Action at International Airports: An Analysis of the Airport Carbon Accreditation Programme. In G. Nhamo, & D. D. Chikodzi (Eds.), *Sustainable Development Goals for Society Vol. 2. Sustainable Development Goals Series* (pp. 237–251). Springer. doi:10.1007/978-3-030-70952-5_16

Dube, K. (2021b). Climate Action at International Airports: An Analysis of the Airport Carbon Accreditation Programme. In G. Nhamo, D. Chikodzi, & K. Dube (Eds.), *Sustainable Development Goals for Society* (pp. 237–251). Cham: Springer. doi:10.1007/978-3-030-70952-5_16

Dube, K. (2022). Nature-based tourism resources and climate change in Southern Africa: Implications for conservation and sustainable development. In L. Stone, M. Stone, P. Mogomotsi, & G. Mogomotsi (Eds.), *Protected Areas and Tourism in Southern Africa: Conservation Goals and Community Livelihoods* (pp. 160–173). New York: Routledge.

Dube, K., & Nhamo, G. (2021). Sustainable development goals localisation in the tourism sector: Lessons from Grootbos Private Nature Reserve, South Africa. *GeoJournal*, *86*(5), 2191–2208.

Dube, K., Nhamo, G., & Mearns, K. (2020). & Beyond's Response to the Twin Challenges of Pollution and Climate Change in the Context of SDGs. In G. Nhamo, G. Odularu, V. Mjimba, G. Nhamo, G. Odularu, & V. Mjimba (Eds.), *Scaling up SDGs Implementation. Sustainable Development Goals Series* (pp. 87–98). Cham: Springer.

Fan, Y., Ullah, I., Rehman, A., Hussain, A., & Zeeshan, M. (2022). Does tourism increase $CO_2$ emissions and health spending in Mexico? New evidence from nonlinear ARDL approach. *The International Journal of Health Planning and Management*, *37*(1), 242–257.

Guix, M., Ollé, C., & Font, X. (2022). Trustworthy or misleading communication of voluntary carbon offsets in the aviation industry. *Tourism Management*, *88*, 104430.

Hall, C. M., Lundmark, L., & Zhang, J. J. (2020). *Degrowth and Tourism*. Oxfordshire: Routledge.

Hallak, R., Brown, G., & Lindsay, N. J. (2012). The place identity–performance relationship among tourism entrepreneurs: A structural equation modelling analysis. *Tourism Management*, *33*(1), 143–154.

Hashmi, G. (2015). *Lancaster London Hotel Glamour and Sustainability: Thriving on a Strong Sustainability Culture as a Pioneer in Hospitality*. Business School Lausanne, Switzerland: Greenleaf Publishing.

IPCC. (2021). *AR6 Climate Change 2021:The Physical Science Basis*. IPCC. Retrieved January 25, 2022, from https://www.ipcc.ch/report/ar6/wg1/#FullReport

Korstanje, M., Strang, K., & Tzaneli, R. (2019). Tourism Security after Climate Change. In M. Korstanje (Ed.), *The Anthropology of Tourism Security* (pp. 27–48). New York: Nova Science Publishers.

Ladkoo, A. D. (2019). Sensitization of Tourists about Climate Change and Its Associated Impacts on the Tourism Sector. The Case of Mauritius. In Dogan Gursoy & Robin Nunkoo (Eds.), *The Routledge Handbook of Tourism Impacts* (pp. 327–337). London: Routledge.

Lengieza, M. L., Swim, J. K., & Hunt, C. A. (2021). Effects of post-trip eudaimonic reflections on affect, self-transcendence and philanthropy. *The Service Industries Journal, 41*(3–4), 285–306.

Liburd, J., Duedahl, E., & Heape, C. (2020). Co-designing tourism for sustainable development. *Journal of Sustainable Tourism*, 1–20. doi:10.1080/09669582.2020.1839473

Liedong, T., Ajide, O. E., & Osobajo, O. A. (2022). Tackling Climate Change in Africa Through Corporate Social Responsibility. In U. Idemudia, F. X. Tuokuu, & T. Liedong (Eds.), *Business and Sustainable Development in Africa Medicine or Placebo?* Routledge.

Mabibibi, M. A., Dube, K., & Thwala, K. (2021). Successes and challenge in sustainable development goals localisation for host communities around Kruger National Park. *Sustainability, 13*(10), 5341.

Michelson, E. S. (2021). Science philanthropy, energy systems research, and societal responsibility: A match made for the 21st century. *Energy Research & Social Science, 72*, 101886.

Morena, E. (2017). Follow the money: Climate philanthropy from Kyoto to Paris. In J. F. Stefan, & C. Aykut (Eds.), *Globalising the Climate* (pp. 95–115). London: Routledge.

Nhamo, G., Chikodzi, D., & Dube, K. (2021). *Sustainable Development Goals for Society* (Vol. 2 ed.). Springer International Publishing. Doi:10.1007/978-3-030-70952-5

Nhamo, G., Dube, K., & Chikodzi, D. (2020). *Counting the Cost of COVID-19 on the Global Tourism Industry*. Springer Nature. Doi:10.1007/978-3-030-56231-1

Peña-Miranda, D. D., Guevara-Plaza, A., Fraiz-Brea, J. A., & Camilleri, M. A. (2021). Corporate social responsibility model for a competitive and resilient hospitality industry. *Sustainable Development*. Doi: 10.1002/sd.2259

Pérez-Morón, J., & Marrugo-Salas, L. (2021). CORSIA Evolution: A Global Scheme for a Sustainable Colombian Aviation Industry. In L. M.-S. James Pérez-Morón (Ed.), *Environment and Innovation Strategies to Promote Growth and Sustainability* (pp. 120–150). CRC Press. Doi:10.1201/9781003 136712-8

Pill, M. (2021). Linking solidarity funds and philanthropic giving to finance loss and damage from climate change related slow-onset events. *Current Opinion in Environmental Sustainability, 50*, 169–174.

Pill, M. (2022). Towards a funding mechanism for loss and damage from climate change impacts. *Climate Risk Management, 35*, 100391.

Rasoolimanesh, S. M., Ramakrishna, S., Hall, C. M., Esfandiar, K., & Seyfi, S. (2020a). A systematic scoping review of sustainable tourism indicators in relation to the sustainable development goals. *Journal of Sustainable Tourism*, 1–21. Doi:10.1080/09669582.2020.1775621

Rasoolimanesh, S. M., Sundari, R., Michael Hall, K., Esfandiar, K., & Seyfi, S. (2020b). Systematic review of indicators of sustainable tourism in relation to the goals of sustainable development. *Journal of Sustainable Tourism*, 1–21. Doi:10.1080/09669582.2020.1775621

Reid, S., Johnston, N., & Patiar, A. (2017). Coastal resorts setting the pace: An evaluation of sustainable hotel practices. *Journal of Hospitality and Tourism Management, 33*, 11–22.

Ritchie, B. W., Kemperman, A., & Dolnicar, S. (2021). Which types of product attributes lead to aviation voluntary carbon offsetting among air passengers? *Tourism Management, 85*, 104276.

Salim, E., Ravanel, L., & Deline, P. (2022). Does witnessing the effects of climate change on glacial landscapes increase pro-environmental behaviour intentions? An empirical study of a last-chance destination. *Current Issues in Tourism*, 1–19. Doi:10.1080/13683500.2022.2044291

Santos, M. C., Veiga, C., Santos, J. A., & Águas, P. (2022). Sustainability as a success factor for tourism destinations: A systematic literature review. *Worldwide Hospitality and Tourism Themes*, *14*(1), 20–37.

Scott, D. (2021). Sustainable tourism and the grand challenge of climate change. *Sustainability*, *13*(4), 1966.

Scott, D., & Gössling, S. (2021). Destination net-zero: what does the international energy agency roadmap mean for tourism? *Journal of Sustainable Tourism*, 1–18. Doi:10.1080/09669582.2021. 1962890

Sibitane, Z., Dube, K., & Lekaota, L. (2022). Global warming and its implications on Nature Tourism at Phinda Private Game Reserve, South Africa. *International Journal of Environmental Research and Public Health*, *19*(9), 5487.

Sinclair-Maragh, G. (2016). Climate change and the hospitality and tourism industry in developing countries, climate change and the 2030 Corporate Agenda for Sustainable Development. *Advances in Sustainability and Environmental Justice*, *19*, 7–24.

Smith, M. K., Smit, I. P., Swemmer, L. K., Mokhatla, M. M., Freitag, S., Roux, D. J., & Dziba, L. (2021). Sustainability of protected areas: Vulnerabilities and opportunities as revealed by COVID-19 in a national park management agency. *Biological Conservation*, *255*, 108985.

Stanković, V., Batrićević, A., & Joldžić, V. (2021). Legal aspects of ecotourism: Towards creating an international legislative framework. *Tourism Review*, *77*(2), 503–514.

Tasci, A. D., Fyall, A., & Woosnam, K. M. (2021). Sustainable tourism consumer: Socio-demographic, psychographic and behavioral characteristics. *Tourism Review*, *77*(2), 341–375.

Trojanowska, M. (2022). Climate change mitigation and preservation of the cultural heritage—A story of the Municipal Park in Rumia, Poland. *Land*, *11*(1), 65.

UN. (2021). *Meetings Coverage and Press Releases Secretary-General Calls Latest IPCC Climate Report 'Code Red for Humanity', Stressing 'Irrefutable' Evidence of Human Influence*. Retrieved March 12, 2022, from https://www.un.org/press/en/2021/sgsm20847.doc.htm#:~:text=Today's%20 IPCC%20Working%20Group%201,of%20people%20at%20immediate%20risk

UNWTO, UNEP. (2008). *Climate Change and Tourism Responding to Global Challenges*. World Tourism Organization and United Nations Environment Programme. Doi:10.18111/9789284412341.

van Zyl, C. (2021a). Introduction. In V. Katsoni, & C. van Zyl (Eds.) *Culture and Tourism in a Smart, Globalized, and Sustainably World*. Cham, Switzerland: Springer.

van Zyl, C. (2021b). Responsible Ocean Governance Key to the Implementation of SDG 14. In Walter Leal Filho et al. (Eds), *Encyclopaedia of the UN Sustainable Development Goals. Life Below Water*. Cham, Switzerland: Springer.

Vousdoukas, M. I., Clarke, J., Ranasinghe, R., Reimann, L. K. N., Duong, T. M., & Simpson, N. P. (2022). African heritage sites threatened as sea-level rise accelerates. *Nature Climate Change*, *12*, 256–262.

# 5 Baby boomers and sustainable tourism

## The need for a new research agenda

*Ian Patterson and Adela Balderas-Cejudo*

## Introduction

In an ageing world, there is a growing need for a more sustainable and balanced relationship between the planet and its inhabitants. As a result, sustainable tourism has become an increasingly popular field of research since the late 1980s and is now the new "buzz" word for tourism operators in the global economy in the 21st century. Poon (1989) described the emergence of "new tourists" who are ecologically aware, demand more environmental resource-based experiences, and are becoming sensitive to the actual environmental quality of destinations.

This is because sustainable tourism is helping to provide a balance between the three dimensions of tourism development – environmental, economic, and socio-cultural – which aims to guarantee the long-term sustainability of tourism through the maintenance of cultural integrity, important ecological processes, biodiversity, and life support systems (UNWTO, 2005). Sustainable tourists are generally opposed to traditional mass tourism or as it is termed "overtourism" because of the negative effects that it causes to the health of the environment. That is, pollution of beaches, environmental damage to marine life, and uncontrolled waste management. This is becoming of increasing concern as the global population continues to grow, resulting in climate change driven impacts that are now beginning to accelerate (Intergovernmental Panel on Climate Change, 2013). Furthermore, local communities are prevented from being able to enjoy the natural environment because resort developers may have purchased the beach areas solely for the use of guests. Social impacts include increased congestion and pollution caused by growing numbers of tourists, as well as an increase in house prices; sometimes criminality in the local area is common (Weeden & Boluk, 2014).

The aim of this chapter is to undertake a systematic review of the academic literature to determine how the behaviour of older tourists, and destination management organisations (DMOs), have changed in a post-COVID era, and to predict whether older tourists are undertaking new and different types of travel behaviour that are specifically promoting sustainable tourism practices.

## Our ageing world

Not only is the global population growing, but it is also ageing. The United Nations stated in its latest report *World Population Ageing* (2019) that ageing is a global phenomenon and that, by 2050, one in six people or 16% will be over the age of 65, which is a rise from one in eleven or 9% in 2019. Globally, the number of older adults is expected to double from

DOI: 10.4324/9781003291763-6

2000, 2015, 2030 AND 2050

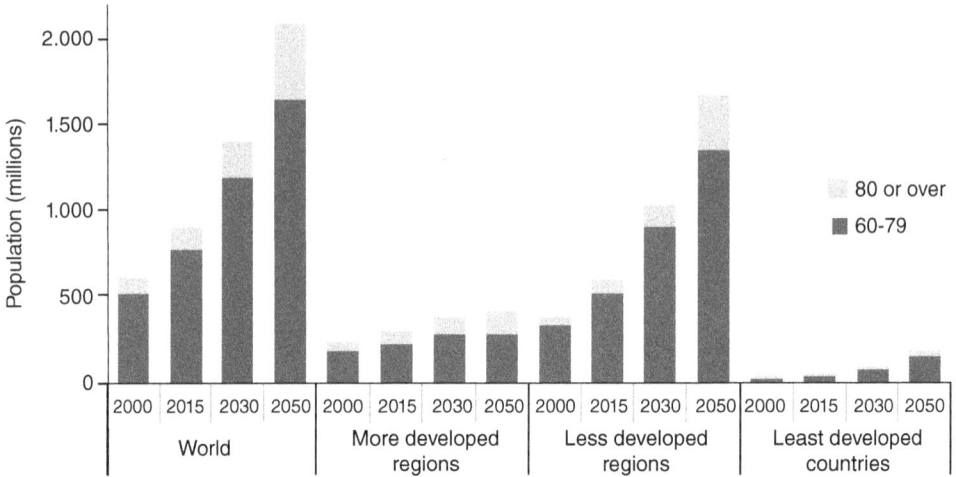

*Figure 5.1* Population aged 60–79 and 80 years and over, by development group.

*Source*: United Nations (2015).

around 617 million to 1.6 billion by 2050 (He, Goodkind, & Kowal, 2016). According to UN estimates (UN, 2020) the number of older people aged 65 and up will rise by 16% in terms of the resident population between now and 2050. This has been described as a tsunami which will mean that, for the first time in history, older people will outnumber children younger than five years of age in the United States.

As a result of this huge demographic shift, the United Nations (2015) has recognised that the numbers of older adults who are travelling are growing rapidly, with baby boomers now accounting for a greater share of all tourism spending than previous cohorts of older travellers. "Baby boomers" was originally coined as a marketing term that described the segment of the population who were born between the years 1946 and 1964 (aged between 56 and 76 years in 2022). Baby boomers are being increasingly targeted by marketers and travel companies as an expanding niche market. UNWTO (2005) has predicted that, by 2050, international travellers who will be aged 60 years and over will exceed two billion trips per annum, compared to 593 million in 1999. In the USA, it was projected (AARP Research, 2018) that baby boomers will expect to take four to five leisure trips in 2019, with about only half travelling within the country, and about half travelling domestically and abroad. They plan on spending over $6,600 on their 2019 travels. This generation continues to redefine the traditional ideas of ageing with the firm belief that the 50s and 60s are now part of middle age (Patterson, Balderas et al., 2021). This is primarily attributed to longer life expectancies and significant improvements in their overall health and wellbeing (Cavanaugh & Blanchard-Fields, 2019).

## Sustainable tourism

The UNWTO described sustainable principles as referring to the environmental, economic, and socio-cultural aspects of tourism development, and a suitable balance must be established between these three dimensions to guarantee the long-term sustainability of tourism. This is because the health of the environment is a major concern as the global population continues to grow, resulting in climate change driven impacts that are accelerating. In addition, global urbanisation rates will continue to rise, resulting in a growing realisation that an unhealthy environment can have an adverse impact on individual and community health (IPCC, Climate Change, 2013).

Therefore, sustainability is more than just looking after the natural environment; it is about considering the social and economic impacts of what we do and how we do it. Thus, the pillars of sustainable tourism are environmental integrity, social justice, and economic development. Sustainable tourists are generally opposed to traditional mass tourism. They respect and support the integrity of local cultures by purchasing local goods and participating with small, local businesses. They favour businesses that conserve cultural heritage and traditional values and support local economies that conserve resources that are environmentally conscious through the use of the least possible amount of non-renewable resources (McClimon, 2019).

In simpler terms, sustainable tourism is committed to creating a low impact on the environment and to local cultures, while contributing to the generation of income and employment for the local population. Sustainable tourism interventions require an understanding of the social and demographic trends that influence traveller behaviour, and how well tourism products and service providers are able to cater for these needs and expectations (Haddouche & Salomone, 2018; Robinson & Schänzel, 2019). Let us now discuss the relationship between baby boomers and sustainable tourism.

**Sustainable tourism and baby boomers**

Many older boomers are decisive, confident travellers that are not constrained by budget issues. They often prefer active, outdoor exploration and sightseeing, and are more willing to venture off the beaten path, to engage in adventure travel that, for some, involves strenuous hiking and staying in rustic accommodation, tracking wild animals, and witnessing tribal ceremonies in exotic lands (Lehto et al., 2008). Lehto et al. (2008) concluded that baby boomers often prefer to participate in holiday activities that include long-haul adventure trips, discovery and cultural trips, and volunteering holidays. Boomers generally find pleasure and escape in the outdoors, and they want to share that with their children and grandchildren. They believe their best years are ahead of them, and they emphasise family time and exploration. Nature tourism is a popular pastime as boomers relax outside or explore the natural world in recreation vehicles (RVs), or through a range of different camping experiences.

The lived experiences of older outdoor enthusiasts have become a common narrative. They openly express their real and enduring love for the outdoors and adventure tourism in blogs on the Internet. One older woman narrated her outdoor experience with a Colorado-based company, Walking the World:

> I made arrangements through the company for a seven-day hiking tour of the Canadian Rockies, specifically Banff and Jasper National Parks. I was in a group of six women and two men and two guides, a man, and a woman. All of us were 60 years and older, nevertheless everyone was fit and had some hiking experience. Our guides were expert naturalists and planned daily walks that varied from four to six hours and took us to elevations of 2,800 feet ... we all enjoyed the trip very much. I thought many times

during the trip that travelling with my contemporaries increased my enjoyment. The vistas were the same but the pace was more leisurely. My group of "elderlies" outwalked many younger people, and good spirits and fitness carried us further on the trail than some other groups have ventured.

(Harnik, 1998, p. 42)

Another adventurer, Mary Ballard, a 50-year-old resident of Rochester, New York, who is visually impaired, signed up with Wilderness Inquiry Inc., a non-profit organisation based in Minneapolis, Minnesota that runs trips with, not for, people with disabilities. Mary Ballard described her first sled trek into northern Minnesota three years ago. She recalled vividly the running of the dogs:

"There's the wind in your face," she said. "There's a feeling of freedom, complete peace and wellbeing, skimming over the snow behind a team of dogs." And then there was the night that Ms. Ballard, "jumped into a hole in a lake when it was close to zero degrees, just to do it," she said. "I stood on the edge of the ice hole in my bathing suit. It was 11 o'clock at night in mid-February," she recalled. "Now, this I didn't think I would do. But everybody in the group assumed that everybody else would. So, I did. We had been in a sauna first. We went back into the sauna, came out and jumped in the lake again."

(Hevesi, 5 July, 1987, p. 10)

Several adventure travel agencies are now promoting baby boomer travel by including blogs on their websites about the lived experiences of boomers that promote the importance of nature based and adventure tourism:

Our trip was excellent in every respect – trip leader, local guides, drivers, staff at estancias. Obviously, a great deal of effort was spent on planning the hikes and other activities, travel, accommodations. Our trip leader Rob Noonan and all the guides educated us in every aspect of Buenos Aires and Patagonia – history, geology, meteorology, flora, and fauna. The hikes were strenuous, but we could set our own paces, so, even I, at age 74, was able to complete all the hikes without overtaxing myself.

(Wilderness Travel, 2022)

The positive effects of connecting to the environment are diverse and influences all levels of individual wellbeing and alleviates feelings of social isolation. In addition, studies support that being outdoors reduces stress by lowering the stress hormone cortisol (Gidlow, Randall, et al., 2016). The amount of green space in the neighbourhood and access to a garden or allotment were found to be significant predictors of stress. In addition, regular physical activity, the frequency of visits to green space in winter, and pleasant views from the home were also predictors of positive health outcomes (Ward et al., 2016) (Figure 5.2).

## Trends and issues: Key contributors

Several of the earlier studies of older travellers (Backman, Backman, & Silverberg, 1999; Moisey & Bichis, 1999; Cleaver & Muller, 2002a) used the following terms in the late 1990s and early 2000s: "nature-based tourists", "ecotourists", and "outdoor and adventure tourists" before "sustainable tourism" became the more popular term that is used in the current literature.

| Environmental Factors | Emotional Responses | Behavioural Outcomes |
|---|---|---|
| ★ Physical Space<br><br>★ Natural Environmental Settings | ★ Satisfaction/Dissatisfaction<br><br>★ Excitement/Non-excitement<br><br>★ Responsibility/Negligence | ★ Acceptance and problem resolution<br><br>★ Avoidance |

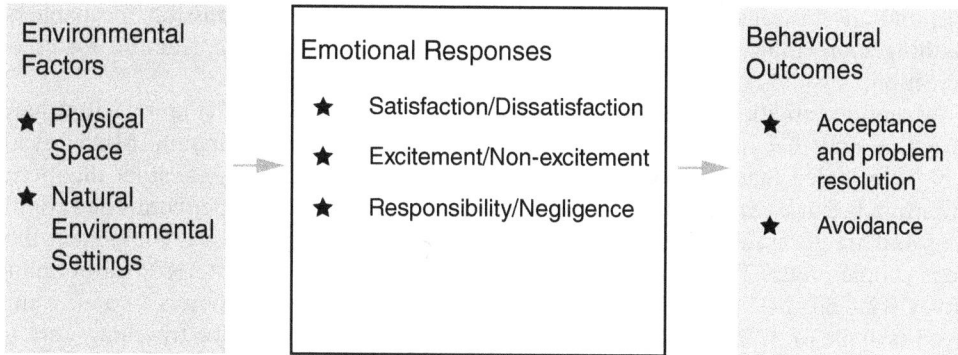

*Figure 5.2* Importance of green space and the social environment.

*Source*: Ward et al. (2016).

Backman et al. (1999) used a commercial mailing list to survey a sample of 334 older adults (36% response rate) who were aged 55 years and older in Carolina and Georgia, USA, travelled regularly, and had an interest in wildlife photography. They further stated that "the senior market was an important sub-segment of the nature-based travelling public" (p. 13). Their study supported the contention that psychographic segments are useful to differentiate the senior nature-based market. Variables such as age, income, gender, education, and health are important to consider as they are based on different aspects of older people's lifestyles. Backman et al. (1999) also revealed that there are distinct segments within the senior market that are very useful to marketers of nature-based travel products. They found that younger respondents in the 55–64 years age group displayed a stronger conservation/protection attitude than the consumptive attitude of the older age group. Furthermore, the younger, older tourists were less interested in the educational/nature benefits than older tourists, and more interested in the benefits associated with camping such as relaxation.

Moisey and Bichis (1999) noted that nature tourism is one of the fastest growing segments of the tourism industry and were interested in examining the motivations and benefits that seniors sought from nature tourism involvement. A total of 421 visitors to the Katy Trail agreed to participate in the survey during the summer of 1998 at Rocheport Missouri, USA. Their results confirmed that the Trail was a challenge for the senior market who were seeking a destination that provides a physical fitness experience in a natural environment. The researchers stated that this is exactly what the nature-oriented baby boomer market is seeking – activities that promote good health and physical fitness, an understanding of nature, and personal enrichment.

Cleaver and Muller (2002) used the term "ecotourists" rather than "nature-based tourists" to describe tourists that are more active, and who try to have a positive impact on the destination. Ecotourists also exhibit a higher level of interest and involvement in conservation than do other tourist types. However, ecotourism is not common to all members of the baby boomer generation, and what they termed the new wave of ecotourists were found to consist primarily of well-educated, environmentally responsible, and socially aware baby boomers.

*Major travel trend: Soft adventure travel*

Baby boomers see themselves as a youthful generation (Fitzpatrick, King, et al., 2013) who seek to learn new things while travelling (Cleaver & Muller, 2002). Some are very active, able, and adventurous travellers, while others are frail, inactive, or possibly incapable of anything more than passive tourism, preferring more relaxing and tranquil forms of recreation.

Adventure activities have been categorised as either soft adventure (Figure 5.3) or hard adventure and that remain at opposite ends of the adventure continuum. Muller et al. (2000) concluded that baby boomers showed a clear preference for soft rather than hard adventure because soft adventure activities are usually conducted under controlled conditions and are generally led by trained guides that supply the educational component that older people prefer. The American Association for Retired People (AARP) travel online survey (Gelfeld, 2017) found that almost half (49%) of the baby boomers who were surveyed (sample of 1,724) stated that the most common motivations for travelling were to spend time with their family and friends (57%), to relax and rejuvenate (48%) (up from 38% in 2017), while 47% were looking for a getaway from everyday life (up from 39%). Further research has revealed that older generations such as 68% of baby boomers and 65% of Generation X, people born between the mid-1960s and the early-1980s (Gen X), are likely to choose sustainable travel options when planning and booking their travel (https://globalnews.booking.com/gen-z-and-the-future-of-sustainable-travel/).

Wilson et al. (2017) were interested in determining what types of adventurous pursuits baby boomer visitors were interested in participating in: soft versus hard. They used a large

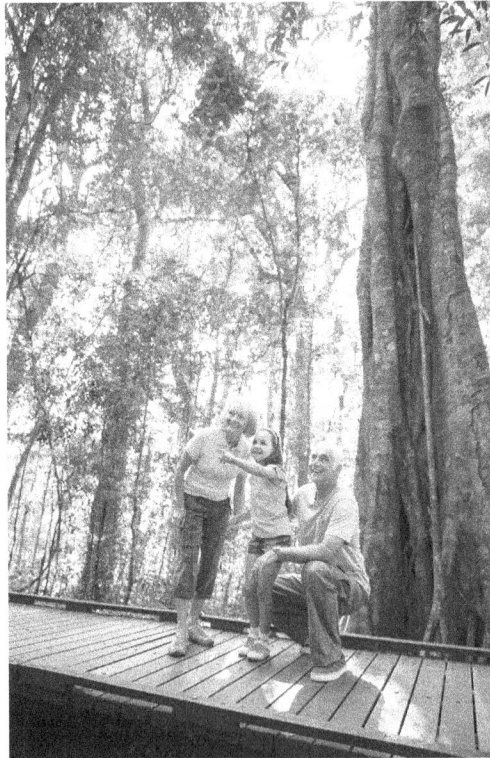

*Figure 5.3* Soft adventure travel experiences for baby boomers.

*Source*: Tourism & Events Queensland.

sample of 403 boomers in Sequoia and Kings Canyon National Parks, USA. They found that, overall, baby boomer visitors were much more active than the typical visitor to national parks, with a higher level of engagement in many different outdoor recreation activities. At the same time, these baby boomer visitors admitted that they were aware that, because they were ageing, they were less prepared for some of the activities that they sought out. The researchers concluded that baby boomers were mainly adventurous, but still thoughtful and careful when contemplating or choosing risky outdoor activities.

### Major travel trend: Mixed hard and soft tourism

It has been noted that baby boomers often choose destinations that offer hard tourism that requires a lot of walking, as well as soft tourism activities that require restful attractions such as good cafes, parks, pleasant views, or local museums and shops to spend the day strolling or relaxing (Scott, 2019). Scott goes on to state that while some baby boomer travellers are super-active and looking for a hard core trek to Everest Base Camp or an intense hiking journey in the Dolomites (a mountain range located in north-eastern Italy), these experiences are not attractive or attainable for everybody.

Most of the adventurous travel destinations, such as hiking in Patagonia, Guatemala, Tasmania, and Spain's Camino de Santiago, and even learning to surf and scuba dive, have become popular in the last decade. DMOs are now catering for older travellers who generally prefer to take things more slowly, such as two-night stays rather than one-night stayovers, and to choose more scheduled free time when they can rest or take in attractions at their own pace compared to younger travellers (Patterson, 2018). By including other types of entertainment such as sing-alongs, games, or informative talks on local bird life, tourist service providers can ensure that less active or soft tourism travellers are well catered for in their programmes (Figure 5.4).

### Major travel trend: Special interest tourism

Streimikiene et al. (2021) noted that older tourists often plan their trips around their hobbies, such as communication and finding of friends, searching for romance, and interest in historical places and events. Rest and calmness are significant, while religion, safety, and health are significant for others. Seniors were more likely to choose destinations where there is calmness, silence, or little noise, concluding that they found these types of experiences more frequently in the natural environment. Zielińska-Szczepkowska (2021) also found that the main reason why older respondents (average age 68.4 years) went on holiday was to enjoy rest and silence. Safety, nature, visiting historical sites, quality of services, and easy transportation connections were found to be the top five attraction factors for seniors when choosing a destination.

### To escape the stress of city life

City living is generally associated with a stressful social environment as well as an increased risk of mental illness, such as the prevalence of mood and anxiety disorders, which is higher in urban areas than in rural areas and where the incidence of schizophrenia is greater in people born and brought up in cities (Yates, 2011). Many older tourists like to choose destinations where there is calmness, silence, or little noise to escape the overcrowding, noise, and pollution of the big cities, which is more frequently found in the natural environment.

*Figure 5.4* Baby boomer guide providing interpretation.
*Source*: Tourism & Events Queensland.

### A strong preference for green hotels

Recently, the term "green tourism" has also been used to link sustainable issues that are associated with tourism (Lu & Nepal, 2009). Specifically, green hotels focus on sustainability in business practices and are very similar to eco-friendly hotels. They differentiate themselves from other hotels with a strong focus on reducing carbon emissions, water usage, waste reduction, and electricity usage. Research in the green hotel sector has confirmed that sustainability can help hotels build a positive image (Lee et al., 2010), enhance guest satisfaction, stimulate consumer behavioural intentions to stay at hotels (Xu & Gursoy, 2015), and to even increase the consumer willingness to pay extra premiums. Businesses are aiming at becoming greener by either participating in certification programmes or by maintaining memberships in green hotel associations. Green hotels are generally built from sustainable materials that aim to leave a low or no carbon footprint, and many are also carbon zero.

### Greater demand for experiential or immersion travel

Here there is a focus on the importance of experiencing a destination at a deeper level, allowing travellers to immerse themselves in the local culture rather than at a superficial level. There is little doubt that experiential travel and the adventure activity segment have gained momentum over the past few years, especially before the outbreak of COVID-19. Soft adventure tourists want personal fulfilment and more enriching lives through travel activities, experiences, and learning. Experiential travel (also called immersion travel) is a type of tourism where visitors prioritise experiencing a destination by engaging with its local customs, culture, and cuisine. This is because many tourists seek authenticity, novelty, and human connection. Of course, self-contained experiences such as resorts and cruises will still remain popular, but many are increasingly preferring travel that promotes hands-on learning experiences. Experiential or immersion travel has been discussed by different travel blogs; however, there is a need to conduct further research on this new niche area of research.

## Future predictions

In 2021 Booking.com undertook a large scale survey of consumers in multiple source markets, including 29,000 travellers across 30 countries and territories, producing a Sustainable Tourism Report. It found that 83% of global travellers thought that sustainable travel is vital, with 61% stating that the pandemic had made them want to travel more sustainably in the future. Travellers revealed that, while on vacation in the past 12 months, 45% made a conscious decision to turn off their air conditioning/heater in their accommodation when they were not there, 43% took their own reusable water bottles, and 33% did activities that supported the local community. Among their other findings, 79% reported wanting to use more environmentally friendly modes of transport, such as walking, cycling, or public transport, rather than taxis or rental cars, while almost three-quarters (73%) wanted to have authentic experiences that are representative of the local culture when they travelled.

Sustainability is more than merely participating in outdoor activities – it also involves caring for the natural environment. According to the United Nations Climate Change Conference (COP26) held in 2021, countries around the world have been asked to secure global net-zero emissions by mid-century to ensure a sustainable future. Hence, sustainable tourism is helping to reduce the carbon footprint, support local communities, and improve the impact that tourism has on the planet, which will be a travel trend for the future.

Figure 5.5 shows the future of sustainable tourism as it applies to the three key pillars of the economic, social, and environmental impacts as well as addressing the needs of host communities.

## Environmental effects of tourism: Opportunities and challenges

In the future, travelling to warmer climates for holidays will remain popular; however, studies are suggesting that many travellers are demanding more responsible holidays (Goodwin, 2021) and to visit new exotic destinations in their search for memorable experiences (Tung & Ritchie, 2011). As a result, it is becoming more likely that in the future the popularity of beach holidays will drop due to the damaging effects of the sun and increased rates of skin cancer. In the past, about half of all tourism took place in coastal areas, with beaches and coral reefs among the most sought-after places. But what will be the effect of global warming with a rise in sea levels by at least 25 metres (82 feet) by the year 2100 (Jarratt & Davies, 2020)? In the USA, almost 30% of the population lives in relatively high-population-density coastal areas where the sea level plays a role in flooding, shoreline erosion, and hazards from storms (Climate.gov, 2020).

New tourist trends are now pushing beach tourism in the opposite direction, through the development of complementary hinterland attractions, including agritourism, and gastronomic and wine tourism. Another alternative strategy by developers is to acquire relatively

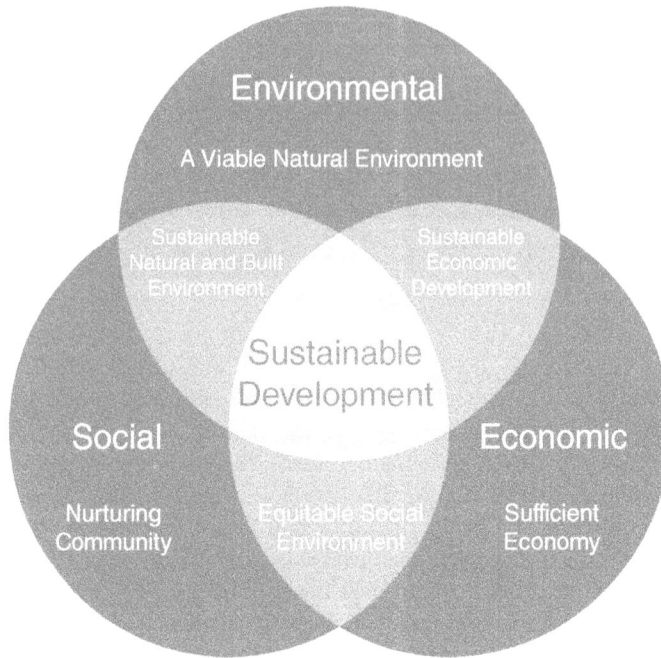

*Figure 5.5* Venn diagram showing how sustainable development ties together a concern for the carrying capacity of natural systems with the social and economic challenges faced by humanity.

underdeveloped coastal areas so as to introduce a wide range of experiences that are not singularly focused on the pleasure beach but offer a different range of activities and attractions that are aimed at more discerning tourists (Picken, 2018).

Another alternative is to bring "the beach to the city" by creating sandy areas in towns and cities by importing sand onto concrete areas. There may also be artificial pools and fairground rides. The most famous urban beach is the Paris Plage, where Parisians and summer tourists have been able to lounge under palm trees on the banks of the river Seine. This cost over €2 million to create and has since been extended due to its popularity (Davies et al., 2020). This will also place less pressure on sand dunes and beach erosion from overtourism.

### Socio-cultural effects: Opportunities and challenges

Sustainable tourist studies need to promote and respect the socio-cultural authenticity of host communities so as to conserve their built and living cultural heritage and traditional values, and to contribute to intercultural understanding and tolerance. In addition, there is a need for an increased awareness of the consequences of the establishing of new hotels and resorts in exotic locations on the society and the people who live there and respecting the cultural and natural heritage of these new and often exotic destination areas.

It is also likely that trips will substantially increase for the purposes of educational and/ or cultural tourism, soft adventure holidays, visiting heritage sites, and volunteering holidays. This is because tourists generally prefer to take holidays where they can learn something new and/or embark on different historical and cultural hands-on experiences

(Patterson, 2018). Educational (including volunteer) tourism is becoming more popular for many who commit themselves to learning about a different culture, and future studies need to highlight the lived experiences of tourists who volunteer to help residents from less developed countries, often in times of crisis.

### Economic effects: Opportunities and challenges

To enable the future growth and development of sustainable tourism it is necessary to provide economic security and funds, as well as to support the economic wellbeing of the population and the local community. However, care must be taken to ensure that economic interests do not become the most important and to respect the needs of the preservation of the environment, and the care of the social progress and lives of the local community (Čepić, 2017).

Future studies need to consider the economic benefits for all stakeholders of host communities in the development of new and exotic destinations. These include increased employment and income-earning opportunities as well as the provision of improved social service payments and public services, such as education and healthcare to support and help to alleviate poverty levels.

### Conclusion

The use of terminology to define sustainable tourism has changed over the past 30 years. To the average tourist, such terms as "nature-based tourism", "ecotourism", or "adventure soft tourism" are perceived to be identical and are often used interchangeably. However, although these terms have great similarities, they also have subtle differences. Overall, they generally refer to practices that minimise the negative impact of visitors on the natural environment, while preserving the local biodiversity and respect for local culture. Sustainable tourism is based around the three key pillars of the economic, social, and environmental impacts as well as addressing the needs of host communities. Terms such as "nature-based" and "ecotourist" have mainly focused on reducing the environmental impacts of tourism, with little consideration for their economic and social impacts.

Another area of future research is the interesting trend toward experiential or immersion travel that has been linked to volunteer tourism as a means of supporting sustainable principles of supporting the integrity of local cultures and traditional values, sometimes in situations of great crisis due to natural disasters. These studies need to show through qualitative research studies the importance of the lived experiences of travellers, rather than reading about them in the travel blogs of newspaper articles and the Internet sources of travel agencies promoting their different trip agendas.

There is no doubt that the terms "sustainable development" and "sustainable tourism" are gaining greater respect and attention from DMOs. As a result, future research studies must carefully consider the social and economic impacts of what we do and how we do it, and to seriously weigh up the impact on the three pillars of sustainable tourism that relate to environmental integrity, social justice, and economic development. In addition, there is a need for a new research agenda that explores the positive relationship between how sustainable tourism contributes to the health and wellbeing of older tourists. Most of the more recent studies have used samples of tourists more generally and have not considered older tourists to any great extent.

## References

AARP (2018). Age 65+ Adults Are projected to outnumber children by 2030. Retrieved at: https://www.aarp.org/home-family/friends-family/info-2018/census-baby-boomers-fd.html

Backman, K. F., Backman, S. J., & Silverberg, K. E. (1999). An Investigation into the psychographics of senior nature-based travellers. *Tourism Recreation Research*, *24*(1), 13–22.

Booking.com (2021). Booking.com's 2021 sustainable travel report affirms potential watershed moment for industry and consumer. https://globalnews.booking.com/bookingcoms-2021-sustainable-travel-report-affirms-potential-watershed-moment-for-industry-and-consumers/

Cavanaugh, J. C., & Blanchard-Fields, F. (2019). *Adult Development and Aging* (8th ed). London: Wadsworth.

Čepić, N. (2017). Economic Effects of Sustainable Tourism. In *10th International Scientific Conference, "Science and Higher Education in Function of Sustainable Development"*, 06–07 October 2017, Mećavnik – Drvengrad, Užice, Serbia.

Cleaver, M., & Muller, T. E. (2002a). I want to pretend i'm eleven years younger: Subjective age and seniors' motives for vacation travel. *Social Indicators Research*, *60*, 227–241.

Cleaver, M., & T. E. Muller. (2002b). "I Want to Pretend I'm Eleven Years Younger: Subjective Age and Seniors' Motives.

Cleaver, M., & T. E. Muller. (2002c). "I Want to Pretend I'm Eleven Years Younger: Subjective Age and Seniors' Motives.

Climate.gov (2020, August 14). Climate change: Global sea level. Retrieved at: https://www.climate.gov/news-features/understanding-climate/climate-change-global-sea-level

Fitzpatrick, M., King, C., & Davey, J. (2013). Getting the focus right: New Zealand baby boomers and advertisements for glasses. *Health Marketing Quarterly*, *30*(3), 281–297.

Gelfeld, V. (2017). *AARP Travel Research: 2018 Travel Trends*. Washington, DC: AARP Research, November 2017. https://doi.org/10.26419/res.00179.001

Gidlow, C. J., Randall, J., Gillman, J., Silk, S., & Jones, M. V. (2016). Hair cortisol and self-reported stress in healthy, working adults. *Psychoneuroendocrinology*, *63*, 163–169. doi:10.1016/j.psyneuen.2015.09.022

Goodwin, H. (2021, June 22). *Consumers are demanding more Responsible Tourism*. Retrieved at: http://traveltomorrow.com/consumers-are-demanding-more-responsible-tourism/

Haddouche, H., & Salomone, C. (2018). Generation Z and the tourist experience: Tourist stories and use of social networks, *Journal of Tourism Futures*, *4*(1), 69–79. 10.1108/JTF-12-2017-0059

Harnik, E. (1998, May 18). Seniors seek adventure hiking Canadian Rockies. *Insight on the News*, *14*, 41

He, W., Goodkind, D., & Kowal, P., (2016). *An Aging World: 2015*. Washington, DC: United States Government Publishing Office.

Hevesi, D. (1987). A sense of adventure, *Travel Insider*, June 10, 2012. *The New York Times*. Retrieved at: November 22, 2021, http://www.smh.com.au/travel/advanced-sense-of-adventure-20120607-1zxvt.html

Intergovernmental Panel on Climate Change. (2013). IPCC publishes full report Climate Change 2013: The physical science basis. Retrieved at: https://www.ipcc.ch/2013/01/30/ipcc-publishes-full-report-climate-change-2013-the-physical-science-basis/

Jarratt, D., & Davies, N. J. (2020). Planning for climate change impacts: Coastal tourism destination resilience policies. *Tourism Planning & Development*, *17*(4), 423–440.

Lee, J. S., Hsu, L. T., Han, H., & Kim, Y. (2010). Understanding how consumers view green hotels: How a hotel's green image can influence behavioural intentions. *Journal of Sustainable Tourism 18*, 901–914.

Lehto, X. Y., Jang, S. S., Achana, F. T., & O'Leary, J. T. (2008). Exploring tourism experience sought: A cohort comparison of baby boomers and the silent generation. *Journal of Vacation Marketing*, *3*, 237–252.

Lu, J., & Nepal, S. K. (2009). Sustainable tourism research: An analysis of papers published in the *Journal of Sustainable Tourism*. *Journal of Sustainable Tourism*, *17*(1), 5–16.

McClimon, T. J. (2019, December 9). The future of preservation is sustainable tourism. *Forbes*. Retrievedat:https://www.forbes.com/sites/timothyjmcclimon/2019/12/09/the-future-of-preservation-is-sustainable-tourism/?sh=186875d16ea5

Moisey, R. M., & Bichis, M. (1999). Psychographics of senior nature tourists: The Katy nature trail. *Tourism Recreation Research*, *24*(1), 69–74.

Muller, T., & Cleaver, M., (2000). Targeting the CANZUS baby boomer explorer and adventure segments. *Journal of Vacation Marketing*, *6*(2), 154–169.

Naidoo, P., Ramseook-Munhurrunb, P., Seebaluck, N. V., & Janvier, S. (2015). Investigating the motivations of baby boomers for adventure tourism. *Procedia – Social and Behavioral Sciences*, *175*, 244–251

Patterson, I. (2018). *Tourism and Leisure Behaviour in an Ageing World*. UK: CABI Oxfordshire.

Patterson, I., Balderas, A., & Pegg, S. (2021). Tourism preferences and perceptions of senior travellers and their impact on healthy ageing. *Anatolia*, *32*(4). DOI: 10.1080/13032917.2021.1999753

Picken, F. (2018). *Beach tourism. The SAGE International Encyclopedia of Travel and Tourism*. Thousand Oaks: Sage.

Poon, A. (1989). Consumer Strategies for a New Tourism. In C. Cooper (ed.), *Progress in Tourism, Recreation and Hospitality Management* (pp. 91–102). London: Bellhaven.

Robinson, V., & Schänzel, H. (2019). A tourism inflex: Generation Z travel experiences. *Journal of Tourism Futures*, *5*(2), 127–141. DOI:10.1108/JTF-01-2019-0014

Scott, S. (2019, February 22). Trend watch – baby boomer travel. Retrieved at: https://www.travel stride.com/blog/trend-watch-baby-boomer-travel. Accessed December 18, 2021.

Streimikiene, D., Svagzdiene, B., Jasinskas, E., & Simanavicius, A. (2021). Sustainable tourism development and competitiveness: The systematic literature review. *Sustainable Development*, *29*(1), 259–271.

Tung, V. W. & Ritchie, J. R. (2011). Exploring the essence of memorable tourism experiences. *Annals of Tourism Research*, *38*(4), 1367–1386.

United Nations, Department of Economic and Social Affairs, Population Division (2015). *World Population Prospects*. New York: United Nations, Department of Economic and Social Affairs, Population Division.

United Nations Department of Economic and Social Affairs (2019). *World Population Ageing 2019. Highlights*. Retrieved at: https://www.un.org/en/development/desa/population/publications/pdf/ageing/WorldPopulationAgeing2019-Highlights.pdf

United Nations. (2019). World population aging. Retrieved at: https://www.un.org/en/development/desa/population/publications/pdf/ageing/WorldPopulationAgeing2019-ypdf

United Nations (2020). World population prospects 2019. Retrieved at: https://population.un.org/wpp/DataQuery/

United States Census. (2018, March 13, 2021). The graying of America: More older adults than kids by 2035. Retrieved at: https://www.census.gov/library/stories/2018/03/graying-america.html. Accessed December 24, 2021.

UNWTO (2005). Sustainability tourism development guidelines. Retrieved at: https://www.unwto.org/sustainable-development

Wang, T-C., Cheng, J-S., Shih, H-Y., Tsai, C-L., Tang, T-W., Tseng, M-L., & Yao, Y-S. (2019). Environmental sustainability on tourist hotels' image development. *Sustainability*, *11*, 2378, doi: 10.3390/su11082378

Ward Thompson, C., Aspinall, P., Roe, J., Robertson, L., & Miller, D. (2016). Mitigating stress and supporting health in deprived urban communities: The importance of green space and the social environment. *International Journal of Environmental Research and Public Health*, *13*(4), 440. 10.3390/ijerph13040440

Weeden, C. & Boluk, K. (2014). *Managing Ethical Consumption in Tourism*. London: Routledge.

Wilderness Travel. (2022). Retrieved at: https://www.wildernesstravel.com/

Wilson, D., Hallo, J., Sharp, J., Mainella, F., & McGuire, F. (2017). Activity selection among baby boomer national park visitors: The search for a sense of adventure. *Journal of Outdoor Recreation and Tourism*, *19*, 37–45.

World Tourism Organization. (2001). *Tourism 2020 Vision: Global Forecasts and Profiles of Market Segments* 7. Madrid, Spain: World Tourism Organization.

Xu, X., & Gursoy, D. (2015). Influence of sustainable hospitality supply chain management on customers' attitudes and behaviors. *International Journal of Hospitality Management, 49*, 105–116.

Yates, D. (2011). The stress of city life. *Nature Reviews Neuroscience, 2*, 430.

Zielińska-Szczepkowska, J. (2021). What are the needs of senior tourists? Evidence from remote regions of Europe. *Economies, 9*(4), 148.

# 6 How big is your tourism footprint?

## In search of the sustainable tourist

*Daisy Kanagasapapathy*

### Introduction

How much longer do we have left on Earth? "The Earth has a deadline," screams Climate-Clock (2022). When the Climate Clock hits zero, it is estimated that the world's carbon budget will be depleted, and the likelihood of devasting global climate impacts will be very high; NASA (2022) predicts the world has until 2030 to make significant reductions in global carbon emissions before the catastrophic damage happens. Notably, excess use and misuse of environmental resources are rising, and consequently the Earth's vital resources are shrinking to an alarming level.

The tourism sector is highly vulnerable to climate change and, at the same time, contributes to the emission of greenhouse gases. $CO_2$ emissions from tourism are forecasted to increase by 25% by 2030 from 2016 levels (UNWTO/IFT, 2019). Tourism also has a range of well-documented negative environmental consequences (Juvan & Dolnicar, 2016; UNWTO, 2022).

Under the banner of "last chance tourism", tourists flock to and are endangering the most vulnerable destinations, such as the Great Barrier Reef, the Everglades of Florida, Mount Kilimanjaro, the Galapagos Islands, visiting rhinoceros in Africa, Machu Picchu, and the Maldives (Abrahams et al., 2022). The effects of this are the world's fastest-melting glaciers, rapidly disappearing corals, water pollution, overcrowding, vanishing islands, and endangered habitats. At this rate, the world will face climate consequences even before the date specified in the original drafting of the Paris Agreement (UN, 2022).

Simultaneously, sustainability continues to grow as an issue for researchers and destinations in the field. Sustainability has traditionally been regarded as a critical factor for competitiveness in tourism destinations; 2017 was the International Year of Sustainable Tourism for Development by The United Nations (UN) 70th General Assembly. This emphasised the importance of government policies, business practices, and consumer behaviour for creating a more sustainable tourism sector that can contribute to the Sustainable Development Goals (SDGs). The concept of sustainability now appears to be present in many government tourism policies. UNWTO constructed tourism as a catalyst for positive change. Yet, the question remains: Why are we not sustainable when we travel? The role of the tourist in sustainable travel can hardly be underestimated as all tourism stakeholders, including tourists, are responsible for creating a sustainable destination (ETC, 2021; UNWTO, 2022).

Thus, this chapter explores tourists' sustainable consumption behaviour by seeking to understand how tourists adopt more sustainable tourism lifestyles. Specifically, the chapter

DOI: 10.4324/9781003291763-7

provides an overview of Malaysian tourism and unsustainable consumption behaviours, addressing the following: How sustainable are you when you travel?

## Understanding sustainable tourism

Sustainable tourism is a phrase we often hear, an idea we rarely publicly question, and a goal we have yet to achieve. In 1962, Rachel Carson catalysed the global environmental movement with *Spring Silent*. She alerted us to the dangers of chemical pesticides, which ultimately led to the creation of the U.S. Environmental Protection Agency (EPA) (Carson, 1962) and the sustainability movement. Following that, the Brundtland Report was released in 1987, where the notion of sustainable development was defined for the first time. The report outlined sustainability as "development that meets the needs of the present without comprising the ability of future generations to meet their own needs" (WCED, 1987). Despite ongoing sustainability efforts, the birth of the SDGs in 2015 by the UN (2019) provided a new direction for safeguarding environmental resources, economic growth, and social equity. Sustainability may have begun as a political movement, but today it is a need. Sachs (2015) points out that sustainable development is central to our age for solving global problems. Sustainability is no longer a niche topic.

Tourism is an environment-dependent industry. It is all about the environment. Mathieson and Wall (1982) remind us that "in the absence of an attractive environment, there would be little tourism. From the basic attractions of sun, sea, and sand to the undoubted appeal of historic sites and structures, the environment is the foundation of the tourist industry" (p. 97). Every place in the world is a tourist destination; as such, sustainability is crucial due to the rapid growth of tourism.

UNWTO (2021) describes sustainable tourism as "tourism that takes full account of its current and future economic, social and environmental impacts, addressing the needs of visitors, the industry, the environment, and host communities". A sustainable approach recognises that whilst tourism can provide many advantages, it also creates burdens that, if not identified and managed, will result in the futures of certain destinations becoming at risk (ETC, 2021). Sustainable tourism practices are, therefore, about planning, developing, and managing tourism that safeguards the latter by:

- Minimising negative impacts and maximising positive effects;
- Conserving natural resources and biodiversity;
- Respecting, celebrating, and preserving cultural traditions and heritage;
- Strengthening local economies and livelihoods;
- Enhancing communities' wellbeing and quality of life and involving them in tourism decision-making.

Sustainable tourism is not a product, a niche, a market proposition, or even a form of tourism – all types of tourism can be made more sustainable. Mowforth and Munt (2016) observe that some businesses and governments deliberately use "sustainable" and allied terms such as "ecotourism" in their marketing as a form of greenwashing to give a false impression of environmental and social responsibility to the public. Interestingly, the lockdown response to the pandemic cleared air pollution and allowed destinations to rest and rejuvenate. In India, the air pollution around the Himalayan Mountain ranges reduced tremendously. The residents living in the Indian state of Punjab were in awe when they could see the Himalayas clearly for the first time in 30 years (Picheta, 2020). The same happened in Nepal: the reduced air pollution cleared the Kathmandu Valley, making it possible to see Mount Everest from some 200 kilometres. The COVID-19 pandemic presents a unique opportunity for change to reduce negative impacts. ETC (2021) and OECD (2020) echo the same reaction. OECD (2020) affirms this as "a once-in-a-lifetime opportunity to move towards more sustainable and resilient models of tourism development".

### Sustainable consumption behaviour

The notion of sustainable consumption behaviour (SCB) has been at the forefront due to its impact on the economy, society, and the environment. The Oslo Symposium in 1994 proposed a working definition of sustainable consumption as

> the use of goods and services that respond to basic needs and bring a better quality of life while minimising the use of natural resources, toxic materials and emissions of waste and pollutants over the life cycle, so as not to jeopardise the needs of future generations.
> (IISD, 2018)

Based on The Oslo Symposium, the transition to sustainable consumption and production patterns must be more than just enabling consumers to buy somewhat sustainable products (Nekmahmud et al., 2022). "Sustainable consumption" is an umbrella term that connects several crucial issues, such as enhancing the quality of life, improving resource efficiency, increasing renewable energy sources, minimising waste, taking a life cycle perspective, and taking the equity dimension into account. Integrating these parts is the central question of how to provide the same or better services to meet the basic requirements of life and the aspirations for improvement for current and future generations while continually reducing environmental damage and risks to human health.

Sustainable tourism should also deliver tourists a high level of satisfaction, presenting them with a meaningful and culturally rich experience, and raising their awareness concerning sustainability practices (Tasci et al., 2022). The sustainable consumption spectrum goes beyond direct consumption. It includes individuals' consumption patterns and emphasises improving their quality of life without focusing on materialistic gains.

Bremner and Dutton (2021) pointed out that the Scandinavian countries and the European region are sustainable travel leaders and setting the pace. The European Union is driving a strong sustainability agenda through its European Green Deal (EC, 2022). The USA lies in 35th position, followed by the UK in 40th position in the Sustainable Travel Index Ranking, while Pakistan, India, Mauritius, Vietnam, the Philippines, Singapore, and Malaysia are at the bottom of the chart for being the least sustainable.

### Tourism footprint

The tourism footprint family comprises the ecological tourism footprint (TEF), the tourism carbon footprint (TCF), and the tourism water footprint (TWF) (Wang et al., 2017). TEF provides a complete valuation of the impact of tourism activities on the environment. TCF focuses on carbon emissions from tourism activities on climate change. Finally, TWF analyses the effects of water consumption from tourism activities on water resources. Tourism has one of the most extensive effects on climate change and global warming. Water resources are overused for hotels, water parks, swimming pools, golf courses, and personal water use by tourists (UNWTO, 2022).

**A fresh look at tourist attitudes: The case of Malaysia**

Tourists' willingness to embrace more responsible and sustainable practices when travelling is a critical factor in determining the success of sustainable tourism. Tourists' attitudes play a crucial role in shaping how the concept of sustainability in travel is marketed, how readily businesses are prepared to implement sustainable practices as part of their daily operations, and ultimately whether a change in practices is successful and achieves real impact.

Malaysia was chosen as the study location to understand SCB from a least sustainable country. Malaysia sits 85th in the Sustainable Travel Index Ranking (Bremner & Dutton, 2021). Located in the heart of the Asia Pacific region, the country had the tallest buildings in the world until 2004, the Petronas Twin Towers in Kuala Lumpur. The country is also home to four World Heritage Sites. Tourism plays a crucial role in Malaysia's economy, where it is the second-largest foreign exchange earner after manufacturing. On average, each international tourist spends about USD715 (TM, 2021). Malaysia is known for its multi-ethnic and multi-cultural society, which are capitalised on by the tourism tagline "Malaysia Truly Asia". The key tourist attractions are its rich natural beauty, cultural heritage, uniqueness of cuisine, and friendliness of its people. To take advantage of these abundant tourism resources and attractions, Malaysia has developed and promoted different kinds of tourism: cultural tourism, ecotourism, education, medical, golf tourism, and MICE (meeting, incentive, conference, exhibition). These products were developed to serve the needs and preferences of a wide range of tourists.

The Malaysian Government declared that sustainable development would remain central to its policies and strategies during the Malaysia Sustainable Development Goals (SDGs) Summit 2019 (FMT, 2019). In 2018, the Ministry of Energy, Science, Technology, Environment & Climate Change (MESTECC) was established with its mission to

> (1) manage energy resources strategically to optimise renewable energy and energy efficiency services as well as to ensure a continuous supply of electricity that is affordable and sustainable; (2) explore, develop and capitalise science and technology based on the country's economic interests through R&D&I strategic collaboration with the industry as a measure to increase commercialisation of technology and labour productivity; and (3) preserve the environment through education, awareness and enforcement towards the pollution-free environment as well as leading climate change adaptation and mitigation measures to ensure country's resilience and create new growth opportunities.
>
> (MESTECC, 2020)

The Malaysian government also has various policies that ideally should aid sustainable tourism development:

1. National Green Technology Policy (NGTP);
2. Malaysia's Roadmap Towards Zero Single-Use Plastics 2018–2030;
3. Green Technology Master Plan (GTMP) Malaysia 2017–2030;
4. National Solid Waste Management Policy 2016;
5. National Energy Efficiency Action Plan (Neeap) 2016–2025;
6. National Policy on the Environment (NPE);
7. National Policy on Climate Change (NPCC).

Despite the country's richness in nature and cultural beauty and the ongoing efforts of the government, Malaysia still faces a significant sustainability challenge. Malaysia's development is now more than ever threatened by its residents' and tourists' unsustainable consumption and production patterns. The government is also facing challenges in encouraging participation to foster sustainable behaviour. Malaysia is ranked eighth among the top ten worst plastic polluters in the world (WWF, 2018). Most plastics are dumped, a small portion burnt, and a tiny fraction recycled. These plastics take approximately 400 years to decompose, equivalent to 16 generations. Plastic has polluted the soil, rivers and oceans. Every year, many sea creatures such as dugongs, turtles, dolphins, and whales die from ingesting plastic in Malaysia. For sea turtles, eating even a single piece of plastic can be deadly. Indirectly, plastic pollution also threatens smaller marine life species, such as fish and clams, and is among protein sources in the human food chain. Malaysia is also known for its haze, air pollution, and rising carbon emissions from coal, oil, and natural gas, which directly affects climate change (BP, 2019).

A crucial question looms: How will Malaysia sustain its present-day resources for future generations if Malaysians and tourists continue these consumption patterns? Much research is still needed concerning tourists' consumption behaviour in the country. Thus, this research aspires to investigate why tourists in Malaysia are reluctant to engage in sustainable tourism practices.

The city of Penang in Malaysia was an ideal location to deploy this research. Penang, the "Pearl of the Orient", is a robust civil society, its state leadership is by the national opposition party, and it has an ethnically and racially diverse population. Penang offers a unique setting for research related to decentralised service delivery, the role of civil society, participatory planning, challenges related to intra- and intergovernmental coordination, tourism as an economic growth strategy, historic preservation, solid waste management, and water management. Penang is also home to a World Heritage Site, Georgetown, which was celebrated in 2008 for its history as a trading and cultural exchange link between the East and West in the Straits of Malacca for over 500 years (UNESCO, 2012). Penang is home to the iconic Penang Bridge (which connects mainland Penang to the island Penang), Penang Hill (with its funicular train), Penang National Park, beautiful beaches, unique architecture, a cultural townscape, botanical gardens, food tourism, and the oldest cross-channel ferry service in the country (Figures 6.1 and 6.2). Penang is one of the top destinations for domestic and international tourists.

**Research methodology**

This research adopts an inductive phenomenological research philosophy to improve the understanding of the sustainable consumption behaviours of tourists. Using ethnographic methods of participant observations and semi-structured interviews, the study was conducted between June 2021 and December 2021. Participant observation is an ethnographic method that seeks to understand the context of everyday life. This method allowed us to inductively build and guide explanations of the behaviour of tourists (Creswell & Plano-Clark, 2018; Stone, 2012). The second stage of the research utilised semi-structured interviews, which drew on a convenient sample of 15 tourists visiting the city of Penang. The respondents were from the UK, Australia, Singapore, and various states within Malaysia – Kedah, Kelantan, Kuala Lumpur, and Selangor. They were eight females and six males. Interviews were conducted within the spirit of "co-authored narratives" and characterised by an appreciation for the interviewee's responses as a "joint social creation" (Kvale & Brinkmann, 2009).

*Figure 6.1* Penang Island.

Photo: Courtesy, Unsplash.com

*Figure 6.2* Penang Hill and its funicular trail.

Photo: Courtesy, Unsplash.com

## Deciphering sustainable consumption behaviour

### *Exploring the unawareness*

The level of SCB in Malaysia is relatively low. From a tourist's decision to take a holiday comes the choices of sustainable tourism, starting with the location, transport, and accommodation; none of these mattered to tourists arriving in Penang due to a lack of awareness of the options provided there. The city is a desired destination for food tourism, beach tourism, and weekend gateways for neighbouring states, which has made it overcrowded, congested, littered, water polluted, and occasionally air polluted.

Most interviewees had heard the word "sustainable" but did not understand how it works in tourism. The interviewees from Kelantan and Selangor said:

> To me, sustainable tourism is new. I am not aware of it. I know ecotourism and homestays. I have not seen any promotions in the hotel about it. This is new and interesting to me. Knowing that I can make a change for my country. I might be willing to try. What can I do? Where can I start it?
>
> Sustainability, yes – I heard about it. But I am unsure what it is about. I don't understand what the need for it is. We are living here fine. The government is building more new attractions. Did you see – they are claiming the sea for a new project in the city? I guess all is all right then.

### *Resistant to change*

Most interviewees are not interested in sustainability because it is too expensive or there is not enough information about it. Interestingly, international tourists are well aware of sustainable tourism yet choose not to practise it. Achieving sustainable tourism consumption behaviours is going to be challenging. An interviewee from the USA mentioned:

> I am well-aware of sustainable tourism. I read an article on how we travellers can be sustainable in our travels. I would love to try some of them someday, but right now, it is inconvenient, and I am not interested. Plus, I love travelling around Asia; there aren't known sustainable practices here.

The lack of understanding of sustainable tourism among tourists may result from the lack of education, especially for domestic tourists. However, it was observed that the younger generation of tourists is willing to engage in sustainable tourism practices. They are seen to practise bringing their reusable water bottles and bags and throwing their waste into recycling bins. This echoed with three of the interviewees from Singapore and Kedah:

> When I travel, I bring my water bottle with me. I encourage my parents and friends to do so. I guess starting small is better than nothing.
>
> I would like to start recycling, but I don't have the opportunity. There aren't any facilities made available for us to use by the local council in the holiday areas.
>
> I turn off my air-conditioner when I leave my hotel room. That's a start. But the rest, I am not used to doing it.

Sustainability is about balancing the environmental, economic, and social impacts in ways that do not burden future generations. Still, the tourists are reluctant to take up this obligation, seriously preferring to assume that technological advances will solve this problem. A domestic tourist from Kedah stated:

> It is a good effort, and we have many apps that can solve the problems. Grab, for example, for transportation and food delivery. I think they are sustainable, which means I am also sustainable in a way.

There was a similar response from a Kuala Lumpur interviewee:

> Sustainability is big hype back in Kuala Lumpur as well; I love coming to Penang for holidays. I try to litter in the bin; sometimes, it is difficult to find one. I don't know – maybe creating an app or robot that solves these sustainability issues for us – why not, right? Malaysia can.

## Political agenda

Sustainability has also been seen as a political agenda that has increased the reluctance of tourists to engage in sustainability practices. As a tourist, they are willing to be involved in policymaking by providing their ideas for consideration. A male interviewee from Selangor commented:

> A joke. Waste of my time. I don't want to be involved in politics. All they do is have a budget for it, and it's all gone. Everything is a con. Because they come with all these policies, they should talk to the public and understand what is happening. Instead of creating policies that they think will look good on paper and expenditure.

This sentiment was echoed by interviewees from the UK and Selangor, who felt:

> Sustainability, if done for the right cause, will definitely change our world. However, I don't see the United Nations' campaign as impactful as Americans are still unsustainable. I don't believe it makes a difference. It looks at government paper, though.
>
> I think it is a way for the government to use the funds easily; I see advertisements all over reminding us to recycle. But how do we recycle – when everything is in one bin when the collection day arrives? Even worse for a tourist, those green hotel owners increase their prices. The way I see it, sustainability is a hocus pocus.

## Keeping with sustainable tourism: Where is it heading?

The consensus is that sustainable consumption is needed, essential, and vital; nevertheless, these positive attitudes do not necessarily translate into sustainable consumption behaviours (Deloitte, 2022). One of the most significant challenges for sustainable tourism development is encouraging tourists to act and minimise environmental impacts. Accordingly, abundant literature on tourists' ecological behaviour, investigating factors influencing their attitudes, and even strategies to educate them on environmental concerns are available. Nevertheless, why are we still seeing unstainable consumption behaviours among tourists?

For sustainable tourism to flourish, firstly it must be acknowledged that the internal barriers preventing tourists from engaging in sustainable practices comes from individuals' lack of knowledge; and the ability to understand the consequences of their unsustainable practices needs to be addressed. Secondly, tourists are also influenced by external aspects related to the accessibility of tourism-related green products, the convenience of accessing them, and the belief that one person cannot make a difference (Nikolic et al., 2021). Sustainable tourism should be affordable for all for changes to happen. A person who spends money on sustainable tourism products does not necessarily become more sustainable. Lopez-Sanchez and Pulido-Fernandez (2015) underline that any tourist that achieves a higher level of commitment, attitude, knowledge, and behaviour toward sustainability becomes more sustainable. Finally, it is time for policymakers to listen to why tourists tend to stick to their unsustainable behaviours despite all the ongoing SDG campaigns. Perhaps, it is time for that change. It was also observed that poor guidelines and governance could be a barrier to sustainable tourism. The future of sustainable tourism depends on understanding and re-educating tourists of the consequences of their actions.

What is needed right now? System change or behaviour change? It needs to be both. Day (2021) and ETC (2021) also echo this argument. When and how do sustainable tourism consumption behaviours strengthen or weaken subsequent behaviours? When do sustainable tourism behaviours lead to a positive spillover, or does it allow a tourist to engage in potentially unsustainable consumption behaviours in the future? Sustainable intelligence might be the key. The ability of a tourist and provider to apply their experience and knowledge concerning the impacts of tourism on the environment where it is practised, and developing proactive behaviour towards sustainable tourism, from the consumption and production viewpoints, are desirable (Goleman, 2006). A tourist with a greater extent of sustainable intelligence empathises with sustainable tourism development in the area where they enjoy their holiday. As a result, these tourists have an intellectual awareness of sustainability, making it easier to incorporate sustainability into tourism's production and consumption processes. Developing sustainable intelligence is critical for tourists to modify their motivations, expectations, and behaviours to embrace a more reasonable and responsible attitude toward the destination. Sustainable tourism is not a goal but a process that needs to be embedded into our daily lives (Figure 6.3). Sustainable tourism as a concept has been an enormous success, but we still have a long way to go to realise its full potential in our tourism ecosystems. The ETC's (2021) efforts can be a model for us to follow. Scandinavia and Europe are setting the pace for sustainable travel. To reiterate, all stakeholders (policymakers, businesses, destination management authorities, tourism operators, local communities, and visitors) play a part in developing a sustainable destination.

How successful will sustainable travel be in 2032 and the years beyond? Sustainable tourism demands the informed participation of all stakeholders and solid political guidance, securing wide-scale involvement and support for sustainable tourism to thrive worldwide. To reach such goals, a continuous process must be enforced to monitor impacts continually and elaborate on preventive and corrective actions. By 2052, the world will have more high sustainable intelligence tourists with the aid of re-educating and sound policies. These tourists will be the ones to move the engine by showing empathy toward sustainable tourism development in the destination where they enjoy their holidays. They will proactively take measures to facilitate the incorporation of sustainability in the processes of production and tourism consumption.

The majority of tourists seek conscious and transformative holidays as a trend where responsible travelling will be automatically embedded in their travel plans. They will want

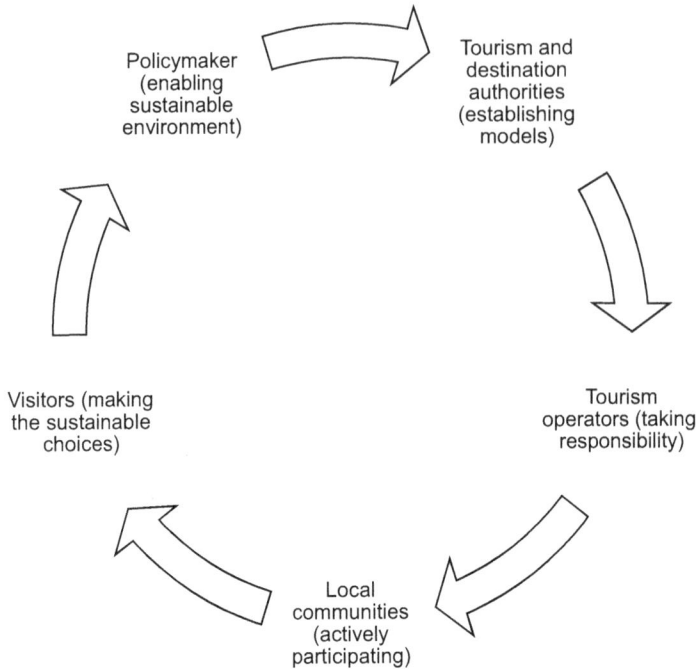

*Figure 6.3* Sustainable tourism as a process.

to embrace meaningful experiences that will help them develop personally and collectively. The future of tourism will be filled with highly educated transformative tourists that want to reinvent themselves and the world.

## Conclusion

This chapter has offered some insights into sustainable tourist consumption and how it contributes to their action of acting sustainably. From the practitioners' perspective, the chapter has provided valuable input to the policymaking, planning, and implementation of sustainable travel.

## References

Abrahams, Z., Hoogendoorn, G., & Fitchett, J. M. (2022). Glacier tourism and tourist reviews: An experiential engagement with the concept of "Last Chance Tourism". *Scandinavian Journal of Hospitality and Tourism 22*, 1–14.

BP. (2019). *BP Statistical Review of World Energy 2019*. T. B. P. C. plc. https://www.bp.com/content/dam/bp/business-sites/en/global/corporate/pdfs/energy-economics/statistical-review/bp-stats-review-2019-full-report.pdf

Bremner, C., & Dutton, S. (2021). *Top Countries for Sustainable Tourism*. London: Euromonitor International.

Carson, R. (1962). *Silent Spring*. Houghton Mifflin Company. https://doi.org/978-0618249060

ClimateClock. (2022). *The Most Important Number in the World*. ClimateClock.world. https://climateclock.world

Creswell, J., & Plano-Clark, V. (2018). *Designing and Conducting Mixed Methods Research*. London: Sage.

Day, J. (2021). Sustainable Tourism in Cities. In A. M. Morrison & J. A. Coca-Stefaniak (Eds.), *Routledge Handbook of Tourism Cities* (pp. 52–64). London: Routledge.

Deloitte. (2022). *Shifting Sands: Are Consumers Still Embracing Sustainability? Changes and Key Findings in Sustainability and Consumer Behaviour in 2021.* Deloitte. https://www2.deloitte.com/uk/en/pages/consumer-business/articles/sustainable-consumer.html

EC. (2022). *A European Green Deal: Striving to be the First Climate-Neutral Continent.* European Commission. https://ec.europa.eu/info/strategy/priorities-2019-2024/european-green-deal_en

ETC. (2021). *Sustainable Tourism Implementation.* E. T. Commission.

FMT. (2019). Malaysia committed to sustainable development, says. *Dr M. FMT News.* https://www.freemalaysiatoday.com/category/nation/2019/11/06/malaysia-committed-to-sustainable-development-says-dr-m/

Goleman, D. (2006). *Social intelligence.* New York: Bantam Books.

IISD. (2018). *Oslo Rountable on Sustainable Production and Consumption.* International Institute for Sustainable Development (IISD). https://enb.iisd.org/consume/oslo004.html#top

Juvan, E., & Dolnicar, S. (2016). Measuring environmentally sustainable tourist behaviour. *Annals of Tourism Research 59*, 30–44.

Kvale, S., & Brinkmann, S. (2009). *InterView.* Copenhagen: Hans Reitzels Forlag.

Lopez-Sanchez, Y., & Pulido-Fernandez, J. I. (2015). In search of the pro-sustainable tourist: A segmentation based on the tourist "sustainable intelligence". *Tourism Management Perspectives 17*, 59–71.

Mathieson, A., & Wall, G. (1982). *Tourism: Economic, Physical, and Social Impacts.* London: Longman.

MESTECC. (2020). *About Us.* MESTECC. Retrieved 16 February from https://pvms.seda.gov.my/pvportal/news/mestecc-is-ministry-of-energy-science-technology-environment-and-climate-change/

Mowforth, M., & Munt, I. (2016). *Tourism and Sustainability: New Tourism in the Third World.* London: Routledge.

NASA. (2022). *Is It Too Late to Prevent Climate Change?* NASA's Jet Propulsion Laboratory. https://climate.nasa.gov/faq/16/is-it-too-late-to-prevent-climate-change/

Nekmahmud, M., Ramkinssoon, H., & Fekete-Frakas, M. (2022). Green purchase and sustainable consumption: A comparative study between European and non-European tourists. *Tourism Management Perspectives 43*, 100980.

Nikolic, T., Pantic, S., Paunovic, I., & Filipovic, S. (2021). Sustainable travel decision-making of Europeans: Insights from a household survey. *Sustainability 13*, 1960.

OECD. (2020). *Rebuilding Tourism for the Future: COVID-19 Policy Responses and Recovery.* O. F. E. C. A. Development. https://www.oecd.org/coronavirus/policy-responses/rebuilding-tourism-for-the-future-COVID-19-policy-responses-and-recovery-bced9859/

Picheta, R. (2020). People in India can see the Himalayas for the first time in 'decades' as the lockdown eases air pollution. *CNN Travel.* https://edition.cnn.com/travel/article/himalayas-visible-lockdown-india-scli-intl/index.html

Sachs, J. D. (2015). *The Age of Sustainable Development.* New York: Columbia University Press.

Stone, P. R. (2012). Dark Tourism as 'Mortality Capital' : The Case of Ground Zero and the Significant Other Dead. In R. Sharpley & P. R. Stone (Eds.), *Contemporary Tourist Experience* (pp. 71–94). London: Routledge.

Tasci, A., Fyall, A., & Woosnam, K. M. (2022). Sustainable tourism consumer: Socio-demographics, psychographics and behavioural characteristics. *Tourism Review 77*, 341–375.

TM. (2021). *Malaysia Tourism Statistics in Brief.* Tourism Malaysia. https://www.tourism.gov.my/statistics

UN. (2019). *The Sustainable Development Agenda.* United Nations. https://www.un.org/sustainabledevelopment/development-agenda/

UN. (2022). *The Paris Agreement.* United Nations Framework Convention on Climate Change https://unfccc.int/process-and-meetings/the-paris-agreement/the-paris-agreement

UNESCO. (2012). *Melaka and George Town, Historic Cities of the Straits of Malacca.* UNESCO. http://whc.unesco.org/en/list/1223

UNWTO. (2021). *Sustainable Development*. https://www.unwto.org/sustainable-development
UNWTO. (2022). *Sustainable Development of Tourism*. UNWTO. https://tourism4sdgs.org/
UNWTO/IFT. (2019). *Transport-related CO2 Emissions of the Tourism Sector – Modelling Results*. https://doi.org/10.18111/9789284416660
Wang, S., Hu, Y., He, H., & Wang, G. (2017). Progress and prospects for tourism footprint research. *Sustainability 9*(1847), 1–17.
WCED. (1987). *Our Common Future*. New York: O. U. Press.
WWF. (2018). *Living Planet 2018: Aiming Higher*. Washington, DC: W. W. Federation. https://s3.amazonaws.com/wwfassets/downloads/lpr2018_summary_report_spreads.pdf

# 7 Whale-watching tourism

## Future sustainability trends

*Chaitanya Suárez-Rojas, Carmelo J. León, Javier de León
and Yen E. Lam-González*

### Introduction

Whale watching involves direct and close encounters with whales, dolphins, and other species of cetaceans in their natural environment, with the intention of providing an educational experience for tourists while at the same time promoting wildlife conservation (Suárez-Rojas & Lam-González, 2022). The magnificence and uniqueness of these charismatic species has led cetaceans to become the central attraction of marine ecotourism, positioning whale watching as one of the most in-demand recreational activities in the world (Buultjens et al., 2016). Whale watching first emerged in the 1950s as a non-extractive, wildlife, conservation-oriented ecotourism activity to counteract the worldwide decline in whale populations caused by whaling (Duffus & Dearden, 1993; Wakamatsu et al., 2018).

In light of this, the sustainable management of the activity is related to how operators carry out whale tours to reduce animal disturbance, thereby ensuring the preservation of whale habitats and marine environments (Amerson & Parsons, 2018). In addition, operators face the challenge of providing satisfactory whale-watching experiences to tourists, so contributing to the economic growth of the business and the whale-watching destinations as a whole (Suárez-Rojas & Lam-González, 2022).

However, inappropriate human recreational interactions with cetaceans negatively impact whales' and dolphins' ecology and welfare and constrain the required quality standards of the tourist experience (Finkler & Higham, 2020; Parsons, 2012). Despite the efforts of decision-makers to design regulations and guidelines, the rapid growth rate of the sector has consistently challenged the activity's operations within the parameters of sustainable practices (Wearing et al., 2014).

Academics have been concerned for decades about leading the sustainability paradigm towards practice. Research efforts have focused on enacting reliable insights to ensure an environmentally respectful, ethical, and socially responsible whale-watching tourism activity (Amerson & Parsons, 2018; Constantine & Bejder, 2008; Higham et al., 2016). The current research debate still deals with this issue.

This chapter reviews the major issues approached in the research field to progress towards new trends that yield responses for sustainable development pathways in whale-watching tourism. We will explore the last 30 years of scientific literature on whale-watching tourism to assess the evolution and the leading topics, relate these to some industry milestones, and identify the research gaps of the field of study. Towards this aim, a co-word analysis was conducted with the 1,042 keywords identified in the 343 publications retrieved from WoS and Scopus, the leading citation databases.

DOI: 10.4324/9781003291763-8

The chapter is structured as follows. The next section is devoted to presenting a general overview of the evolution of publications in the field and the main outcomes of the last 30 years (1990–2020), and how they have related to global socio-political milestones and the growth of the industry. In the third section, we identify challenges and opportunities for sustainability with implications for academics, industry operators, consumers, and policy-makers. The fourth section is dedicated to outlining the principal contribution of the chapter, from the theoretical and practical perspectives, and providing additional remarks.

## Major trends and issues in whale-watching tourism, policy, and research

As with other research fields, a whale-watching literature has evolved and been influenced and underpinned by its intellectual and political environments. Figure 7.1 shows the tendency of the number of publications in the research field and the density view of the keywords by decade. The lighter keywords are those with a higher occurrence density in the literature (Van Eck & Waltman, 2010). This overview focuses on the last 30 years (1990–2020). Notably, the 1990s corresponded with the first boom in scientific publications, coinciding with an explosive growth of whale-watching tourism – whale-watching destinations multiplied by a factor of five between the 1980s and 1990s (Hoyt, 1996).

In the early 1990s, the activity was recognised as a safeguard for whales as part of the worldwide effort to end commercial whaling (O'Connor et al., 2009), ecotourism being one of the top-occurring keywords within the research field. Whales were recovering from years of uncontrolled hunting, thanks to the commercial whaling moratorium[1] and the growth of whale watching as an ecotourism conservation-oriented activity (Hoyt, 2001). Likewise, Canada was also shown as an important research destination. Whale watching in the country was already considered a somewhat mature sector and an economically profitable alternative to whaling (Hoyt, 2001). However, by the mid-1990s, recreational harassment appeared as a new factor impacting whales' welfare, and whale watching experimented with its most intensive growth. Thus, while Duffus and Dearden (1993) pointed out that whale watching needed to be managed to avoid resource degradation and optimise the recreational experience, the activity had come to be considered just another form of harmful marine-wildlife exploitation by the end of the decade (Orams, 2000).

The following decade (the 2000s) supposed a period of mind-change concerning the principles of whale watching. Whale-watching destinations had increased to over 100, and consumer demand, along with the socio-economic benefits, continued to grow (O'Connor et al., 2009). This activity's explosive growth led scholars to start turning their gaze towards the environmental impacts of whale watching and the management requirements to address them.

As shown in Figure 7.1, the number of studies nearly tripled. This period witnessed the opening up of the research field to include a broader range of species, led by bottlenose dolphins, killer whales, and humpback whales. These three species are the best-understood cetaceans owing to their wide-ranging distribution, easy accessibility, and straightforward identification (Weinrich, 2001). However, management began to show a high occurrence density. Indeed, whale-watching management was still considered inadequate or utterly lacking (Constantine & Bejder, 2008), despite Duffus and Dearden's (1993) recommendations and years of extensive academic debate. In addition, tour boats, human disturbance, and behavioural responses, among other keywords regarding the impacts of the activity and the ecological responses of targeted species, were frequently occurring keywords. By the end of the decade, the International Whaling Commission underlined the need for

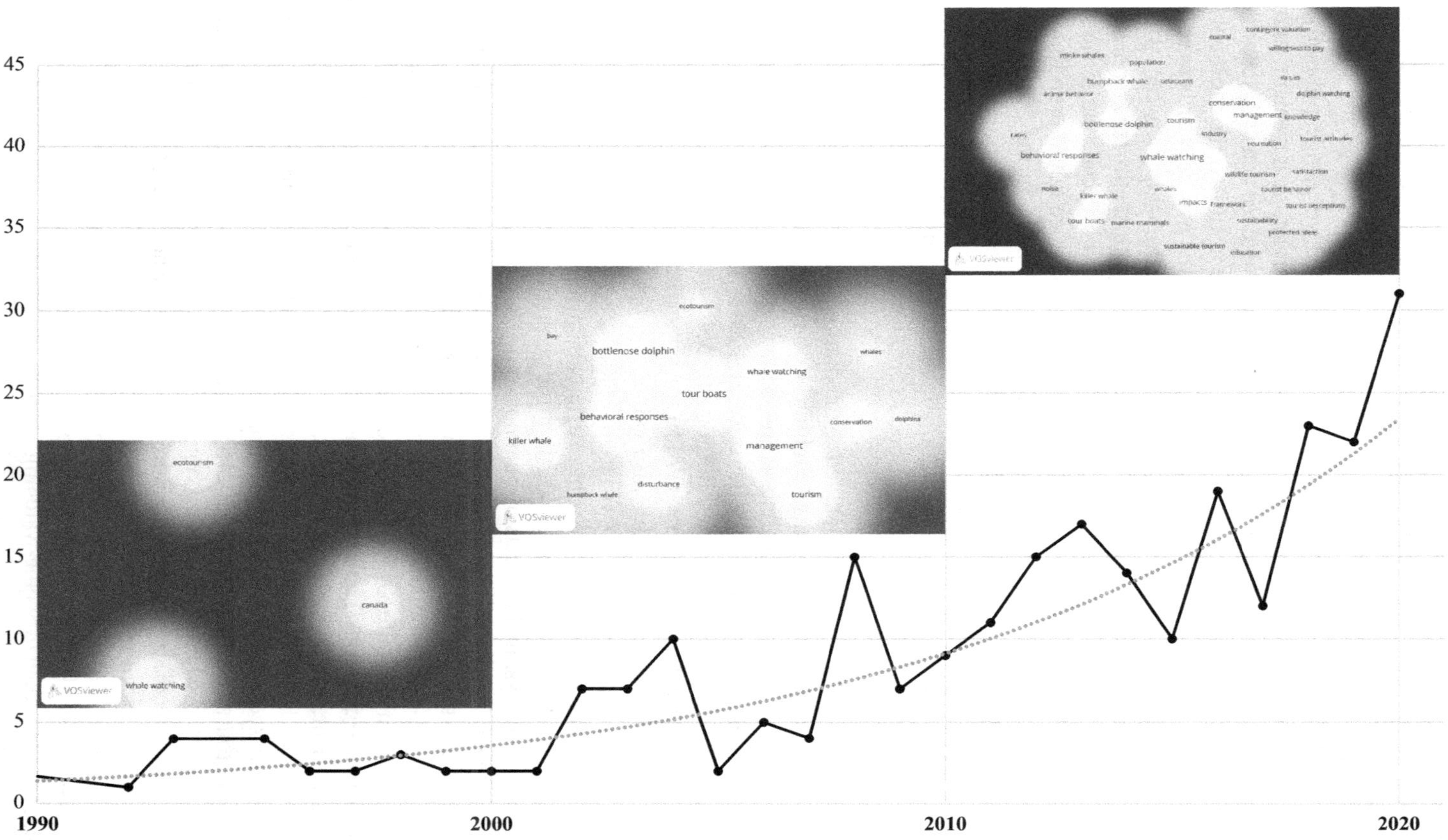

*Figure 7.1* The evolution of whale-watching tourism research in terms of number of publications and keyword occurrence.

more precise research efforts in regard to the long-term impacts of whale watching (O'Connor et al., 2009). Interestingly, tourism seems to eclipse the more environmentally friendly concept of ecotourism. According to Malcolm and Duffus (2008), whale watching was no longer considered a benign activity carried out by environmentally respectful (eco-) tourists, as it initially was.

From 2010 to 2020, with whale watching widely consolidated worldwide, publications in the research field grew exponentially. During this decade, tourism, management, and conservation were closely co-occurring keywords with a high occurrence density. According to Parsons (2012), management guidelines for whale watching constituted the most common strategy introduced to mitigate the impacts of the activity. Many new keywords were added concentrically to the keyword network explaining the relationship between watching impacts and whale responses (Figure 7.1). Research and analytical efforts extended to the tourism demand for whale watching. Orams (2000) had asked a decade earlier for more effort to understand the effect of whale-watcher motivations, since it had been found that these were rarely as simple as getting close to cetaceans. Thereby, in the 2010s, scholars firmly recognised the importance of understanding consumer behaviour and how this affects the industry's sustainability (Bentz et al., 2016; García-Cegarra & Pacheco, 2017). Thus, the determinants of value and satisfaction, focusing on tourist perceptions, attitudes, and behaviour, were identified as crucial aspects for harmonising the industry's development with natural resource conservation.

Education and knowledge were included as mediators, and thereby as a management solution for ensuring sustainable whale watching. Robust education programmes seemed to encourage pro-environmental awareness and behaviour and foster the compliance of management guidelines by operators (Bentz et al., 2016; Cornejo-Ortega et al., 2018; García-Cegarra & Pacheco, 2017). Methodological tools such as contingent valuation to elicit whale watchers willingness to pay (WTP) for improved management solutions and determine the activity's economic value also gained importance. As Cheung et al. (2019) pointed out, understanding tourists' WTP may encourage service quality and provide higher economic benefits to achieve feasible, sustainable tourism.

Between 2020 and 2021, research concerning the ecological impacts of the activity has mainly focused on analysing the effects of boat noise and the impacts of the swim-with activity due to the increasing number and size of whale-watching and swim-with-cetacean boats worldwide, and the still unquantified effects on species' behaviour (Arranz et al., 2021). For instance, with the aim of responding to these issues, de Freitas et al. (2021) investigated the potential of a navigation route tool to assess the management of the activity and the compliance with current legislation. The authors concluded that to promote responsible environmental management, policy-making should focus on the boat operators and tourists (de Freitas et al., 2021). From the social side of research, studies have focused on how empathy and knowledge-based activities positively improve policy compliance and the quality of whale-watching tours (Cárdenas et al., 2021; Villalba-Briones et al., 2021). Cárdenas et al. (2021) concluded that when operators accomplish regulations and provide adequate information about whale ecology, tourists are more satisfied than when operators do not comply with sustainability guidelines. Scholars have also focused on providing further insights into the social preferences and economic value of improved solutions for the industry's sustainability. Suárez-Rojas et al. (2021) underlined the market potential for promoting a green and socially responsible model of whale-watching tourism.

**Challenges and opportunities for sustainability**

Concerns in the literature about the responsible environmental management and sustainability challenges in whale watching are growing in parallel to the industry numbers and figures (Soto-Cortés et al., 2021). According to the 2030 Agenda, genuinely transformative actions will still be required in the next decade to encourage marine tourism consumption without neglecting future generations' needs and wellbeing (United Nations, 2015). Good governance implies the monitoring of impacts, raising environmental awareness, and more capacity-building actions (United Nations, 2015).

The development of the whale-watching tourism industry calls for an imminent management shift in which the socio-ecological relationships need to be addressed from a more holistic approach, as well as encouraging a collaborative atmosphere between public and private stakeholders and tourists (Hooper et al., 2021; Suárez-Rojas & Lam-González, 2022). There is a need to take advantage of the scientific knowledge in order to understand the long-term ecological impacts and the needs of the different actors (Mallard, 2019).

As studies have extensively reported, the existing regulations have largely failed in the attempt to ensure sustainable development at whale-watching destinations around the world. Collective interests have been ignored for years, the building of trust relationships has been unsuccessful, and scientific evidence has not been completely exploited – probably due to the existing gap between what academics provide and what the industry needs (Garrod & Fennell, 2004; Higham et al., 2014). Researchers have underlined the fact that operators do not always follow the rules appropriately, as they are mainly concerned with maximising their profits, which has a negative impact on cetaceans and the tourist experience (Soto-Cortés et al., 2021). Hooper et al. (2021) pointed out that this happens even though operators are aware of the existing guidelines and regulations, which is a measure of the lack of understanding of the serious implications that non-compliance has on their own long-term prosperity.

Academics have also found that, on the one hand, consumers are willing to pay more for responsible experiences and, on the other hand, that the transmission of information about cetaceans (ecology, impacts, etc.) increases their satisfaction with the experience and their concerns about wildlife conservation (Figure 7.2). This indicates the potential influence tourists may have in driving operators to behave according to good practice (Cárdenas et al., 2021; Finkler & Higham, 2020). However, an existing limitation is that tourists usually ignore the guidelines to recognise the violations perpetrated by skippers (Hooper et al., 2021).

To deal with these issues, there are two main priorities that whale-watching tourism stakeholders may have to face in the following years: first, academics should be more effective in translating scientific insights into understandable information for the different actors – operators, tourists, and policy-makers – and in promoting co-creation initiatives between them. Second, the various actors should cooperate in awareness raising and empowering the most responsible consumers and operators that are able to guide whale watching towards a genuine sustainability shift (Soto-Cortés et al., 2021). As shown in Figure 7.3, stakeholders need to be connected to guaranteeing whale-watching sustainability according to the principles of trust and reciprocity, establishing the input of researchers as an opportunity in this avenue.

*Figure 7.2*  Whale watching in Iceland.

Photo: Courtesy, Unsplash.com.

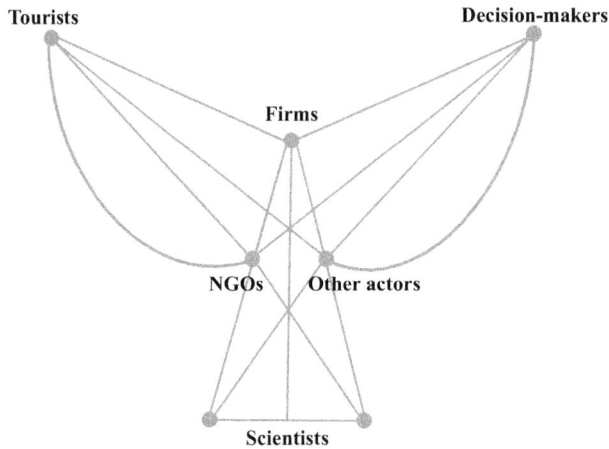

*Figure 7.3*  The whale-watching stakeholder sustainable network.

*Source*: Adapted from Suárez-Rojas & Lam-González (2022).

### Catalysts and research needs

Academics from different disciplines should collaborate more in this field to reorient study goals and methodologies within a multidisciplinary perspective. This is of paramount importance, for instance, in the current climate emergency we are living through. The projected impacts – such as changes in seawater temperatures – that provoke cetacean losses and displacements to new breeding and feeding sites for 2030–2100 will worsen the sustainability challenges of the sector. This could be exacerbated due to operators' level of competence in keeping their businesses alive and the consequent implications for the image and attractiveness of destinations (Albouy et al., 2020; Richards et al., 2021). Academia is called to strengthen research on the vulnerability and risk that future climate effects pose to whale-watching development and the actions needed to deal with them, thus promoting smart adaptation based on win–win mitigation strategies (Albouy et al., 2020).

It is undeniable that data collection is highly costly, particularly with regard to the study of cetaceans' ecology. Species longevity and migratory patterns make monitoring a challenging task (Burnham et al., 2021). By 2030, research efforts need to be directed to: (i) provide open-source data; (ii) work towards data sampling standardisation worldwide; (iii) implement recent technological advances to reduce effort on tracking and, for example, decoding whales' stress signals during encounters; and (iv) implement more robust analytical methods, such as in statistical modelling (Burnham et al., 2021). Data and knowledge sharing, as well as strengthening links between scholars from different research sites, would be adequate for the assessment of ecological impacts, for example, by comparing the various forms of pressure and animal responses between the different study cases in diverse destinations – and the design of common, science-based, adaptive management schemes for the whale-watching tourism sector (Higham et al., 2016).

Harmonising the industry's development with cetacean and marine habitat conservation is subjected to a better understanding of whale-watchers' preferences, values, and concerns. However, comprehending how the whale-watching activity meets tourist expectations and leads to a behavioural change is a challenging hurdle. Some studies have underlined the fact that whale-watching consumers are heterogeneous in their interests and preferences, although they do share the motivation of watching cetaceans in their natural environment (Malcolm & Duffus, 2008). Further research should be directed towards developing proactive educational approaches based on emotions or interpretative tools. These kinds of strategies have the potential to temper consumer expectations towards pro-environmental attitudes, behavioural changes, and long-term intentions to engage in wildlife conservation actions (Finkler & Higham, 2020). For instance, Finkler and Davis (2022) found that the projection of videos about sustainable whale watching before the tour invokes positive changes in consumers' attitudes and behavioural intentions to learn about responsible practices.

There is still a need to bring forth reliable insights into how to reconcile the various demands with responsible, sustainable practices and how much should be invested in more environmentally friendly actions. An in-depth understanding of tourists' preferences and estimation of the economic value of improved solutions for compensating the less environmentally friendly preferences is fundamental for bridging these gaps. This will provide the industry with trustworthy, empirical insights to invest with greater financial security in an innovative, ethical, and responsible form of whale watching.

### Industry and policy implications

The limited or ineffective communication between the science, policy, and business spheres does not make the 2030 Agenda Sustainability Goals (Finkler & Higham, 2020; Higham et al., 2016) any easier to reach. While researchers and policy-makers state that operators still violate the existing regulations, firms are involved in a highly fragmented regulatory context which is sometimes not well understood by them or does not respond to the particular conditions of their areas of operation (Garrod & Fennell, 2004).

Cooperation is needed among these stakeholders to develop bottom-up solutions that bridge the existing communication gap (Higham et al., 2014). Even though some good examples of self-regulating whale-watching management have been evidenced, such as in the El Vizcaíno Biosphere Reserve (Mexico) and Península Valdés (Argentina), there is a need for further effort to extrapolate these successful models to different worldwide destinations (IWC, 2020; Mayer et al., 2018). Actions should be directed towards: (i) building trust relationships and highlighting the common interests that the different actors share; (ii) making the scientific outcomes and recommendations for the industry more comprehensible; (iii) encouraging academia and policy-makers to attend to the industry's factual needs; (iv) involving entrepreneurs in policy-making; and (v) promoting co-creation in a way that gives to operators the opportunity to have more input in regards to their valuable gained experience, level of information, and innovation capabilities.

Co-creation is also well-established as an opportunity for whale-watching tourists to be involved in decision-making (Mayer et al., 2018). According to the International Whaling Commission (2020), tourists are more willing to comply with regulations when they feel privileged to be at a whale-watching destination, declared under a multi-stakeholder collaborative process. Co-creation on this side of the whale-watching stakeholder network (Figure 7.3) is a process with the potential to encourage more socially responsible behaviour on the part of firms and consumers (Judge et al., 2020; Xie et al., 2020).

With regard to this last issue and considering that there is market potential to be engaged in ethical practices without compromising economic returns, corporate social responsibility approaches are needed (Suárez-Rojas et al., 2021). This requires the design of new incentive schemes, in which operators are better informed with tailored cost–benefit analyses that clarify how the costs associated with the desired change will be compensated (Mayer et al., 2018). An understanding of site-specific idiosyncrasies and consideration of all the possible scenarios are also required for this aim (Pacheco et al., 2021).

**Major contributions**

Whale-watching tourism is involved in a complex scenario of ecological, socio-cultural, economic, and political dimensions. After reviewing the last 30 years of literature on whale-watching tourism through a co-word analysis, this chapter has provided an overview of the major trends and issues within the research field, underlining how the scientific literature has aimed to respond to industry and political milestones.

This chapter has shown that from the moment the presumption of innocence regarding the benefits of the activity was lost – recognising that whale watching was causing damage to the marine environment – the ecological impacts on whales due to human disturbance have focused academia's research efforts. Wildlife welfare and conservation concerns have led academics to strongly orient their works towards proposing management solutions for sustainable whale watching. This moment was crucial and allowed for new knowledge with genuine usefulness for policy-makers. Research in the last ten years has broadened its scope towards understanding why tourists engage in whale watching and what determines their preferences, behaviour, and perceived value in regard to the activity, thereby underlining the existing potential to promote more environmentally friendly experiences that meet tourists' expectations, thus ensuring higher levels of satisfaction.

From the policy and managerial perspectives, the remaining challenges rely on: (i) the effective use of existing valuable information and knowledge to lead behavioural change; (ii) the building of new socio-ecological relationships to reorient management practices towards a more integrative approach based on scientific breakthroughs and collaborative stakeholder networks; and (iii) the promotion of further knowledge and the empowerment of the society, residents, operators, and tourists calling for a shift to redirect the future of the industry's development and research efforts. This requires close cooperation between the whale-watching tourism industry, authorities, and other private and public stakeholders, posing a challenge for marine ecotourism governance. Managing human–cetacean interactions is fundamentally about managing people. In light of this, particular attention should be given to the potential of co-creation initiatives and participatory governance. We will see a real and timely shift towards sustainability in whale-watching tourism if we address the current shortcomings of research and the weaknesses of socio-ecological relationships.

## Note

1 The International Whaling Commission (IWC) is a global body of 88 member governments from countries all over the world, charged with the conservation of whales and responsible for setting catch limits for commercial whaling. In 1982, the IWC decided that there should be a pause in commercial whaling on all species and populations between 1985 and 1986, which remains in place today.

## References

Albouy, C., Delattre, V., Donati, G., Frölicher, T. L., Albouy-Boyer, S., Rufino, M., ... Leprieur, F. (2020). Global vulnerability of marine mammals to global warming. *Scientific Reports*, 10(1), 1–12.

Amerson, A., & Parsons, E. C. M. (2018). Evaluating the sustainability of the gray-whale-watching industry along the pacific coast of North America. *Journal of Sustainable Tourism*, 26(8), 1362–1380.

Arranz, P., de Soto, N. A., Madsen, P. T., & Sprogis, K. R. (2021). Whale-watch vessel noise levels with applications to whale-watching guidelines and conservation. *Marine Policy*, 134, 104776.

Bentz, J., Lopes, F., Calado, H., & Dearden, P. (2016). Enhancing satisfaction and sustainable management: Whale watching in the Azores. *Tourism Management*, 54, 465–476.

Burnham, R. E., Duffus, D. A., & Malcolm, C. D. (2021). Towards an enhanced management of recreational whale watching: The use of ecological and behavioural data to support evidence-based management actions. *Biological Conservation*, 255, 109009.

Buultjens, J., Ratnayke, I., & Gnanapala, A. (2016). Whale watching in Sri Lanka: Perceptions of sustainability. *Tourism Management Perspectives*, 18, 125–133.

Cárdenas, S., Gabela-Flores, M. V., Amrein, A., Surrey, K., Gerber, L. R., & Guzmán, H. M. (2021). Tourist knowledge, pro-conservation intentions, and tourist concern for the impacts of whale-watching in las perlas archipelago, Panama. *Frontiers in Marine Science*, 8, 627348.

Cheung, L. T., Ma, A. T., Chow, A. S., Lee, J. C., Fok, L., Cheng, I. N., & Cheang, F. C. (2019). Contingent valuation of dolphin watching activities in South China: The difference between local and non-local participants. *Science of The Total Environment*, 684, 340–350.

Constantine, R., & Bejder, L. (2008). Managing the whale-and dolphin-watching industry: Time for a paradigm shift. In *Marine Wildlife and Tourism Management: Insights from the Natural and Social Sciences*. Oxfordshire, UK: CABI Publishing, 321–333.

Cornejo-Ortega, J. L., Chavez-Dagostino, R. M., & Malcolm, C. D. (2018). Whale watcher characteristics, expectation-satisfaction, and opinions about whale watching for private vs community-based companies in Bahía de Banderas, Mexico. *International Journal of Sustainable Development and Planning*, 13(5), 790–804.

de Freitas, D. C., dos Santos, J. E. A., da Silva, P. C. M., de Oliveira Lunardi, V., & Lunardi, D. G. (2021). Are dolphin-watching boats routes an effective tool for managing tourism in marine protected areas? *Ocean & Coastal Management*, 211, 105782.

Duffus, D. A., & Dearden, P. (1993). Recreational use, valuation, and management, of killer whales (Orcinus orca) on Canada's Pacific coast. *Environmental conservation*, 20(2), 149–156.

Finkler, W., & Davis, L. S. (2022). Filmmaking, affective communication, and the construction of tourism imaginaries: Putting the wow into sustainable whale watching. *Tourism Culture & Communication*, 22(2), 205–217.

Finkler, W., & Higham, J. E. (2020). Stakeholder perspectives on sustainable whale watching: A science communication approach. *Journal of Sustainable Tourism*, 28(4), 535–549.

García-Cegarra, A. M., & Pacheco, A. S. (2017). Whale-watching trips in Peru lead to increases in tourist knowledge, pro-conservation intentions and tourist concern for the impacts of whale-watching on humpback whales. *Aquatic Conservation: Marine and Freshwater Ecosystems*, 27(5), 1011–1020.

Garrod, B., & Fennell, D. A. (2004). An analysis of whalewatching codes of conduct. *Annals of Tourism Research*, 31(2), 334–352.

Higham, J. E., Bejder, L., Allen, S. J., Corkeron, P. J., & Lusseau, D. (2016). Managing whale-watching as a non-lethal consumptive activity. *Journal of Sustainable Tourism*, 24(1), 73–90.

Higham, J., Bejder, L., & Williams, R. (Eds.). (2014). *Whale-Watching: Sustainable Tourism and Ecological Management*. Cambridge, UK: Cambridge University Press.

Hoarau, H., & Kline, C. (2014). Science and industry: Sharing knowledge for innovation. *Annals of Tourism Research*, 46, 44–61.

Hooper, L. K., Tyson Moore, R. B., Boucquey, N., McHugh, K. A., & Fuentes, M. M. (2021). Compliance of dolphin ecotours to marine mammal viewing guidelines. *Journal of Sustainable Tourism*, 1–19.

Hoyt, E. (1996). Whale watching: A global overview of the industry's rapid growth and some implications and suggestions for Australia. In *Encounters with Whales, 1995 Proceedings*. Canberra, Australia: Australian Nature Conservation Agency (pp. 31–36).

Hoyt, E. (2001). *Whale Watching 2001: Worldwide Tourism Numbers, Expenditures, and Expanding Socioeconomic Benefits*. Yarmouth Port, USA: International Fund for Animal Welfare.

IWC (2020). *Case Study*. Argentina: Península Valdés, Chubut. https://wwhandbook.iwc.int/en/case-studies/argentina-ptagonia (accessed December 2, 2021).

Judge, C., Penry, G. S., Brown, M., & Witteveen, M. (2020). Clear waters: Assessing regulation transparency of website advertising in South Africa's boat-based whale-watching industry. *Journal of Sustainable Tourism*, 29(6), 964–980.

Malcolm, C., & Duffus, D. (2008). Specialization of whale watchers in British Columbia waters. In *Marine Wildlife and Tourism Management: Insights from the Natural and Social Sciences*. Oxfordshire, UK: CABI, 109–129.

Mallard, G. (2019). Regulating whale watching: A common agency analysis. *Annals of Tourism Research*, 76, 191–199.

Mayer, M., Brenner, L., Schauss, B., Stadler, C., Arnegger, J., & Job, H. (2018). The nexus between governance and the economic impact of whale-watching. The case of the coastal lagoons in the El Vizcaíno Biosphere Reserve, Baja California, Mexico. *Ocean & Coastal Management*, 162, 46–59.

O'Connor, S., Campbell, R., Cortez, H., & Knowles, T. (2009). *Whale Watching Worldwide: Tourism Numbers, Expenditures and Expanding Economic Benefits, a Special Report from the International Fund for Animal Welfare*. Yarmouth, MA: Economists at Large.

Orams, M. B. (2000). Tourists getting close to whales, is it what whale-watching is all about? *Tourism Management*, 21(6), 561–569.

Pacheco, A. S., Sepúlveda, M., & Corkeron, P. (2021). Whale-Watching Impacts: Science, Human Dimensions and Management. *Frontiers in Marine Science*, 8, 1126.

Parsons, E. C. M. (2012). The negative impacts of whale-watching. *Journal of Marine Biology*, 2012.

Richards, R., Meynecke, J. O., & Sahin, O. (2021). Addressing dynamic uncertainty in the whale-watching industry under climate change and system shocks. *Science of the Total Environment*, 756, 143889.

Soto-Cortés, L. V., Luna-Acosta, A., & Maya, D. L. (2021). Whale-watching management: Assessment of sustainable governance in Uramba Bahía Málaga National Natural Park, Valle del Cauca. *Frontiers in Marine Science*, 8, 71.

Suárez-Rojas, C., González Hernández, M. M., & León, C. J. (2021). Do tourists value responsible sustainability in whale-watching tourism? Exploring sustainability and consumption preferences. *Journal of Sustainable Tourism*, 1–20.

Suárez-Rojas, C., & Lam-González, Y. E. (2022). Whale-watching tourism. In *Encyclopedia of Tourism Management and Marketing* (pp. 1–3). Edward Elgar Publishing.

United Nations. (2015). *70/1. Transforming our World: The 2030 Agenda for Sustainable Development*. Resolution adopted by the General Assembly on September 25, 2015.

Van Eck, N. J., & Waltman, L. (2010). Software survey: VOSviewer, a computer program for bibliometric mapping. *Scientometrics*, 84(2), 523–538.

Villalba-Briones, R., González-Narvaez, M. A., & Vitvar, T. (2021). How empathy-based sensitisation and knowledge reinforcement affect policy compliance: A case study of dolphin watching, Ecuador. *Australian Journal of Environmental Education*, 37(3), 285–305.

Wakamatsu, M., Shin, K. J., Wilson, C., & Managi, S. (2018). Exploring a gap between Australia and Japan in the economic valuation of whale conservation. *Ecological Economics*, 146, 397–407.

Wearing, S. L., Cunningham, P. A., Schweinsberg, S., & Jobberns, C. (2014). Whale watching as ecotourism: How sustainable is it? *Cosmopolitan Civil Societies: An Interdisciplinary Journal*, 6(1), 38–55.

Weinrich, M. (2001). Cetacean societies: Field studies of dolphins and whales by J. Mann, R. C. Connor, P. L. Tyack, and H. Whitehead. *The Journal of Wildlife Management*, 65(2), 366–367.

Xie, J., Tkaczynski, A., & Prebensen, N. K. (2020). Human value co-creation behavior in tourism: Insight from an Australian whale watching experience. *Tourism Management Perspectives*, 35, 100709.

# 8 Fair pricing in tourism

## From profitability towards sustainability

*Tomasz Napierała and Adam Pawlicz*

## Introduction

A combination of two words, "fair" and "pricing", may seem, at first glance, to be an oxymoron. Price is the primary component of company revenues and profits, while fairness, according to the *Cambridge Dictionary* (2022), is defined as "the quality of treating people equally or in a way that is right or reasonable". Fairness as a notion is associated by the general public with sport (fair play) or social policy (fair access of resources), rather than with the market economy. Still, when customers perceive the price they pay is unfair, they may sanction all kinds of businesses in the future. The notion of price fairness may relate not only to the price itself but also to the general policy of a company. An extreme example would be the situation of the market of alimentary goods in times of unrest when a seller may charge customers a prohibitively high price as they have few alternatives. The company may temporarily raise their profits, but their long-term prospects are bleak. Another example could be Western European countries buying gas from the Russian state company Gazprom. The price is usually competitive but involves supporting a regime that is non-democratic according to the buyers' standards.

The tourism market usually does not involve such drastic examples but still is a place where price fairness plays an important role. Its intrinsic characteristics, like high-income elasticity, the intangibility of the product, information asymmetry, the predominance of intermediaries, and advance bookings, make this market sensitive to price fairness. Tourism companies often rely on their brands, seek a long-term relationship with customers, and look for a positive image, rather than maximise profits from one-off transactions. Hence, it is essential for them to be perceived as a fair company that charges fair prices for quality products. The importance of price fairness is positively correlated with company size. With major tourism providers, the issue of price fairness is not limited to customer perspective but encompasses the whole spectrum of social and environmental responsibility.

This chapter is organised as follows. Following this introduction, the notion of price fairness is presented. Next, a customer perspective and the methods of measurement are discussed. Later we present the consequences for non- and for-profit organisations and the problems of ethics in pricing. Finally, we describe sustainability issues and the future of fair pricing in tourism.

## The notions of price fairness and fair pricing

The concept of price fairness was introduced by Huppertz et al. (1978). Subsequently, Okun (1981) argued that a price increase is accepted as fair when justified by an increase in

DOI: 10.4324/9781003291763-9

cost, but practices of increasing price following an increase in demand would be seen as unfair. However, he listed some exceptions, such as transport companies and hotels (Okun, 1981), which are of particular interest in this chapter. Chapuis (2012) and Kahneman et al. (1986) used the notion of price fairness to emphasise that enterprises are social constructs rather than legal or economic only. Recently, scientific and managerial discussions on price fairness have become vital in the field of business and marketing (Malc et al., 2016).

It is common knowledge that maximising short-run profit cannot be the only focus of companies. Such an approach might lead to customers' feeling regret as they find themselves victims of an unfair price increase. They interpret such a decrease in satisfaction as a consequence of a loss in real income when they must pay more for the same good which was priced lower before (Rotemberg, 2011). Such inequality of prices perceived by customers might have two very different contexts: advantageous, when a customer feels guilty for gaining by the unfairly low price, or disadvantageous, when a customer feels angry due to an unfairly high price which stimulates their sense of loss (Oh, 2003). Alternatively, reputation and long-term profits result from customers' goodwill and employees' engagement which are triggered by price fairness (Kahneman et al., 1986).

Price fairness judgement is based not only on knowledge (a cognitive aspect of price fairness), but also on emotions (the affective aspect) of buyers (Xia et al., 2004). This has resulted in the application of various psychological theories as a theoretical framework of studies on price fairness. For example, price fairness was the subject of the adaptation-level theory. Atypical price endings are used by enterprises to inform customers that prices they set are fair (Kinard et al., 2013). The range-frequency theory allows us to discuss changes in price structures as leading to the contrast and assimilation of price judgements. Price fairness judgement depends on strategies and techniques applied by marketers targeting various groups of customers, with discrimination or generalisation as the goal (Cunha & Shulman, 2011).

Chapuis (2012) and Schuitema et al. (2011) explained that fair pricing emphasises rules and mechanisms applied to set the prices, in contrast to price fairness which relates only to the paid price. Fair pricing leads to price fairness (Chapuis, 2012). While price fairness belongs to the domain of distributive justice, pricing fairness is the concept related to procedural justice (Gielissen et al., 2008). Fair pricing, related as well to participative pricing, allows tourism enterprises to save significant amounts of time and other resources to fix prices; the co-creation of the product (by tourism enterprises and tourists) might be an inspiration for a co-decision on pricing (Adhikari, 2019).

Some factors determining consumers' perceptions of prices should be mentioned. Bolton et al. (2003) suggested that fair prices are the output of the comparison of prices with microeconomic factors like previous prices and estimated costs for the seller, reference competitors' prices, but also macroeconomic determinants like inflation. A fair price is the result of the self-interest bias of customers, and the motives of sellers as perceived by buyers (Gielissen et al., 2008). Price fairness is influenced by the characteristics of other transactions experienced, noticed, or considered by a customer (Bolton et al., 2003), as well as a buyer's disposable income (Malc et al., 2016).

**Customers' perceptions of pricing**

The retail price is a piece of information that potential buyers consider when making a purchase decision. Hotel managers use the same number to estimate their revenues and by competitors to set their pricing policies. A historical record of prices along with occupancy allows bookkeepers to calculate profits and salient indicators such as revenue per available room (RevPAR). From the price fairness perspective, how customers perceive prices is more important than raw numbers. Perception is defined as a process by which people select, organise, and interpret information inputs to construct a meaningful portrait of the world (Akaegbu, 2013). Price perception is a process in which consumers interpret prices and attribute value to goods and services (Byun & Sternquist, 2010). Previous research argues that price perception may consist of positive and negative cues to customers and as such is of the utmost importance from a marketing perspective.

Numerous factors influence price perception but the most important is a reference price, defined as a price against which buyers compare the offered price of a service or tangible product (Lichtenstein et al., 1993). There are two types of reference prices: external and internal. The external one is provided by a producer, while the internal one is formed by a customer. Academic attention is given mostly to the latter but both have certain effects over the final monetary perception.

Previous experience is used as a primary indicator to establish an internal reference price, that is, a customer develops an internal reference price using the prices of their previous purchases and other advertised prices. It has been mostly conceptualised as a medium price or weighted log-mean, the most frequently encountered (mode) price, the lowest price seen, the last price seen, the highest willingness to pay, or even a future price when customers believe that these future prices will be higher (Chandrashekaran & Jagpal, 1995; Pedrajaiglesias & Guillén, 2000). If a customer chooses a new tourist destination, their internal reference price is built on prices of similar products, for example, a price for a week-long holiday package in Turkey is compared against a similar product from Greece. Alternatively, customers may rely on word-of-mouth or base the price perception on general economic information about a country of origin, for example, Switzerland is known for high-quality expensive watches, so high quality and expensive hospitality services are expected as well. The more unique the experience, the fewer the comparison possibilities.

Apart from the reference price, there are other factors impacting price perception. These are related to cultural values like uncertainty avoidance, value consciousness, price consciousness, sale proneness, price-quality schema, and prestige sensitivity. Factors related to culture are vital to the international tourism industry, where the same products are sold to customers in different countries. In cultures where uncertainty avoidance is valued, the high price may imply high quality, especially in situations where quality assessment is costly (Meng, 2011). This principle can be easily applied to the tourism market, which is characterised by high information asymmetry, as the product is purchased long before its consumption. One of the key customer values of tourism intermediaries is to reduce information asymmetry between potential prospects and tourism providers.

Price perception may have an impact on customer reaction to price changes, loyalty, and customer satisfaction. This is more salient in service industries, where each experience is different and there exists a high variability in product delivery.

## Implications for an organisation's policy and image

Positive consumer perception of the prices of products is vital for entrepreneurs, as the feeling of being charged an unfair price might push the customers to sanction a business by not returning to it, spreading negative word-of-mouth or simply making formal complaints. This is also accompanied by an overall lower customer satisfaction and represents a real problem for a company's long-term profit. Some customers are even willing to invest a considerable amount of time and money to punish the seller for unfair treatment, especially when strong negative emotions occur with the perception of price unfairness. Although this choice is irrational from the economic point of view, the psychological benefits prevail as customers feel that this is the way for them to prevent others from being exploited (Xia et al., 2004). Moreover, the increase in customers' perception of price fairness positively impacts repurchase intention, the value of a brand, customer loyalty, and trust.

As stated above, the concept of price fairness is built on customer price evaluations that are based on the relationship between the input and output ratios. People compare prices they pay against other customers, against what they are used to pay, or what they think the price should be (Xia et al., 2004). Numerous factors, such as image, trust between buyer and seller, and customer's meta-knowledge of the marketplace, impact the perception of price fairness. Previous research has shown that customers are much more willing to accept a price surge if a seller has a brand with a positive image, while price increases from a company perceived as exploitative of consumers will be deemed unfair (Jin et al., 2016). Similar effects are observed when there exists a trust between transaction parties, which often results from previous transactions. The rise of knowledge about market practices also impacts a customer's understanding of price changes, which has already been shown in yield management practices in the 1990s (Kimes, 1994). A price increase in the fossil fuels market or inflation may well be used to justify to customers the rising prices of tourism services.

Another factor that may increase the fairness of price policies is its association with collective considerations like social and environmental values. Still, as shown in the study by Schuitema et al. (2011) conducted in the Netherlands, the environmental justice considerations (the protection of future generations and nature) were evaluated more positively than those that promoted social justice (everyone was equally affected). Those social and nature-related concerns are closely related to considering the moral values of customers and tourism providers. Tourism businesses anticipate the problem and aim to minimise the negative impact on company reputation by a communication policy that either tries to explain a price surge or a price differentiation.

When customers have only a limited possibility to compare company prices with others, the standard technique to minimise the negative perception of price lift was to increase the rack rate which, at that time, was widely used as a reference price. For example, Kimes (1994) showed that 95% of airline passengers had received discounts. Nowadays, the ability to screen prices offered by numerous suppliers allows the customer to build their own image of the reference price. Another common method is associated with the creation of product bundles that are often used as a means to convey a price surge and as a method to lower the price without adverse effects on the brand image and deterioration of long-term customer relationships (Mitra, 2020).

The very foundation of yield management raises fairness questions, as customers are charged different prices for the same product. The wide use of yield management practices results in customers seeing a constantly changing price. A tourist may see a different price each time they access the company or e-travel agent site. What businesses call inventory optimisation may be viewed by customers as opportunistic behaviour aimed solely at profit maximisation and exploiting the ignorance (due to information asymmetry) of clients. A company thus needs to explain price differences to customers. Mauri (2007) suggests techniques to reduce the conflict arising from the use of yield management that is based on the privacy of pricing and on setting price fences acceptable from the customer's point of view. Still, the construction of fences must be socially acceptable. Setting different prices based on age (e.g., discounts for seniors or children), time of purchase, and seasons is acceptable, but little understanding could be expected when fences are based on gender, nationality, and race. As shown by Mattila and Choi (2005), if customers are reminded of hotel pricing policy (i.e., they know that advance purchase means lower prices) during the booking process, they are much more willing to accept fluctuations in price. Regardless, there remains a perception that yield management is something that is done to clients rather than for them. Creating mutual value in the customer–company relationship should be one of the primary aims of any organisation (McMahon-Beattie, 2011).

**Ethics of pricing in tourism**

The ethics of pricing in tourism refers to the ethics of mechanisms applied to set up prices for tourism services, though initially to the ethics of prices per se. Two questions should be addressed: "Are the prices in tourism fair ( the ethics of price functions)?" and "Are the strategies, techniques, and tactics of pricing fair (the ethics of revenue management)?" The ethics of pricing in tourism should be considered from the perspective of moral, political, and economic governance (Fennell, 2019). This raises a further two questions: "How are the values of price fairness embedded in the stakeholders of tourism industries, including individuals, organisations, and communities?" and "How is price fairness regulated by the law and customs?"

From the perspective of ethics, two functions of prices considered by neoliberal economics (Friedman, 1966) should be emphasised: distribution and supply. These refer to the social and environmental responsibility of enterprises when making price decisions. Although omitted from neoliberal theory, the exclusive function of price should be considered. It is argued that neoliberal ideology and the capitalism-based free-market culture are the fundamentals of the ethics of tourism industries rather than the principles of culture which underscore the commodification and the promoting of socio-cultural and environmental values (Fennell, 2019).

Social responsibility in tourism demands asking a question about pricing incentives for all stakeholders of tourism enterprises: owners (prices vs profit); managers (prices vs performance); employees (prices vs salaries); tourists (prices vs value or experience); suppliers (prices vs costs), including intermediaries (prices vs commissions); and other groups directly or indirectly dependent on tourism industries, such as local communities. The ethics of pricing leads to a comparison between prices and all the above-mentioned indicators of stakeholders' interests. It should also be emphasised that all those interests must be given the same amount of attention.

As a social phenomenon, tourism is inherently related to: the issue of justice: equity and solidarity between visitors and those visited (justice tourism); and rights: the human right to travel (egalitarian tourism) (Fennell, 2019). The exclusive function of prices is omitted from neoliberal theory, but it is paradoxically the result of capitalism. The airline or hotel industry's loyalty programmes are associated with customers' status symbols and privileges resulting from the amount of money spent on tourism services. Wealthy tourists are encouraged to consume more and to pay more to keep their status and privileges, while less affluent customers are outside the "velvet rope", as Schwartz (2020) calls it. Regarding social responsibility, fair prices in tourism should benefit local communities similarly to other stakeholders of tourism development (the problem of justice) and make tourism affordable for everyone (the case of human rights).

The environmental responsibility of tourism industries requires special attention since climate change and decreasing bio- and geodiversity are the most substantial threats to the planet. It is argued that capitalism has applied the strategy of cheap nature which means, by analogy with labour, employing nature for a very low salary (Moore, 2016). Nature is one of the most significant assets for tourism and leisure, and one of most impacted by tourism. Although pricing in tourism mostly fails to target the problem of using environmental assets, some initiatives to solve that issue have already been undertaken, such as taxing the airlines' $CO_2$ emissions (Hofer et al., 2010), or tourist destinations' ecological impact (Palmer & Riera, 2003). However, these solutions have been widely criticised for failing to solve the problem itself but merely limiting the use of scarce natural resources to wealthy tourists (Fennell, 2019).

The ethics of revenue management are the last consideration of this section. The fairness of revenue management practices, including differential pricing, depends on the following aspects: the ethical character of the practice itself, the fairness of profits resulting from the practice applied, the ethics of price presentation, and, finally, the perception of prices (Hayes & Miller, 2011). Ethical issues include the lack of transparency in differential pricing, and the links between differential pricing and unethical aims or attitudes, including overcharging practices for particular market targets, such as international tourists or particular times of travel (Elegido, 2011; Hayes & Miller, 2011; Keating, 2009). The ethical dilemma of revenue management is whether strategies, techniques, and tactics of pricing can solve the environmental and social problems of mass tourism, and the injustice resulting from elitist tourism.

## Sustainable tourism pricing

It is expected that the tourism industry should address all sustainable development goals (UNWTO, 2018). Local embeddedness characteristic in micro- and small enterprises allow that part of the tourism industry to meet the needs of the local environment, society, and economy much more efficiently than large multinational tourism corporations (Kc et al., 2021). However, tourism prices at the local level, when significantly influenced by commissions and pricing strategies of intermediaries, as well as pricing patterns of international tourism brands, might lead to the deterioration of local economic assets, a decrease in competitiveness of the local tourism sector, as well as a decline in net exports at the national level. For example, the high cost of access to many developing tourism destinations forces them to compensate for that cost by lower prices of services offered by local tourism industries (Kc et al., 2021; Napierała, 2013).

As Alpízar (2006) suggested, the limited budget of public agencies governing tourism attractions, such as cultural heritage sites or protected areas of natural heritage, push the institutions to charge tourists. The other argument for valuing such attractions is to reduce tourism flows within their areas to make them more equally distributed across time and space. Pricing is seen as contributing simultaneously to local economic prosperity and a protection of culture and natural heritage (Alpízar, 2006). However, some limitations of the fee-paying policy should be indicated. Charging tourists for access to the most attractive places or areas might just change the locations with congestion problems from overcrowded attractions to overcrowded entrance stations (Voltaire, 2017).

When considering factors influencing the increase of tourism enterprises' awareness of the environmental contexts of their operations, including pricing, the following categories of determinants should be indicated: (1) supply-based – the need to preserve local environmental assets as substantial elements of tourism attractiveness; (2) demand-based – growing demand for environmentally friendly tourism services; (3) cost-based – the opportunity to reduce operating costs; and (4) legal-based – the enforcement by authorities of environmental regulations (Bohdanowicz, 2006). It has been empirically confirmed that tourists value sustainability, and so are willing to accept higher prices (García-Pozo et al., 2013). Most of the environmental activities undertaken by tourism enterprises belong to the cost-based category, namely energy conservation, responsible waste management, and water conservation (Bohdanowicz, 2006). In consequence, higher prices and lower costs enable higher profits for sustainable tourism industries. Sustainability is consistent with the core business priorities of tourism enterprises (Woodland & Acott, 2007).

The very problem of the missing effectiveness of pricing in tourism in terms of sustainable development results not from the pricing itself, but from the neoliberal fundamentals of the capitalist system (Higgins-Desbiolles, 2006; Stroebel, 2015). Even in the field of tourism research, the focus is on how tourism impacts economic growth, rather than the agenda of sustainable development (Kronenberg & Fuchs, 2021). The limitations of studies on tourism pricing should be kept in mind, as contemporary researchers follow mainstream neoliberal ideology rather than modern economic theories and concepts.

**The future of fair pricing**

The contemporary understanding of prices as constantly evolving due to demand fluctuations, weather, time of the day, and so on is greater, especially if those factors are properly communicated. It can be predicted that tourism providers that previously hesitated with the implementation of yield management techniques would more freely introduce them. This applies to small enterprises, tourism attractions, and non-government organisations that usually get most of their revenues from sources other than ticket sales, such as donations, subsidies, and grants. Greater acceptance of price variations implies that customers' perceptions of price fairness would also be impacted, as the number of reference prices would be limited.

Another trend relates to the so-called fair-trade programmes that were successfully introduced as early as in the 1970s in coffee markets. Although scholars argue how fair is fair trade and whether this model is superior to free trade or protectionism from the point of view of goods suppliers (Maseland & de Vaal, 2002), it has been proven that the fair-trade label may increase consumption, and consumers infer greater health benefits of foods containing such labels (Berry & Romero, 2021). A corresponding term in the tourism context would be sustainable tourism (rather than "fair tourism") which has great prospects in the market due to the rise of environmental awareness and an increase of disposable income that allows consumers to choose fair products which are usually more expensive. The number of tourists that value the relationship between companies' environmental or social policies and prices will grow. Still, as the total tourism numbers are growing, it is unclear whether the percentage of tourists willing to pay a premium for green products actually increases. It is difficult to expect environmental considerations from those undertaking their first international journey or from those who can only afford an international trip once in a few years. We must not forget that tourism is still a luxury good in many developing countries.

From the research point of view, an increasing contribution of human geography and economic sociology to the knowledge of pricing is expected in the future. It has already been discussed in this chapter that the ethics of fair pricing demands that we consider the

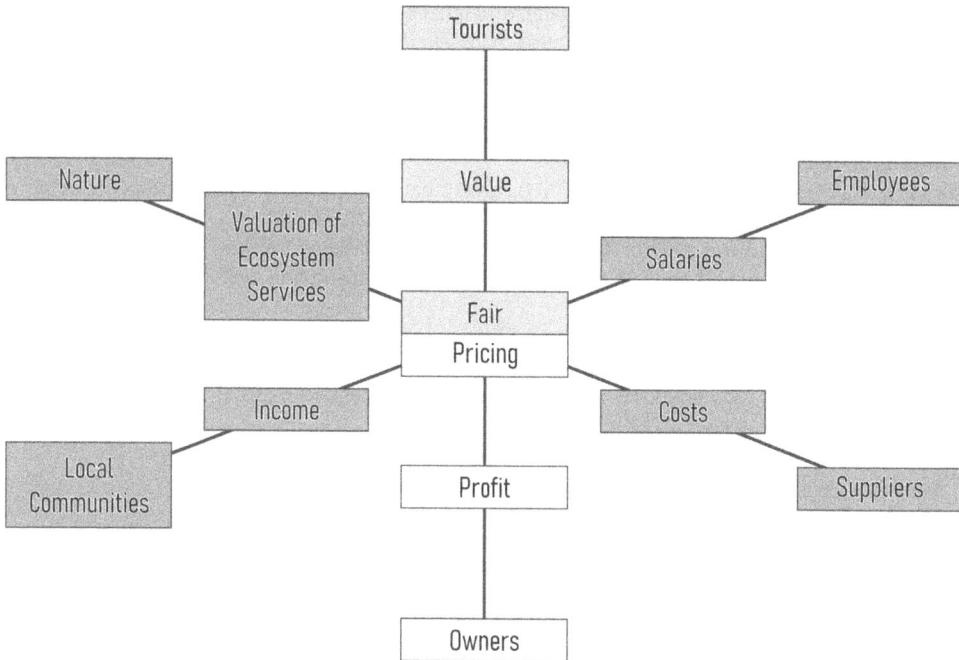

*Figure 8.1* Visual abstract: Fair pricing.

interests of all stakeholders of tourism industries. The focus should be on local communities, organisations, and local actors, rather than tourists and tourism enterprises only. A better understanding of the tourism relations identified in places, localities, and regions becomes a necessity for tourism management, including revenue management.

The idea of the competitiveness of destinations will collapse as the concept of sustainable tourism breaks ties with the neoliberal ideology of growth. The shift from territorial competitiveness towards territorial value is expected (Jeannerat & Crevoisier, 2022). The future of fair pricing is seen as more place-related, more democratic, and more participatory. As traditional revenue management was developed mainly within the tourism industry and with the contribution of computer engineering, future fair pricing will become a domain of the humanities and social sciences, including the above-mentioned fields of human geography and economic sociology. Depending on social innovations rather than technological ones, fair pricing will contribute to community building and collective empowerment to overcome local social issues and challenges.

A visual abstract of this chapter is depicted in Figure 8.1.

## References

Adhikari, A. (2019). Effect of reference price in PWYTF pricing in tourism sector. *Theoretical Economics Letters*, 9(4), 555–562. https://doi.org/10.4236/tel.2019.94038

Akaegbu, J. B. (2013). An exploratory study of customers' perception of pricing of hotel service offerings in Calabar Metropolis, Cross River State, Nigeria. *International Journal of Business and Social Science*, 4(13), 7.

Alpízar, F. (2006). The pricing of protected areas in nature-based tourism: A local perspective. *Ecological Economics*, 56(2), 294–307. https://doi.org/10.1016/j.ecolecon.2005.02.005

Berry, C., & Romero, M. (2021). The fair trade food labeling health halo: Effects of fair trade labeling on consumption and perceived healthfulness. *Food Quality and Preference*, *94*, 104321. https://doi.org/10.1016/j.foodqual.2021.104321

Bohdanowicz, P. (2006). Environmental awareness and initiatives in the Swedish and Polish hotel industries—Survey results. *International Journal of Hospitality Management*, *25*(4), 662–682. https://doi.org/10.1016/j.ijhm.2005.06.006

Bolton, L. E., Warlop, L., & Alba, J. W. (2003). Consumer perceptions of price (un)fairness. *Journal of Consumer Research*, *29*(4), 474–491. https://doi.org/10.1086/346244

Byun, S.-E., & Sternquist, B. (2010). Reconceptualization of price mavenism: Do Chinese consumers get a glow when they know? *Asia Pacific Journal of Marketing and Logistics*. https://doi.org/10.1108/13555851011062232

Cambridge University Press. (2022). *Fairness*. Cambridge Dictionary. https://dictionary.cambridge.org/pl/dictionary/english/fairness

Chandrashekaran, R., & Jagpal, H. (1995). Is there a well-defined internal reference price? *ACR North American Advances*, *22*, 230–235.

Chapuis, J. M. (2012). *Price fairness versus pricing fairness* (SSRN Scholarly Paper ID 2015112). Social Science Research Network. https://papers.ssrn.com/abstract=2015112

Cunha, M., & Shulman, J. D. (2011). Assimilation and contrast in price evaluations. *Journal of Consumer Research*, *37*(5), 822–835. https://doi.org/10.1086/656060

Elegido, J. M. (2011). The ethics of price discrimination. *Business Ethics Quarterly*, *21*(4), 633–660. https://doi.org/10.5840/beq201121439

Fennell, D. A. (2019). The future of ethics in tourism. In E. Fayos-Solà & C. Cooper (Eds.), *The Future of Tourism* (pp. 155–177). Springer International Publishing. https://doi.org/10.1007/978-3-319-89941-1_8

Friedman, M. (1966). *Essays in Positive Economics*. Chicago: The University of Chicago Press.

García-Pozo, A., Sánchez-Ollero, J.-L., & Marchante-Mera, A. (2013). Environmental sustainability measures and their impacts on hotel room pricing in Andalusia (Southern Spain). *Environmental Engineering and Management Journal*, *12*(10), 1971–1978.

Gielissen, R., Dutilh, C. E., & Graafland, J. J. (2008). Perceptions of price fairness: An empirical research. *Business & Society*, *47*(3), 370–389. https://doi.org/10.1177/0007650308316937

Hayes, D. K., & Miller, A. A. (2011). *Revenue Management for the Hospitality Industry*. Hoboken, NJ: Wiley.

Higgins-Desbiolles, F. (2006). More than an "industry": The forgotten power of tourism as a social force. *Tourism Management*, *27*(6), 1192–1208. https://doi.org/10.1016/j.tourman.2005.05.020

Hofer, C., Dresner, M. E., & Windle, R. J. (2010). The environmental effects of airline carbon emissions taxation in the US. *Transportation Research Part D: Transport and Environment*, *15*(1), 37–45. https://doi.org/10.1016/j.trd.2009.07.001

Huppertz, J. W., Arenson, S. J., & Evans, R. H. (1978). An application of equity theory to buyer-seller exchange situations. *Journal of Marketing Research*, *15*(2), 250. https://doi.org/10.2307/3151255

Jeannerat, H., & Crevoisier, O. (2022). From competitiveness to territorial value: Transformative territorial innovation policies and anchoring milieus. *European Planning Studies*, 1–21. https://doi.org/10.1080/09654313.2022.2042208

Jin, N. (Paul), Line, N. D., & Merkebu, J. (2016). The effects of image and price fairness: A consideration of delight and loyalty in the waterpark industry. *International Journal of Contemporary Hospitality Management*, *28*(9), 1895–1914. https://doi.org/10.1108/IJCHM-03-2015-0094

Kahneman, D., Knetsch, J. L., & Thaler, R. H. (1986). Fairness as a constraint on profit seeking: Entitlements in the market. *The American Economic Review*, *76*(4), 728–741. https://doi.org/10.1017/CBO9780511803475.019

Kc, B., Dhungana, A., & Dangi, T. B. (2021). Tourism and the sustainable development goals: Stakeholders' perspectives from Nepal. *Tourism Management Perspectives*, *38*, 100822. https://doi.org/10.1016/j.tmp.2021.100822

Keating, B. (2009). Managing ethics in the tourism supply chain: The case of Chinese travel to Australia. *International Journal of Tourism Research, 11*(4), 403–408. https://doi.org/10.1002/jtr.706

Kimes, S. E. (1994). Perceived fairness of yield management: Applying yield-management principles to rate structures is complicated by what consumers perceive as unfair practices. *Cornell Hotel and Restaurant Administration Quarterly, 35*(1), 22–29. https://doi.org/10.1177/001088049403500102

Kinard, B. R., Capella, M. L., & Bonner, G. (2013). Odd pricing effects: An examination using adaptation-level theory. *Journal of Product & Brand Management, 22*(1), 87–94. https://doi.org/10.1108/10610421311298740

Kronenberg, K., & Fuchs, M. (2021). Aligning tourism's socio-economic impact with the United Nations' sustainable development goals. *Tourism Management Perspectives, 39*, 100831. https://doi.org/10.1016/j.tmp.2021.100831

Lichtenstein, D. R., Ridgway, N. M., & Netemeyer, R. G. (1993). Price perceptions and consumer shopping behavior: A field study. *Journal of Marketing Research, 30*(2), 234–245. https://doi.org/10.1177/002224379303000208

Malc, D., Mumel, D., & Pisnik, A. (2016). Exploring price fairness perceptions and their influence on consumer behavior. *Journal of Business Research, 69*(9), 3693–3697. https://doi.org/10.1016/j.jbusres.2016.03.031

Maseland, R., & de Vaal, A. (2002). How fair is fair trade? *De Economist, 150*(3), 251–272. https://doi.org/10.1023/A:1016161727537

Mattila, A. S., & Choi, S. (2005). The impact of hotel pricing policies on perceived fairness and satisfaction with the reservation process. *Journal of Hospitality and Leisure Marketing, 13*(1), 25–39. https://doi.org/10.1300/J150v13n01_03

Mauri, A. G. (2007). Yield management and perceptions of fairness in the hotel business. *International Review of Economics, 54*(2), 284–293. https://doi.org/10.1007/s12232-007-0015-4

McMahon-Beattie, U. (2011). Trust, fairness and justice in revenue management: Creating value for the consumer. *Journal of Revenue and Pricing Management, 10*(1), 44–46. https://doi.org/10.1057/rpm.2010.42

Meng, J. G. (2011). Understanding cultural influence on price perception: Empirical insights from a SEM application. *Journal of Product & Brand Management.* https://doi.org/10.1108/10610421111181831

Mitra, S. K. (2020). An analysis of asymmetry in dynamic pricing of hospitality industry. *International Journal of Hospitality Management, 89*, 102406. https://doi.org/10.1016/j.ijhm.2019.102406

Moore, J. W. (2016). The rise of cheap Nature. In J. W. Moore (Ed.), *Anthropocene or Capitalocene? Nature, History, and the Crisis of Capitalism* (pp. 78–115). Oakland, CA: PM Press.

Napierała, T. (2013). *Przestrzenne zróżnicowanie cen usług hotelowych w Polsce* [Spatial Volatility of Hotel Prices in Poland]. Krakow: Wydawnictwo Uniwersytetu Łódzkiego.

Oh, H. (2003). Price fairness and its asymmetric effects on overall price, quality, and value judgments: The case of an upscale hotel. *Tourism Management, 24*(4), 387–399. https://doi.org/10.1016/S0261-5177(02)00109-7

Okun, A. M. (1981). *Prices & Quantities: A Macroeconomic Analysis.* Oxford: Basil Blackwell Publisher.

Palmer, T., & Riera, A. (2003). Tourism and environmental taxes. With special reference to the "Balearic ecotax". *Tourism Management, 24*(6), 665–674. https://doi.org/10.1016/S0261-5177(03)00046-3

Pedrajaiglesias, M., & Guillén, M. J. Y. (2000). The role of the internal reference price in the perception of the sales price: An application to the restaurant's services. *Journal of Hospitality & Leisure Marketing, 7*(3), 3–22. https://doi.org/10.1300/J150v07n03_02

Rotemberg, J. J. (2011). Fair pricing. *Journal of the European Economic Association, 9*(5), 952–981. https://doi.org/10.1111/j.1542-4774.2011.01036.x

Schuitema, G., Steg, L., & van Kruining, M. (2011). When are transport pricing policies fair and acceptable? *Social Justice Research, 24*(1), 66–84. https://doi.org/10.1007/s11211-011-0124-9

Schwartz, N. (2020). *The Velvet Rope Economy: How Inequality became Big Business* (1st edition). New York: Doubleday.

Stroebel, M. (2015). Tourism and the green economy: Inspiring or averting change? *Third World Quarterly*, *36*(12), 2225–2243. https://doi.org/10.1080/01436597.2015.1071658

UNWTO. (2018). *Tourism for SDGs*. https://tourism4sdgs.org/

Voltaire, L. (2017). Pricing future nature reserves through contingent valuation data. *Ecological Economics*, *135*, 66–75. https://doi.org/10.1016/j.ecolecon.2016.12.032

Woodland, M., & Acott, T. G. (2007). Sustainability and local tourism branding in England's South Downs. *Journal of Sustainable Tourism*, *15*(6), 715–734. https://doi.org/10.2167/jost652.0

Xia, L., Monroe, K. B., & Cox, J. L. (2004). The price is unfair! A conceptual framework of price fairness perceptions. *Journal of Marketing*, *68*(4), 1–15. https://doi.org/10.1509/jmkg.68.4.1.42733

# 9 Tourism sustainability is a big problem in the development of marine tourism in Indonesia

*Ahmad Bahar*

## Introduction

Tourism based on the coastal and marine environment (marine-based tourism) in Indonesia has experienced very rapid development over the last two decades. This is due to the increasing demand for tourists to visit various destinations in the coastal and marine environment. Labuan Bajo, Raja Ampat, and Wakatobi are among the top priority destinations and have a rapidly increasing number of visits. The opening of these ten priority destinations also provides foreign tourists with more diverse choices, in the form of biophysical objects of the coastal environment, the diversity of ecosystems, community culture, as well as the uniqueness of marine biota organisms.

Domestic and foreign tourist visits to various marine tourism destinations in Indonesia will certainly have an impact on increasing the country's foreign exchange, opening up new jobs and opportunities for doing business in the community. This also causes this sector to become one of the leading ones to be developed, including several coastal districts/cities which have excellent attractiveness to making marine tourism a priority sector. In particular, the Ministry of Maritime Affairs and Fisheries of the Republic of Indonesia has issued Ministerial Regulation Number 93/PERMEN-KP/2020 concerning Marine Tourism Villages (Dewi Bahari). The hope is that coastal villages that have the potential to attract marine tourism can be developed using village funds from the government or partnering with investors. Coastal tourism villages are now popping up everywhere. The development of marine tourism has had an economic impact on the government, local communities, and private sector entrepreneurs as well as the parties involved in the marine tourism sector.

The development of marine tourism that is not carried out responsibly can also have negative impacts. Therefore, the development of this sector requires serious attention from the government so that the positive impact it has can continue to be sustainable. One thing that needs serious attention is the use of marine resources as a tourist attraction, which are currently being continuously degraded or even damaged. In fact, various works in the literature reveal that the degradation or decrease in the quality of tourist attraction objects will reduce tourist satisfaction which can ultimately reduce the number of tourist visits. A significant reduction in the number of tourist arrivals can kill the tourism business (Hall, 2001; Harriott, 2002).

Another problem is that the contribution of the tourism sector to Indonesia's GDP is only 10.4% (WTTC, 2019) with the contribution from the marine tourism sector only around 10%. This low income is also felt by several tourist-receiving districts such as Wakatobi Regency, which initially made the marine tourism sector a leading one for regional

DOI: 10.4324/9781003291763-10

development but is now turning to the fisheries sector. One of the causes is the low regional income from marine tourism.

Therefore, in addition to promotional efforts in an effort to increase the number of tourist visits, it is also important for the Indonesian government to establish policies or regulations so that the use of marine biological resources as objects or attractions for tourism can be sustainable. The policies or regulations issued by the government are based on studies or technical frameworks that have been prepared by the government, academics, or non-governmental organisations such as World Wildlife Fund, The Nature Conservancy, or Conservation International. Up to 2019, the government had targeted 20 million foreign tourists, four million of whom were to visit marine destinations, of which ten new ones were to be opened. Although the target of visiting numbers has not been fully achieved, it turns out that the impact on resources in several marine tourism destinations has resulted in significant degradation.

### History of the development of marine tourism in Indonesia

The city of Jakarta, as the nation's capital, made Ancol Beach the best marine tourism destination in 1970, though not many activities could be engaged in because recreational facilities at sea were still very limited at that time. The most common activities were bathing, swimming, boating, walking and playing in the sand on the beach, or just sitting around waiting for the sunset. But they were enough to relieve stress for the residents of Jakarta and its surroundings at that time. Ancol Beach was quickly recognised in the country and became a marine tourism destination for those visiting the city.

In the early 1980s, tourists who like to dive into the sea have looked at the Thousand Islands as a new destination that offers calm and beautiful beaches and coral reef ecosystems with diverse fish. This period was the beginning of the development of marine tourism in Indonesia along with the discovery of snorkels as a tool for activities in the sea, including enjoying the beauty of the underwater world. Despite the snorkel-only facilities, visits to the Thousand Islands continued to increase until the late 1990s.

In the early 1990s, a new marine tourism destination emerged on Bunaken Island, North Sulawesi Province. Similar to the Thousand Islands, Bunaken Island is also a National Park that sets aside part of its territory to be used for marine tourism activities. The difference is that Bunaken Island is closer to Manado City. This fairly easy access caused Bunaken to quickly develop as a marine tourism destination that presents the beauty of coral reefs and the steep cliffs of the island. Although there is much diving equipment provided by dive operators, many tourists still choose snorkelling, boats, and catamarans to enjoy the underwater beauty of Bunaken because diving is still limited. The destination was increasingly successful under the control of the Bunaken National Park Management Board (DPTNB) before finally fading away due to the increasing amount of garbage from Manado City covering coral reefs and the emergence of new destinations elsewhere.

In 2005, the PT Wakatobi Dive Resort (WDR) started operating on Tomia Island, offering the beauty of Wakatobi's underwater world to the international community. Wakatobi was later known worldwide as one of the best destinations. Tourists who want to dive in Wakatobi via WDR must wait for three to four months. When Ir. Hugua led Wakatobi Regency, he made tourism a leading sector. The new Matahora airport was initiated so that access from Jakarta to Wakatobi was shortened from three days to one day. With the slogan "Wakatobi, a real underwater paradise" that Hugua was promoting everywhere, he overtook the attractiveness of other marine tourism destinations in Indonesia. Due to the limited infrastructure that supports the tourism sector in Wakatobi, many tourists felt unsatisfied with their travels. Wakatobi was then less resonant, especially when Ir. Hugua was replaced by Arhawi, SE as the Regent of Wakatobi, the fisheries sector overtook tourism as the leading sector.

Exploration of the beauty of the underwater world then continued to shift to the east. Publications of international researchers later attracted many tourists with a special interest in visiting Raja Ampat. As noted by Dr. Gerald R. Allen who found 283 types of reef fish in Tanjung Kri in one dive. "An excellent (fertile) water condition that is rarely found in any marine waters in the world," wrote the Australian marine biologist. Around 2010, tourists began to flock to the Bird's Head Islands in small groups according to their respective interests. There are those who want to see the beauty of coral reef ecosystems, view the seascapes in Wayag, see manta rays on Arborek Island, or just see the birds of paradise in Saporkren, Waisai. This destination continues to grow, although not too fast because it is quite far from Jakarta.

The beauty of the coral reef ecosystems in Raja Ampat and Wakatobi may not be enough for travellers. For those who like unique species or extremely strong currents, they visit Labuan Bajo and Komodo National Park. Marine biota tourism objects such as sharks and manta rays are excellent in destinations that have started to develop since 2015. Compared to previous destinations, Labuan Bajo is growing rapidly. In 2015 the number of visits was only 50,000; two years later it was 130,000. As a result, many tourists are now starting to complain about the density of visits to dive spots.

The development of marine tourism in Labuan Bajo has also led to many operators using live-aboards. This is different from previous destinations. The presence of this live-aboard accelerates the movement of tourists to the east of Indonesia, such as in the Morotai Islands, North Maluku, or to the Banda Neira Islands in the Banda Sea.

## Sustainability threats due to degradation of marine tourism resources and plastic waste

### *Coral reef degradation*

One of the most attractive coastal ecosystems for tourists is the coral reef. This ecosystem has a high marine biodiversity because it is associated with various types of fish and other marine biota so that it is attractive and very beautiful to look at. The condition of coral reefs with high biodiversity is certainly very attractive as a marine tourism destination, especially for diving.

However, the condition of coral reef ecosystems in Indonesia is currently experiencing a decline in quality. Data from the Indonesian Institute of Sciences (LIPI) released in 2018 showed only 6.39% of coral reefs were in very good condition, with 23.4% in good condition, 35.06% in adequate condition, and 35.15% in bad condition. These results were taken from 108 locations and 1,064 observation stations throughout Indonesian waters. The reasons are many fishing activities that damage coral reefs, coastal reclamation, bleaching of corals due to global warming, and irresponsible marine tourism activities.

Maritime tourism, if managed properly, will deliver welfare to the nation. However, if this sector is not developed carefully, it will have negative impacts, social and ecological. One of the reasons for the shift in marine tourism destinations in Indonesia is the decline in the quality of the attractiveness of resources due to the increasing rate of damage to coral reefs (Figure 9.1). This has a simultaneous impact on decreasing the level of tourist

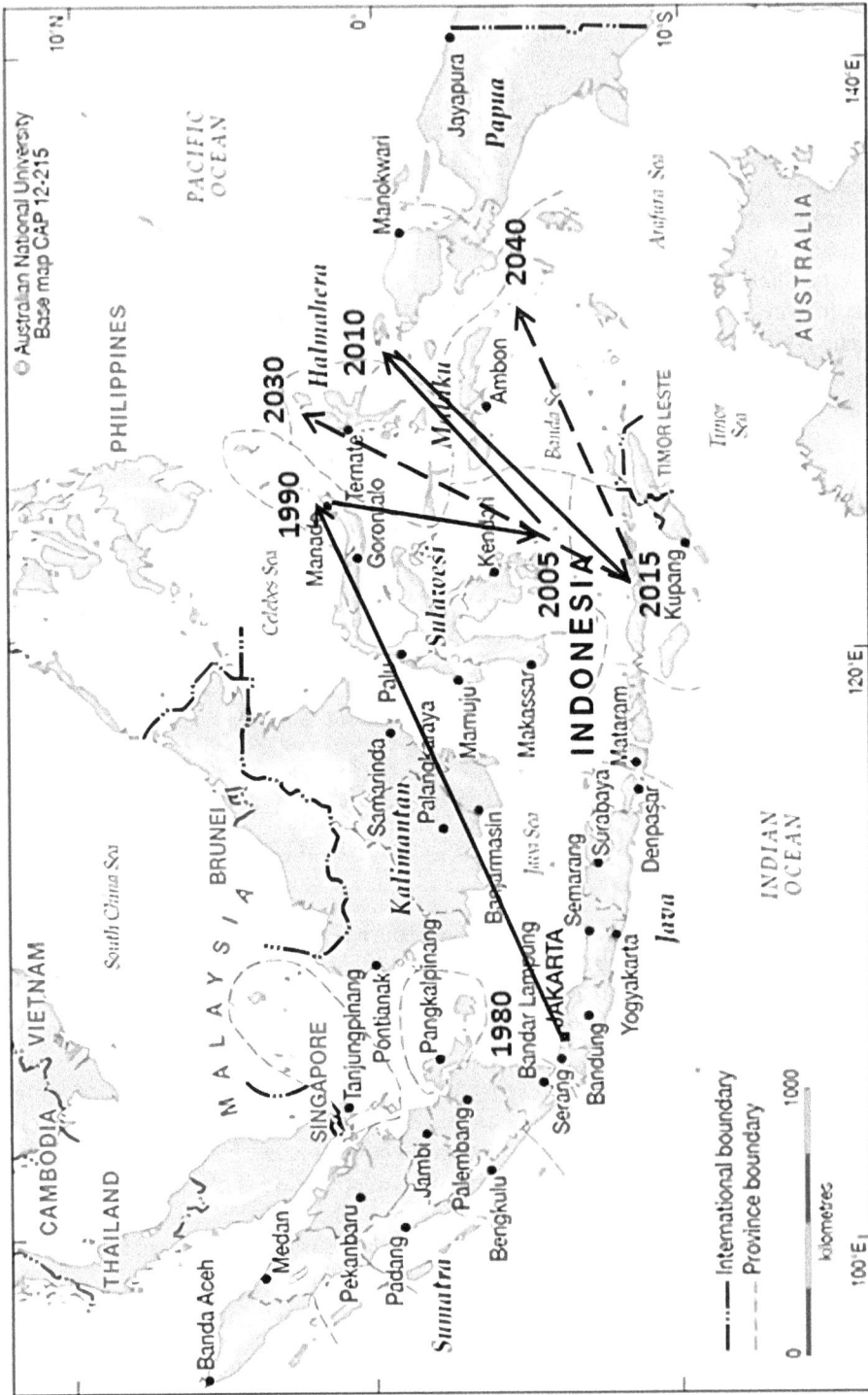

*Figure 9.1* Movement of favourite marine tourism destinations in Indonesia.

**Hard Corals**

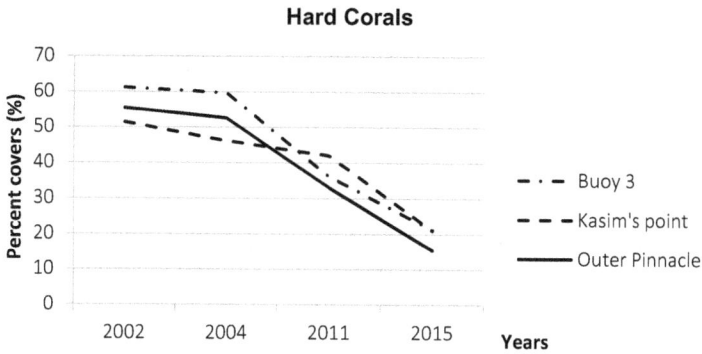

*Figure 9.2* Coral reef degradation due to diving tourism activities on Hoga Island, Wakatobi Regency, 2002–2015.

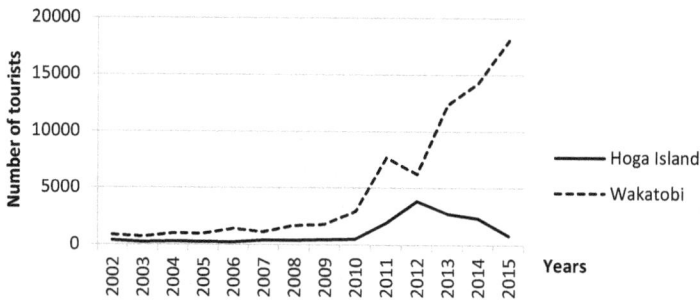

*Figure 9.3* Domestic and foreign tourist visits to Hoga Island and Wakatobi, 2002–2015.

*Source*: Wakatobi Tourism Office and Wakatobi National Park Office.

satisfaction. Another cause is the desire of tourists to find new attractions or unique tourist resources.

Research conducted on Hoga Island, one of the best dive sites in Wakatobi Regency, which sampled three dive points in the Tourism Utilization Zone (where fishing is prohibited), namely Buoy 3, Kasim's Poin, and Outer Pinnacle, showed a significant decrease in hard coral cover of 33.6% over 13 years (Figure 9.2) (Bahar, 2017). In fact, marine tourism on Hoga Island has implemented the principles of responsible tourism, managed by Operation Wallacea (Opwall) in collaboration with the local community (the Wakatobi Foundation), where visitors are generally tourists and foreign researchers with high education and Professional Association of Diving Instructors (PADI) certified diving skills. Therefore, you can imagine the impact of damage to coral reefs for diving tourism destinations that are freely open, without any signs for visitors made by tourism managers. Currently, tourist visits to Hoga Island, domestic and foreign, are decreasing (Figure 9.3).

### Interaction with marine life

In addition to the beauty of coral reefs, the emergence of rare and unique marine life also attracts many tourist visits to Indonesia. Several destinations to see the attractions of the whale shark (*Rhyncodon typus*) have been developed, such as the Cenderawasih Bay National

Park in West Papua Province, Saleh Bay in West Nusa Tenggara Province, Botubarani Beach in Gorontalo Province, and Tali Sayan on Derawan Island, East Kalimantan. Tourist destinations to see manta ray attractions have developed on Arborek Island in Raja Ampat Regency and Labuan Bajo in Manggarai Regency. Dive sites to see shark and dolphin attractions have also spread.

The results of research by Mustika et al. (2015) show that the behaviour of boats in Lovina to witness dolphin attractions is not in accordance with best practices such as is done in Australia. Although we cannot prove a causal effect of this practice violation, we can conclude from other studies that this situation is likely to adversely affect the dolphins in Lovina. Tourist responses also according to Mustika et al. (2013) indicate tourist dissatisfaction with the driving behaviour of boatmen, the excessive number of boats approaching animals, and interactions that are too close to dolphins and which can cause behaviour change (Stensland & Berggren, 2007; Christiansen et al., 2010). Things that cause visitor dissatisfaction could disrupt the sustainability of the economic impact of Lovina Beach marine tourism. A better understanding of tourism is essential in designing sustainable marine wildlife tourism in developing countries. A good experience by tourists will tend to promote trips for others (Mustika et al., 2013).

The disappearance of the whale shark, which is a tourist attraction on Botubarani Beach, has also occurred, starting from the end of 2017 (Figure 9.4). Many tourists were disappointed because they did not see the whale shark on subsequent visits. The number of visits is not limited, the number of boats that interact directly with the animals is very large and at very close range, sometimes even crashing into them. Whale shark monitoring

*Figure 9.4* Explosive and chaotic tourist visits at the beginning of the appearance of whale sharks in 2016 at Botubarani Beach, Gorontalo Province.

carried out by the Makassar Coastal and Marine Resources Management Center (BPSPL) of the Ministry of Maritime Affairs and Fisheries (KKP) during the period March to June 2019 on 16 whale sharks obtained five wounds on the mouth, three wounds on the fins, two wounds on the body, and six tails were not injured.

Djunaidi et al. (2021) suggested the importance of cooperation between the government and the local community in managing whale shark tourism in Botubarani so that marine tourism attractions are sustainable. Local governments should take a role in educating the public, managers and visitors. The reduced number of tourist visits during the COVID-19 pandemic led to a recovery in the presence of whale sharks in this area. Monitoring by BPSPL in Makassar in 2020 detected the appearance of 18–21 whale sharks per month. This is in accordance with Israngkura's research (2022) that during the pandemic, the condition of the resources and objects of marine tourism attractions was restored due to reduced tourist visits.

*Plastic waste*

The threat of sustainable tourism in Indonesia also comes from plastic waste pollution. This type of pollution has a wide impact, including human health, the economy, tourism, and beach aesthetics (Thompson et al., 2009). The results of research by Hayati et al. (2020) on Tidung Island, a small area in Jakarta Bay, show that tourism is one of the largest sources of waste generation. Plastic waste is the largest part of solid waste, accounting for 83.86% of the total. Unmanaged tourism waste causes the beach index to be included in the "very dirty" category which causes a decrease in the visitor acceptance index. Such conditions will make tourism unsustainable.

Stranded plastic waste observed by Suteja et al. (2021) on 14 tourist beaches on the island of Bali could threaten the sustainability of coastal tourism. The abundance of plastic waste of 0.36 items/m$^2$ with a weight of 3.8 g/m$^2$, which is dominated by plastic bags, straws, and plastic cups, is very disturbing for tourists who seek recreation on the beach, especially during the rainy season where plastic waste stranded from the sea is significantly abundant.

The decline in the number of tourist visits to the Bunaken Marine National Park in North Sulawesi was also triggered by plastic pollution. Poor waste management from the mainland of Manado City causes plastic bag waste through rivers to enter the sea and cover coral reefs. The peak occurs during the rainy season which not only brings plastic waste, but also sedimentation from higher land which causes the death of coral reefs because sediment covers the pores. The revenue of the Bunaken National Park Management Board (DPTNB) from year to year continues to experience a drastic decline.

## Economic impact of marine tourism

It is undeniable that tourist visits to some of the destinations mentioned above have had an economic impact on foreign exchange for the government, the private sector, and income for local communities. Data from the tourism office in 2014 states that the tourism sector has contributed to a foreign exchange income of IDR120 trillion by providing job opportunities to 11 million residents. Meanwhile, data from the World Travel and Tourism Council (WTTC) in 2014 stated that tourism contributed 9% of GDP or IDR946.1 trillion. And up to the end of 2019 the tourism sector contributed 10.4% of GDP (WTTC, 2019).

Compared to other archipelagic countries that rely on the marine tourism sector, such as Madagascar, Hawaii, and the Maldives, Indonesia is still lagging behind. For example, in the Maldives, 74% of its GNP is supported by the marine tourism sector (Raina &

Agarwal, 2004). Yet the potential of Indonesia's marine tourism objects is not inferior to these countries. For example, for manta ray diving, according to 2015 Manta Watch data, although there are more Pari Manta spots in the Maldives, the chances of encountering these protected biota are three times higher in West Manggarai. It is not surprising therefore that O'Malley et al. (2013) estimate the economic potential of Indonesian manta tourism at USD15 million per year. That, just from the Pari Manta tour alone.

## References

Bahar, A. (2017). Degradasi ekosistem terumbu karang akibat aktifitas wisata selam di Pulau Hoga, Taman Nasional Wakatobi. In Okto Irianto & Nova V. Pati (Eds.), *Pengelolaan Taman Nasional Laut Bunaken sebagai destinasi wisata laut dunia*. Manado: Kementerian Koordinator Kemaritiman dan Universitas Sam Ratulangi.

Christiansen, F., Lusseau, D., Stensland, E., & Berggren, P. (2010). Effects of tourist boats on the behaviour of Indo-Pacific bottlenose dolphins off the south coast of Zanzibar. *Endangered Species Research*, 11, 91–99.

Dimitrovski, D., Lemmetyinen, A., Nieminen, L., & Pohjola, T. (2021). Understanding coastal and marine tourism sustainability-A multi-stakeholder analysis. *Journal of Destination Marketing & Management*, 19, 100554.

Djunaidi, A., Jompa, J., Nadiarti, N., Bahar, A., & Tilahunga, S. D. (2021, May). Benefit sharing from whale shark tourism in Botubarani, Gorontalo and Labuhan Jambu, Teluk Saleh. In *IOP Conference Series: Earth and Environmental Science* (Vol. 763, No. 1, p. 012063). IOP Publishing.

Hall, C. M. (2001). Trends in ocean and coastal tourism: the end of the last frontier?. *Ocean & Coastal Management*, 44(9–10), 601–618.

Harriott, V. J. (2002). *Marine tourism impacts and their management on the Great Barrier Reef* (No. 46). Townsville: CRC Reef Research Centre.

Hayati, Y., Adrianto, L., Krisanti, M., Pranowo, W. S., & Kurniawan, F. (2020). Magnitudes and tourist perception of marine debris on small tourism island: Assessment of Tidung Island, Jakarta, Indonesia. *Marine Pollution Bulletin*, 158, 111393.

Israngkura, A. (2022). Marine resource recovery in Southern Thailand during COVID-19 and policy recommendations. *Marine Policy*, 137, 104972.

Mustika, P. L. K., Birtles, A., Everingham, Y., & Marsh, H. (2013). The human dimensions of wildlife tourism in a developing country: Watching spinner dolphins at Lovina, Bali, Indonesia. *Journal of Sustainable Tourism*, 21(2), 229–251.

Mustika, P. L. K., Birtles, A., Everingham, Y., & Marsh, H. (2015). Evaluating the potential disturbance from dolphin watching in Lovina, north Bali, Indonesia. *Marine Mammal Science*, 31(2), 808–817.

O'Malley, M. P., Lee-Brooks, K., Medd, H. B. (2013). The global economic impact of Manta Ray Watching Tourism. *PLoS One* 8(5). doi: 10.1372/journal.pone.0065051.

Raina, K., Agarwal, S. K. (2004). *The Essence of Tourism Development: Dynamics, Philosophy, and Strategies*. New Delhi: Saurup & Sons. p. 414.

Stensland, E., & Berggren, P. (2007). Behavioural changes in female Indo-Pacific bottlenose dolphins in response to boat-based tourism. *Marine Ecology Progress Series*, 332, 225–234.

Suteja, Y., Atmadipoera, A. S., Riani, E., Nurjaya, I. W., Nugroho, D., & Purwiyanto, A. I. S. (2021). Stranded marine debris on the touristic beaches in the south of Bali Island, Indonesia: The spatiotemporal abundance and characteristic. *Marine Pollution Bulletin*, 173, 113026.

Thompson, R. C., Swan, S. H., Moore, C. J., & Vom Saal, F. S. (2009). Our plastic age. *Philosophical Transactions of the Royal Society B: Biological Sciences*, 364(1526), 1973–1976. https://doi.org/10.1098/rstb.2009.0054

WTTC. (2014). *World Travel and Tourism Council*. wttc.org

WTTC. (2019). *World Travel and Tourism Council*. wttc.org

# Part II

# Planning and development

In Chapter 10, **Annemarie Nicely, Shweta Singh, Dan Zhu, Chutong Jiang**, and **Jihon Choe** discuss *Trader harassment of visitors: A global problem in need of a global response*. The goals of this chapter are to demonstrate how widespread the problem of trader harassment of visitors was before the COVID-19 pandemic, to note a key pattern to the phenomenon, and to posit ideas on how global tourism leaders may assist in reducing the phenomenon globally. To accomplish this, the authors share the results of a 2019 study. They reviewed 1,489 documents (blog posts, journal articles, master's theses, online magazine and newsletter articles, news reports, newspaper articles, project reports, speeches, and press releases), and the response to an open-ended question taken from 659 surveys. From the analysis, the research team found 80 countries and territories where significant trader harassment of visitors was reported between the period 2010 and 2019. Among the states were developed and developing nations. Most states found were in the equatorial climate zone (the Caribbean, South East Asia, and Southern Asia). The chapter provides suggestions on measures that regional and global tourism leaders should take to curtail the problem going forward.

*Cross-border tourism: The case of the Greater Bay Area* is Chapter 11's topic by **Jinah Park, Haiyan Song**, and **Yeung (William) Kong**. Tourism has played a critical role in regional development and integration for cross-border destinations. Although improved infrastructural interconnectivity and technologies facilitate cross-border mobilities, there are some inherent and external roadblocks to cross-border tourism such as institutional complexity and the current immobility situation caused by the COVID-19 pandemic. By taking the case of the Guangdong–Hong Kong–Macao Greater Bay Area (GBA), which spans geographical and institutional borders, this chapter traces the policy-based networking and development phases of the GBA and identifies people's changing perceptions of the GBA destination image before and during the pandemic. On the basis of interviews and longitudinal surveys, the authors found four pillars of cross-border tourism collaboration and a sense of local belonging in adjusting marketing strategy during and after the pandemic. The GBA case provides important implications for cross-border tourism planning and management in the post-pandemic era by addressing how a cross-border destination has been affected by the pandemic but also remains resilient.

**Jens Thraenhart** and **Alastair M. Morrison** elaborate on *Trends and issues with regional tourism partnership formation* in Chapter 12. This chapter commences by identifying recent trends in regional tourism partnerships based on a GUESS framework (growth, universality, e-marketing, sustainable development and sustainable tourism, and scope). Sixteen specific examples of regional tourism partnerships are highlighted. Then, there is a discussion

DOI: 10.4324/9781003291763-11

of the reasons and catalysts for the increasing interest in regional tourism partnerships. Two successful partnership cases are presented (the Wild Atlantic Way and the Experience Mekong Collection). Next, the prerequisites, barriers, and challenges for regional tourism partnerships are identified. The roles of stakeholders in creating and maintaining partnerships are discussed. The chapter then reviews the potential contributions of regional tourism partnerships in achieving the UN's Sustainable Development Goals (SDGs) and presents six expected future trends for regional tourism partnerships. The chapter ends by outlining the potential contributions of the work.

Chapter 13 is *Disseminating small island destinations in the face of global challenges: A strategic analysis* by **Eduardo Parra-López, Almudena Barrientos-Baez**, and **María de los Ángeles Pérez Sánchez**. There is a lack of a comprehensive understanding of how island tourist destinations can improve social, economic, cultural, and territorial issues. A need to support decision-making through an intelligent framework and rigorous research is the focus of this chapter. The proposal revolves around four key elements: size, tourism specialisation, economic growth, and tourism competitiveness. The construct of tourism specialisation relates to tourism supply and demand, while the construct of competitiveness corresponds to demand. The concept of smallness is an island's ability to strategise a prosperous path of sustainable economic growth through tourism. The research proposal emerges from this route and towards the empowerment of island destinations. The aim is to achieve improvement for residents and to explore opportunities arising in a post-pandemic era and results in improvements for the lives of residents. The chapter presents a rigorous transversal analysis of the literature to explain the relationship between tourism and growth.

**Huu Nghia Le** in Chapter 14 focuses on *Island tourism development for inclusive growth*. Island tourism has always been one of the most popular types of tourism in the world. Yet, it is difficult to analyse the tourism growth on an island because it is a dynamic entity with distinctive characteristics. In recent years, island tourism development has dealt with problems like climate change, environmental degradation, economic downturns, cultural loss, and diseases like the COVID-19 pandemic. Thus, tourism for inclusive growth is key to ensuring the destination can keep up with demand. The author chose the SLIQ concept because it was a new way of looking at an old problem and picked the Hai Tac islands in Vietnam as a case study and conducted in-depth interviews with five island people. The findings demonstrate that local authorities should govern the island as a living entity, which needs a unique approach to contribute to the SDGs. The chapter also affirms that sustainable economic growth, environmental protection, and human fairness should be the three goals of public policy when planning and managing inclusive tourism development on islands.

Chapter 15, *Community-based rural tourism development: An intersectional exploration of possibilities and challenges*, is authored by **Neha Nimble**. Community-based tourism has established itself as an advanced, sustainable, and collaborative form of tourism that is aimed at the development of host communities. The last two decades have seen India's Ministry of Tourism developing and implementing a number of community-based rural tourism development programmes in different parts of the country. While community-based tourism is definitely a better alternative to mass tourism practices, it also has its own disadvantages, if not implemented with a critical awareness of the power equations in the community. Drawing from PhD research conducted in a rural tourism site, Naggar, in the Himalayan state of Himachal Pradesh in India, this chapter explores the various ways in which community-based rural tourism constructs and transforms the unequal power relations of gender, class, and caste in the community. Located within a theoretical framework drawn from feminist political thought (feminist materialism, feminist topography, and an intersectionality perspective), the author finds that the communities are not homogeneous and that their members experience tourism and its benefits as well as costs differently if the existing power relations are not weaved into the structure of the tourism development activities. The study adopted a phenomenological methodology to describe the varied lived experiences for people of tourism development in their community. Tourism processes and activities (introduced as part of the state's tourism development for livelihood generation in the community) reflect the prevailing social relationships and confirm existing power relations while also creating space for some transformation. Through an analysis of the cultural, social, and material lives of women in their productive and reproductive spheres, the study specifically reveals that a person's socio-spatial location in the community in terms of her gender, caste, and class shape her experiences of tourism and also determine the share of its costs and benefits for her.

**Paul Tully** and **Neil Carr** in Chapter 16 discuss the ***Evolving position of stray domestic animals in tourism***. Contemporary and historic examples abound of the removal and termination of stray domestic animals from the holiday space to spare tourists from encounters with such animals and the negative social constructions surrounding them. Yet such actions are increasingly meeting resistance driven by the changing social perceptions of animals as sentient beings with rights and welfare needs. As a result, and through tourism, we are witnessing a re-imagining of stray animals, from undesirables to new trendy attractions. This chapter discusses the historic construction of stray animals as pests in the holiday environment and how they have traditionally been dealt with. In the light of changing societal views about animal rights and welfare, the chapter discusses changing trends in how stray animals in the tourism experience are imagined and dealt with. The chapter highlights the need for the tourism industry to re-imagine its relationship with stray domestic animals to ensure its own sustainability.

# 10 Trader harassment of visitors

## A global problem in need of a global response

*Annmarie Nicely, Shweta Singh, Dan Zhu, Chutong Jiang and Jihon Choe*

### Introduction

Trader harassment of visitors (THV), simply defined, is an unwanted or undesired selling behaviour directed toward visitors by vendors and micro-business operators at destinations (Nicely & Ghazali, 2014; Nicely & Selvia, 2019; de Albuquerque & McElroy, 2001; Griffin, 2003). While not all taxicab drivers, craft traders, informal tour guides, and street performers interacting with visitors engage in harassment, unfortunately many do. Three common ways micro-traders harass visitors include duping them into overpaying for goods and services, quoting them inflated prices, or multiple traders approaching visitors one after the other (Nicely, Singh, & Zhu, 2020).

While the origin of the phenomenon is unclear, studies on the topic began emerging in leading scholarly journals on tourism during the early 2000s, such as *Current Issues in Tourism, Tourism Management*, and *Annals of Tourism Research* (Chambers & Airey, 2001; Chaudhary, 2000; de Albuquerque & McElroy, 2001). However, the issue has not caught the attention of global tourism organisations such as the UNWTO. One factor that may explain why the phenomenon has fallen below the radar is that tourism leaders suppress reports of the practice at their location (Overseas Security Advisory Council, 2016). With the emergence of social media in the early 2000s (Ortiz-Ospina, 2019), travellers have begun to share directly with the public their experiences of the practice. In 2022, there were numerous reports on the phenomenon on popular travel websites, such as Tripadvisor (M, 2017; NoSuchPerson, 2010) and video-sharing platforms like YouTube (The Other Side, 2022).

Some believe THV only occurs in developing nations. Scholars who believe the problem is more widespread had little evidence to support this (McElroy et al., 2007a). So, in 2019 a group of us embarked on a study to determine how widespread the problem of THV was. The research team had three goals: to determine the number of countries and territories with significant trader harassment of visitors (STHV), those that were developed states, and the geographic regions where the phenomenon dominated.

While several factors may be fuelling the phenomenon, to date researchers have only been able to agree on one determinant. That is, poor socioeconomic conditions at the destination (Ajagunna, 2006; Chepkwony & Kangogo, 2013; de Albuquerque & McElroy, 2001; Dunn & Dunn, 2002a, 2002b; Griffin, 2003; Kozak, 2007; McElroy et al., 2007a; Nicely & Selvia, 2019). However, individual visitor, micro-trader, tourism stakeholder, societal, and natural environmental factors may be at play as well. These are yet to be proven through research.

DOI: 10.4324/9781003291763-12

*Table 10.1* Types of documents (archives) used

| Types of archives | Descriptions |
| --- | --- |
| Blog posts | Comments noted on a discussion website, such as below an online news article or on the discussion board of a travel website. The comments chronicle the author's experience with trader harassment at a destination. Example: Tripadvisor post. |
| Journal articles and master's theses | *Journal articles*: studies, scholarly opinion, and concept papers published in academic journals. *Master's theses*: detailed reports of studies conducted by graduate students as the final requirement for their graduate programme. Example: *Annals of Tourism Research*. |
| Magazine and newsletter articles | *Magazine articles*: stories published in online periodicals, usually for a wide audience. *Newsletter articles*: stories published on a website for a small audience, usually persons who share a common interest such as travel. Neither are usually published daily. Examples: World Nomads article; Condé Nast Traveler article. |
| News reports and newspaper articles | Publications by a media house (such as by a television or newspaper company) on local and international happenings. Published daily whether online or in paper format. Examples: Geo News (Pakistan); *The Independent* (UK). |
| Official speeches and press releases | *Speeches*: written presentations delivered by public or private officials. *Press releases*: written communication distributed to the media by a public or private body. Examples: speeches by a minister of government; press release from local tourism governing authority. |
| Project reports | Official documents, from a reputable body (such as a non-governmental or governmental organisation), outlining programmes executed and the results or describing a situation in a country. Example: Travel Foundation Report; Crime and Safety Report for a particular state. |

This study is important for several reasons. It is the first known attempt at enumerating countries and territories with STHV at one or more of their tourist locations. A list of countries and territories could prove useful to researchers and practitioners, such as in regional collaborations. It would provide evidence that the phenomenon is indeed global and hence should be examined and ways found to address it by global tourism leaders and bodies.

The period of focus for the study was 2010 to 2019, immediately before the COVID-19 pandemic. The focus of the study was on THV, not on other forms of visitor harassment such as institutional harassment (e.g., the bothering of visitors by police officers and military personnel) or sexual harassment (i.e., advances of a sexual nature). These should be examined later.

STHV is unwanted or undesired selling practices by micro-traders when engaging with visitors and which demand or have the attention of authorities at the destination. States with STHV are countries and territories where there is documentary evidence (e.g., studies, newspaper articles, blog posts) of the phenomenon occurring at one or more locations in the destination and a team of researchers agrees (after reviewing the documents) that the practice is at noteworthy levels. Listed in Table 10.1 are the descriptions of the types of archives used in the study. Finally, a state is a country or territory.

## Methodology

To accomplish the goals of the study, the research team used a mixed methods approach to data gathering and analysis. The unit of analysis was the state. Hence, the target population for the study was all countries and territories globally (United Nations Statistics Division, 2019b). The research team attempted a census.

The team used two data sources: archives and a previously administered survey. The archives included journal articles, master's theses, project reports, speeches, press releases, news reports, newspaper articles, magazine articles, newsletter articles, and blog posts (Table 10.1) describing THV during the target period.

To gather the archives, a group of research assistants searched the Internet. They used online sources such as: the search engine, Google; the library databases, Factiva and Newspaper Source Plus; and the popular travel forum website, Tripadvisor. The assistants used the following keywords: "tourist harassment", "visitor harassment", "tourist scam", and the name of each state (or their popular tourism cities or locations). They also searched newspaper and national tourism body websites using key terms such as "tourist harassment", "visitor harassment", "tourist scam", and/or the local names of micro-trading groups. The research team collated the responses to an open-ended question describing the respondents' THV experience at a destination. The period of focus for the survey was THV which occurred between 2010 and 2016, within the period of focus for the present study. The search for archives started in the summer of 2016 and ended in the summer of 2019.

### Data collection

#### Step 1: Archives classified according to type

The first step was gathering and reviewing the archives. The aim of the research team here was to collate and classify all the archives gathered by type. To accomplish this, one leading member of the research team collated the archives (not the surveys), reviewed them, and created an archive-type codebook. The principal investigator (PI) then assigned each archive a number and randomly selected 10% for review. Then two other members of the research team coded and recoded the sample of archives separately, and the convergence rate was ascertained. Once the minimum convergence rate for an interrater reliability of 0.75 (Stephanie, 2016) was achieved, the pair then coded the remaining archives, again separately. So each archive was coded twice, once by each coder. A leading member of the research team (i.e., the second author of this chapter) then identified conflicts between the coders and resolved them.

#### Step 2: First flagging of states with significant trader harassment using reviewers' feedback

The next step was the first flagging of states with STHV. A leading member of the research team gave an international team of researchers electronic copies of the archives organised in folders by state and a Microsoft Excel spreadsheet containing important details on each state, namely (1) the cities in the state where THV was reported in the archives and surveys; (2) the number of archives found by type on the state (such as the number of blog posts or newspaper articles found) and the number of surveys with THV information on the state; (3) the period covered in each group of archives (such as the period covered in the blog posts, journal articles, and theses on the state); and (4) the World Press Freedom Ranking Scores (WPFRS) for the state for the years 2010 to 2018. Excluded were WPFRS for 2019, which were not available at the time of the study. WPFRS are produced annually by the group Reporters Without Borders (2019). They indicate the extent to which each state has a reputation for withholding information from the public. The international reviewers that participated in the first flagging of the archives were primarily graduate students at a leading research university in the midwestern region of the USA. They used the period covered in the archives to determine the pervasiveness of the phenomenon during the target period. Meanwhile, they used the WPFRS to make a judgement on the sufficiency of the evidence provided.

The PI then asked the reviewers to do the following for the states assigned. Read the archives for the states. Review the information provided in the Excel spreadsheet for those states. Then indicate, after their review, the states with STHV. A leading member of the research team then gathered the reviewers' spreadsheets and entered their decisions into a master spreadsheet. Hence, the information on each state was independently reviewed by three people. The PI then examined the decisions of the reviewers and, where two or more agreed that a state had STHV at one or more of their tourist locations, flagged the state as such.

*Step 3: Second flagging of states with significant trader harassment using archival evidence score*

The next major step was determining archival evidence scores (AES) for each state. To accomplish this, the research team did the following. A leading member administered a survey to a diverse panel comprising ten people. The survey ascertained the believability of the types of archives gathered (Table 10.2). In the survey, the research team asked the respondents to: "Indicate the extent to which each (of the archival types named) would be a reliable source of information on the existence of STH at a tourist destination." A ten-point scale of 1 "I would not believe" to 10 "I would be convinced" was used. The mean and median scores for each archive type were then determined. Listed in Table 10.2 are the mean and median scores per archive type.

The research team then used each archive type's believability mean and median scores to determine each archive's weight. Archives where the mean and median believability scores were between 7.5 and 10 (in the top 25 percentile) (i.e., journal articles and master's theses), the team assigned a weight of 0.3. For archives where the mean and median believability scores were between 5.5 and 7.4 (in the second top 25 percentile) (in particular, project reports, news and newspaper articles, speeches and press releases, magazine, and newsletter articles), the team assigned a weight of 0.2. And those with mean and

*Table 10.2* Archive rankings

| Archive type | Mean | Median | Percentile | Weighting for each |
| --- | --- | --- | --- | --- |
| Journal articles and master's theses | 8.6 | 9.0 | Top 25% (0.75–1.00) | 0.30 |
| *Surveys | 7.9 | 8.0 | Second 25% (0.55–0.74) | 0.20 |
| Project reports | 6.9 | 7.0 | | 0.20 |
| News reports and newspaper articles | 6.3 | 6.0 | | 0.20 |
| Speeches and press releases | 6.1 | 6.0 | | 0.20 |
| Magazine and newsletter articles | 5.6 | 5.5 | | 0.20 |
| Blog posts | 5.5 | 5.0 | Third 25% (0.25–0.54) | 0.10 |

* Individual surveys were treated like blog posts and assigned a weight of 0.10.

median believability scores of between 2.5 and 5.4 the team assigned a weight of 0.1. Since blog posts had a mean and median score of 5.5 and 5.0, respectively, the team assigned a weight of 0.1. No archive type had mean and median believability scores of less than 5.0.

Surveys reported mean and median believability scores of 7.9 and 8.0, respectively. Since the surveys were not treated as a collective but rather individually, the team assigned them a weight of 0.2 (Table 10.2). Hence, the team counted the single open-ended response from each survey as one archive, because of its similarity to a blog post. Of the seven archive types, journal articles and master's theses were thought to be the most believable source of information on THV at a destination; blog posts were viewed as believable, but the least so, of the seven archive types (Table 10.2).

For a state to be flagged as possibly having STHV, its AES had to be one or above. For example, there must a minimum of four journal articles and master's theses describing the phenomenon at the destination. The team used the below formula to compute the AES for each state using the software Microsoft Excel 2010.

**AES per state** = (weighting for archival type 1 X number of archival type 1) + (weighting for archival type 2 X number of archival type 2) … (weighting for archival type $x$ X number of archival type $x$)

*Step 4: List of states with significant trader harassment of visitors finalised*

The team then used two factors to determine the final list of states with STHV: the reviewers' feedback (Step 2) and the state's AES score (Step 3). Hence, states identified by two or more reviewers as having STHV and had an AES of 1 or more were noted as having STHV.

However, before the list was published, the PI conducted another review using four sources of information: her travels to the destinations during the target period, YouTube videos, other archives sourced, and the original archives themselves. After the review, some states were removed from the list, but none were added.

### Reliability, validity, and data analysis

The team took steps to ensure the accuracy of the research findings. For example, individuals with diverse backgrounds participated in the search for archives and the document review. The 17 undergraduate and graduate research students and one professor who participated in the gathering of the archives were from countries in Asia, Europe, and the Americas. Nine graduate students and two undergraduate research students reviewed the archives for appropriateness. The archives and summary data for each state were reviewed by three experts, each from a different region of the world. To further minimise bias, none of the reviewers who participated in the archive and summary data review (Step 2) were from the state or geographic region assigned. For example, the PI did not assign a reviewer from Asia to review the archives and summary data for a state in Asia. The 11 reviewers who participated in the archive and summary data review were from states in Asia, Europe, the Americas, and Africa. There were no reviewers from the Oceania region.

To ensure the credibility of the findings, the research team used three factors to determine states with STHV: the number of archives and surveys describing THV in the state, the believability of the archives, and expert opinion of scholars from different parts of the world. In addition, the research team used three levels of review to determine the list of states with STHV: (1) expert review of the archives sourced; (2) the research team's review of each state's AES; and (3) the final review of the list of states by the PI for the project.

The creators of the survey also took steps to ensure the accuracy of the qualitative data the survey generated. For example, they asked the respondents not to give their names in the survey and ensured the confidentiality of their responses at the start.

### *Data analysis*

The research team used three data analysis techniques. To determine the number of archives gathered by type, the team used frequency analysis. To ascertain the states with STHV, the team used narrative and interpretive phenomenological analyses. To ascertain probable geographic patterns to TH globally, the team used cross-tabulation analysis. The above was accomplished using the software Statistical Package for Social Sciences (SPSS) 26.0 and Microsoft Excel 2010.

### Findings

The research team analysed 1,489 and 659 archives and surveys, respectively, for the study. Of the documents analysed, most were blog posts (41.8%) as well as news reports and newspaper articles (19.7%) (Table 10.3).

The research team attempted to gather data on 245 states. For some states, no documents and surveys describing THV were found. The team found 80 states with STHV at one or more of their tourist locations. The team also found that the practice was not unique to developing nations. Of the 80 states with STHV during the period 2010–2019, 16 were developed states. For example, the team found STHV in developed states such as the USA, Israel, and Italy (Table 10.4).

Another noteworthy finding was the high ratio of states with STHV in five geographic regions: the Caribbean, South East Asia, Southern Asia, Central America, and Northern Africa, namely 67.9, 63.6, 55.6, 50, and 42.9% of the states in these regions, respectively (Table 10.3). Hence, it could be concluded that at minimum 67.9, 63.6, 55.6, 50, and 42.9% of the states in the Caribbean, South East Asia, Southern Asia, Central America, and

*Table 10.3* Documents analysed

| Documents | Count | % |
|---|---|---|
| Blogs | **897** | **41.8** |
| News reports and newspaper articles | **423** | **19.7** |
| Journal articles and theses | **44** | **2.0** |
| Magazine and newsletter articles | **90** | **4.2** |
| Project reports | **34** | **1.6** |
| Speeches and press releases | **1** | **0.0** |
| *Surveys* | **659** | **30.7** |
| TOTAL | **2,148** | **100.0** |

*Table 10.4* States with significant trader harassment of visitors during the period 2010 to 2019

| Geographic regions | Total states by region | States not examined by region | Number of states found with STHV by region | States found with STHV by region (%) | States with STHV by region |
|---|---|---|---|---|---|
| Northern Africa | 7 | Western Sahara | 3 | 42.9 | Egypt<br>Morocco<br>Tunisia |
| Eastern Africa | 22 | British Indian Ocean Territory<br>French Southern Territories | 6 | 27.3 | Ethiopia<br>Kenya<br>Mauritius<br>The United Republic of Tanzania (including Zanzibar)<br>Zambia<br>Zimbabwe |
| Middle Africa | 9 | — | 0 | 0 | — |
| Southern Africa | 5 | — | 1 | 20.0 | South Africa |
| Western Africa | 17 | | 4 | 23.5 | Cabo Verde<br>The Gambia<br>Ghana<br>Nigeria |
| Caribbean | 28 | — | 19 | 67.9 | Anguilla<br>Antigua & Barbuda<br>Aruba<br>Bahamas<br>Barbados<br>Cuba<br>Dominica<br>Dominican Republic<br>Grenada<br>Haiti<br>Jamaica<br>St. Kitts & Nevis<br>St. Lucia<br>St. Vincent & the Grenadines<br>St. Maarten (Dutch Part)<br>St. Martin (French Side)<br>Trinidad & Tobago<br>Turks & Cacaos Islands<br>United States Virgin Islands |
| Central America | 8 | — | 4 | 50.0 | Belize<br>Costa Rica<br>Mexico<br>Nicaragua |
| South America | 16 | Bouvet Island<br>South Georgia and the South Sandwich Islands | 6 | 37.5 | Argentina<br>Brazil<br>Chile<br>Colombia<br>Ecuador<br>Peru |

*(Continued)*

*Table 10.4* (Continued)

| Geographic regions | Total states by region | States not examined by region | Number of states found with STHV by region | States found with STHV by region (%) | States with STHV by region |
|---|---|---|---|---|---|
| Northern America | 5 | — | 1 | 20.0 | The United States of America* |
| Central Asia | 5 | — | 0 | 0.0 | — |
| Eastern Asia | 6 | — | 3 | 50.0 | China<br>Republic of Korea (South Korea)<br>Taiwan[†] |
| South-eastern Asia | 11 | — | 7 | 63.6 | Cambodia<br>Indonesia<br>Lao People's Democratic Republic<br>Malaysia<br>Philippines<br>Thailand<br>Viet Nam |
| Southern Asia | 9 | — | 5 | 55.6 | Bangladesh<br>India<br>Nepal<br>Pakistan<br>Sri Lanka |
| Western Asia | 18 | — | 4 | 22.2 | Bahrain<br>Israel*<br>Jordon<br>Turkey |
| Eastern Europe | 10 | — | 2 | 20.0 | Bulgaria*<br>Hungary* |
| Northern Europe | 15 | Åland Islands<br>Svalbard and Jan Mayen Islands | 3 | 20.0 | Latvia*<br>Sweden*<br>United Kingdom of Great Britain & Northern Ireland* |
| Southern Europe | 16 | — | 6 | 37.5 | Gibraltar*<br>Greece*<br>Italy*<br>Portugal*<br>Serbia*<br>Spain* |
| Western Europe | 9 | — | 2 | 22.2 | France*<br>Germany* |
| Australia and New Zealand | 6 | Christmas Island<br>Cocos (Keeling) Islands<br>Heard Island and McDonald Islands<br>Norfolk Island | 1 | 16.7 | Australia* |

*(Continued)*

*Table 10.4* (Continued)

| Geographic regions | Total states by region | States not examined by region | Number of states found with STHV by region | States found with STHV by region (%) | States with STHV by region |
|---|---|---|---|---|---|
| Melanesia | 5 | — | 1 | 20.0 | Fiji |
| Micronesia | 8 | United States Minor Outlying Islands | 0 | 0.0 | — |
| Polynesia | 10 | Pitcairn | 2 | 20.0 | Samoa Tonga |
| TOTAL | 245 | 13 | 80 | | |

* Developed states and list of countries and areas globally (United Nations, 2018; United Nations Statistics Division, 2019a).
† Examined but not in the United Nations list of states around the world.

Northern Africa, respectively, had STHV at one or more of their tourist locations during the period 2010 to 2019 (Table 10.4). Most of the regions just mentioned are within the equatorial climate zone (Kottek et al., 2006).

Also, of the 80 states identified as having STHV, the research team identified 48 states to watch as they too may have had STHV during the target period. For these states, either the reviewers thought they had STHV, or their AES was greater than 1. So there could be as many as 128 states with STHV at one or more of their tourist locations during the period 2010 to 2019, or 52.2% of the countries and territories worldwide.

**Discussion**

The 2019 study found 80 states with STHV at one or more of their popular tourist locations. There may be more. Of the 80 states, 16 were developed countries and territories. The researchers also discovered that most states found with STHV were in the Caribbean, South East Asia, and Southern Asia, largely in the equatorial climate zone.

This chapter has filled an important gap in the literature. At the time the research was conducted in 2019, it was the only one of its kind to paint a picture of how widespread the problem of THV was immediately before the pandemic.

It is predicted that the problem will re-emerge to pre-COVID-19 levels if action is not taken. Here is why. Studies have found an indirect connection between socioeconomic conditions and THV intensity levels at destinations. When local socioeconomic conditions decline, THV intensity levels increase (Ajagunna, 2006; Chepkwony & Kangogo, 2013; de Albuquerque & McElroy, 2001; Dunn & Dunn, 2002a, 2002b; Griffin, 2003; Kozak, 2007; McElroy et al., 2007b; Nicely & Selvia, 2019). The pandemic has hurt socioeconomic conditions at destinations and has wrecked economies and education systems worldwide. Tackling socioeconomic issues fuelling THV globally would assist in the attainment of the United Nations Sustainable Development Goals 1 (no poverty), 4 (quality education), and 8 (decent work and economic growth).

Second, rising outdoor temperatures, due to global warming, are also likely to result in negative visitor/ micro-trader interactions in outdoor and quasi-outdoor spaces. Studies on

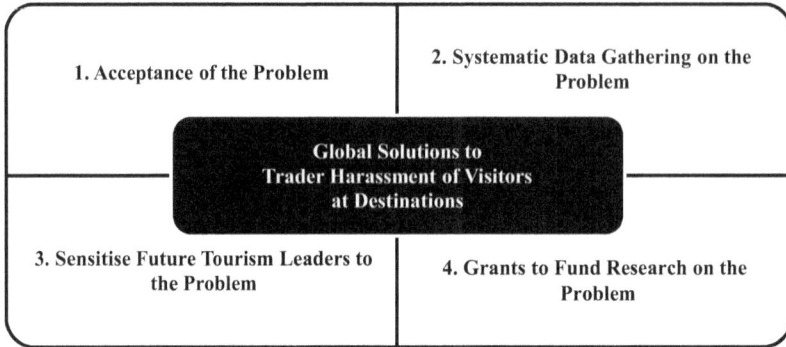

*Figure 10.1* Four global solutions to the problem of trader harassment of visitors at destinations.

atmospherics, such as temperature and precipitation levels on customer service, have confirmed that these factors play a role in negative service provider/customer interactions in spaces (Howarth & Hoffman, 1984; Kolb, Gockel, & Werth, 2012).

## Recommendations

We recommend the following. First, acceptance among world tourism leaders that THV is a global issue in need of a global response (Figure 10.1).

Second, systematic data gathering on THV intensity levels at destinations is required. UNWTO is best poised to take on this important task. That is, gather and make available to the public (including researchers and tourism leaders at destinations) periodic data on THV intensity levels in various cities around the world. Researchers globally would then use the data to do the needed work in this area, such as determining the myriad of causes of the phenomenon and effective ways to address them. Meanwhile, destination management organisations could use this data to monitor and enhance the effectiveness of their mitigation programmes.

Third, there is a need to sensitise the next generation of tourism leaders and change agents in this sector to this critical global problem. Again, UNWTO could be the catalyst for this initiative.

Fourth, grants to fund THV research are required. Funding research in this area would fast-track the science tourism practitioners needed to create highly effective mitigation programmes, and again UNWTO could assist here (Figure 10.1).

## Limitations and future research

This research was not without limitations. First, the methodology used was not effective at capturing all states with STHV. Second, the approach used was better at capturing bothersome micro-trading, and not so much intimating, illegal, and immoral micro-trading. The next logical step is to review and improve the current research protocol for identifying states with STHV to one involving robust quantitative measures and analysis.

## References

Ajagunna, I. (2006). Crime and harassment in Jamaica: Consequences for sustainability of the tourism industry. *International Journal of Contemporary Hospitality Management, 18*(3), 253–259.

Chambers, D., & Airey, D. (2001). Tourism policy in Jamaica: A tale of two governments. *Current Issues in Tourism, 4*(2–4), 94–120. doi:10.1080/13683500108667884

Chaudhary, M. (2000). India's image as a tourist destination – A perspective of foreign tourists. *Tourism Management, 21*(3), 293–297. Retrieved from https://ac.els-cdn.com/S0261517799000539/1-s2.0-S0261517799000539-main.pdf?_tid=f93c861e-a421-42c7-8869-3a3036a61583&acdnat=152047 4294_5fdf5e86d4774e11cb5c1ffec6ca1718

Chepkwony, R., & Kangogo, M. (2013). Nature and factors influencing tourist harassment at coastal beach of Mombasa. *International Research Journal of Social Sciences, 2*(11), 17–22.

de Albuquerque, K., & McElroy, J. L. (2001). Tourist harassment: Barbados survey results. *Annals of Tourism Research, 28*(2), 477–492. doi:10.1016/S0160-7383(00)00057-8

Dunn, H., & Dunn, L. (2002a). *People and Tourism*. Kingston, Jamaica: Arawak Publications.

Dunn, H., & Dunn, L. (2002b). Tourism and popular perceptions: Mapping Jamaican attitudes. *Social and Economic Studies, 51*(1), 25–45.

Griffin, C. (2003). *The Caribbean Tourism Integrated Standards Project: Analysis and Policy Recommendations re Harassment of and Crime Committed by and Against Tourists in the Caribbean*. Retrieved from San Juan, Puerto Rico.

Howarth, E., & Hoffman, M. S. (1984). A multidimensional approach to the relationship between mood and weather. *British Journal of Psychology, 75*(1), 15–23.

Kolb, P., Gockel, C., & Werth, L. (2012). The effects of temperature on service employees' customer orientation: an experimental approach. *Ergonomics, 55*(6), 621–635. doi:10.1080/0014013 9.2012.659763

Kottek, M., Grieser, J., Beck, C., Rudolf, B., & Rubel, F. (2006). World Map of the Köppen-Geiger climate classification updated *Meteorologische Zeitschrift, 15*(3), 259–263. doi:https://doi.org/10.1127/0941-2948/2006/0130

Kozak, M. (2007). Tourist harassment: A marketing perspective. *Annals of Tourism Research, 34*(2), 384–399. Retrieved from http://www.sciencedirect.com/science/article/pii/S0160738306001381

M, M. (2017). Average beach with annoying venders nagging at you constantly. Tamarindo and the roads-absolutely filthy and littered.

McElroy, J., Tarlow, P., & Carlisle, K. (2007a). Tourist harassment: Review of the literature and destination responses. *International Journal of Culture, Tourism and Hospitality Research, 1*(4), 305–314.

McElroy, J. L., Tarlow, P., & Carlisle, K. (2007b). *Tourist Harassment and Responses*. Oxfordshire, UK: CAB International.

Nicely, A., & Mohd Ghazali, R. (2014). Demystifying visitor harassment. *Annals of Tourism Research, 48*, 266–269. Retrieved from http://doi.org/10.1016/j.annals.2014.05.011

Nicely, A., & Selvia, A. (2019). *Socioeconomic Predictors of Visitor Harassment: Why Knowing Vendor Migration Patterns is Important*. Visitor Harassment Research Unit Yellow Paper Series. West Lafayette, Indiana: School of Hospitality & Tourism Management. Purdue University.

Nicely, A., Singh, S., & Zhu, D. (2020). Visitor (trader) harassment further explained. *Journal of Quality Assurance in Hospitality & Tourism, 22*, 1–16. doi:10.1080/1528008X.2020.1848748

NoSuchPerson. (2010). *My Experience in Tobago* (Vol. 2019). Hayesville, NC: TripAdvisor.

Ortiz-Ospina, E. (2019). The rise of social media. *Our World in Data*. Retrieved from https://our worldindata.org/rise-of-social-media

Overseas Security Advisory Council. (2016). *Barbados 2016 crime and safety report*. Retrieved from Washington DC:

Reporters Without Borders. (2019). Reporters without border: For the freedom of information. Retrieved from https://rsf.org/en/media-center

Stephanie. (2016). Inter-rater reliability IRR: Definition and calculation. *Statitics How To*. Retrieved from https://www.statisticshowto.datasciencecentral.com/inter-rater-reliability/

The Other Side (Producer). (2022). Don't visit egypt until you watch this - Pyramids BEWARE.

United Nations. (2018). *World Economic Situation and Prospects 2018*. Retrieved from New York.

United Nations Statistics Division. (2019a). Developed regions. *Methodology*. Retrieved from https://unstats.un.org/unsd/methodology/m49/

United Nations Statistics Division. (2019b). Geographical regions. *Methodology*. Retrieved from https://unstats.un.org/unsd/methodology/m49/

# 11 Cross-border tourism

## The case of the Greater bay area, China

*Jinah Park, Haiyan Song and Yeung (William) Kong*

## Introduction

Cross-border tourism collaboration aims to create more diversified experiences by integrating different tourism products and forming destination networks across borders. Given the enormous scale of cross-border flows and collaborations, cross-border tourism initiates functional interlinkages in transport, business, education and research, the environment, and tourism. During the process of cross-border collaboration, destinations must strategically optimise tourism resources to achieve strategic objectives for all stakeholders (Kozak & Buhalis, 2019; Kirillova et al., 2020) and organise relevant measures to respond to internal and external events (Park et al., 2022). Although advances in transportation and information technology have streamlined cross-border collaboration at the urban, regional, and national levels, various negative events have generated roadblocks to this collaboration, aside from its challenges such as the complexities of spatial and institutional structures. Intrinsically, given that cross-border integration encompasses multiple governmental systems and organisational variations in the respective regions, the process presents multiple challenges that require formal and informal innovative solutions (Park et al., 2022; Perkmann, 2007). Exogenous shocks, in particular the COVID-19 pandemic, have triggered a serious crisis in cross-border tourism because of the resulting border closures worldwide and restricted mobility and travel. The collaboration of regular dialogues and multilateral interactions for joint marketing plans has largely stopped. The COVID-19 pandemic and border lockdowns have generated disturbing impacts on tourism destinations in border regions, and these areas must learn how to work together again to revive cross-border tourism. This timely debate is important in prompting concerns regarding new methods to revitalise cross-border tourism collaborations and achieve sustainable tourism mobility in and between destinations (Park et al., 2022). By using the real-world scenario of the Guangdong–Hong Kong–Macao Greater Bay Area (GBA), this chapter highlights the process of developing a new cross-border tourism destination from its early initiatives (top-down policy formulation by the central government) and extension (bottom-up feedback and involvement from local actors) to the current immobility situation caused by the pandemic. This chapter also discusses some market predictions for cross-border tourism based on consumers' perspectives.

In the empirical analysis, a mixed-method design was applied, including a qualitative approach for in-depth interviews and secondary data and a quantitative approach for longitudinal surveys. The in-depth interviews were conducted with tourism-related experts from government, academia, and industry positions from different sides of the border (including Guangdong, Hong Kong, and Macao). Secondary sources of information were

DOI: 10.4324/9781003291763-13

also gathered to supplement the interview data and to identify the policy-based networking and development phases of the region from the era of the Pearl River Delta (PRD) and the Pan-PRD to the GBA. The interviews outlined the key insights of a variety of stakeholders involved in the collaborative planning and marketing and identified collective responses and resilience to crises in the cross-border tourism destinations. Regarding the quantitative part of this study, longitudinal surveys were conducted in 2019 and 2021 with Mainland Chinese individuals living in and out of GBA cities. The results from the longitudinal surveys compare the situation before and during the pandemic and examine the heterogeneity and changes between in GBA and out of GBA tourists (local and non-local domestic tourists) over time.

## Driving factors of cross-border mobilities

Since the GBA project was proposed by the central Chinese government in 2018 and its development plan was launched in 2019 (Greater Bay Area, 2019), cross-border collaboration and governance have been put on the agenda to achieve and co-create a competitive tourism destination (BrandHK, 2019; Park & Song, 2021). Given that the project aims to promote the intra-regional mobility of people, goods, services, capital, and information across the internal borders between Guangdong, Hong Kong, and Macao in the region (Xinhua, 2019), this state-led top-down regionalisation increasingly supports cross-border tourism collaborations at the local level. Accordingly, member cities in the region devised downstream measures, which are presented in Table 11.1. The collaboration between formal actors at the local level encompasses various tourism-related development strategies, from city clusters, service trades, and the development of multi-site tourist products to tourism cooperation and joint tourism. More recently, the GBA-related new initiatives among local actors in the region, such as xx organisations (DMOs), have also involved tripartite cross-border collaboration among Guangdong, Hong Kong, and Macao, and culture is regarded as an ideal vehicle for connectivity within the GBA. The driving factors of cross-border mobilities can be indexed into five dimensions.

The first is improvements in infrastructural interconnectivity across borders. Tourist mobility is a result of the interplay of various drivers (Williams, 2013), and among them, travel time and cost of transportation are the main factors driving travel choices (Masiero et al., 2022). In the GBA case as well, cross-border mobilities have been facilitated by significant improvements in infrastructural interconnectivity among cities across the borders (Park et al., 2022); examples of these include the Hong Kong–Zhuhai–Macao bridge, the Guangzhou–Shenzhen–Hong Kong high-speed railway, and Greater Bay Airlines. This growing interconnectivity is instigated by the long-term strategic development plan, in force since the 1990s, and which has had a major influence on cross-boundary travel time and accessibility, as well as the economic growth of cities in the Pan-PRD (Hou & Li, 2011). Along with physical connectivity, immigration convenience is also an issue in the case of tourist mobility in cross-border regions (Park et al., 2022). Prominent examples include the streamlined exit–entry procedures for cross-border tourists through the immigration and customs clearance services for vehicular traffic and passengers on the Hong Kong–Zhuhai–Macao bridge and the provision of e-channel services for regional residents and visitors. Advances in information technology and smart GBA initiatives have also facilitated cross-boundary connections among cities and systems by co-building a digital ecosystem and e-governance (Park et al., 2022).

*Table 11.1* The landmarks of cross-border tourism collaboration in the Greater Bay Area

| Year | Landmark initiatives or constructions |
| --- | --- |
| 1994 | The cities in Guangdong were first integrated under the concept of the PRD, which the Guangdong provincial government officially introduced to promote economic renaissance and urban agglomeration. |
| 2003 | Two special administrative regions (Semi-Autonomous Regions or SARs, including Hong Kong and Macao) were included later under the Pan-PRD Regional Cooperation. |
| 2008 | The Outline of the Plan for the Reform and Development of the Pearl River Delta (2008–2020) was announced by the State Development and Reform Commission. It called for closer collaboration between Guangdong, Hong Kong, and Macao in infrastructure and city planning and supported collaborative planning for priority projects along the PRD. |
| 2011 | The Tourism Marketing Organization of Guangdong, Hong Kong, and Macao, which was jointly launched by the Tourism Administration of Guangdong Province, HKTB, and MGTO, led regional cooperation and joint promotional plans. |
| June 2015 | The China National Tourism Administration (CNTA) and the Macao SAR government signed an agreement to establish a joint working committee to support Macao in building a world tourism and leisure centre. |
| March 2016 | The idea of a city cluster in Southern China is mentioned in the country's 13th Five-Year Plan (2016–2020), and the concept of the GBA was introduced in the Outline of the 13th Five-Year Plan for National Economic and Social Development of the People's Republic of China. |
| 2017 | Based on CEPA (Mainland and Hong Kong Closer Economic Partnership Arrangement and Mainland and Macao Closer Economic Partnership Arrangement), which had been signed in 2003, the mainland and Hong Kong signed the Agreement on Trade in Goods under CEPA to consolidate and update the commitments on liberalisation and facilitation of the trade of goods and services between the mainland (Guangdong) and Hong Kong. |
| July 2017 | The Framework Agreement on Deepening Guangdong–Hong Kong–Macao Cooperation in the Development of the Bay Area was signed. |
| December 2017 | The Tourism Federation of Cities in Guangdong, Hong Kong, and Macao Bay Area was established under the Agreement on Deepening Guangdong–Hong Kong–Macao Cooperation in the Development of the GBA. The Guangdong–Hong Kong–Macao Bay Area Travel Trade Cooperation Summit was held in Hong Kong with the aim of displaying development opportunities and promoting the development of multi-site tourist products. |
| January 2018 | The 2018 National Tourism Work Conference was held, announcing guidelines for tourism development in the GBA, in which the concept of tourism cooperation and joint tourism brands was emphasised. |
| February 2018 | The 78th Work Meeting of the Tourism Marketing Organization of Guangdong, Hong Kong, and Macao was held to discuss the joint promotional projects domestically and overseas. |
| April 2018 | The first General Membership Meeting of the Tourism Federation of Cities in Guangdong, Hong Kong, and Macao Bay Area was held in Guangzhou to discuss and propose tactics to develop tourism in the GBA, which include developing a united image and tagline and implementing regular promotion and advertising. |
| February 2019 | The Outline Development Plan for the Guangdong–Hong Kong–the Central Committee of the Communist Party of China and the State Council published Macao Greater Bay Area. The central government introduced eight policy measures (March) and 16 additional policy measures (November). |
| April 2019 | The Construction Plan for Hengqin International Recreation Island was published. |

(*Continued*)

*Table 11.1* (Continued)

| Year | Landmark initiatives or constructions |
| --- | --- |
| September 2019 | The Implementation Measures (Trial) for Hong Kong and Macao Tour Guides and Tour Escorts practising in Hengqin New Area, Zhuhai was issued by the Human Resources and Social Security Department and the Culture and Tourism Department of Guangdong Province in September 2019. This measure allows Hong Kong and Macao residents to apply for and obtain a tour guide qualification certification in the Hengqin New Area. In 2020, more than 400 Hong Kong and Macao tour guides passed the examination and obtained special tour guide certificates. |
| 2020 | Based on CEPA and other policies, each of the five wholly owned travel agencies in Hong Kong and Macao have been approved as the pilot operation of outbound tourism business (excluding Taiwan), offering services to mainland residents. |
| May 2020 | The Department of Culture and Tourism of Guangdong Province announced the following five trail themes, with 27 physical trails and a GBA trail identification system: Sun Yat Sen Cultural Heritage Trail, the maritime Silk Road Cultural Heritage Trail, the overseas Chinese Cultural Heritage Trail, the ancient Post Road Cultural Heritage Trail, and the Haiphong historical sites Cultural Heritage Trail. |
| September 2020 | The agreement on the GBA 9+2 Urban Tourism Market Joint Supervisory Authority was signed to establish a governmental body of tourism market supervision and cooperation. Under the agreement, local DMOs in the GBA could jointly build working mechanisms such as information notification, joint inspection, joint responses, and emergency and crisis management. |
| October 2020 | The 2020 Work Plan of the Framework Agreement on Hong Kong and Guangdong Cooperation. The Work Plan includes tourism collaboration measures for "fostering cooperation in modern service industries (cooperation in the financial services sector and professional services, as well as cultural and tourism cooperation)." |
| November 2020 | With the "Beautiful China–Experience with Heart & Eyes" online promotion hosted by the Ministry of Culture and Tourism of Guangdong Province, "Journey to the GBA" was jointly promoted to the regional and global markets. |
| December 2020 | The Ministry of Culture and Tourism of the People's Republic of China published the Cultural and Tourism Development Plan of the GBA to jointly build the GBA as a cultural and leisure bay area and a quality circle for living, working, and travelling. |
| January 2021 | According to the Chief Executive's 2020 Policy Address of Hong Kong, the Greater Bay Area Youth Employment Scheme was launched in 2021 to encourage enterprises to employ local university graduates in Hong Kong and deploy them to stations and work in mainland GBA cities. In 2022, a total of 1,091 graduates in Hong Kong have been employed by cross-border enterprises in the GBA under the scheme. |

*Source*: Hong Kong and Macao Affairs Office, the People's Government of Guangdong Province (2017); GovHK (2022); Greater Bay Area (2019); KPMG (2020); Ministry of Culture and Tourism of the People's Republic of China (2020); and data provided from interviewees.

The second factor is cross-border collaborations between local DMOs to create a new development path. As Kozak and Buhalis (2019) also indicated, organisational collaboration among different tourism boards and governments is a critical success factor in cross-border tourism marketing which aims to co-create value and be differentiated towards the local marketing of each member city. In the GBA, the Macao Government Tourism Office (MGTO) is currently collaborating with member cities in Guangdong to diversify its tourism products for attracting new visitors beyond the existing casino markets. Our interviews with government officials in Macao indicated that the city is taking full advantage of the GBA project and co-working with several member cities in Guangdong. For example, Macao is co-promoting culinary tourism resources with Foshan (a city in Guangdong province) and heritage tourism with Jiangmen (another city in Guangdong province). From 2018 to 2019, the MGTO promoted multi-destination travel in GBA products through the cooperation of various market representatives and local travel agencies. According to the data provided by an interviewee, the total sales volume of this promotion was more than 210,000 people. The interviewees from Hong Kong stated that the Hong Kong Tourism Board (HKTB) is also taking advantage of the multi-destination concept to enhance Hong Kong's attractiveness, maintain current markets, and attract new markets. An example of this initiative is bleisure destination marketing (Lichy & McLeay, 2018), which combines business in Hong Kong with leisure in the GBA. Policy support at a supra-regional level is key to the success of these collaborations, such as the agreement on the 144-hour visa-exemption transit for overseas group tours in the GBA. This agreement was established based on the support of existing policies and cooperative actions since it was signed by Guangdong and Hong Kong under the Agreement Concerning Amendment to the Mainland and Hong Kong Closer Economic Partnership Arrangement (CEPA) Agreement on Trade in Services.

The third factor is the emergence of new actors to coordinate and monitor the collective governance of cross-border mobilities. By echoing existing findings that indicated political relations as a vital component of cross-border tourism development in the cases of the Greater Mekong sub-region (Sofield, 2006) or the China–Myanmar border (Su & Li, 2021), cross-border openness and flows are greatly affected by collaborative interactions between institutional actors in the GBA. To deepen their collaborations, local actors have also jointly launched a new cross-border agency, called the GBA 9+2 Urban Tourism Market Joint Supervisory Authority, as a strategic actor. Hence, they have established collaborative mechanisms for co-working, problem-solving, monitoring, and supervising the GBA tourism market. The role of these new strategic actors has become even more important because of the COVID-19 pandemic – clearly, current cross-border collaborations without a powerful supra-regional agency have shrunk or paused, including joint tourism marketing and branding, cross-border employment for youth and entrepreneurial support, and regular inter-city dialogues.

The fourth key factor is regional education and enlightenment to prompt integration. An interviewee from academia proposed to:

> think of the Erasmus scheme, which was a way of education, getting students to move around a lot, all kinds of European countries in the foundation of the European economy. It was the EEC [European Economic Community] which became the EU. They, over many years, took different steps to prompt integration in hard measures and soft measures … [to] build the cross-border identity. Now, so, industry representatives, delegates [in the GBA or other cross-border destinations], they can have conversations that are quite important fairly early on, but there's also the more organic side, which is the younger generation and their interests and engagement. … You can have school groups, primary groups, secondary [groups], plus industry, plus governments. You probably need all of these things, but then some prioritization: what is it that comes first? Is it driven by economics? … Whereas in Europe, they put quite a lot of emphasis on building cultural understanding. And in Europe, you've got so many different languages. [Some programmes should be developed to reduce/eliminate these differences, and] to build some shares and get to know one another. There wasn't really a particularly strong message about "go and join" or "go and sign up". It was more like just getting to understand who the other people are.

As such, following the construction of connective infrastructure (e.g., the Hong Kong–Zhuhai–Macao Bridge, high-speed railways, and customs facilities on the other side of the borders), the GBA has also focused on nurturing regional youth talent which carries a global mindset and a sense of GBA connectedness like the kind the EU has long paid attention to (European Union, 2016). For instance, fostering mobility in the GBA for younger generations was one of the main policy areas to enhance the understanding of opportunities in the regions; a HKD300 million grant was arranged for the Youth Development Fund to enhance young peoples' cross-border mobility within the region (Greater Bay Area, 2019). However, from more than 20,000 applications, the Greater Bay Area Youth Employment Scheme matched only 1,091 fresh graduates to 3,000 job vacancies under the Scheme (GovHK, 2022). Since the breakdown of these cross-border mobility plans during the COVID-19 pandemic, collective responses and resilience to the crisis have become more evident recently, as these are not only engaged in rekindling cross-border mobilities but also encouraging the younger generation to take an active role in cross-border regional integration.

The fifth key factor is common-pool resources that can be shared, jointly investigated, and co-utilised by local actors across borders (Park et al., 2022). In the GBA case, the Hengqin New District (located in Zhuhai, the cross-border city of Hong Kong and Zhuhai) serves as a showcase for the experiments of cross-border tourism collaborative initiatives. This echoes the previous findings on synergetic tourism collaboration and governance that focused on various stakeholders' exchanges and interactions to accomplish collective goals more effectively and efficiently (Borges et al., 2014). Since the Construction Plan for Hengqin International Recreation Island was published in 2019, Hengqin has been developed into a venue for exchanges and interactions, and a series of Hengqin experiments have been launched, such as trial measures for an integrated certification system for tour guides. Similar models within the GBA can also be found in the Qianhai Shenzhen–Hong Kong Modern Service Industry Cooperation Zone (Shenzhen Government Online, 2021), the "one zone, two parks" concept in the Shenzhen–Hong Kong I&T Cooperation Zone of the Loop (cross-border areas around Lok Ma Chau and San Tin, Hong Kong), and large-scale regional development plans, such as the Northern Metropolis and the Lantau Tomorrow Vision (Chief Executive's 2021 Policy Address, 2021).

**Challenges in collaborations: A longitudinal approach to GBA image change**

Challenges in cross-border tourism collaboration were noted in differences that pertained to the existence of borders, such as differences in bureaucratic prominence, administrative arrangements, business environments, social development, and resident attitudes and their engagement (Kozak & Buhalis, 2019; Park et al., 2022). Furthermore, immobility during the COVID-19 pandemic and related changes in people's perceptions of cross-border regions would be the major challenges for local DMOs and tourism businesses when they can work together to rekindle cross-border tourism and promote new joint products (Antwi et al., 2022; Ju et al., 2022; Park et al., 2022).

To gauge people's responses to the GBA initiatives and the current challenges border regions face from the pandemic, we examined data from longitudinal surveys conducted in 2019 and 2021, before and during the pandemic. A total of 310 and 1,010 responses were collected in 2019 and 2021, respectively. For the data from 2019 ($n = 310$), 155 responses were from people living in one of the GBA member cities, and 155 responses were from people living in non-GBA cities in Mainland China. For the data from 2021 ($n = 1,010$),

514 responses were from people living in the GBA, and 496 responses were from Chinese people living in non-GBA cities. To identify the image of a destination region on a megalopolitan scale, we developed a comprehensive measurement scale of 24 items based on existing literature on destination image, regional and cross-border destination marketing, and the governing documents and consultancy reports of the GBA project. As shown in Table 11.2, the 24 items include regional tourism planning, destination attractiveness for leisure and business travel, and destination development environment (Park & Song, 2021). The items are measured on a five-point scale (from 1 = strongly disagree to 5 = strongly agree).

To assess changes in GBA destination image before and during the pandemic, a paired sample t-test was performed to examine these changes among in-GBA and out-of-GBA residents in 2019 and 2021 (before and during the pandemic). The results regarding the destination image changes from 2019 to 2021 are shown in Figure 11.1. The results of the paired sample t-test for before and during the pandemic are shown in Table 11.2. Significant differences existed in several image items among in-GBA and out-of-GBA residents between 2019 and 2021. Generally, GBA destination image tended to improve for in-GBA and out-of-GBA residents, despite their intra-regional mobilities being severely affected by the pandemic. Some items of destination image directly related to the border lockdown measures were significantly worse, including transportation accessibility, efficient travel route planning, wide choice of multi-destination travel products, and economic development. We also found that the perceived image of smart tourism destination was negatively affected by the pandemic. This observation is well supported by the definition of smart mobility in megalopolitan regions that emphasises smart, connected, and safe mobility (CIVAS, 2020). As Mantero (2022) indicated, after experiencing the pandemic, sustainable, smart, and safe mobility has become even more important for destinations, local communities, and tourists. The positive aspect of our results is that cross-border tourism destinations may not be hard hit by the pandemic, although methods to restart tourism and mobility should be discussed carefully and circumspectly. Local tourism that targets cross-border travellers within the region would be the most feasible marketing opportunity during and after the pandemic, as the results showed that local people (in-GBA residents) tended to have a more positive image in most aspects. The difference between their image perceptions became more evident for the following items: nightlife, hospitable host community in two SARs, good for a family trip, quality of tourism service, good partnership among member cities, distinctive role sharing of member cities, and difference in political and economic systems. In-GBA residents' perceptions tended to be more favourable toward those items than those of out-of-GBA residents.

### A rosy prediction for cross-border tourism destinations in the post-pandemic era

The GBA case addresses how a cross-border destination has been affected by the pandemic, but it also remains resilient. For governments and tourism organisations, the COVID-19 pandemic has highlighted constraints that impede long-term collaboration and the sustainable development of cross-border tourism destinations. In border regions, several joint actions could help to create sustainable synergistic mechanisms. Specifically, an active coalition of local DMOs could be a prerequisite for rekindling cross-border tourism and mobility more effectively. The DMOs can promote more cross-border partnership projects for the public and private sectors and also establish a collective governance system by introducing a new strategic agency consisting of local DMOs and different stakeholders from member cities across borders. Another point highlighted by our results is that

*Table 11.2* Paired sample t-test before and during the pandemic, 2019 versus 2021

| Items | Year | In-GBA | | | Out-of-GBA | | |
|---|---|---|---|---|---|---|---|
| | | Mean | f | Sig. | Mean | f | Sig. |
| Unique lifestyle | 2019 | 4.38 | 1.029 | | 3.94 | 1.444 | ** |
| | 2021 | 4.37 | | | 4.16 | | |
| Appealing local cuisine and beverages | 2019 | 4.17 | 0.163 | ** | 3.75 | 3.548 | *** |
| | 2021 | 4.38 | | | 4.18 | | |
| Cleanliness | 2019 | 3.90 | 9.727 | *** | 3.75 | 3.996 | *** |
| | 2021 | 4.21 | | | 4.10 | | |
| Safety | 2019 | 4.00 | 17.810 | *** | 3.57 | 0.121 | *** |
| | 2021 | 4.35 | | | 4.07 | | |
| Nightlife | 2019 | 4.14 | 6.225 | | 4.04 | 7.235 | |
| | 2021 | 4.23 | | | 3.95 | | |
| Hospitable host community (nine mainland cities) | 2019 | 3.80 | 0.059 | *** | 3.54 | 0.622 | *** |
| | 2021 | 4.23 | | | 4.04 | | |
| Hospitable host community (two SARs) | 2019 | 3.43 | 15.999 | *** | 3.30 | 0.329 | *** |
| | 2021 | 4.33 | | | 4.07 | | |
| Good for a family trip | 2019 | 3.80 | 0.032 | *** | 3.77 | 0.738 | *** |
| | 2021 | 4.32 | | | 4.12 | | |
| Tourism infrastructure and facilities | 2019 | 4.07 | 14.978 | *** | 3.93 | 9.294 | *** |
| | 2021 | 4.48 | | | 4.40 | | |
| Quality of tourism service | 2019 | 3.78 | 1.171 | *** | 3.74 | 0.127 | ** |
| | 2021 | 4.29 | | | 3.94 | | |
| Variety of cultural attractions and entertainment | 2019 | 4.13 | 16.302 | *** | 3.92 | 6.911 | *** |
| | 2021 | 4.39 | | | 4.26 | | |
| Environment for business, exhibitions, and conferences | 2019 | 4.14 | 10.662 | *** | 3.97 | 4.595 | |
| | 2021 | 4.27 | | | 4.07 | | |
| Outdoor recreation activities | 2019 | 4.01 | 20.997 | *** | 3.70 | 0.561 | *** |
| | 2021 | 4.40 | | | 4.33 | | |
| Economic development | 2019 | 4.43 | 10.243 | *** | 4.12 | 19.644 | *** |
| | 2021 | 4.19 | | | 3.85 | | |
| High level of internationalisation | 2019 | 4.31 | 5.405 | | 4.15 | 2.238 | |
| | 2021 | 4.29 | | | 4.02 | | |
| Transportation accessibility | 2019 | 4.39 | 3.205 | ** | 4.19 | 2.097 | ** |
| | 2021 | 4.22 | | | 3.96 | | |
| Good partnership among member cities | 2019 | 4.09 | 9.810 | * | 3.99 | 4.192 | |
| | 2021 | 4.23 | | | 4.02 | | |
| Distinctive role sharing of member cities | 2019 | 3.94 | 4.617 | *** | 3.82 | 0.129 | ** |
| | 2021 | 4.32 | | | 4.08 | | |
| Difference in political and economic systems | 2019 | 3.84 | 0.306 | *** | 3.81 | 0.146 | *** |
| | 2021 | 4.39 | | | 4.25 | | |
| Smart tourism destination | 2019 | 4.17 | 2.220 | *** | 3.94 | 1.211 | * |
| | 2021 | 4.33 | | | 4.08 | | |
| Efficient travel route planning | 2019 | 3.86 | 2.951 | *** | 3.72 | 1.189 | *** |
| | 2021 | 4.32 | | | 4.14 | | |
| Wide choice of multi-destination travel products | 2019 | 3.89 | 2.624 | *** | 3.77 | 0.484 | *** |
| | 2021 | 4.27 | | | 4.13 | | |
| Leading role of mature destinations in regional reputation | 2019 | 4.05 | 10.067 | *** | 3.94 | 3.325 | *** |
| | 2021 | 4.35 | | | 4.31 | | |
| Government and policy support | 2019 | 4.25 | 3.698 | | 4.09 | 1.700 | |
| | 2021 | 4.19 | | | 4.00 | | |

* $p < 0.1$; ** $p < 0.05$; *** $p < 0.001$.

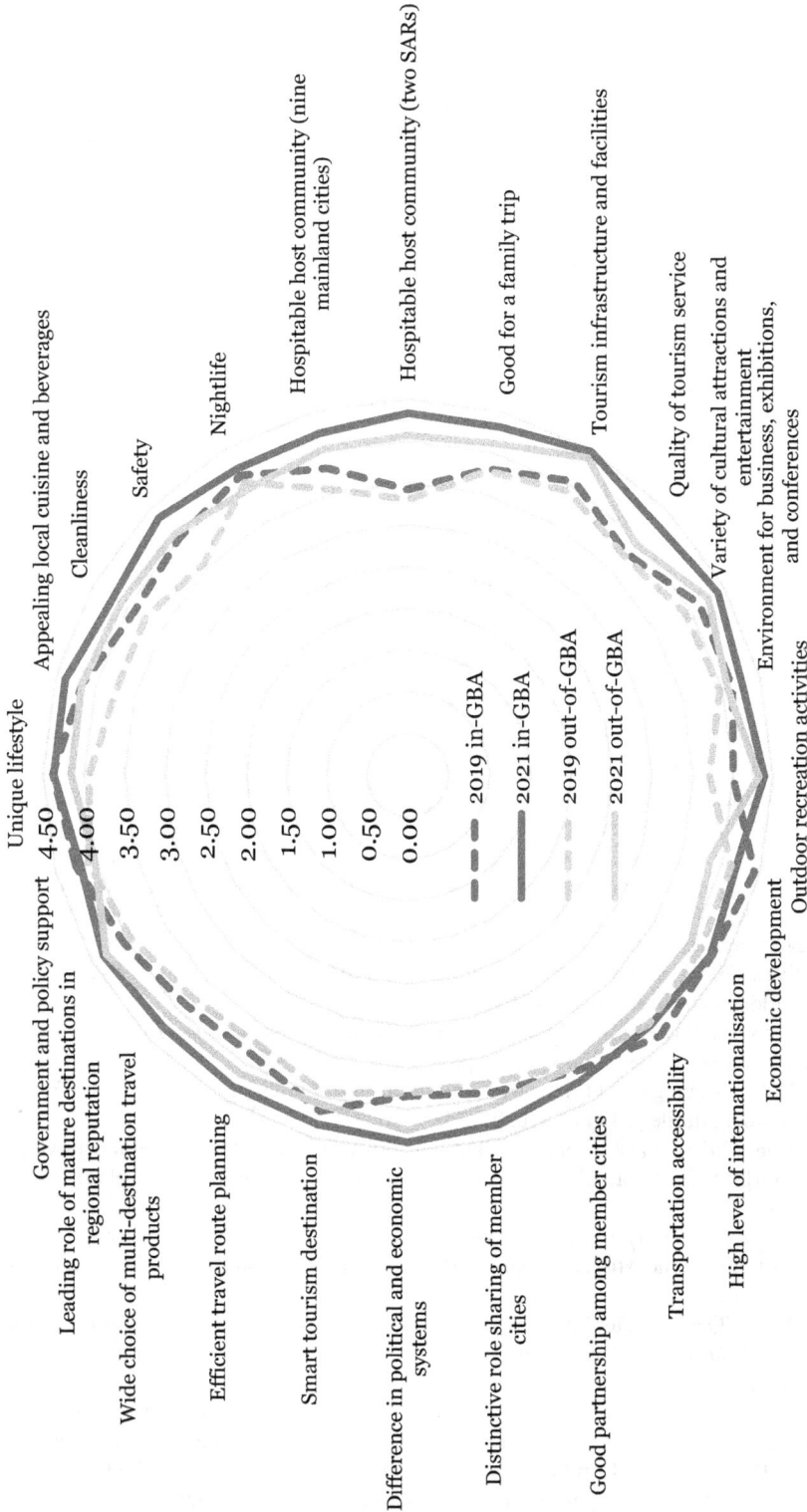

*Figure 11.1* Destination image changes of in-GBA and out-of-GBA residents in 2019 versus 2021.

governments in cross-border regions should build dialogue in many sectors beyond tourism. Through conversations among different sectors and cities, structures and actions can be established to prioritise groups in relevant cities, such as younger generations in the GBA, especially in Hong Kong. Increasing successful cross-border collaborations with deliverable goals would mutually benefit destinations in the border region. As part of this endeavour, cross-border destinations can share and co-utilise common resources to showcase their drastic measures and fruitful outcomes.

In the comparison of GBA destination images before and during the pandemic, we have learned that cross-border tourism destinations are calling to the local market. This is not only because people need a supplement for long-haul travels but also because a sense of local belonging has surfaced, which could rescue the tourism and hospitality industry in the near future (Sharma et al., 2021). Regarding the longitudinal survey results, local residents, who are the potential tourism market for the region, tended to have more positive images and were less affected by the negative pandemic situation. Before inbound travel can resume, local markets across borders will boost the resumption of the tourism industry and cross-border collaborations in the wake of the pandemic. Although pandemic-related travel restrictions and border shutdowns have created severe barriers for tourism destinations, especially those in border regions, the pandemic may open the door for local tourism across borders and enhance peoples' willingness to experience local nature and culture and enjoy high-quality tourism services in the region.

## References

Antwi, C. O., Ntim, S. Y., Boadi, E. A., Asante, E. A., Brobbey, P., & Ren, J. (2022). Sustainable cross-border tourism management: COVID-19 avoidance motive on resident hospitality. *Journal of Sustainable Tourism*. https://doi.org/10.1080/09669582.2022.2069787

Borges, M., Eusébio, C., & Carvalho, N. (2014). Governance for sustainable tourism: A review and directions for future research. *European Journal of Tourism Research*, 7(1), 45–56.

BrandHK (2019). Strategic focus: Greater Bay Area Information Services Department, The Government of the HKSAR (2019). Retrieved from https://www.brandhk.gov.hk/html/en/%20Strategic Focus/GreaterBayArea.html

Chief Executive's. (2020). Policy Address (2020). *Striving Ahead with Renewed Perseverance*. https://www.policyaddress.gov.hk/2020/eng/pdf/PA2020.pdf

Chief Executive's. (2021). Policy Address (2021). *Building a Bright Future Together*. https://www.policyaddress.gov.hk/2021/eng/pdf/PA2021.pdf

CIVAS. (2020). Introducing CIVITAS sustainable and smart mobility for all. Retrieved from https://civitas.eu/sites/default/files/CIVITAS2030%20booklet.pdf

European Union. (2016). What about you: Do you feel you belong to Europe and in what ways? Retrieved from https://europa.eu/youth/nnfe/what-about-you-do-you-feel-you-belong-europe-and-what-ways_en

Greater Bay Area. (2019). The Guangdong-Hong Kong-Macao Greater Bay Area Development Office, Constitutional and Mainland Affairs Bureau. Retrieved from https://www.bayarea.gov.hk/en/outline/plan.html

GovHK. (2022). LCQ20: Greater bay area youth employment scheme. *The Government of the Hong Kong Special Administrative Region*. Retrieved from https://www.info.gov.hk/gia/general/202202/16/P2022021600191.htm

Hong Kong and Macao Affairs Office, the People's Government of Guangdong Province. (2017). Agreement on the liberalization of service trade between the mainland and Hong Kong in Guangdong. Retrieved from http://hmo.gd.gov.cn/CEPAzcfg/content/post_42750.html

Hou, Q., & Li, S. M. (2011). Transport infrastructure development and changing spatial accessibility in the Greater Pearl River Delta, China, 1990–2020. *Journal of Transport Geography*, 19(6), 1350–1360.

Ju, B., Dai, H. M., & Sandel, T. L. (2022). Struggling in im/mobility: Lived experience of Macao's mainland Chinese migrant laborers via WeChat Moments during COVID-19. *Communication, Culture and Critique*. https://doi.org/10.1093/ccc/tcac022

Kirillova, K., Park, J., Zhu, M., Dioko, L. D., & Zeng, G. (2020). Developing the coopetitive destination brand for the Greater Bay Area. *Journal of Destination Marketing & Management, 17*. https://doi.org/10.1016/j.jdmm.2020.100439

Kozak, M., & Buhalis, D. (2019). Cross–border tourism destination marketing: Prerequisites and critical success factors. *Journal of Destination Marketing & Management, 14*. https://doi.org/10.1016/j.jdmm.2019.100392

KPMG (2020). Keys to success in the Greater Bay Area: Third annual survey on drivers for growth. *KPMG China, HSBC and the Hong Kong General Chamber of Commerce (HKGCC)*. Retrieved from https://assets.kpmg/content/dam/kpmg/cn/pdf/en/2020/01/keys-to-success-in-the-greater-bay-area.pdf

Lichy, J., & McLeay, F. (2018). Bleisure: motivations and typologies. *Journal of Travel & Tourism Marketing, 35*(4), 517–530.

Mantero, C. (2022). Sustainable, smart and safe mobility at the core of sustainable tourism in six European islands. In T. Tsoutsos (ed.), *Sustainable Mobility for Island Destinations* (pp. 1–18). Cham: Springer.

Masiero, L., Hrankai, R., & Zoltan, J. (2022). The role of intermodal transport on urban tourist mobility in peripheral areas of Hong Kong. *Research in Transportation Business & Management*. https://doi.org/10.1016/j.rtbm.2022.100838

Ministry of Culture and Tourism of the People's Republic of China. (2020). 2020 Hong Kong and Macao "Beautiful China Experience with Heart & Eyes" online promotion activity ended. Retrieved from https://www.mct.gov.cn/gtb/index.jsp?url=https%3A%2F%2Fwww.mct.gov.cn%2Fwhzx%2Fwhyw%2F202011%2Ft20201124_902880.htm

Park, J., & Song, H. (2021). Variance of destination region image according to multi-dimensional proximity: A case of the Greater Bay Area. *Journal of Destination Marketing & Management, 20*. doi:https://doi.org/10.1016/j.jdmm.2021.100600

Park, J., Tse, S., Mi, S. D., & Song, H. (2022). A model for cross-border tourism governance in the Greater Bay Area. *Journal of China Tourism Research*, 1–25. doi:https://doi.org/10.1080/19388160.2022.2036664

Perkmann, M. (2007). Policy entrepreneurship and multilevel governance: A comparative study of European cross-border regions. *Environment and Planning C: Government and Policy, 25*(6), 861–879.

Sharma, G. D., Thomas, A., & Paul, J. (2021). Reviving tourism industry post-COVID-19: A resilience-based framework. *Tourism Management Perspectives, 37*. https://doi.org/10.1016/j.tmp.2020.100786

Shenzhen Government Online. (2021). Qianhai Shenzhen-Hong Kong Modern Service Industry Cooperation Zone. Retrieved from http://www.sz.gov.cn/en_szgov/news/infocus/Qianhai/content/post_8707179.html

Sofield, T. H. B. (2006). Border tourism and border communities: An overview. *Tourism Geographies, 8*(2), 102–121.

Su, X., & Li, C. (2021). Bordering dynamics and the geopolitics of cross-border tourism between China and Myanmar. *Political Geography, 86*. https://doi.org/10.1016/j.polgeo.2021.102372

Williams, A. M. (2013). Mobilities and sustainable tourism: Path-creating or path-dependent relationships? *Journal of Sustainable Tourism, 21*(4), 511–531.

Xinhua. (2019). *Development plan for Guangdong–Hong Kong–Macao Greater Bay Area*. Beijing: The State Council of the People's Republic of China. Retrieved from http://www.gov.cn/zhengce/2019-02/18/content_5366593.htm#1; http://english.www.gov.cn/policies/latest_releases/2019/02/18/content_281476527605892.htm

# 12 Trends and issues with regional tourism partnership formation

*Jens Thraenhart and Alastair M. Morrison*

## Introduction

A tourism region is

> a geographically-delineated area where there is an agreement to develop and/or market tourism, or to engage in other forms of tourism partnerships. The area may be within an individual country or span multiple countries (cross-border tourism). The regional tourism agreement can be through legislation or inter-governmental treaties or may be a more informal cooperation approach.
>
> (Morrison, 2022, pp. 4–5)

The research on partnering stretches back to the 1980s, although practical partnering applications pre-date academic publishing.

Why are regional development and regional cooperation required in tourism? First, when considering tourism development within one individual local tourism region the reason is usually to grow the economy and improve the social conditions of local people. This is particularly important for rural and other economically disadvantaged areas. When the development covers multiple regions or countries, there is a need to coordinate and integrate tourism policy, planning, development, and marketing. Second, there are several reasons for and benefits from regional cooperation in tourism development, marketing, and other partnerships. The synergy that results provides the participants with more than they would have by going it alone. The following are eight specific reasons for regional tourism partnerships:

- Adding to budgets;
- Enhancing images and branding;
- Giving better customer service;
- Having a shared presence;
- Increasing market appeal;
- Providing new target market opportunities;
- Sharing research and other information;
- Tackling social responsibilities.

In addition to these benefits, there are other motivations for seeking to cooperate across borders. For example, Kozak and Buhalis (2019) suggest that "cross-border collaboration has become more valuable particularly for destinations that have had long-standing disputes in political relations but now seek peace. They are often forced to collaborate and

DOI: 10.4324/9781003291763-14

develop economic and political bridges, developing friendships on the way". These two authors analysed potential partnering between Greece and Turkey. Other authors have examined regions where there were prior political and civil differences (e.g., Kennell et al., 2019; Lagiewski, 2004).

Despite the many reasons for regional cooperation, there are issues and challenges in forming such alliances and there have been failed attempts to partner and situations where the results were disappointing. It is worthwhile, therefore, to determine the common barriers and challenges to regional cooperation.

There are substantial contributions to the literature on partnering and collaboration in business in general, although much less has been written about partnering in tourism. The early contributions by Kanter (1994) and Dent (1999) are highlighted later in this chapter. Sullivan and Warner (2017) and Nahm (2021) are more contemporary contributions on strategic alliances and collaborative advantage, respectively.

Having introduced the topic and outlined the influential contributions so far, the principal objectives of this chapter are to: (1) describe the recent growth trends in regional tourism partnerships globally; (2) identify barriers to regional approaches to tourism and elaborate on issues and challenges associated with regionalisation in tourism; (3) examine the roles of various stakeholders in regional tourism collaborations; (4) consider potential contributions of regional tourism partnerships to the achievement of the UN Sustainable Development Goals (SDGs); and (5) elaborate future trends expected in regional tourism partnerships.

## Recent trends and growth in regional tourism partnerships

There are five major recent trends associated with regional tourism partnerships which correspond with the acronym GUESS – growth (G), universal (U), e-marketing (E), sustainability (S), and scope (S).

### Growth

The number of regional tourism partnerships has increased, although that growth has been gradual rather than rapid. Table 12.1 provides a sample of 16 regional tourism partnerships from across the world.

### Universality

Regional tourism partnerships are now present on all continents of the world, so this trend is universal. Table 12.1 provides examples from Africa, Asia, Europe, the Caribbean, North and South America, and Central America; however, there are also partnerships in the Middle East (e.g., through the Gulf Cooperation Council). These types of partnerships are pervasive, and all regions and nations recognise their potential benefits.

### E-marketing

The usage of ICT and e-marketing is another noteworthy trend in regional tourism partnerships. Online collaboration is appropriate given customer preferences and also because it removes barriers of physical geography that can constrain the activities of regional tourism partnerships.

*Table 12.1* Selection of regional tourism partnerships

| Regional tourism Partnerships | Countries and description |
| --- | --- |
| **Asia** | |
| GMS – Greater Mekong Subregion www.GreaterMekong.org | Cambodia, Lao PDR, Myanmar, PR China (Yunnan, Guangxi), Thailand, Viet Nam |
| MTCO – Mekong Tourism Coordinating Office www.MekongTourism.org www.DestinationMekong.com | Set up in 2006 by the GMS member countries with assistance from the Asian Development Bank (ADB). |
| Silk Road – managed by the UNWTO World Tourism Organization https://silkroad.unwto.org | Albania, Armenia, Azerbaijan, Bangladesh, Bulgaria, PR China, Croatia, DPR Korea, Rep. Korea, Egypt, Georgia, Greece, Indonesia, Iran, Iraq, Israel, Italy, Japan, Kazakhstan, Kyrgyzstan, Malaysia, Mongolia, Montenegro, Pakistan, Romania, Russia, San Marino, Saudi Arabia, Spain, Syria, Tajikistan, Turkey, Turkmenistan, Ukraine, and Uzbekistan. The UNWTO Silk Road Programme is a collaborative initiative designed to enhance sustainable tourism development along the historic Silk Road route. |
| **Australia and Pacific** | |
| SPTO – South Pacific Tourism Organisation www.southpacificislands.travel | American Samoa, Cook Islands, Federated States of Micronesia, Fiji, French Polynesia, Kiribati, Nauru, Marshall Islands, New Caledonia, Niue, Papua New Guinea, Samoa, Solomon Islands, Timor Leste, Tokelau, Tonga, Tuvalu, Vanuatu, Wallis & Futuna, Rapa Nui, and the People's Republic of China. Established in 1983 as the Tourism Council of the South Pacific. In addition to its 21 government members, the SPTO has about 200 private sector members. |
| **North, Central and South America, and Caribbean** | |
| Travel South USA https://industry.travelsouthusa.com/ | Alabama, Arkansas, Georgia, Kentucky, Louisiana, Mississippi, Missouri, North Carolina, South Carolina, Tennessee, Virginia, and West Virginia. Travel South USA is the official regional destination marketing organisation for the southern United States. The non-profit organisation promotes travel to and within its member states. |
| ACAT – Atlantic Canada Agreement on Tourism http://acat-etra.ca/ | New Brunswick, Newfoundland and Labrador, Nova Scotia, Prince Edward Island. ACAT drives growth in the sector by promoting travel to Atlantic Canada through research-driven marketing campaigns and activities in key international markets such as the United States, the United Kingdom, Germany, as well as to select markets within Canada. |
| COTAL – Confederación de Organizaciones Turística de América Latina https://www.cotalamerica.org/ | Argentina, Aruba, Belize, Brazil, Bolivia, Chile, Colombia, Costa Rica, Cuba, Dominican Republic, Ecuador, Guatemala, Honduras, Nicaragua, Panama, Paraguay, Peru, Puerto Rico, , Venezuela, and Mexico. COTAL was formed in 1957 and is a non-profit organisation bringing together all of the national travel agency associations in Latin American countries. |

*(Continued)*

*Table 12.1* (Continued)

| Regional tourism Partnerships | Countries and description |
|---|---|
| CATA – Central America Tourism Association www.visitcentroamerica.com/en https://www.catatourismagency.org | Belize, Costa Rica, El Salvador, Guatemala, Honduras, Nicaragua, Panama, and Dominican Republic. CATA is a non-profit international organisation based in Madrid, Spain and with a sub-headquarters in El Salvador. |
| CTO – Caribbean Tourism Organization www.onecaribbean.org | Antigua and Barbuda, Bahamas, Barbados, Belize, Cuba, Dominica, Guadeloupe, Martinique and St. Martin, Grenada, Guyana, Haiti, Jamaica, Aruba, Bonaire, Curacao, St. Eustatius, St. Maarten, St. Kitts and Nevis, St. Lucia, St. Vincent and the Grenadines, Suriname, Trinidad and Tobago, Anguilla, Bermuda, British Virgin Islands, Cayman Islands, Montserrat, Turks and Caicos Islands, Puerto Rico, and the U.S. Virgin Islands. CTO, established in 1989, is the region's tourism development agency. |
| **Africa** | |
| RETOSA – Regional Tourism Office of Southern Africa http://www.retosa.co.za/ | Angola, Botswana, DR Congo, Lesotho, Madagascar, Malawi, Mauritius, Mozambique, Namibia, South Africa, Swaziland, Tanzania, Zambia, and Zimbabwe. The Regional Tourism Organization of Southern Africa (RETOSA) is a Southern African Development Community (SADC) body responsible for the Promotion and Marketing of Tourism in the region. |
| **Europe** | |
| Alp Net www.alp-net.eu | Germany, Switzerland, Austria, and Italy. Ten of the leading tourist organisations in the Alps have decided to collaborate to further develop and boost sustainable Alpine tourism. |
| BTC – Baltic Sea Tourism Center www.bstc.eu | Germany, Denmark, Poland, Lithuania, Finland, and Sweden. BTC was established to jointly improve competitiveness for sustainable tourism in the Baltic Sea Region. |
| ETC – European Travel Commission https://etc-corporate.org | Austria, Belgium, Bulgaria, Croatia, Cyprus, Czech Republic, Denmark, Estonia, Finland, Germany, Greece, Hungary, Iceland, Ireland, Italy, Latvia, Lithuania, Luxembourg, Malta, Monaco, Montenegro, Netherlands, Norway, Poland, Portugal, Romania, San Marino, Serbia, Slovakia, Slovenia, Spain, and Switzerland. Established in 1948, the European Travel Commission is a unique association in the travel sector, representing the National Tourism Organizations of the countries of Europe. Its mission is to strengthen the sustainable development of Europe as a tourist destination. |
| RCC – Regional Cooperation Council https://www.rcc.int/tourism | Albania, Bosnia and Herzegovina, Croatia, Greece, Macedonia, Montenegro, Serbia, and Kosovo. The EU funded (EUR5 million) and RCC implemented Tourism Development & Promotion project works to create joint and internationally competitive cultural and adventure tourism offers in the six Western Balkans (WB6) economies which will attract more tourists to the region, lengthen their stay, increase revenues, and contribute to growth and employment. |

*(Continued)*

*Table 12.1* (Continued)

| Regional tourism Partnerships | Countries and description |
| --- | --- |
| V4 – Visegrad Group https://www.visegradgroup.eu<br>Discover Central Europe – European Quartet<br>www.discover-ce.eu<br>DCC – Danube Competence Center www.danubecc.org | Czech Republic, Hungary, Poland, Slovakia.<br>The European Quartet, otherwise known as the Visegrad Four (V4), has been working together to ensure long-term success in fields of common interest, through continued and reinforced internal cooperation.<br>Germany, Austria, Slovakia, Hungary, Croatia, Serbia, Romania, Bulgaria, Moldova, and Ukraine.<br>The Danube Competence Center (DCC), based in Belgrade, is a Danube focused association of tourism actors for a sustainable and competitive destination Danube, supported by GIZ. |

### Sustainable development and sustainable tourism

Regional partnerships being viewed as a pathway to more sustainable tourism is another recent trend. The Experience Mekong Collection is a particularly good example of this trend: "The Experience Mekong Collection (EMC) is a curation of small businesses and social enterprises that display travel experiences that are sustainable and responsible in the Greater Mekong Subregion (GMS)" (Thraenhart, 2022, p. v).

Several scholars have analysed the potential contributions of partnerships to sustainable development and sustainable tourism. For example, Graci (2013) studied multi-stakeholder collaboration with certain islands in Indonesia that resulted in the creation of an NGO, the Gili Eco Trust. Dunets et al. (2019) considered the contribution of tourism partnerships to the sustainable development of a transborder mountain region in Russia, Mongolia, China, and Kazakhstan. El-Khadrawy et al. (2022) examined several cases in Botswana, Egypt, Jordan, Mexico, and Saudi Arabia on how partnerships were contributing to the sustainable tourism development of cultural heritage sites.

### Scope

It is accurate to say that the traditional focus on tourism partnerships has been from the perspectives of marketing and branding (Kozak & Buhalis, 2019; Morrison, 2019), and economic growth and development (Hampton, 2009). However, the scope of regional tourism partnerships is continuously expanding. Shared interests are the core of the attraction of forming partnerships. Morrison (2019) outlined several bases (or shared interests) on which regional partnerships can be grounded, and these include: (1) similar resources; (2) contiguous or similar locations; (3) similar markets; and (4) shared challenges and problems (Figure 12.1). Greater attention of late has a focus on shared challenges and problems. For example, Jiang and Ritchie (2017) discussed disaster collaboration in the context of cyclones in Australia. Shrestha and Decosta (2021) examine how partnerships and collaboration in Nepal can help in overcoming the problems caused by COVID-19.

There is increasing attention given to regional tourism partnerships based on similarities of locations and resources in the context of Indigenous peoples and ethnic minorities. Gao et al. (2019) looked at cross-border tourism and partnerships based on the ethnic minorities in Yunnan Province and Myanmar. Tham et al. (2020) reviewed the partnerships linking Indigenous and ethnic minority peoples across Asia and the Pacific.

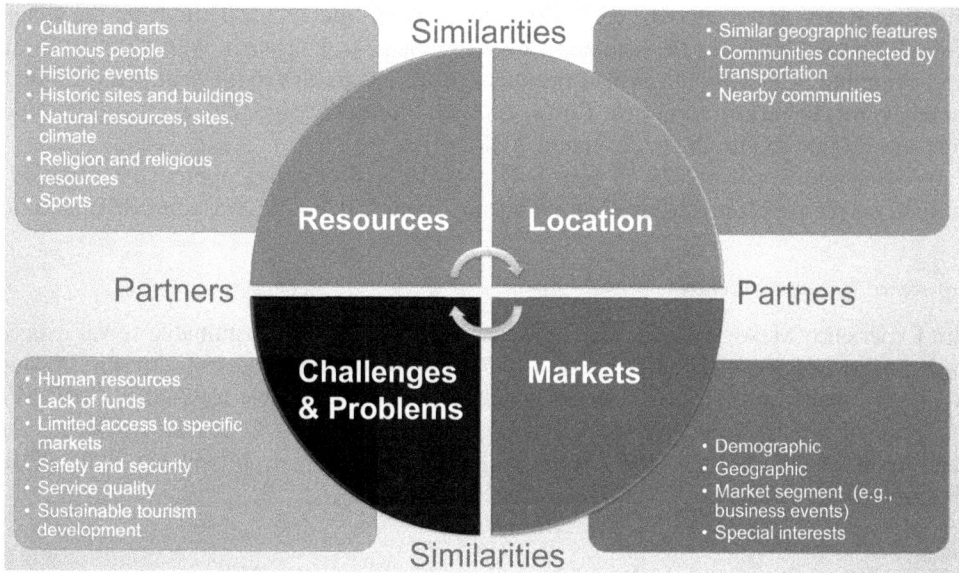

Figure 12.1 Partner identification wheel.

*Source*: Updated from Morrison (2019).

Other evidence of the increasing scope for regional tourism partnerships include joint efforts in the better handling of solid waste in Thailand (Jotaworn et al., 2021) and how improved child welfare can be achieved through Indigenous tourism partnerships in Canada (Huneault & Otomo, 2020).

## Catalysts of regional tourism partnerships

The chapter introduction cited eight specific reasons for forming regional tourism partnerships, and these are among the main catalysts. The ability to conveniently accomplish joint objectives and tasks online has also pushed greater regional tourism cooperation through shared websites and social media marketing. Government agencies, NGOs, and multilateral development banks are advocates of regional tourism partnerships and are providing the financial and technical support to sustain these collaborative efforts.

In addition to the business case for partnering and the influence of technology and institutional support, recent crises and disasters have convinced partners to form closer alliances. This was certainly the case in 2020–2022 when everyone found themselves in the same predicament of severely reduced business volumes.

## Successful cases in regional tourism partnerships

There are many successful regional tourism partnerships around the globe as can be attested by their longevity.

### Wild Atlantic Way

The Wild Atlantic Way (WAW) encompasses the coastline and hinterland of the nine coastal counties of the West of Ireland – Donegal, Leitrim, Sligo, Mayo, Galway, Clare,

Limerick, Kerry, and Cork. The route itself stretches for almost 2,500 km, from the village of Muff on the Inishowen Peninsula in County Donegal to Kinsale in West Cork. In addition, a number of urban centres have been identified as gateways to the Wild Atlantic Way, namely Cork, Killarney, Limerick, Ennis, Galway, Westport, Sligo, Donegal, and Letterkenny, as accommodation hubs (Sweeney, 2018).

WAW was initiated in 2010–11 primarily as a regional branding approach by Ireland's national government tourism development agency, Fáilte Ireland. WAW is a self-driving route.

### *Experience Mekong Collection*

The Experience Mekong Collection showcases responsible and sustainable travel experiences in the Greater Mekong Subregion (GMS). Our Mekong Tourism Advisory Group (MeTAG), made up of tourism professionals active in responsible tourism in the GMS, endorses all nominations. The region is comprised of the six countries bordering the Mekong River: Cambodia, PR China (Guangxi and Yunnan provinces), Lao PDR, Myanmar, Thailand, and Vietnam (Mekong Tourism, 2017).

## Prerequisites, barriers, and challenges for regional tourism partnerships

Kozak and Buhalis (2019) identified 12 prerequisites for cross-border destination marketing. These were trust, politics, product, marketing, distribution, accessibility, organisational cooperation, facilitation, planning, education, economic benefits, and socio-cultural benefits. McComb et al. (2017, p. 291) suggest conditions for effective stakeholder collaboration in tourism that include identifying all legitimate stakeholders, involving stakeholders throughout the partnering process, active participation of all stakeholders, and stakeholders believing their participation has the potential to influence decisions.

These prerequisites and conditions can also present potential barriers and challenges to forming regional partnerships. The presence and absence of mutual trust is particularly crucial in this respect (McComb et al., 2017). There are macro-level and individual-level barriers and challenges to the formation of regional tourism partnerships, and some of these are now discussed.

Politics and recent histories of conflict can pose significant barriers to regional tourism cooperation. For example, Kennell et al. (2019) reviewed regional tourism partnering in the WB6 countries (Albania, Bosnia and Herzegovina, Kosovo, Montenegro, North Macedonia, and Serbia). Among the partnering issues that were found within this region were the following three (Kennell et al., 2019, p. 6):

• Political issues, illegal migration, and terrorism have already made an impact on the tourism industry worldwide, and especially in the WB6 region.
• The region is historically known for many unresolved political issues that have been the source of tensions between neighbouring economies. Moreover, since 2015, many WB6 economies have also attracted negative attention due to civil protests against governments.
• From 2015, the region became one of the main migratory paths into the European Union, known as the Western Balkan Route. However, the number of illegal border crossings on this route has been falling steadily.

Apart from macro-level barriers, competitiveness and independence are two major barriers to cooperation and collaboration in tourism, and they are deeply ingrained in this economic sector. According to Kirillova et al. (2020) "tourism collaboration in its pure form is rarely observed because destinations tend to be excludable and rivalrous". The authors have personally experienced failed attempts to form regional partnerships for these reasons and others. Often, this was because stakeholders had other agendas and partnering was not a high enough priority for them.

Azazz et al. (2021) found that the main stakeholder opposition (in Egypt) to public–private partnerships was a result of the lack of satisfactory values or benefits, limited relational strength among stakeholders (mistrust, lacking commitment and support, and an unfriendly environment), and lack of technical knowledge and expertise.

For partnering to occur, there must be a willingness at the organisational and individual decision-maker levels. This point is not always obvious from reviewing the scholarship on tourism partnerships; however, it is certainly noticeable on the ground in the real world of tourism. Often, the major benefits from partnering do not emerge rapidly; they require longer-term commitments in order to be fully realised. Also, there can be perceptions that uneven levels of benefits will accrue from partnerships and that some partners will gain significantly more than others (e.g., smaller players will gain more than larger ones). Leaders of organisations may be more interested in building legacies in the shorter term or may be unwilling to enter a partnership with perceived uneven benefits. It also needs to be acknowledged that individual personalities influence readiness to engage in partnerships and other long-term relationships (Dent, 1999).

## Stakeholder interests and roles

The collaboration of various stakeholders is essential for successful regional tourism partnerships (McComb, Boyd, & Boluk, 2017). All stakeholders should share an interest in establishing such partnerships as each has the potential to benefit from collaboration. In principle, everyone should win, and no one should lose, through partnering.

### Tourists

Tourists undoubtedly are the beneficiaries of effective regional tourism partnerships as these groupings generate new offers such as experiences, packages, themed trails and routes, travel information, and access. However, it is even more desirable that tourists be invited to co-create travel experiences and other activities, products, and services.

### Tourism industry

Partnerships can bring new customers and revenues for tourism businesses, and this represents a major incentive for partnership participation. The active involvement of the tourism industry in regional tourism partnerships is usually a must, as they deliver the experiences, services, and products to visitors.

### Community residents

The involvement of social enterprises, such as in the case of the Experience Mekong Collection, and the promotion of community based tourism (CBT) projects are two ways in which regional tourism partnerships benefit community based residents. As with tourists, residents should also be involved in co-creating new experiences and activities, services, and products.

*Government*

Government agencies are often the instigators of regional tourism partnerships as was the case with Failté Ireland with the Wild Atlantic Way. The public sector also is frequently a major funder of such partnerships (Zapata & Hall, 2012). Governments are called upon to facilitate cross-border travel and infrastructure provision, establish regional standards and quality assurance schemes, facilitate capacity development and training, support tourism planning efforts, and harmonise policies. Governmental destination management organisations (DMOs) usually are involved in the marketing and branding of regional tourism partnerships as well.

*NGOs and the environment*

NGOs are often involved in the formation of regional tourism partnerships. For example, the Danube Competence Centre was supported by GIZ, the German international development agency. Multilateral development banks, including the Asian Development Bank, are other agencies sometimes called in to provide monetary support for regional tourism partnerships, as ADB did in the Mekong partnership.

Although a "silent" stakeholder, the environment usually plays a significant role in regional tourism partnerships by providing a focal point for collective priority and action. This is demonstrated in the following description of how partnerships can potentially contribute to the UN's Sustainable Development Goals (SDGs).

**Potential contributions to SDGs**

There is a substantial amount of attention being paid to the potential contribution of tourism partnerships to the achievement of the SDGs. A 2019 conference held in New Zealand concluded there was a "need for diverse actors to work in partnership to achieve the SDGs" (Scheyvens & Cheer, 2021). Several scholars have taken up this call; for example, Ferrer-Roca et al. (2020) examined how cross-border partnerships in a region of Spain and France could help achieve the SDGs there. Movono and Hughes (2020) examined the SDGs from the perspective of two community-focused tourism businesses in Fiji, focused on SDG 17 to explore how partnerships between tourism businesses and local community stakeholders supported local development. Hughes and Scheyvens (2021), using Fiji as their case study, considered how partnerships among hotels, their guests, and local communities helped with SDG attainment.

These scholars have considered the contributions of partnerships to a variety of SDGs and particularly related to environmental (e.g., SDGs 13, 14, and 15) and social (e.g., SDGs 1 and 5) ones. From the authors' own experiences, these five are valid expected contributions, and partnerships also materially enhance the following two goals:

SDG 8: Promote sustained, inclusive, and sustainable economic growth, full and productive employment, and decent work for all.
SDG 17: Strengthen the means of implementation and revitalise the global partnership for sustainable development.

**Future trends**

It is expected that regional tourism partnerships will continue to grow and be more diverse in the decades ahead.

*Lesser individual resources: Powering up through collaboration*

The period 2020–2022 has been one for downsizing in tourism and within tourism organisations. It is likely that this situation will persist for more years and that the benefits of partnering will be even greater in the time leading up to 2030. Five future trends are anticipated with regional tourism partnerships.

*Technological advances: Gluing future regional tourism partnerships*

An increasing emphasis on e-marketing was one of the earlier mentioned recent trends. Regional tourism partners will increasingly value the benefits of sharing new technological advances, including extended reality (augmented and virtual reality), artificial intelligence (AI), and the Internet of Things (IoT).

*Sustainability surge: Getting closer to 2030*

The pressure on tourism organisations to positively contribute to the achievement of the SDGs will intensify in the next seven to eight years.

*Climate change: Global warming knows no boundaries*

The Glasgow Declaration on climate change action in tourism has committed all stakeholders to the following:

> We declare our shared commitment to unite all stakeholders in transforming tourism to deliver effective climate action. We support the global commitment to halve emissions by 2030 and reach Net Zero as soon as possible before 2050. We will consistently align our actions with the latest scientific recommendations, so as to ensure our approach remains consistent with a rise of no more than 1.5°C above pre-industrial levels by 2100.
> (One Planet Network, 2022)

Climate change actions provide a platform for future tourism partnerships as climate knows no boundaries and must be addressed regionally and globally.

*Creativity emphasis: Openly innovating and co-creating*

There can be little doubt that tourism will be quite different in the future and destinations will continuously have to innovate and find new ways to capitalise on opportunities and address issues and challenges. Rather than operating as closed shops, regional tourism partnerships will increasingly have to embrace open innovation (Chesbrough, 2003) and engage in co-creation with all stakeholders. Tourists and community residents need to be involved in these processes, as do tourism industry businesses.

*Changes ahead: Embracing organisational structure changes*

Tourism organisations, especially DMOs, will have to change in the future as a result of new realities in the marketplace and society (Reinhold, Beritelli, & Grünig, 2019). These authors (p. 1135) state that "the need and legitimacy of destination management organisations (DMOs) are increasingly questioned".

**Contributions and future research priorities**

Although the research studies and other scholarship efforts on regional tourism partnerships are on the increase, this is still a neglected part of the literature. This chapter has identified recent and expected future trends in regional tourism partnerships, which constitutes an important contribution. Suggestions for future research priorities are also provided in the following paragraphs.

Partnering in tourism is still without solid theoretical foundations. There are such frameworks available for business in general including the concept of collaborative advantage that is attributed to Kanter (1994). She proposed the '8 Is that make successful we's' and they were individual excellence, importance, interdependence, investment, information, integration, institutionalisation, and integrity.

There is a need to develop a partnering readiness scale that specifically addresses regional tourism partnerships. Outside of tourism, there are several long-standing scales for such measurement including Dent's (1999) concept of "partnering intelligence" and measuring an individual's "partnering quotient". MIT's Readiness Checklist is a tool to measure an organisation's readiness to enter into partnerships (Massachusetts Institute of Technology. (2022).

Success stories in regional tourism partnerships are well recognised and covered; however, failures are not given the same attention. For example, Ghanem et al. (2022) document the failure of a public–private destination management system (DMS) project in Egypt after the public sector agency decided to withdraw from the collaborative project.

Thraenhart (2022) has made a noble effort to catalogue and describe many of the regional tourism partnerships on a worldwide basis. However, it is acknowledged that this inventory is incomplete and more future work is needed to provide a comprehensive listing and categorisation of such partnerships.

**Conclusion**

Regional tourism partnerships, if effectively implemented, are multiple win propositions. They are a global phenomenon of which the benefits are universally recognised. Despite the many catalysts and reasons for creating regional tourism partnerships, there are also significant barriers and challenges in forming and implementing such collaborations.

The nature and scope of regional tourism partnerships will continually change in the future. There is a need in the years to come for these partnerships to become a more mainstream topic of tourism research.

**References**

Azazz, A. M. S., Elshaer, I. A., & Ghanem, M. (2021). Developing a measurement scale of opposition in tourism public-private partnerships projects. *Sustainability*, *13*, 5053, https://doi.org/10.3390/su13095053

Chesbrough, H. W. (2003). *Open Innovation: The New Imperative for Creating and Profiting from Technology*. Cambridge, MA: Harvard Business Review Press.

Dent, S. M. (1999). *Partnering Intelligence: How to Profit from Smart Alliances*. Palo Alto, CA: Davies-Black Publishing.

Dunets, A. N., Ivanova, V. N., & Poltarykhin, A. L. (2019). Cross-border tourism cooperation as a basis for sustainable development: A case study. *Entrepreneurship and Sustainability Issues*, *6*(4), 2207–2215.

El-Khadrawy, R. K., Attia, A. A., & Rashed, R. (2022). Partnerships for sustainable tourism development in the cultural heritage sites. In A. Versaci et al. (eds.), *Conservation of Architectural Heritage, Advances in Science, Technology & Innovation*. Cham, Switzerland: Springer Nature Switzerland AG. https://doi.org/10.1007/978-3-030-74482-3_5

Ferrer-Roca, N., Guia, J., & Blasco, D. (2020). Partnerships and the SDGs in a cross-border destination: The case of the Cerdanya Valley. *Journal of Sustainable Tourism*. https://doi.org/10.1080/09669582.2020.1847126

Gao, J., Ryan, C., Cave, J., & Zhang, C. (2019). Tourism border-making: A political economy of China's border tourism. *Annals of Tourism Research*, *76*. https://doi.org/10.1016/j.annals.2019.02.010

Ghanem, M., Elshaer, I., & Saad, S. (2022). Tourism public-private partnership (PPP) projects: An exploratory-sequential approach. *Tourism Review*, *77*(2), 427–450.

Graci, S. (2013). Collaboration and partnership development for sustainable tourism. *Tourism Geographies*, *15*(1), 25–42.

Hampton, M. P. (2009). *The Socio-Economic Impacts of Singaporean Cross-Border Tourism in Malaysia and Indonesia*. Working paper. Kent Business School, University of Kent, Canterbury.

Hughes, E., & Scheyvens, R. (2021). Tourism partnerships: Harnessing tourist compassion to 'do good' through community development in Fiji. *World Development*, *145*, 105529, https://doi.org/10.1016/j.worlddev.2021.105529

Huneault, G., & Otomo, M. (2020). From unlikely to likely partnerships for change - Child welfare and indigenous tourism in Canada. *Journal of Sustainable Tourism*. https://doi.org/10.1080/09669582.2020.1817047

Jiang, Y., & Ritchie, B. W. (2017). Disaster collaboration in tourism: Motives, impediments and success factors. *Journal of Hospitality and Tourism Management*, *31*, 70–82.

Jotaworn, S., Nitivattananon, V., Kusakabe, K., & Xue, W. (2021). Partnership towards synergistic municipal solid waste management services in a coastal tourism sub-region. *Sustainability*, 13, 397, https://doi.org/10.3390/su13010397

Kanter, R. M. (1994). Collaborative advantage. *Harvard Business Review*, *72*(4), 96–108.

Kennell, J., Chaperon, S., Šegota, T., & Morrison, A. M. (2019). *Western Balkans Tourism Policy Assessment and Recommendations*. London: University of Greenwich.

Kirillova, K., Park, J., Zhu, M., Dioko, L., & Zeng, G. (2020). Developing the coopetitive destination brand for the Greater Bay Area. *Journal of Destination Marketing & Management*, *17*. https://doi.org/10.1016/j.jdmm.2020.100439

Kozak, M., & Buhalis, D. (2019). Cross-border tourism destination marketing: Prerequisites and critical success factors. *Journal of Destination Marketing & Management*, *14*, 100392. https://doi.org/10.1016/j.jdmm.2019.100392

Lagiewski, R., & Revelas, D. (2004). Challenges in cross-border tourism regions. Accessed from https://scholarworks.rit.edu/other/551

McComb, E. J., Boyd, S., & Boluk, K. (2017). Stakeholder collaboration: A means to the success of rural tourism destinations? A critical evaluation of the existence of stakeholder collaboration within the Mournes, Northern Ireland. *Tourism and Hospitality Research*, *17*(3), 286–297.

Massachusetts Institute of Technology. (2022). Readiness checklist: How ready are we to initiate the partnership? http://d-lab.mit.edu/sites/default/files/inline-files/12.%20Readiness%20Checklist_0.pdf

Morrison, A. M. (2019). *Marketing and Managing Tourism Destinations*, 2nd ed. London: Routledge.

Morrison, A. M. (2022). *Tourism Marketing in the Age of the Consumer*. London: Routledge.

Movono, A., & Hughes, E. (2020). Tourism partnerships: Localizing the SDG agenda in Fiji. *Journal of Sustainable Tourism*. https://doi,org/10.1080/09669582.2020.1811291

Nahm, J. (2021). *Collaborative Advantage: Forging Green Industries in the New Global Economy*. Oxford, England. Oxford Scholarship Online.

One Planet Network. (2022). Glasgow declaration: Climate action in tourism, https://www.oneplanetnetwork.org/programmes/sustainable-tourism/glasgow-declaration

Reinhold, S., Beritelli, P., & Grünig, R. (2019). A business model typology for destination management organizations. *Tourism Review*, *74*(6), 1135–1152.

Scheyvens, R., & Cheer, J. M. (2021). Tourism, the SDGs and partnerships. *Journal of Sustainable Tourism*. https://doi.org/10.1080/09669582.2021.1982953

Shrestha, R. K., & Decosta, P. L. (2021). Developing dynamic capabilities for community collaboration and tourism product innovation in response to crisis: Nepal and COVID-19. *Journal of Sustainable Tourism*. https://doi.org/10.1080/09669582.2021.2023164

Sullivan, R., & Warner, M. (2017). Introduction. In M. Warner, & R. Sullivan (Eds.), *Putting Partnerships to Work: Strategic Alliances for Development between Government, the Private Sector and Civil Society* (pp. 1–12). London: Routledge.

Sweeney, P. (2018). An Irish entrepreneurial state's success: The Wild Atlantic Way. *Progressive Economy*. https://www.tasc.ie/blog/2018/04/25/an-irish-entrepreneurial-states-success/

Tham, A., Ruhanen, L., & Raciti, M. (2020). Tourism with and by Indigenous and ethnic communities in the Asia Pacific region: A bricolage of people, places and partnerships. *Journal of Heritage Tourism*, *15*(3), 243–248.

Thraenhart, J. (2022). How to strengthen cross-border tourism integration and resilience? An analysis of the Greater Mekong Subregion with an emphasis on post-COVID-19 recovery. Unpublished doctoral thesis. The Hong Kong Polytechnic University.

Zapata, M. J., & Michael Hall, C. (2012). Public–private collaboration in the tourism sector: Balancing legitimacy and effectiveness in local tourism partnerships. The Spanish case. *Journal of Policy Research in Tourism, Leisure and Events*, *4*(1), 61–83.

# 13 Island destinations in the face of global challenges

## A strategic analysis

*Eduardo Parra-López, Almudena Barrientos-Báez
and María de los Ángeles Pérez Sánchez*

### Introduction

Tourism within island destinations (IDs) has been dependent on international tourism, intermediation, and air and maritime connectivity. Significant price competition and infrastructure development, especially in accommodation, have seen investment focused on coastal tourism, with value-added resorts and the consolidation of an all-inclusive product. IDs specialising in mass tourism at a global level require a wide variety of actors with multiple inputs. Further research is needed to understand how IDs with multiple actors can enhance their economic sustainability (Randall, 2021).

Our research has established a set of standardised tourism services that makes local culture invisible and difficult to distinguish among destinations. This means IDs are readily interchangeable, but not mutually exclusive. Climate tends to be a key attraction, together with destination security. These conditions, together with engaged markets, political uncertainty, and other factors in competing destinations, have favoured an increase in international tourists over the past five years.

IDs are destinations desired and demanded by millions of tourists, projecting their future as a preferred choice, consolidated and specialised, towards a need for gradual renewal and diversification. Self-confident, but without neglecting to look towards other markets and competitors, IDs worldwide maintain employment (direct and indirect) amidst criticisms of precariousness and drive the economy in and from the tourism system by increasing quality levels.

While there are no prescriptive formula or procedures to respond to the challenges faced by IDs, we aim to reflect on a strategic process of learning to encourage a commitment to improve and make the most of resources and infrastructure.

### Resilience effect in island destinations: Concepts and factors

To address the challenges of the past, tourism must adapt. This process should minimise tensions between societal behaviours, thoughts, and actions, which can be learned and developed by individuals and public and private institutions. This approach requires a balanced construction between all actors in the ID tourism system to ensure stimulus, health, social and economic security, and resilience within the sector (McLeod et al., 2022).

The main factors associated with the resilience effect in tourism are (Booth et al., 2020; Klein et al., 2003; Coghlan & Prideaux, 2009; Pelling, 2011; Parra-López et al., 2010; Parra-López & Martínez-González, 2018):

1. The capacity to introduce realistic tourism plans and follow the steps to implement these plans;

DOI: 10.4324/9781003291763-15

2. To possess a positive view of ourselves as a society and have confidence in our strengths and abilities as a tourism industry;
3. To develop skills in communication and problem solving within tourism;
4. The ability to manage the feelings and impulses of our society in relation to the will-power of all those involved with ID tourism.

The United Nations (UNISDR, 2012) advanced that the development of resilient cities requires the identification of threats and weaknesses through training and development in conflict management. These factors are equally relevant for IDs aiming to enhance their resilience. Strategic planning through effective leadership, based on the implementation of solutions and the reinforcement of strengths, should not depend on individuals, but on the synergies established to develop resilient communities and markets.

Sustainable ID development can be defined as that which is capable of meeting the needs of the present without compromising the ability of future generations to meet their own needs. Sustainable development based on the Sustainable Development Goals (SDGs) requires concerted effort to build an inclusive, sustainable, and resilient future for people and islands. Therefore, it is essential to harmonise the three basic elements: (a) economic growth, (b) social inclusion, and (c) environmental protection. These elements are interrelated and are all essential for the wellbeing of individuals and society (Table 13.1).

The implementation of tourism development programmes must show continuity and internal coherence, through strong administration (smart governance), maximising the resources and strengths of the destination, promoting renewal, and enhancing the creation of decently paid jobs. These tourism reactivation processes should arise from the recognised needs of different stakeholders who would collaborate in their development and process improvement, as is expressed in Table 13.1.

*Table 13.1* How to develop more resilient islands based on the SDGs

| Prevention | Solution | Retrieval |
|---|---|---|
| • 2030 Agenda as a clear, measurable, and participatory definition process for citizens, academia, public and private companies, public administrations, and the third sector, to address these challenges in a consensual manner and leave no one behind.<br>• Take urgent action to combat climate change and its effects.<br>• Reduce poverty and enhance food security.<br>• Ensure the availability of water and its sustainability.<br>• Protect and promote the responsible use of our natural resources.<br>• Strengthen the meaning of the implementation and revitalisation of a global pact of islands for sustainable development. | • Crisis planning business continuity management.<br>• Effective solution-based leadership.<br>• Achieve gender equality and the empowerment of women.<br>• Improve the productivity and reduce the consumption of firms.<br>• Conserve and make the oceans and seas more sustainable.<br>• The energy transition and its sustainability in the islands. | • Crisis learning.<br>• Negative consequences assumption.<br>• Reinforcement of strengths. |

*Source*: Adapted from UNISDR, United Nations, 2012. Handbook for Local Government Leaders. (See also: Small Island Developing States |Department of Economic and Social Affairs (un.org))

Forms of governance adopted by the market in recent decades, which emphasised economic growth and competitiveness, favoured public–private partnerships and networks as a form of organisation, and the search for consensus (Muñoz Mazón et al., 2012; Santana Talavera, 2016) may no longer be sufficient. The enhanced use of e-communications can enhance win-to-win dynamics, with synergies that responsibly project economic and emotional gains for the parties involved (Kotler et al., 2011). It is possible to project a future for tourism in which IDs follow the principles of legitimacy, transparency, participation, shared knowledge, co-responsibility, efficiency, and equity (OECD, 2004). In doing so, they would contribute to the basis for improvement in the quality of life for citizens and tourists, without ignoring a need for economic returns for business.

It is also recognised that the reduction of risk would be greater, with a higher degree of resilience achieved. The tourist destination, as the first organisational level, which encompasses all critical infrastructure, would be the line of formation that places one next to the other, as well as the axis that manoeuvres and allows the maximum result of the strategies implemented to be obtained with a minimum of vulnerability.

**Specialisation as an imperative of a new reality: Knowledge and analysis**

IDs globally need analytical and strategic thinking more than ever. Reasonable and reflective thinking are needed to address current and future challenges. Ennis (1996) and Lipman (1991) have emphasised that in contexts of uncertainty it may be necessary to redirect the behavioural aspects of business and public decision-making. The purpose of decisions should focus on agility and efficiency, considering how they can be achieved and the self-correcting nature of processes, rather than decisional passivity.

The nature of strategic thinking in tourism is complex and defies synthesis into a single definition. However, important characteristics include skills, disposition and foresight, reasoned judgement of the situation in question, and the self-correction of strategies that may be proposed. Halpern (1998) suggests that strategic thinking must be purposeful, reasoned, and directed towards new objectives. It should be noted that to think analytically is to evaluate, not only the outcome of the thought processes, but also to look at the reasoning that led to the conclusions or the factors that have led to a decision. Therefore, strategic thinking in tourism should involve the evaluation of island specific solutions, with the aim of providing useful and accurate feedback so as to improve outcomes.

The development of sustainable tourism proposes a view of tourism activity that embraces all elements of sustainable development and not just economic outcomes (Mauss, 2004; Santana Talavera, 2016; Bauza Martorell & Melgosa Arcos, 2020). This would include the capacity to structure responses for territories and populations, products and consumption, as well as supporting social constructs and ways of life.

A holistic approach that takes into consideration the agents involved in the contexts of production and reproduction is required to implement the principles of sustainable development. It is an open system, which is intrinsically and reciprocally related to other systems (economic, political, environmental) and is linked by feedback (Díaz Rodríguez et al., 2012). This openness can create uncertainties, with many nuances that are dependent on external circumstances over which IDs have little or no control. It is against this backdrop that planning should reflect critical and boom times. The global tourism system, as has been seen during the pandemic, is capable of reacting in a coordinated manner and responding to complexities with extreme dynamism.

The success of specialisation lies in knowledge and its adaptive capacity to foster innovation and renewal (in markets, processes, infrastructure, business models, products, and the workforce). The mantra of competitiveness, quality, satisfaction, sustainability/responsibility, and profitability is repeated with devotion in the business literature. However, while often cited it is seldom reflected in reality. Tourism processes are the cause and effect of many different actors operating in synchrony (Donaire et al., 2014; Santana Talavera, 2016). IDs have been trapped in deadlines, agendas, and results, and the true value of learning has been lost. When destinations embark on learning, they are learning and discovering, they are reconnecting with reality.

### Economic growth and tourism competitiveness: Seeking shared island leadership

Although there is no proven methodology for designing tourism scenario modelling for islands (Senge, 2001; McCabe et al., 2012), indicators – such as population growth (especially an analysis of the behaviour of the middle classes at origin), the number of tourists, price elasticity, the type of tourist, tourist expenditure, supply, activities carried out, direct and indirect employment, dependency rates, competitiveness between destinations, efficiency of the marketing developed, and others of a more qualitative nature combining variables related to perceptions, relationships, and decision-making at different levels – are commonly used.

The why of scenario modelling is to be answered with a simple example based on the principle of communicating vessels. The total number of potential tourists will be limited and with slow (or very slow) replenishment rates. They make up, more or less geographically distributed by groupings of destinations of majority choice (about which they already have stereotypes), those who will be tourists in the global tourism system of the future.

In principle, all IDs will start out with equal potential, with the easing of pandemic restrictions. Market imposed laws and destinations will compete to capture the market. Internal (economic, political, social, and health) and external pressures (pandemics, terrorism, severe environmental changes, natural disasters) of competing destinations and the capacity of destination managers will influence customers and intermediaries (online travel agencies or OTAs, tour operators, and travel agents) between IDs (or geographical areas). The approach aims to reconfigure tourism planning from scratch. It requires destinations to analyse and learn from other experiences to solve problems or improve processes, utilising a bottom-up approach.

Island territories globally have had diversified and consolidated offerings which can be reconfigured through inter-enterprise and citizen collaboration. By identifying demand, the combination of products to form collaborative products, which are linked and sustained in a circular economy (European Environment Agency, 2018), may contribute to qualitative renewal, reinforce sensations, and evoke emotions in the tourist. According to Barrientos-Báez et al. (2019) these qualities are essential to increase satisfaction and the perception of quality experiences.

This manifestation of resilience (adaptive change) (Adger, 2000; Larsen et al., 2011; Roca & Villares, 2016; Dávila-Lorenzo & Saladrigas-Medina, 2020) requires the exercise of decisive leadership, communication, regulatory and bureaucratic ease, and economic support for entrepreneurship, talent, and training, such as shared leadership. Internal competitiveness can favour the creation and integration of products with high added value, which is transmitted by the tourist on social networks and virtual intermediation channels (Tripadvisor, Booking, or Google Review).

Strategic modelling should also embrace the multiplier effect of investment in technology, which has implications for the future of island tourism and unemployment. For the destination, technology may provide tools that aim to minimise economic risks and maximise benefits.

Technology may provide for the improvement of services and the quality of life for the direct and indirect actors in the system. The process of digital transformation, as an application of technological capabilities to tourism processes, may allow the reworking of products and assets. The outcomes of improved efficiency, customer experience through personalisation, the management of post-COVID-19 risks, and the discovery of new opportunities for income generation may benefit a destination. According to Parra-López et al. (2020), global digitalisation has resulted in social and economic transformation unlike any other in modern history. To facilitate these outcomes requires coordinated behaviour that allows interaction between local and global networks. Each individual becomes an agent of change, providing sensitivity to the environment, analysing perceptions, and proposing actions to generate infrastructure that creates tourist satisfaction.

Unemployment and the creation of jobs on the IDs is also a critical issue. Job creation leads to social improvement and is a critical consideration in planning and destination models. Well-defined roles and responsibilities contribute to efficiency, and all positions in the system, with their specific functions and activities, are relevant. Each worker contributes to product development, and ensuring that tasks are dignified, recognised, and valued will enhance quality.

Specific training (not only formal), commitment, and responsibility for employee professional development will be necessary. It is a matter of nurturing aptitude (training) and attitude (willingness) towards the company and with the tourist company, by being a participant and assuming an active role in the co-creation and communication of the product. By doing so the employee may enhance tourist satisfaction and work towards a more responsible and sustainable company (Figure 13.1).

This process is utopic if not accompanied by measures that enhance employee standards of living and job satisfaction, provides intellectual benefits (expansion and provision of basic knowledge), and has the potential to improve the medium-term prospects of their family unit, focusing above all on the future of their children. These aspired socio-economic changes and employment trends, some of which are already in play, force IDs to adapt tourism strategies to address these challenges. It is suggested that the values that function as attractors for tourists (stereotypes that make the destination attractive) will also appeal to innovative (risk-taking) foreign companies and a higher-skilled external workforce (Liu & Wall, 2005; Lundberg et al., 2009; Rosentraub & Joo, 2009). This has been reflected in other stages of island tourism development (Tiago & Borges-Tiago, 2022; Vyline-Cuffy, 2022). Strength-based tourism and teamwork, together with specialisation, will be essential.

Making decisions based on data, accurate information, and predictive analytics that allow the development of possible future scenarios is essential. Digital based scenarios convert tourism processes into a set of data, directing the objectives of firms towards tourists and their level of experience. The use of data driven strategic visions will make businesses more powerful within the island value chain process.

Rethinking traditional assumptions related to island tourism strategy is anchored in a process of shared leadership. It also leads to a different perspective of business models of value, where the tourism workforce is a driving force. An automated digital experience based on continuous improvement can be used by high-performance collaborative teams with a focus on the value of tourism. It calls for us to turn our back on short-termism

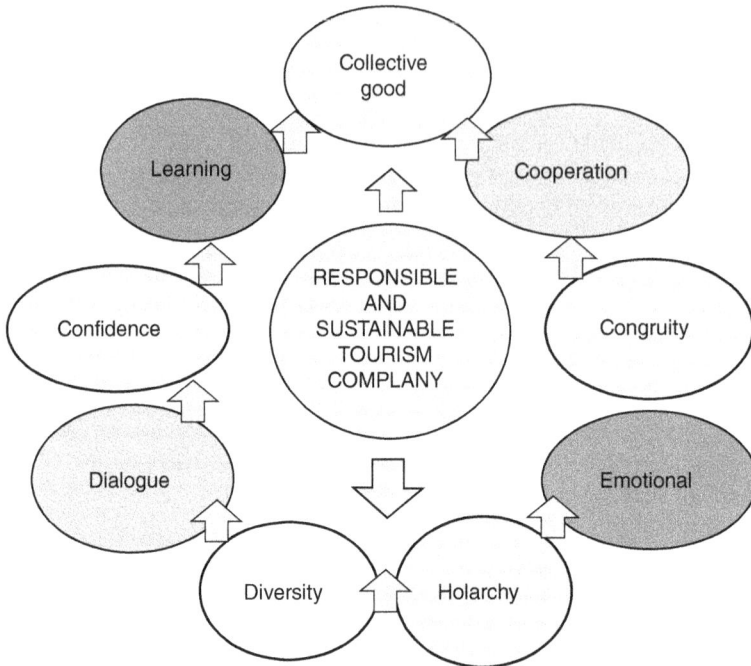

*Figure 13.1* Values and principles for new tourism firms with a focus on SDGs.

*Note*: This proposal of values and principles for island destinations is applicable to any island destination regardless of territorial size or geographic area.

which has dominated thinking in recent years. The process of digital transformation cannot be ignored, as many aspects of tourism marketing, innovation, and quality are already evolving.

The challenge for IDs is not to return to what they were but to bring about the renewal of values and a change of perceptions through the generation of new opportunities and supportive social behaviour. Tourism organisations on islands must adjust quickly to meet the expectations of key players in a highly competitive and constrained market.

### Opportunities and challenges

Descriptive and reflective analysis has highlighted the opportunities and challenges facing IDs. Areas of consensus and possible future contributions have been explored and indicate a need for agility to bring the desired outcomes to fruition. The development of mechanisms, plans, and operational frameworks for the coordination of IDs at a global level is required. These systems should provide support for public–private decision-making (coordination between agents, risk communication, transparency, and technical governance). Similarly, there is a need to activate and prioritise research, knowledge, and innovation in tourism by researchers, academics, and experts from the islands. Developing a communication network between agents and people, which generates a probabilistic routing that successfully reinforces a path forward and highlights the network of island products and services and people, is needed. This approach reinforces the logic of more efficient time parameters, generating constant corrective actions, making random behaviour visible, and achieving greater anticipation in the response.

Governments of IDs should provide a stimulus to accelerate recovery. A financial stimulus is essential and may take the form of favourable tax policies, the lifting of travel restrictions as soon as the health emergency allows, prompt visa facilitation, or safe tourist corridors. Other strategies may include an increase in marketing campaigns to promote tourist confidence in the destination. Tourism must be placed at the centre of the recovery policies and action plans of IDs.

We must prepare ourselves for a change of focus where the capabilities of ID tourism are an engine for local and inward growth. Strategies should also work towards examining the capacity to achieve value generation.

Looking towards tourism in 2030 and in concert with the SDGs the proposed approach encourages joint contributions, sustainable development and resilience building, the application of learnings from the pandemic crisis, and support from the experience and collective intuition of all actors in the system. This approach needs to create the team before seeking tourism diversity and develop an awareness process that generates natural growth, structuring spaces, and autonomies. These factors will allow innovation to address ID challenges. According to Barrientos-Báez et al. (2020), change must start with small initiatives that show that it is possible and achievable.

There is also a call to ID tourism agents to ensure there are training plans for employees. Agents can also take advantage through the transition to the circular and regenerative economy, the knowledge economy, and through digital transformation.

Incorporating society more broadly to contribute to strategy in the local context will allow the complexities and difficulties of tourism to emerge. These contributions may form a foundation to face the current pandemic crisis and future challenges. From the global to the local, geolocation (Robertson, 1994) is a process that takes an economic and cultural perspective, which allows tourism development to adapt to the unique qualities of each island, enabling greater differentiation and avoid business and political tensions.

Transparency may also pose a challenge as an information process in the face of uncertainties and the digital age. It requires efficient communication of the measures adopted and should provide tourists with the measures that are being carried out. IDs will need to become influencers within the value chain.

For the local community it may be helpful to encourage a knowledge of self and to abandon hopelessness, resignation, or passivity towards tourism that often emerges in island populations. Change is a natural process and only by joining this process will the community be able to grow and face future challenges. The most adaptable and flexible species change and survive (Darwin, The Origin of Species, 1859/1995).

IDs are in a time of stress and perseverance, and this is a key challenge for them up to 2030. The capacity to overcome this state of stress depends on the ability to decide and order their behaviour and the idea of carrying out a process as agents of change. Tourist destinations face complex decisions that should not incapacitate managers, nor weaken decision-making, which is known as the paradox of choice. This paradox, the learning–knowledge binomial, is an asynchronous relationship that explores new forces, lays the foundations for a new reality, and will respond to the tourism system: why me and now? IDs globally are embarking on a path where there will be errors and deviations. Learning from these mistakes will help to manage difficulties and provide new skills.

To conclude, we must continue to focus on innovative methodologies and business models such as customer journey maps, user experience, design thinking, artificial intelligence, disruptive innovation, and the plan–do–check–act Deming cycle. Competition is intensifying, and it will be the capacity of IDs for innovation and invention that will determine the survival of companies.

## References

Adger, W. N. (2000). Social and ecological resilience: Are they related? *Progress in Human Geography*, *24*(3), 347–364. https://doi.org/10.1191/030913200701540465

Barrientos-Báez, A., Barquero-Cabrero, M., & Rodríguez-Terceño, J. (2019). La educación emocional como contenido transversal para una nueva política educativa: El caso del grado de turismo. *Revista Utopía y Praxis Latinoamericana*, *24*(4), 147–165. https://produccioncientificaluz.org/index.php/utopia/article/view/29796

Barrientos-Báez, A., Caldevilla-Domínguez, D., Cáceres Vizcaíno, A., & Sueia Val, E. G. (2020). Sector Turístico: Comunicación e Innovación sostenible. *Revista de Comunicación de la SEECI*, *53*, 153–173. https://doi.org/10.15198/seeci.2020.53.153-173

Bauza Martorell, F. J., & Melgosa Arcos, F. J. (2020). *El turismo después de la pandemia global, análisis, perspectivas y vías de recuperación*. Asociación Española de Expertos Científicos en Turismo (AECIT). https://bit.ly/3InrlGx

Booth, P., Chaperon, S. A., Kennell, J. S., & Morrison, A. M. (2020). Entrepreneurship in island contexts: A systematic review of the tourism and hospitality literature. *International Journal of Hospitality Management*, *85*. https://doi.org/10.1016/j.ijhm.2019.102438

Coghlan, A., & Prideaux, B. (2009). Welcome to the wet tropics: The importance of weather in reef tourism resilience. *Current Issues in Tourism*, *12*(2), 89–104. https://doi.org/10.1080/13683500802596367

Darwin, C. (1859/1995). *On the Origin of Species*. Boston: Harvard University Press.

Dávila-Lorenzo, M., & Saladrigas-Medina, H.-M. (2020). Modelo de gestión de comunicación pública del patrimonio. *Revista Latina de Comunicación Social*, *77*, 329–356. https://doi.org/10.4185/RLCS-2020-1461

Díaz Rodríguez, P., Santana Talavera, A., & Rodríguez Darias, A. J. (2012). Selección patrimonial: Del consumo cotidiano al consumo turístico. *Revista Andaluza de Antropología*, *2*, 86–107. https://doi.org/10.12795/RAA.2012.i02.05

Donaire, J. A., Camprubi, R., & Galí, N. (2014). Tourist clusters from Flickr travel photography. *Tourism Management Perspectives*, 11, 26–33. https://doi.org/10.1016/j.tmp.2014.02.003

Ennis, R. H. (1996). *Critical Thinking*. Upper Saddle River: Prentice Hall.

European Environment Agency. (2018). *Trends and Projections in Europe 2018. Tracking Progress towards Europe's Climate and Energy Targets*. Brussels: Publications Office of the European Union.

Halpern, D. F. (1998). Teaching critical thinking for transfer across domains: Disposition, skills, structure training, and metacognitive monitoring. *American Psychologist*, *53*(4), 449–455. https://doi.org/10.1037/0003-066X.53.4.449

Klein, R. J. T., Nicholls, R. J., & Thomalla, F. 2003. Resilience to natural hazards: How useful is this concept? *Environmental Hazards*, *5*(1–2), 35–45. https://doi.org/10.1016/j.hazards.2004.02.001

Kotler, P., Bowen, J., Makens, J. C., García, J., & Flores, J. (2011). *Marketing Turístico*. Madrid, España: Editorial Pearson Educación.

Larsen, R. K., Calgaro, E., & Thomalla, F. (2011). Governing resilience building in Thailand's tourism-dependent coastal communities: Conceptualising stakeholder agency in social-ecological systems. *Global Environmental Change*, *21*(2), 481–491. https://doi.org/10.1016/j.gloenvcha.2010.12.009

Lipman, M. (1991). *Thinking in Education*. Cambridge: MA, Cambridge University Press.

Liu, A., & Wall, G. (2005). Human resources development in China. *Annals of Tourism Research*, *32*(3), 689–710. https://doi.org/10.1016/j.annals.2004.10.011

Lundberg, C., Gudmundson, A., & Andersson, T. D. (2009). Herzberg's two-factor theory of work motivation tested empirically on seasonal workers in hospitality and tourism. *Tourism Management*, *30*(6), 890–899. https://doi.org/10.1016/j.tourman.2008.12.003

Mauss, M. (2004). *Manual de etnografía*. Mexico City: Fondo Editorial Económico.

McCabe, D., Treviño, L. K., & Butterfield, K. D. (2012). *Cheating in College: Why Students Do It and What Educators Can Do About It*. Baltimore: The Johns Hopkins University Press.

Mcleod, M., Dodds, R., & Butler, R. (2022). *Island Tourism Sustainability and Resiliency*. London: Routledge.

Mekong Tourism. (2017). Experience Mekong Collection, https://mekongtourism.org/experience-collection/

Muñoz Mazón, A. I., Fuentes Moraleda, L., & Fayos-Solà, E. (2012). Turismo como instrumento de desarrollo: Una visión alternativa desde factores humanos, sociales e institucionales. *PASOS. Revista de Turismo y Patrimonio Cultural, 10*(5), 437–444. http://www.redalyc.org/articulo.oa?id=88124507001

Naciones Unidas [UNISDR]. (2012). Cómo desarrollar ciudades más resilientes: Un manual para líderes de los gobiernos locales. Desarrollando ciudades más resilientes. https://www.unisdr.org/files/26462_manualparalideresdelosgobiernosloca.pdf

OECD. (2004). OECD Principles of Corporate Governance. http://www.oecd.org/corporate/ca/corporategovernanceprinciples/31557724.pdf

Parra-López, E., Barrientos-Báez, A., & Martínez-González, J. A. (2020). La transformación digital del turismo. *Revista de Occidente, 464*, 52–66. https://ortegaygasset.edu/producto/revista-de-occidente-no-464-enero-2020/

Parra-López, E., & Martínez-González, J. (2018). Tourism research on IDs: A review. *Tourism Review, 73*(2), 133–155. https://doi.org/10.1108/TR-03-2017-0039

Parra-López, E., Melchior-Navarro, M., & Fuentes-Medina, L. (2010). Dinámicas de transformación de un destino turístico maduro. In Hernández, R., Santana, A. (Eds.), *Destinos Turísticos Maduros ante el Cambio. Reflexiones desde Canarias*. La Laguna, Spain: Universidad de la Laguna, pp. 217–232.

Pelling, M. (2011). *Adaptation to Climate Change: From Resilience to Transformation*. London: Routledge.

Randall, J. (2021). *An Introduction to Island Studies*. Rowman & Littlefield Publishing Group. https://books.google.co.jp/books?id=WOGqzQEACAAJ

Robertson, R. (1994). Globalization or glocalization? *Journal of International Communication, 1*(1), 33–52. https://doi.org/10.1080/13216597.1994.9751780

Roca Bosch, E., & Villares Junyent, M. (2016). La integración del cambio climático en la planificación de los riesgos ambientales en el litoral catalán. In Olcina Cantos, J., Rico Amorós, A. M., & Moltó Mantero, E. (Eds.), *Clima, Sociedad, Riesgos y Ordenación del Territorio*. Universidad de Alicante. Asociación Española de Climatología, pp. 595–601. https://doi.org/10.14198/XCongreso AECAlicante2016-56

Rosentraub, M. S., & Joo, M. (2009). Tourism and economic development: Which investments produce hains for regions? *Tourism Management, 30*(5), 759–770. https://doi.org/10.1016/j.tourman.2008.11.014

Santana Talavera, A. (2016). *Planteamientos a medio plazo al turismo en Canarias. En ¿Existe un modelo turístico canario?*, Simancas-Cruz, M., & Parra-López, E. (Coord.) Las Palmas: Promotur Turismo de Canarias, pp. 187–203.

Senge, P. (2001). *Peter Senge and Learning Organization*. The Encyclopaedia of Informal Education. Retrieved on June 7, 2022 from http://www.infed.org/thinkers/senge.htm

Tiago, F., & Borges-Tiago, T. (2022). Small ID (Chapter). In Buhalis, Dimitrios (Ed.), *Encyclopedia of Tourism Management and Marketing*. Cheltenham: Edward Elgar Publishing.

UN. (2023). About small Island developing states, https://www.un.org/ohrlls/content/about-small-island-developing-states

UNISDR. (2012). How to make cities more resilient. A handbook for local government leaders, https://www.unisdr.org/campaign/resilientcities/assets/toolkit/documents/Handbook%20for%20local%20government%20leaders%20%5B2017%20Edition%5D.pdf

Vyline-Cuffy, V. (2022). Resilient Island Tourism. In Buhalis, Dimitrios (Ed.), *Encyclopedia of Tourism Management and Marketing*. Cheltenham: Edward Elgar Publishing.

# 14 Island tourism development for inclusive growth

*Huu Nghia Le*

## Introduction

Since mass tourism has become one of the most popular products, tourism destinations have been significantly altered. Remote islands have become more accessible to tourists as a result of the availability of transportation. Island tourism has consistently ranked among the most appealing types of tourism worldwide. While tourism is one of the most important economic activities for islands, local infrastructure and resources are stressed due to the large seasonal influx of visitors to such areas (UNEP, 2014).

As leisure travel on tropical islands has expanded, the unspoiled paradise tagline has become a marketing cliché for tourism all over the world. Briguglio (2008) found that tourism creates money and employment but also contributes to environmental damage. In some ways, hosting tourists on islands can be a double-edged sword as the considerable economic benefits it brings can also lead to local vulnerabilities such as increased demand for water, food, and energy resources. Water and waste pollution, urbanisation and development of coastal areas, traffic congestion, deterioration of natural resources, and especially climate change are huge concerns. Another stumbling block to tourism development is the effect of diseases such as the COVID-19 pandemic. All these impacts make the local community's wellbeing worse.

Although islands are particularly vulnerable to global change processes, the World Bank predicts that coastal and marine tourism will be the most important added value category in the blue economy by 2030, accounting for 26% of all such categories (Brumbaugh & Patil, 2017). Therefore, it is necessary to develop novel approaches because traditional ways do not effectively deal with current difficulties. This study proposes a SLIQ-based strategy to address the issues raised.

Since 2018, Ha Tien city has increasingly developed tourism activities. Among the many stunning sights that have drawn domestic and foreign tourists, the Hai Tac (Pirate) Islands are one. Even though tourism has brought many benefits to this area, there are still some hidden dangers: an imbalance between the social, economic, and environmental pillars, the risks of climate change, and the unequal distribution of economic benefits among the local population. These challenges are the key reasons this site was chosen for a case study.

In response to the concerns stated by Lee and Jan (2019) and their demand for additional research, I took a qualitative approach to the subject matter. The study employed the SLIQ concept as an innovative method to gain a multidimensional insight into the study of Giannoni and Maupertuis (2007). These authors found that tourism success at island destinations depends on the integrated management of environmental quality, lodging infrastructure, and services that contribute to the overall experience. This research, which

DOI: 10.4324/9781003291763-16

supports the conclusions of Farmaki et al. (2016), continues to delve into residents' perspectives. While the empowerment of local communities is a critical goal of sustainability, there is still a lack of recognition of the influence that residents' perceptions and attitudes have towards their ability to develop tourism sustainably. According to Baldacchino's (1993) suggestion, local people are the significant stakeholders because they play a crucial role as change agents compared to other stakeholders.

## Literature review

### *Tourism and inclusive growth*

The discussion concerning tourism's role in development has shifted from focusing primarily on the sector's involvement in poverty reduction to an approach that emphasises the sector's role in reducing inequality (Hampton et al., 2017; Scheyvens & Biddulph, 2018). While global poverty and inequality levels between countries have decreased, inequality levels inside countries have tended to rise. Inequality within a country can contribute to civil instability, hinder progress, and inhibit development.

Academics and international organisations such as the Asian Development Bank (ADB), World Bank, and United Nations World Tourism Organization (UNWTO) have discussed inclusive growth and development (de Haan, 2015). Inclusive growth deals with policies that allow people from different groups – gender, ethnicity, and religion – and from different sectors – agriculture, manufacturing, and services – to contribute to and benefit from economic growth (de Haan, 2015). Inclusive growth is long-term growth that creates and expands economic opportunities for all citizens (Lee, 2019). Rauniyar and Kanbur (2010) defined it as "progress combined with equal opportunities". In many emerging markets, tourism promotes job creation, improves the population's wellbeing (Snyman, 2012), and gradually opens the door to inclusive growth.

Bakker and Messerli (2017) asserted that when promoting inclusive growth for a place, tourism must generate productive employment and economic opportunities for entrepreneurs while ensuring that all people have equal access to these newly created jobs and opportunities. In light of these considerations, the Sustainable Development Goals (SDGs) contain two goals that specifically address inclusive growth: Goal 8 supports inclusive, sustained, and productive economic growth, while Goal 10 lowers inequality within and between nations.

### *Sustainable island tourism*

Small island governments struggle to compete worldwide in agriculture and production, primarily because they have a limited ability to benefit from economies of scale in these sectors. When it comes to the tourism industry, there are no such restrictions, as many remote islands enjoy a competitive edge because of their natural characteristics, which include a pleasant climate, sandy beaches, and unusual appearance. Many such communities rely primarily on tourism to help them grow their economies and prosper (McElroy, 2003). Even though tourism has helped the economy grow, protected the environment, and made people happier, the industry needs to be aware of and deal with several risks. Many scholars and practitioners have taken an interest in the sustainable approach to tourism in islands as a long-term strategic approach to the industry (Ko, 2005).

In light of the three pillars of sustainable tourism (UNWTO, 2004; UNEP and UNWTO, 2005), environmental, social, and economic considerations appear to be significant in the discussion over island tourism balance. There is no point in having these sustainability qualities unless they are accompanied by the management component of sustainable tourism (Modica, 2015). According to McElroy and Albuquerque (2002), "preserving a steady stream of income by creating an adaptive competitive destination niche market through the continuous guidance of participatory community planning without sacrificing the socio-cultural and natural integrity of the asset base" is the sustainable approach in island tourism.

Sustainable island tourism can therefore be seen as improving the quality of life for island residents, protecting the environment and preserving traditional culture, and designing inclusive and participatory tourism to enhance the community's awareness of sustainable goals.

### Residents' perceptions of sustainability

Sustainable tourism cannot be established without the assistance and involvement of residents, as their opinions on inclusive tourism development are a fundamental condition for sustainability (Woodley, 1993; Gursoy & Rutherford, 2004). Increasing tourism activities on islands can be done with the participation of local people, who benefit financially from tourism, as well as visitors who enjoy high-quality experiences and an increased knowledge of environmental issues (Lee, 2013). Thus, more positive perceptions of tourism among inhabitants will result in more support for community-based tourism growth.

Local attitudes toward tourist development can be influenced by a variety of demographic factors, including age, gender, education, and length of residence (Huh & Vogt, 2008). As a result, tourism planning is seen as a significant influence on their awareness (Choi & Murray, 2010). Environmental sustainability (Choi & Murray, 2010), the local economy (Gursoy, Jurowski, & Uysal, 2002), and the stage of tourist growth (Hunt & Stronza, 2014; Lundberg 2015) are other crucial agents.

### SLIQ approach

First presented by the Vietnam National MAB Committee, the SLIQ technique is used to construct, establish, and manage Biosphere Reserves (BRs). SLIQ stands for Systems Thinking, Landscape Planning, Inter-Sectorial Coordination, Quality Economy. SLIQ provides a scientific framework and technique for creating and deploying BRs in Vietnam to meet its and the UN's SDGs.

- *System thinking (S)*. According to Nguyen et al. (2017), system thinking is the "process of understanding how things influence one another within a whole". It takes into account the intricate and soft relationships that underpin human motivation, behaviour, and emotions to provide a holistic approach to difficult policy and social problems. Systems thinking emphasises cyclical cause and effect relationships rather than linear ones.
- *Landscape planning (L)*. Using the ideas of landscape ecology and systems ecology, Nguyen et al. (2017) explain how landscape planning involves zoning, managing, and employing land, water, and other natural resources in a reasonable manner at a specific locale. Spatially zoned land use and marine resource management are at the core of landscape planning in coastal and island areas. The design method must be based on a

specific geology, geomorphology, soil, biological factors, human factors, traditional use, and Indigenous knowledge.

- *Inter-sectorial coordination (I)*. Stakeholders in management activities are connected through inter-sectorial coordination, which is based on already existing policy structures. Local people's participation from the bottom up, as well as a top-down policy with direction, orientation, and long-standing traditional knowledge, are all part of the solution.
- *Quality economy (Q)*. Conservation-based economies are emerging as a result of the quality economy, which is consistent with the present trend toward green economies and sustainable growth. Other operations are involved, such as trademark registration, marketing, and promoting and enhancing the added value of high-quality local products.

## Methods

### *The context of the case study: The Hai Tac (Pirate) Islands*

The Hai Tac Islands comprise 16 large and small islands and two submerged islands, about 20 km from Ha Tien City and 40 km from Phu Quoc Island. The commune has nearly 500 households, with about 2,000 people living mainly in Hon Tre, Hon Giang, Hon Duoc, and Hon U. The island commune has developed community tourism models, which contribute to Ha Tien City's inclusive growth. With its pristine environment, this place attracts tourists who want to experience, explore, and go on adventures (Figures 14.1 and 14.2).

As a new destination on Vietnam's tourism map, Hai Tac's tourism growth has experienced various obstacles. First, tourism planning is accountable for balancing and enhancing the relationship between social, environmental, and economic pillars. Second, if community members are not adequately informed about the environmental impacts of tourism, their engagement in tourist activities to co-create experiences may harm the environment. Third, climate change affects the existence and development of coastal and island populations.

### *Procedures and respondents*

I selected qualitative research employing a constructionist and interpretive method. The constructionist method supposes that humans are actors who make meaning by interacting with

*Figure 14.1* The landscape of the Hai Tac Islands.

*Source*: Kien Giang Department Of Tourism.

*Figure 14.2* Tourist activities.

*Source*: Kien Giang Province's e-portal

*Table 14.1* Respondent information

| Respondents | Gender | Age | Role/occupation |
|---|---|---|---|
| R1 | Male | Over 30 | Local guide |
| R2 | Male | 27 | Homestay owner |
| R3 | Female | Over 30 | Lodge servant |
| R4 | Female | Over 30 | Motorbike taxi driver |
| R5 | Female | 28 | Homestay owner |

the reality they interpret (Holloway & Hubbard, 2001). In contrast, the interpretive approach prompts consideration of people and their relationships to where they live (Stewart, Hayward, Devlin, & Kirby, 1998). Various approaches, including interviews, participant observations, texts, and discourses, were used to obtain data for this study.

The data collection process was split into three phases. I went to the Hai Tac Islands in 2018 and 2019 to participate in the study as an observer. In 2020, I conducted in-depth interviews with residents. In June 2022, I returned to the Islands to observe and talk with interviewees. As indicated in Table 14.1, I conducted interviews with five tourism workers through open-ended questions. The theoretical foundations of Kantsperger's study (2019) were used to design the interview questions: What do you think of how tourism is growing on the Hai Tac Islands? How does tourism change your daily life, or do you experience its effects? In your opinion, what are the most significant obstacles to the Island's tourism development? What are the most important barriers to local communities getting involved in sustainable tourism? How do locals interact with other stakeholders?

## Findings

### *Challenges for tourism development on islands*

The sustainable island concept ensures long-term economic growth and equity by balancing the environmental, social, and economic pillars. This strategy aims to help the island meet its current and future resource and service requirements without compromising socioecological systems. However, when applying this strategy to the case of the Hai Tac Islands there are many issues that need to be addressed:

> Hai Tac is one of Ha Tien's most picturesque islands, but the locals are poor. Fishing was once the primary means of subsistence here. But recently, tourism has grown. Some families set up homestays. Initially, we had challenges such as feeding travellers and luring them to our homestay. We also had no vacationers due to the rain and storms. Despite the benefits of tourists, we struggled during the off-season .... Even though our place was safe, we lost income due to the COVID-19 pandemic. However, in the next five years, we believe that tourism can be a key sector as the government said .... We heard about sustainable tourism development, and we agreed with this idea. We think that tourism brought money, happiness, and joy, but it also left piles of garbage and water that was not clean. Thus, we need to work together to keep our environment safe and clean.
>
> (R2)

> We agree that sustainable development is the best way for our community. However, we are not clear on how to achieve it. It seems that this concept is not valid if we keep talking about it without action .... These days, tourists come to the Hai Tac Islands a lot, leading to massive waste. We do not have the best way to deal with it yet ... Human resources is also a big challenge for many tiny enterprises like us. We need local government help.
>
> (R5)

People living on islands are confronted with several challenges that limit their chances for long-term development and make them particularly vulnerable to the effects of global warming. They are particularly vulnerable to external financial and economic shocks due to diseconomies of scale, low levels of foreign direct investment, and limited means of harnessing their natural resources sustainably:

> I am proud of my beautiful island, and I would like to share its history with many people. This is the reason I became a tour guide. However, due to water and electricity resources restrictions in daily life, tourism activities are still a little boring ... More rain and storms are affecting tourism activities. More importantly, the COVID-19 pandemic pushed us to leave our job. Though I love my job, it is not stable. There are some risks in the tourism industry, particularly in island areas like Hai Tac, where it is not easy to access ... I hope in the future infrastructure connection will be better. It also helps us to reduce trash.
>
> (R1)

When considering the long-term viability of an island, it is impossible to ignore the complicated scenario created by the COVID-19 outbreak. The economic and societal consequences of it are expected to be catastrophic. Island economies that are heavily reliant on tourism will face significant difficulties. Local governments must consider the policies that affect their inhabitants' livelihoods. Tourism for inclusive growth needs poverty strategies:

> Since our island has more tourists, I did this job. I hosted visitors for some guest houses and earned a little income. This job is easy, but the income is also good. I feel happy when meeting many people. Sometimes, I hosted foreign tourists; although I did not understand their language, I did not feel unfamiliar …. COVID-19 affected our lives, but I thought it would pass, and our daily lives would go back to previous regular days. In the future, Hai Tac will host more guests.
>
> (R3)

> Tourism gives me a meaningful life. Since tourism activities were developed on my island, I have had a plus job to pay my children's educational fees. However, when COVID-19 occurred, my family's life fell into struggle. I have to find another job …. Now, besides working here temporarily, I am still working at another place to have enough money to cover our daily lives.
>
> (R4)

> Thanks to tourism activities, my family economy is gradually increasing. However, tourism cannot help many poor people due to some obstacles from them such as lack of knowledge, lack of will, lack of orientation …. The poor still need things that bring benefits in front of them. They don't see the long-term value.
>
> (R5)

### *Island tourism for inclusive growth*

Oriented to developing sea and island tourism towards eco-leisure, the government expects the Hai Tac Islands to become a prominent destination, creating a difference for local tourism products. Accordingly, the island will be designed with many different subdivisions, including a theme park based on the island's pirate story. It is predicted that, by 2030, the Hai Tac Islands will become a green tourist destination, with completed infrastructure services, particularly waste treatment systems. The number of tourists visiting this place will be controlled based on ecological capacity. By 2040, the island targets being a sustainable tourism hub, and local people will become a green citizenry that consumes and wastes less, achieving a green economy and adapting to sea-level rise.

With the above orientations, tourism has the potential to directly or indirectly contribute to the achievement of the SDGs. Sustainable tourism creates local jobs and revenue that can be used to alleviate poverty, support entrepreneurial endeavours, and empower all women and girls to find their voices. Local people will become suppliers of tourist products that create linkages with other parties. Their connection has the potential to grow the local supply chain and spur more local innovation and new enterprises in the tourism industry, particularly among low and middle-income individuals, leading to the achievement of social equality.

Stakeholders agree that actions taken by households or individuals could help make tourism more sustainable. These actions include proper trash disposal and recycling, clean-up campaigns, and saving energy. All of these are related to Goal 6 (clean water and sanitation), Goal 7 (affordable and clean energy), and Goal 11 (sustainable cities and communities).

Several participants identify the model of community tourism development as a cost-effective approach to sustainable tourism. The Hai Tac Islands have already successfully implemented this. Community tourism can contribute to the majority of SDGs, particularly Goal 11 (sustainable cities and communities) and Goal 12 (responsible production and consumption). The island community seeks to preserve a green environment through active involvement.

### The SLIQ approach to sustainable tourism for inclusive growth

To administer islands effectively, we must think of them as living beings rather than just pieces of land that can be divided into sectors. Public policy should achieve three objectives: economic growth, environmental conservation, and social equity. Therefore, the study looked for a new way to handle existing challenges. According to the findings, the SLIQ concept can assist in tackling future issues.

First, system thinking is critical in planning and managing a destination, particularly island locations, because it enables the development of a holistic understanding of concerns, contributing to Goal 11 (sustainable cities and communities). According to Hjorth and Bagheri (2006, p. 76), one of the most significant benefits of systematic thinking is that:

> meeting sustainability objectives will certainly require an increased understanding of the interactions between nature and society. Issues about sustainability are not merely complicated, they involve subsystems at a variety of scale levels and there is no single privileged point of view for their measurement and analysis. Such problems can neither be captured nor solved by sciences that assume that the relevant systems are simple.

Thus, when maximising the tourism development potential of the Hai Tac Islands, local authorities should prioritise this factor.

Second, landscape planning is closely tied to land resources and the practical usage of their physical and biological surroundings. Furthermore, landscape planning is a way for managing the current status of the environment and its value and sensibility. It is the process of determining long-term development plans for a region, taking into account natural and man-made impacts on the land. The final part of this process is the development of an action plan with recommendations for territorial landscape management. All factors affecting island tourism development like land, climate and meteorology, hydrology and hydrogeology, landscape potential, and socioeconomic conditions should be thoroughly analysed. They are a vital tool for achieving Goal 13 (climate action), Goal 14 (life below water), and Goal 15 (life on land).

Third, cross-sector collaboration is becoming increasingly important. The importance of local cooperation and creativity should be highlighted in tourist planning and growth. Indeed, the local government's desire to regulate and protect their own backyard may create opportunities for sustainable sorts of tourism to help establish innovative and resilient island communities (Hampton & Jeyacheya, 2020). Tourism's alignment with the long-term development of other sectors of an island's economy, as Dodds (2007) points out, is widely documented as a vital part of the overall long-term outcome. Cooperation and coordination across sectors and levels are critical for successfully implementing sustainable tourism. Local governments must encourage inhabitants' participation in the island's development as a whole. This contributes to Goal 8 (decent work and economic growth) and Goal 9 (industry, innovation and infrastructure).

Fourth, the quality economy is a driving force behind an island's regional and worldwide expansion. It is reflected in the concept of a blue economy. According to Hampton and Jeyacheya (2020), the blue economy aims to preserve the benefits of the expanding ocean economy while developing it responsibly to secure the sustainable use of the ocean's resources and increase wellbeing and equity in coastal and island societies. Tourism development plays an essential part in the blue quality economy because it can promote Indigenous peoples' wellbeing and tourist satisfaction, positively affecting Goal 8 (decent work and economic growth) and Goal 10 (reduced inequalities).

## Conclusions and limitations

Tourist growth on an island is difficult to study because of its dynamic nature, including the ecological, sociological, experiential, and mystical features. As a result, island tourism development policies must be founded on sustainable criteria to overcome challenges such as climate change, environmental degradation, economic downturns, and cultural erosion. Four sustainability concepts, including economic growth, environmental protection, cultural promotion, and good governance, should be taken into consideration during the tourism planning process, following Gurung and Seeland (2008). To ensure that the destination can keep pace with demand, sustainability planning is essential (Uysal & Modica, 2016).

Many academics are now looking for fresh and inventive approaches to the difficulties previously addressed in this new setting. In this chapter, I have used SLIQ. There should be a framework for academics and practitioners to critically plan long-term tourist policies based on system thinking, landscape planning, and inter-sectorial collaboration. SLIQ's approach will be critical in dealing with hidden threats in the future and supporting inclusive growth.

Though the SLIQ method has been successfully employed to preserve regions, this is the first time it has been used to build and design tourism islands to address future challenges. Therefore, certain restrictions must be addressed in future projects. First, the method should be thoroughly examined from the perspective of local governments. While considering tourism as a spearhead economy in accordance with national tourism development goals, they must consider how tourism relates to and connects with other sectors. Future research should conduct comprehensive examinations of these interactions. Second, support for sustainable tourism growth should be elicited from communities that rely on tourism activities and those that do not. Third, the host–guest interaction must be evaluated to co-create sustainable tourism experiences. To this end, I advise that future scholars should expand on the issue of inter-sectorial coordination.

## References

Bakker, M., & Messerli, H. R. (2017). Inclusive growth versus pro-poor growth: Implications for tourism development. *Tourism and Hospitality Research*, *17*(4), 384–391.

Baldacchino, G. (1993). Bursting the bubble: the pseudo-development strategies of micro-states, *Development and Change*, *24*(1), 29–51.

Briguglio, L. (2008). Sustainable tourism in small island jurisdictions with special reference to Malta. *Journal of Travel Research*, *1*, 29–39.

Brumbaugh, R., & Patil, P. (2017). Sustainable tourism can drive the blue economy: investing in ocean health is synonymous with generating ocean wealth. *World Bank Blogs*. Available at https://blogs.worldbank.org/voices/Sustainable-Tourism-Can-Drive-the-Blue-Economy

Choi, H. C., & Murray, I. (2010). Resident attitudes toward sustainable community tourism. *Journal of Sustainable Tourism, 18*, 575–594.

de Haan, A. (2015). Inclusive growth: More than safety Nets? *European Journal of Development Research, 27*(4), 606–622.

Dodds, R. (2007). Sustainable tourism and policy implementation: Lessons from the case of Calviá, Spain. *Current Issues in Tourism, 10*, 296–322.

Farmaki, A., Altinay, L., & Yaşarata, M. (2016). Rhetoric versus the realities of sustainable tourism: The case of Cyprus. In *Sustainable Island Tourism: Competitiveness, and Quality-of-Life* ed P. Modica and M. Uysal (Boston, MA: CABI), pp. 35–50.

Giannoni, S., & Maupertuis, M-A. (2007) Environmental Quality and Optimal Investment in Tourism Infrastructures: A Small Island Perspective. *Tourism Economics, 13*(4), 499–513.

Gursoy, D., Jurowski, C., & Uysal, M. (2002). Resident attitudes: A structural modeling approach. *Annal of Tourism Research, 29*, 79–105.

Gursoy, D., & Rutherford, D. G. (2004). Host attitudes toward tourism an improved structural model. *Annal of Tourism Research, 31*, 495–516.

Gurung, D. B., & Seeland, K. (2008). Ecotourism in Bhutan: Extending its benefits to rural communities. *Annals of Tourism Research, 35*(2), 489–508.

Hampton, M. P., Jeyacheya, J., & Long, P. H. (2017). Can tourism promote inclusive growth? Supply chains, ownership and employment in Ha Long Bay, Vietnam. *The Journal of Development Studies, 54*, 359–376.

Hampton, M. P., & Jeyacheya, J. (2020). Tourism-dependent Small islands, inclusive growth, and the blue economy. *One Earth, 2*, 8–10.

Hjorth, P., & Bagheri, A. (2006). Navigating towards sustainable development: A system dynamics approach *Futures, 38*, 74–92.

Holloway, L., & Hubbard, P. (2001). *People and Place: The Extraordinary Geographies of Everyday Life* (Essex: Pearson Education).

Huh, C., & Vogt, C. A. (2008). Changes in residents' attitudes toward tourism over time: A cohort analytical approach. *Journal of Tourism Research, 46*, 446–455.

Hunt, C., & Stronza, A. (2014). Stage-based tourism models and resident attitudes towards tourism in an emerging destination in the developing world. *Journal of Sustainable Tourism, 22*, 279–298.

Kantsperger, M., Thees, H., & Eckert, C. (2019). Local participation in tourism development—Roles of non-tourism related residents of the alpine destination bad reichenhall. *Sustainability, 11*, 6947.

Ko, T. G. (2005). Development of a tourism sustainability assessment procedure: A conceptual approach. *Tourism Management, 26*(3), 431–445.

Lee, N. (2019). Inclusive growth in cities: A sympathetic critique. *Regional Studies, 53*(3), 424–434.

Lee, T. H. (2013). Influence analysis of community resident support for sustainable tourism development. *Tourism Management, 34*, 37–46.

Lee, T. H., & Jan, F. H. (2019). Community-based tourism contribute to sustainable development? Evidence from residents' perceptions of the sustainability. *Tourism Management, 70*, 368–380.

Lundberg, E. (2015). The level of tourism development and resident attitudes: A comparative case study of coastal destinations. *Scandinavian Journal of Hospitality and Tourism, 15*(3), 266–294.

McElroy, J. (2003). Tourism development in Small Islands across the world. *Geografiska Annaler. Series B, Human Geography, 85*, 231–242.

McElroy, J. L., & Albuquerque, K. (2002). Problems for managing sustainable tourism in small islands. In *Island Tourism and Sustainable Development: Caribbean, Pacific, and Mediterranean Experiences* ed Y. Apostolopoulos, & D. G. Gayle (Westport, Connecticut: Praeger Publishers), pp. 15–31.

Modica, P. (2015). *Sustainable Tourism Management and Monitoring. Destination, Business and Stakeholder Perspectives* (Milan: Franco Angeli).

Nguyen, H. T., Nguyen, V. T., Le, T. T., & Vu, T. H. (2017). *Hướng dẫn kỹ thuật: Cách tiếp cận SLIQ và thực hiện vốn xã hội, vốn doanh nghiệp trong các khu sinh quyển thế giới của Việt Nam (Technical guide: SLIQ approach and implementation of social capital, social enterprise in Vietnam's world biosphere reserves)* (Hanoi: MAP).

Rauniyar, G., & Kanbur, R. (2010). Inclusive growth and inclusive development: A review and synthesis of Asian Development Bank literature. *Journal of the Asia Pacific Economy*, *15*(4), 455–469.

Scheyvens, R., & Biddulph, R. (2018). Inclusive tourism development. *Tourism Geographies*, *20*(4), 1–21.

Snyman, S. L. (2012). The role of tourism employment in poverty reduction and community perceptions of conservation and tourism in Southern Africa. *Journal of Sustainable Tourism*, *20*(3), 395–416.

Stewart, E., Hayward, B., Devlin, P., & Kirby, V. (1998). The "place" of interpretation: A new approach to the evaluation of interpretation. *Tourism Management*, *19*, 257–266.

UNEP (United Nations Environment Programme). (2014). *Emerging Issues for Small Island Developing States. Results of the UNEP/UN DESA Foresight Process* (Nairobi: UNEP).

UNEP and UNWTO (United Nations Environment Programme, and World Tourism Organization). (2005). *Making Tourism More Sustainable. A Guide for Policy Makers* (Paris and Madrid: UNEP and WTO), pp. 11–12.

UNWTO. (2004). *Indicators of Sustainable Development for Tourism Destinations: A Guidebook* (Madrid: UNWTO).

Uysal, M., & Modica, P. (2016). Island Tourism: Challenges and Future Research Directions *Sustainable Island Tourism: Competitiveness and Quality of life* ed P. Modica & M. Uysal (UK: CABI), pp. 173–188.

Woodley, A. (1993). Tourism and sustainable development: The community perspective. In *Tourism and Sustainable Development: Monitoring, Planning, Managing* ed J. G. Nelson, R. Butler, & G. Wall (Waterloo: University of Waterloo, Heritage Resources Centre), pp. 135–146.

# 15 Community-based rural tourism development

## An intersectional exploration of possibilities and challenges

*Neha Nimble*

## Introduction

This chapter focuses on the analysis of community-based tourism's (CBT) impact on host communities' livelihoods and explores the various ways in which community-based rural tourism (CBRT) constructs and transforms the unequal power relations of gender, class, and caste in the community. "Who is the community in CBT?" By exploring the challenges and opportunities that CBT presents for the development of various groups in the community, the chapter critiques the uncritical implementation of the principles of CBT and/or CBRT projects in host communities.

The chapter focuses on the implementation of a CBRT programme, the Endogenous Tourism Project (ETP) under the Rural Tourism Scheme (RTS) in Naggar, one of 36 rural tourism sites in India. I build on a conceptual understanding from livelihood discourse and adapt the International Fund for Agricultural Development (IFAD) sustainable livelihoods framework to include gendered and intersectional lenses to capture the role of patriarchy and caste in determining the impacts of tourism on the livelihood assets of community households.

The chapter describes community members' lived experiences of tourism development through ETP and its impact on their livelihoods using a feminist and intersectional phenomenological research paradigm. The analysis is drawn from primary data collected from community members and key implementers of the project through interviews, observation, and a review of government documents.

## Community-based tourism development: Significance and critique

For more than two decades, sustainable forms of tourism development, popularly called "new tourism" (WTTC, 2003), have been encouraged to address the negative socio-economic, cultural, and political impacts of mass tourism activities on host communities. New sustainable forms of tourism call for small-scale, community-based, and community-driven tourism development, especially in rural areas. Even though such tourism leads to less developmental contribution, the costs are less too (Timothy, 2002). Therefore, in the last two decades, tourism planning and policy around the world and in India have adopted the principles of sustainable tourism. By enhancing greater community control and integration of the local economy through the use of local resources, livelihood activities, and services, they seek to achieve the objectives of development, empowerment, and self-reliance of host communities, especially of those marginalised therein (Telfer & Sharpley, 2008).

DOI: 10.4324/9781003291763-17

CBT is a local, new, sustainable tourism approach, and is developed in local communities in innovative grassroots ways by different groups of individuals, and corporate, governmental, or non-governmental organisations (Hatton, 1999). While CBT is promoted primarily for its socially sustainable ways of development, it is also linked to the strengthening of culture, the promotion of employment, and empowerment opportunities for women and the most marginalised in the community (Hatton, 1999).

Responding to the global call for adopting CBT approaches, India adopted CBT for community development with a focus on the development of women and the marginalised, especially for tourism expansion in existing rural tourist destinations. CBT is a desired form of tourism development in India as an increasing number of urban and international tourists are willing to experience the local, rural culture, traditional lifestyles, and environment of unfamiliar communities. Indian tourism policy in the early 2000s put a thrust on small and rural segment tourism. Relatedly, the RTS was conceptualised and implemented in various parts of India with the goals of sustainable tourism development (Ministry of Tourism, Government of India, 2011). As claimed by the Ministry of Tourism, Government of India, the scheme aimed at the economic objectives of creating sustainable livelihoods through community-based actions and also at the empowerment of women, marginalised communities, and gender equality through a convergence of economic and social issues (Ministry of Tourism, Government of India, 2011). The benefits sought included a fair distribution of stable employment and income-earning opportunities and social services among host community stakeholders. Key geographical sites were identified with the potential for tourism to promote rural tourism and spread its benefits to the communities.

Since then, a number of studies, globally, have revealed that sustainable CBT development also has many challenges, especially in developing countries with heterogeneous communities. These include issues of distribution of power, control, ownership, and benefits among community members (Mowforth & Munt, 2009). Studies have found that community participation in most cases has been uncritical. The concerns and needs of the locals (except local elites) have not been sought. The projects have been thrust upon communities after being designed in the centres of power with tourism only creating islands of affluence for the elite while marginalising the majority of others with limited resources and power (Hatton, 1999). Despite the empowerment of women being at the centre of CBT, research has shown that practices empower as well as disempower women through CBT (Nara & Irawan, 2020; McCall & Mearns, 2021).

Overall, the research around new forms of tourism has emphasised that unequal power structures have implications on the sharing of costs and benefits within the community. While CBT's possible negative consequences are increasingly and continuously under scrutiny, the research also continues to appreciate the developmental potential of tourism if developed equitably and with the critical involvement of people.

This chapter brings evidence from India to describe how uncritical implementation of CBRT, without a conscious examination of intersectional concerns of gender and caste inequalities, can further marginalise the disadvantaged and consolidate structural inequalities.

## Tourism development in Naggar

### *History and growth of tourism in Naggar*

Naggar is a Gram Panchayat village in the Kullu district of the state of Himachal Pradesh in India. This is a formal democratic local institution for self-government and village administration in India. A Gram Panchayat village often comprises a group of small villages. It offers a rich, natural, and cultural heritage and is home to many buildings and temples of historical importance, and tourism has been a part of its socio-economic and cultural history for centuries. Naggar was a centre of British administration during the period 1845–1910. Tourists have been coming here since pre-colonial times and it is a preferred destination for foreign tourists who seek to experience the nature, culture, and traditions of the Indigenous community. People are primarily engaged in horticulture, agriculture, livestock rearing, and tourism-related economic activities for their livelihoods.

Naggar is a small Himalayan village with beautiful hills and places of religious importance for Hindus (Figure 15.1). The welcoming nature of the host population, the tradition of hosting, the ethnicity, the peculiarly Himalayan ecology, heritage buildings, and colourful fairs are some of the motivating aspects for tourism in Naggar. It is also at a strategic location, offering a starting point for a number of different treks into the higher Himalayas. Being equidistant and close to the towns of Kullu and Manali, this village offers the services of a town as well as the calm of a rural site. People come to enjoy experiences ranging from paragliding to skiing, river rafting, stone carving, wood carving, basket weaving, silver metal working, weaving, knitting, achar chutney making, and wool making.

*Figure 15.1* Himalayan scenery surrounding Naggar.

Photo: Courtesy of Unsplash.com.

The small-scale tourism in Naggar is basically comprised of informal economic activities and therefore, until now, there has been no documentation of how much tourism contributes to the local economy. Even though almost everybody in the village is directly or indirectly engaged with some or other tourism activity, the strong forward and backward linkages with local economic activities make it difficult to say how much tourism in itself actually contributes to the local economy or to the economies of households.

With its specific rural ways of living, heterogenous community, close association with tourists, and a long history of tourism, Naggar offers a rich site to understand and analyse tourism impacts on the livelihoods of the community from a gender-aware and intersectional perspective.

### Endogenous Tourism Project: State intervention for tourism development in rural tourism sites

The most significant and critical development with respect to tourism development happened in the form of the implementation of ETP. Naggar has a small-scale kind of tourism which ensures maximum contact between tourists and locals, the former who come in search of unexplored cultures and places. Considering its tourism potential, it was declared a rural tourism site and the ETP under RTS was implemented from 2005 to 2010 after which tourism development was handed over to the community.

This has been by far the biggest organised tourism development and livelihood generation intervention from the government in the village. ETP is primarily a locally and community based, non-institutionalised, and low-scale tourism development project. For implementation of ETP, key geographical sites were identified with the potential to promote rural tourism and spread its benefits to the communities. For this, the Ministry of Tourism joined hands with the United Nations Development Programme (UNDP) for funds and capacity building. The major activities under ETP included infrastructure development, skill and capacity building programmes, the formation of an implementing committee, orientation visits to other similar sites, the provision of subsidised loans, and training and other materials for gainful work.

The work of the project was done under two components: hardware and software. The software component focused on capacity building and training and was taken care of by the UNDP (through implementing the partner NGO SAVE). The UNDP acted as a supplier of financial capital and human resources for investment in tourism development in Naggar considering the limited government spending. The work done under this component included the composition of a Village Tourism Development Committee (VTDC), orientation visits to other ETP sites in the country, the formation of Self-Help Groups (SHGs), and gender sensitisation programmes. Many other capacity building and training programmes were provided to local men and women in hygiene, hospitality, history, tradition, heritage, toilet building, guides, trekking, tailoring, weaving, embroidery, and cooking. The second component, hardware, was directly managed by the Department of Tourism, Himachal Pradesh and was implemented by the VTDC. Some work was later deployed to Gram Panchayat as the local administrative body collaborated in the implementation of the project. The work done under this component included the construction of roads, buildings, provision of trekking equipment to youth groups, renovation of heritage buildings and temples of historical and tourist importance, and the implementation of a home-based stay scheme.

## Findings

### *Implementation of ETP: A critical assessment*

Under the RTS-ETP, tourism development is a community-based, community driven initiative. It is based on the principles of CBRT which aim to empower the marginalised and develop the rural community through tourism development. It involves local people in collective action and organised community groups for managing tourism in their community (Ashley, 2000).

Despite such principles of rural tourism within which ETP was conceived and implemented in Naggar, it was found that the project was thrust upon the community and that tourism was not really owned or driven by it. The findings from the field agreed with Fillmore (1994) that communities had not been consulted about the use of their resources and the appropriation of their culture in the new tourism. The process of conception, planning, and implementation was essentially and practically top down, and according to community members, there was no preliminary needs assessment survey or any effort to assess and address the composition, concerns, and expectations of the community.

Interestingly, in Naggar, all residents supported the proliferation of tourism related economic activities as they have traditionally and historically been hosts to a number of foreigners and domestic tourists for more than a century. Tourism development in the community, thus, received very little resistance. This also, again, does not mean that the whole community was in charge of this tourism development, or that the whole community was impacted equally, positively, or negatively, for that matter.

As we seek the truth of the promises made to the community by CBRT in Naggar, an important question to consider is who controls this community-based tourism and whether the benefits from tourism go to the local people and which groups of people. Hall (2000) rightly noted that the emphasis on local, bottom-up approaches as associated with sustainable development seems to derive its value from a flawed assumption of the cohesion of local communities. Communities are not homogeneous. In Naggar, they mean a very heterogeneous group of women and men who belong to different caste and class groups. People have multiple identities and these identities (gender, caste, class) intersect with each other to create specific interconnected identities of people which determine their individual privileges and vulnerabilities. However, the project failed to acknowledge and address these multiple identities and differential consequent impacts that any development programme has on a community.

Accordingly, the impact of ETP on people's lives and livelihoods only consolidated existing structural inequalities. The involvement and participation of the local community was done in a very uncritical manner. The implementing agency NGO SAVE chose VTDC members on the recommendation of two influential (dominant caste and class) people in Naggar who were friends of the representative from the NGO. These two were automatically chosen to be President and Secretary of the committee. Further, out of the total 15 VTDC members, five (four men and one woman from dominant caste groups) were nominated to do most of the work as these are the ones who are politically powerful owing to their social and economic standing in the community.

In addition, there was very little awareness about the implementation of ETP in the community. Barring a very few (about 40 in all), women and men did not know about ETP or VTDC. Of those who knew, most of them were VTDC members and those who had participated in some training programme. People belonging to dominant caste and class groups were better aware of the project. Thus, it cannot truly be called a community-owned initiative. In this community-owned project, the VTDC represents, or it would be better to say means, the community.

Participatory activities and meetings as part of the project were also conducted in a very technical manner, as Mowforth and Munt (1998) had found in the case of community-based development programmes. The participation of women and socially and economically backward people in tourism in Naggar had been uncommon and only for the sake of completing the requirements of the papers. There was a clear caste bias while conducting meetings and in relation to the participation of women and people from the *koli*, *haassi*, *luhar*, and *chamar* castes. The nominal vice president of the VTDC resentfully remembers:

> Inclusion of caste people in any decision-making body or meeting is a farce. Their participation is mostly just for paper work. They feel they are favouring us if they allow us to sit next to them in meetings. One can easily sense their disgust at the forced physical proximity for those hours or minutes. But then, we don't feel humiliated or anything, we are used to this treatment. Let them be.

Women were happy enough to be called for meetings as they have traditionally learnt to be decided for. In VTDC, they were functional members, for they went to support the dominant caste's male view.

Under such participation and involvement, most in the local communities reaped few direct or significant benefits from tourism because they had little control over the ways in which the activities and enterprises were developed. The project was conceptualised in the centres of power in Ministry offices and given to the selected few in the community to run. It is these very few local elites that reaped the rewards of the community and derived gains from such tourism development.

## Impact of CBRT on a community's livelihood assets

This chapter assesses the impact of CBT development through studying the impact of ETP on people's livelihood assets. Households and individuals access a complex range of resources and assets in search of their livelihoods. These assets (financial, physical, human, natural, personal, and social) are the building blocks on which people build their livelihood activities (Ashley, 2000). Tourism through ETP essentially has brought to Naggar a transformation in livelihoods. Due to tourism, the community has seen a change in the availability and use of household assets. The main generalised impacts of tourism on the local community are summarised below.

### *Physical assets*

Improvement in the community's physical assets is a definite, direct, and positive outcome in Naggar. Building and renovation work, street lighting, new road connectivity, and the development of parking spaces have provided a better infrastructure, better connectivity to roads, better transport facilities, and improvement of housing and other buildings. The needs of tourists have resulted in better means of communications. Locals working as tourist guides, in adventure tourism, and as cultural performers have newer tools and equipment. Naggar is now better connected within and outside to the bigger towns of Manali and Kullu. This has positive consequences for the livelihood options of locals.

*Natural assets*

There has been a change in the use and management of natural resources in Naggar due to tourism proliferation. The opening up of small scale but private resorts and hotels has restricted the access of locals to certain parts of land which were earlier open to public use. There is increased competition for access to such natural resources. Despite the CBT development, the management of natural resources is not within the hands of the collective community but rather left to individuals and families in a scattered and unsure manner. There is increased competition and unrest over access to land, making it expensive. There is decreased access to common natural resources such as water from streams.

The most important change in access to natural assets is seen in the case of access to land resources in Naggar. With the increase in land prices, people have also been converting their land into other forms of productive asset like homestays, hotels, and restaurants. At the same time, this has negative implications for people with small land holdings as they find it difficult to buy or rent land to gain benefits from new tourism businesses.

*Social assets*

While tourism development under ETP focuses on physical assets, the development of social assets is a by-product. However, the extent to which tourism development has strengthened the social assets of Naggarians is under doubt, considering the unequal power relations and unequal sharing of the benefits of tourism opportunities. Social assets have been affected positively as well as negatively.

Despite caste divides, Naggar has traditionally had strong social networks. People rely upon access to networks of relationships within their caste and other informal organisations outside their households to arrange their livelihoods. For example, for arranging family functions, people rely on temple-based organisations for the provision of materials. The new form of economic engagement in tourism has impacted social networks and community organisation, its rules, norms, and other aspects. The older social relationships and thus related networks are changing and have become weak in terms of reciprocity and mutual exchange.

The principle of community participation in tourism development, despite its uncritical implementation, has also changed and improved people's social assets in many ways. Newer political relationships are also being formed. For example, a women's self-help group in one of the small villages in Naggar has given them a sense of sisterhood from which they derive not only financial support but also share their life experiences to seek possible solutions to their problems. This collective between these women from the same place is a new thing, a result of ETP.

In an emulation of tourist behaviour and general openness in attitude brought about by tourism, a few young men and women have begun to have informal affiliations with people from other castes. This attitude is also seen in a newer openness to making friends with the *kolis*, *haasis*, *chamars*, and *luhars*. There is also increasing scope for formal and informal organisation among village persons for enhancing tourist arrivals.

Social assets also include culture and pride. Rural tourism promoted in Naggar celebrates the cultural ways of people. Tourism has affected people's culture in two ways. It has celebrated it and attracted people's attention to it. Local people take pride in their culture as they find tourists taking an interest in their clothes, cuisine, and ways of living. At the same time, their culture faces erosion as the modern ways of tourists have begun to colour the way youths live. While traditional crafts and skills continue to be appreciated, people's ways of living are steadily being replaced by the ways of city people. This has threatened the sustainability of the authenticity of the local culture in Naggar.

Change in social assets also manifests itself in negative ways. Implementing ETP, which is essentially a social and economic partnership, has given rise to conflicts between VTDC and Gram Panchayat, and resentment between various groups within the community. This is a result of the way in which ETP has been implemented with unequal and non-transparent practices of provision of products and services. This has led to an increase in distrust as well as social differentiation. With newer social organisations, there are conflicts over new power relations, as in the case of VTDC where a huge amount of discontentment is seen among villagers over the benefits that leadership brings. Tourism has resulted in newer economic and political benefits to some, but also in feelings of jealousy, hatred, and unrest among people over the distribution of these benefits. Newer class formations, changing economic status, and the increasing economic independence of traditionally low income and status groups like scheduled caste (SC) households has led to an increase in unrest between the dominant caste groups and the *kolis*, *haasis*, *chamars*, and *luhars*. This has resulted in a decrease in traditional patronages and mutual trust. VTDC and Gram Panchayat have also suffered conflict over the distribution of funds and work under ETP, thus giving rise to diminishing faith in social organisations. Such a lack of faith in social bodies is a new thing in Naggar where people use to have faith despite and within the strong power relations and their fair/unfair manifestations in their lives.

Cumulatively, certain persons, individually and as part of VTDC, have gained recognition and better social networks in the community. Through a few scattered meetings, people's desire to develop their tourism businesses are becoming more organised and seek support and inspiration from other groups and individuals.

### Financial assets

Tourism money has had a multiplier effect on the local economy. Tourism has definitely increased people's financial assets. While it relies on them, it has helped people gain access to and avail themselves of better financial assets. Due to the additional work opportunities that tourism provides to different earning members in the household, total extra household income has improved in Naggar. Schemes like the Homestay Scheme provides loans at subsidised rates to invest in converting traditional houses into homestays and make livelihoods. Tourism offers newer mechanisms of investing for profit and there is also increased return for the traditional economic activities like selling milk and wool. However, while there is increasing extra money, there are newer financial burdens on the community as a result of tourism development. People have to pay increased prices for land and other commodities of daily use. There is a burden of expensive lifestyles as people emulate tourists and their neighbours.

### Human assets

Various training, skill, and capacity building programmes were provided to enhance the human assets of the local community. Women and young men were trained in many activities that may help them make livelihoods through tourism. Training was given in languages, knitting, weaving, tailoring, trekking, cooking, and hospitality. People also learned through orientation visits to organise themselves and be professional hosts. The trainers were chiefly chosen from the village, thus increasing their leadership and training abilities and confidence. At the same time, with increases in financial assets, there is a significant increase in access to education and health facilities. On the downside, there is decreasing traditional knowledge with respect to wild natural foods, medicines, and weaving. With tourism offering quick money, there is decreasing will among youth to go for higher education after schooling.

### Personal assets

As tourism offers opportunities for economic, political, and social betterment, personal assets become very important for making use of such opportunities. Tourism and tourists have had a definite and positive impact on people's personal assets in Naggar. There is an increased will and determination to live a free life as people see tourists and their worlds through them. With newer opportunities, there are increased desires, dreams, and a faith to pursue them. Despite many cultural norms binding people's participation, women and other marginalised people are pushing for their dreams and have begun to participate in their fulfilment as they participate in tourism. This has meant that certain, though few, individuals, women, and SC men have gained prominent recognition in the society.

Tourism's complex, direct, and indirect impacts are thus not seen only in terms of contributions to cash income activities but also on people's existing livelihood assets. ETP has led to an increase in access to financial, human, personal, and physical assets while social assets and natural assets have been increasingly eroded. People are using a diverse array of assets in different amounts and trying to find a balance which might yield the best situation. These transformations in access to and availability of assets are not uniform and are distributed unevenly within the community.

## Discussion: An intersectional exploration of the community's experiences of ETP

### Gendered and intersectional impacts of CBRT

Naggar, like most rural societies, has inherited historically embedded and strengthened practices and patterns of differences and disadvantages. These have led to the creation of various kinds of exclusionary practices on the basis of caste, class, and gender. Such practices and oppressive relations of caste, class, and gender, created and maintained by the institutions of religion, patriarchy, and new forms of capitalism, have determined a social exclusion and resulted in restricted choices of livelihood strategies to many groups of people in Naggar.

The impacts of tourism have been non-uniform and gender and caste as categories of identity determine the exclusion of many from the positive impacts and benefits of tourism. Gender and caste come out as categories of social exclusion that are based in cultural and economic disadvantages. Both structure people's access to land, labour, rights, entitlements, and occupations in Naggar in favour of upper caste men and determine the distribution of labour, productive and reproductive freedom, and resources in favour of upper caste men. The dominant traditional and cultural values degrade women's and lower caste groups' identity, labour, as well as morality. This has negatively affected these groups' participation in tourism work. The cultural ways of valuing and undervaluing certain gender and caste identities have meant the exclusion of women and 'lower' caste groups from economic resources, assets, opportunities, and occupations.

Tourism mediates between the institutions, structures and processes of patriarchy and casteism, and exacerbates the social exclusion of these marginalised groups in many ways and cases. Tourism processes and activities are based in a gendered and casteist society and relations. They are not only informed by these but also inform them – to construct a gendered and casteist participation of these under-privileged groups. While participation in tourism work has meant only a small extra income for all, it has also manifested as the social exclusion of women and 'low' caste and class groups in Naggar. These include: political exclusion from decision making, unequal access to the market, unequal access to productive and investment resources, gender and caste based occupational segregation, exclusion from free and fair access to training and opportunities which are usually caste and gender based, and exclusion from access to and use of common and natural resources. This has meant that the costs of tourism are shared more by women and disadvantaged groups while upper caste males have shared most benefits of tourism. Tourism does offer potential and the possibility to offset such exclusionary practices but requires critical implementation, which is missing as of now.

Gender is a key category in Naggar that structures women as below men, economically as well as socially. The dominant values of the society often create injustice in access to economic resources and the valuation of women's identity. Anything associated with women in this particular society is considered to be of low value and commanding little appropriation.

Caste is another significant category of structuring economic, cultural, and social exclusion in Naggar. It is rooted in religiously ordered and sanctioned economic disadvantage, ordering, and segregation of occupations. The castes belonging to the lowest rung in the hierarchy of caste in the community are restricted into the least rewarding and most stigmatising occupations which are devalued in society (Kabeer, 2000). They form in Naggar, as in many societies, a socially despised and exploited caste. Practices of social exclusion and discrimination are socially approved. Inter-dining is prohibited as these people belonging to the castes in the lowest rung are considered to cause pollution for others through direct or indirect contact (Nayak, 1994). This has implications for the spatial, geographical, as well as societal segregation that these groups find themselves facing. They are found at the margins of not only the village but also on the margins of progress, growth, and empowerment.

*Tourism and social exclusion*

Tourism is an important source of livelihood for people after horticulture which is the primary source of income in Naggar. With its forward and backward linkages with already existing small economic activities, like transportation, milk and vegetable selling, and souvenir making, people have always taken up tourism very positively and hopefully. The increasing economic diversification is proving to be a good strategy for expanding the already existing livelihood options, though there are definitely only a few new tourism occupations that ETP has introduced into the community for people. Thus, rural tourism may fulfil its promise of providing an additional income stream to rural communities and become a vehicle for the development and support of other rural economic sectors (Sharpley & Sharpley, 1997) in Naggar.

However, it is difficult to say that CBRT has optimised the socio-political and economic benefits to all and in equal measures. Fulfilling its responsibilities in this domain has been inadequate. Certain groups on the basis of their caste, class, and gender continue to be socially excluded from society in general and from the benefits of tourism in particular. The unequal sharing of these benefits is embedded in the unequal socioeconomic and cultural structures and processes of Naggar. The various ways in which women and traditionally disadvantaged caste groups are excluded from the benefits of CBRT are as follows.

*Manifestation of social exclusion in tourism in Naggar.* For all the economic and physical improvements, one thing that can be said about tourism development in Naggar is that the more things change, the more they remain the same. This is to say that while people investing effort, will, and resources in tourism are becoming economically prosperous, the majority of the population, especially women, the *kolis, haasis, chamars,* and *luhars,* and low economic class people, are not affected much apart from the little extra money they make directly or indirectly out of tourism work. With an overemphasis on physical infrastructure, ETP has failed to create any new forms of work in Naggar for the under-developed sections of society who needed tourism to function as a tool for their development and socio-economic inclusion. High return activities like the conversion of traditional homes with subsidised loans could not be adopted by these categories of people who had only small houses to begin with. Similarly, a lack of landholding, especially in the profitable parts of the village, has limited these groups' capacity to build restaurants or shops. This has meant that the benefits of tourism have failed to reach the excluded and thus needy groups.

*Unequal access to productive resources.* A close analysis of ETP implementation suggests that tourism development and livelihood generation through tourism is being done by ignoring the need for an improvement of the social condition of marginalised people in a community as the tourism potential of that community is enhanced to increase the number of tourists. It is not difficult to understand now that tourism activities are based within the current and existing but historically shaped socioeconomic context of that community. Hence, tourism does not affect all groups of people similarly as these groups have varying resources and access to them. The research suggests that tourism has benefitted the already resourceful while maintaining or sometimes strengthening the marginalisation of the already disadvantaged (Wilkinson & Pratiwi, 1995). Traditionally denied access to land and other important investment and productive resources, women, the *kolis, haasis, chamars,* and *luhars,* and low economic class families have not been able to convert those resources into new capital despite enormous opportunities for the same. The core tourism activities that offer bigger economic returns need initial investment, as in the case of a homestay, restaurants, or a bakery. Therefore, the majority of women and people from low class and historically oppressed castes have only found low paid, low skilled, and informal kinds of work in the tourism sector where there is little scope for betterment. This means difficult and unsustainable economic opportunities for them. However, it does not mean that the socially disadvantaged sections have not received any benefits from tourism employment. Participation in tourism livelihoods has improved people's economic conditions and this has made them feel empowered and to have more control of their lives. For example, women might have received few benefits, and the impacts on their lives must seem small for outsiders; but for them, these economic changes are very significant (Benaría & Roldán, 1987). But it is to be stressed that the larger benefits of tourism have been reaped by people who have traditionally had resources to convert into tourism capital.

Along with exclusion from decision making around the ETP work, such unequal socio-economic status has meant that the disadvantaged groups of the community are being turned into low paid workers in tourism. In many cases, with the expansion and monopoly of certain and a few rich people's businesses, tourism is ironically turning entrepreneurs into labourers. Ownership is getting lost as poorer people, especially poor women, find themselves ill equipped to compete. This is more intense as tourism has failed to create enough decision-making spaces for women and other disadvantaged people

*Unequal access to the market.* Without a critical assessment of the people and their needs, there has been no creation of market spaces for the products of most women. ETP failed to create any market linkage for women entrepreneurs and producers of souvenirs and crafts. Women have traditionally been restricted from entering the public space of the market or nearby places like Kullu or Manali which have bigger market spaces for their products. The marketability of crafts made by poorer women is reduced in the absence of ownership of such shops and any linkage with the bigger markets outside of Naggar. While some richer and dominant caste women have developed big souvenir shops selling hand-loom products, their poorer counterparts are unable to make good returns out of the same skill. The creation of such scattered opportunities of earning is creating tension in the community between economically advantaged and disadvantaged groups. This has resulted in a sense of resentment among women of these different groups.

*Gender and caste based occupational segregation.* The tourism industry has long been viewed and used as a strategy for development which is capable of creating work opportunities for underprivileged sections of a society and integrating them into the mainstream economy. However, often it has been found that such efforts have ended up confirming stereotyped sexist ideologies and strengthened the socio-economic stratification that exists in society. Tourism livelihood interventions have been found to create new ethnic and gender divisions of labour while strengthening the existing ones.

As proven in various studies of tourism employment (Chant, 1997; Kinnaird & Hall, 1996; Momsen, 2004; Sinclair, 1997), women and men are found to be horizontally segregated into different occupations. The extent of this segregation is dependent on the nature of the work and existing cultural norms of work for a particular gender in that particular community. Women in Naggar also are concentrated in "female" occupations. These are sometimes skilled but mostly semi-skilled, seasonal, part-time, and with little wages. All the work women find in tourism is primarily an extension of their reproductive roles at home and their cultural role in the community. Women mostly undertake these. The work women do in Naggar in homestays – weaving, knitting, souvenir making, sometimes cleaning – has mostly reinforced gender divisions of labour.

Rao (1995) mentioned that usually men are engaged in marketing while women produce, being involved in cooking, managing tourists at homestays, making souvenirs, or being commodities of entertainment of tourists. Men control major sectors of the tourism economy. The low value and wages accorded to women's work results in reduced appreciation of their work. They are seen as supplementing the family income. The lack of recognition of the proper economic and social worth of women's work, then, has consequences for women's access to resources, freedom, and socio-economic mobility (Momsen, 2004; UNWTO-UN, 2010). According to UNWTO-UN (2010), tourism jobs fail to have as many positive consequences for women due to cultural barriers, traditional gender work norms, lack of government interventions, and the lack of women's agency as a group. These prevent women from achieving top political or social status in society. Fillmore (1994) rightly points out that women are positioned differently than men in society and, from the moment tourism enters a community, these differences rule whoever benefits from the opportunities tourism brings. Women are generally less schooled, less trained, culturally subjugated, physically restrained in movement and spaces, and own and control very few productive resources in Naggar. Women work in the homestays owned by their husbands, thus adding to their workload. They make souvenirs from the traditional skill they learnt as part of becoming women in Naggar. Most of them make little but important extra money.

Thus, work opportunities presented by tourism ghettoise women and are only an extension of their domestic and cultural roles (Kinnaird & Hall, 1996; Chant, 1997; Sinclair, 1997). Women have been included in tourism development within the prevailing capital and patriarchal structures, which only confirms socioeconomic inequalities further. While tourism perpetuates gendered divisions of labour, there is also no computation of the unpaid labour done by women informally (Rao, 1997). The result is that, despite increasing participation of women in tourism work and related economic activities, there is no significant sign of their empowerment. Rather, tourism jobs have led in certain cases to increasing vulnerabilities of women as they struggle to fend for their families in a transition phase without any alteration in their family and community roles (Equations, 2011).

Tourism has also created work along caste lines. Under ETP, equipment and training in dance were provided to *koli* men whose caste occupation is dancing in social gatherings and community functions. The traditional occupation of dancing has stigma attached to it in the community and is considered to be of low value and moral standing. Providing training to young men in dancing has confirmed their caste occupation and denied them the chance of mobility. Such practices of providing skills and materials to support livelihoods under tourism projects have made no effort to break social barriers and have ended up strengthening them.

### Exclusive sharing of costs

As women and other disadvantaged sections face exclusionary practices that hinder their livelihoods, it is they who have to disproportionately bear the costs of tourism development. While tourism livelihoods have brought little extra money to families, it has also led to price hikes in the things of daily use, like food and fuel. Restaurants, big entrepreneurs, businessmen, and tourists pay extra prices and have cornered the market. Women have to negotiate over prices and often the food supply of poor houses are lower than before (Badger, 1993). It is usually women in rural households who eat least and have to bear the brunt of rising prices more than the male members of the community. Family, as a space of gender relations, is being impacted and altered due to women's engagement in tourism related livelihoods (Rao, 1995). Gender relations change too as women have to move out of their private domains to public ones to earn from tourism and the little financial autonomy that these earnings lead to have the potential to alter the gender relations within family and community, even though on a small scale. There is an ever increasing double workload on women as earners for the family and homemakers. This role change and reversal leads to tensions within family relationships. Citing from her study in Goa, Fillmore (1994) says that tourism is a double-edged sword for women. In Naggar, it benefits and harms them. Being a systematically and socially excluded group, it is women who have to pay more costs for than gain benefits from tourism. While women are improving their status in the community by becoming earning and contributing members in the economy, they are also being confirmed in feminised work. The women from the families of *kolis, haasis, chamars,* and *luhars* are also being labelled as morally inferior as they move out to find all kinds of work opportunities in tourism. Thus, tourism has the potential to degrade and improve women's status (Fillmore, 1994) and thus to include as well as further exclude them from full participation in family and society.

After seeing such effects, it is important to mention here that gender (Kabeer, 2000) and caste in themselves do not unproblematically translate into exclusion from society. Rather, they often mediate and exacerbate disadvantages and social exclusion. Poverty is harsher for women and SCs with certain added disadvantages associated with the identities of gender and caste for them. Control, restrictions, and denial of access are more common in these categories. However, importantly, this does not mean that access and exclusion cannot be altered. Newer institutional domains like the market may offset the disadvantages of socially excluded groups. Tourism offers such possibility through the potential of creating spaces for more inclusive, social, economic, and political participation.

Though this CBT development has led to small or bigger economic benefits, the distribution of these financial gains does not benefit the entire community. Bigger benefits are reaped by the local elite who already have the productive resources to invest in tourism enterprises. Few benefits reach the marginalised in local communities.

## Conclusion

Tourism approaches women in a gendered and casteist system and in most cases confirms these caste and gender relations as women and men play out their traditional roles and responsibilities even in the ambit of work and livelihoods. Tourism processes and activities reflect the prevailing social relationships and result in differential experiences for women which are based on their socio-spatial locations. Women and men experience tourism development differently. The experiences of women are also differential and reflect the implications of caste and class relations too. The patriarchal system works with the capitalist mode of tourism development to construct gender and caste specific experiences for most, while also leaving some windows open for transformation.

## References

Ashley, C. (2000). The impacts of tourism on rural livelihoods: Namibia's experience. *ODI Working Paper 128*. London: ODI.
Badger, A. (1993). Why not acknowledge women. Tourism in Focus. Winter issue. No. 10. pp. 2–5.
Benaría, L., & Roldán, M. (1987). *The Crossroads of Class and Gender*. Chicago: University of Chicago Press.
Chant, S. (1997). Gender and Tourism Employment in Mexico and the Philippines. In T. Sinclair (ed.) *Gender, Work and Tourism*, London: Routledge.
Equations. (2011). *Women in Tourism: Unfulfilled Promises, Continuing Myths*. Bangalore: Equations.
Fillmore, M. (1994). *Women and Tourism: Invisible Hosts, Invisible Guests*. Bangalore: Equations.
Hall, C. M. (2000). *Tourism Planning: Policies, Processes and Relationships*. Harlow: Prentice Hall.
Hatton, M. (1999). *Community-Based Tourism in the Asia-Pacific*. Toronto: Canadian Tourism Commission, Asia-Pacific Economic Cooperation and Canadian International Development Agency.
Kabeer, N. (2000). Social exclusion, poverty and discrimination: Towards an analytical framework. *IDS Bulletin* 31(4), 83–97.
Kinnaird, V., & Hall, D. (1996). Understanding tourism processes: A gender aware framework. *Tourism Management* 7(2), 96–102.
McCall, C. E., & Mearns, K. F. (2021). Empowering women through community-based tourism in the Western Cape, South Africa. *Tourism Review International* 25, 157–171.
Ministry of Tourism, Government of India. (2011). Revised guidelines for rural tourism, https://tourism.gov.in/sites/default/files/2019-10/011120120217173_0.pdf
Momsen, J. (2004). *Gender and Development*. London: Routledge.

Mowforth, M., & Munt, I. (1998). *Tourism and Sustainability: Development, Globalisation and New Tourism in the Third World*. London: Routledge.

Mowforth, M., & Munt, I. (2009). *Tourism and Sustainability: Development, Globalization and New Tourism in the Third World*, 3rd ed. London: Routledge, London.

Nara, V., & Irawan, N. (2020). Managing women empowerment through participation in sustainable tourism development in Kampong Phluk, Siem Reap, ambodia. *International Journal of Economics, Business and Accounting Research* 4(2), 262–269.

Nayak, P. (1994). Economic Development and Social Exclusion in India. In *Social Exclusion and South Asia Labour Institution and Development Program DP/77/1994*, Geneva.

Rao, N. (1995). Commoditization and commercialization of women in tourism: Symbols of victimhood. *Contours*, 7(1), 30–32.

Rao, N. (1997). Stark realities. Paper presented at the *Workshop on Women and Tourism organized by Institute of Management in Government*, Thiruvananthapuram held from 21–23 July, 1997 at Kochi, Kerala.

Sharpley, R., & Sharpley, J. (1997). *Rural Tourism: An Introduction*. London: Thomson Business Press.

Sinclair, M. T. (1997). Issues and Theories of Gender and Work in Tourism. In T. Sinclair (ed.) *Gender, Work and Tourism*, London: Routledge.

Telfer, D. J., & Sharpley, R. (2008). *Tourism and Development in the Developing World*. New York: Routledge.

Timothy, D. J. (2002). Tourism and Community Development Issues. In R. Sharpley & D. Telfer. (eds.) *Aspects of Tourism: Tourism and Development, Concepts and Issues* (pp. 149–166). Clevedon, England: Channel View Publications.

UN Women & UNWTO. (2010). *The Global Report on Women in Tourism*. Madrid: WTO. Retrieved from http://www2.unwto.org/sites/all/files/pdf/folleto_global_report_on_women_in_tourism-corregido.pdf

Wilkinson, P. F., & Pratiwi, W. (1995). Gender and Tourism in an Indonesian Village. *Annals of Tourism Research* 22(2), 283–299.

World Travel & Tourism Council (WTTC). (2003). *Blueprint for New Tourism*. London: WTTC. Retrieved from http://www.ontit.it/opencms/export/sites/default/ont/it/documenti/archivio/files/ONT_2003-09-18_00146.pdf on 02 March 2015

# 16 Stray domestic animals and tourism

*Paul Tully and Neil Carr*

## Introduction

The technological revolution of the 21st century, combined with rising social concerns for the welfare of animals (Carr & Broom, 2018), is highlighting the precarious position of stray domestic animals in tourism like never before. In combination, these issues are transforming the position and management of stray domestic animals in tourist destinations into an important topic related to the sustainability of tourism. Failure to address this will have potential negative consequences for the attractiveness of destinations. Online environments shed light on and raise global awareness of the cruelty that is inflicted on stray domestic animals. Such behaviour often involves the eradication of them from the holiday space, exemplified by news websites communicating the killing of stray dogs to protect Mauritius's image as a paradise island destination (Stanton, 2015). Alongside this, animal advocacy organisations look to rally support with claims about animal abuse happening in destinations, such as the mass poisoning of stray cats before the peak tourist season (Greek Cat Welfare Society, 2021). Elsewhere online social media provides a place for individuals to share stories and graphic images that can spark outrage at destination tactics employed to deal with stray domestic animals (e.g., Facebook, n.d.; Twitter, 2014). These examples speak of the human domination of sentient animals, how stray domestic animals are seen as problematic for tourism activity, and the extent of concern for the welfare of these animals. Depictions of this animal abuse have the potential to turn tourists, who are increasingly concerned with animal welfare, away from destinations (Shaheer & Carr, 2021).

This chapter explores the circumstances around the domination of stray domestic animals as sentient beings to explain why they are viewed as a problem for tourism activity and how this sits alongside the growing concern of ensuring the welfare of animals. The discussion considers how the dominant narrative of the stray in society has influenced the position of these animals ahead of their needs and wants. In other words, it discusses how a social narrative leads to the disregarding of animal sentience. As ideas of animal-centrism and welfare grow in society, this chapter presents a possible future for stray domestic animals in the holiday space. This is a future based on the scientific study of animal sentience, and this has relevance for all involved with tourism, given the shared human responsibility that exists towards all animals.

## What is a stray domestic animal?

Animals classed, by people, in the domestic category are those subject to human control and dominance. Over thousands of years, people have manipulated the breeding of

DOI: 10.4324/9781003291763-18

certain animal species to match humanity's requirements. As DeMello (2012) explains, practices of selective breeding for behavioural and physical traits have purposely manipulated certain animals. For example, generations of human-initiated modifications of cat characteristics have facilitated close human–cat bonds (Atkinson, 2018). This has been driven by, and continues to drive, in an example of a dominance feedback loop, perceptions of them as a perfect pet. Recognising that not all domestic animals are equal in how they are treated and perceived by people, it is important to identify the divide between pets and other domesticated animals. Humans have strongly manipulated all domestic animals, but it is clear that pets have a unique, close bond with humans. This has meant they are not only domesticated in a breeding sense but also in terms of their access to the human home. Hence, pets are increasingly living inside the human house as part of the human family (Carr, 2014).

Humans' dominance over domesticated animals influences their physical, psychological, and social life experiences. People have the power to determine if domestic animals live their best quality of life or not. Regarding pets, people are also in control of the closeness of the human–animal bond. Humans, therefore, have direct and ancestral responsibilities for how these animals live (Norlock, 2017). Yet human action continuously results in animals straying from this domesticated way.

The existence of stray domestic animals is a failure of human responsibility that highlights the precarious position of these animals in a human-centric world. They are unowned or abandoned domestic animals (either via their personal or ancestors' domestication) that freely roam amongst human activity (Jarvis, 2020). These animals could be a pet dog, a farmed pig, a carriage horse, or a racing pigeon, among many other human ideals, if action towards them was different. Through breeding, stray domestic animals are manipulated by humans to serve humanity, but via the actions (and inactions) of people are left to survive by themselves as if wild, something for which they are often poorly prepared for by human breeding and conditioning. One single label does not exist for these animals, as people can refer to them as stray, free-roaming, semi-owned, roaming, semi-feral, or feral, among others. This creates added complexity as these terms are all loaded with implicit meaning, are often used interchangeably, and differ across cultures and species. A spectrum can begin to highlight differences between these animals, as they can fall between those highly socialised with humans (those newly strayed) to completely non-socialised (those fully feral). Each labelled animal fits within these points (Figure 16.1), though, no matter which label is attached, each indicates that an animal's domestication has not gone as intended. They designate a domestic animal which has strayed from where it should be, namely under human dominance and control. In this scenario, all stray animals can be identified as pests, a problem to be resolved. As such, they do not belong in a human-centric world. They lack the social value ascribed to wild and domestic animals as they do not belong to either human-defined category. The feral animal is not ascribed with the label "wild", and the stray animal is bereft of the title "pet" or "domesticated animal". In this way, the pet is loved, the domesticated animal is valued, and the wild animal is venerated. In contrast, stray and feral animals are pests, unwanted and unloved. Yet even here there are degrees of difference, with some more unwanted and unloved than others depending on the species in question, their location, and the cultural values of the humans they interact with.

Human treatment of stray domestic animals is shaped by their social identity. As DeMello (2012) discusses, animals live surrounded by social constructions that communicate understandings to people. The cognitive and emotional lived experiences of animals are unimportant in such human-centric thinking. The label of "stray" (and stated variations) connects to

| e.g., a dog abandoned to the streetafter recently beinga pet will be a highly social stray. | e.g., a catroaming with no fixed home but receiving food and care from a volunteer may be closer to semi-feral. | e.g., a free-roaming hog, born to feral parentage with domestic ancestors, but never under human control will be fully feral. |

**Stray**   **Feral**

Degree of socialisation with humans

**100%**   **0%**

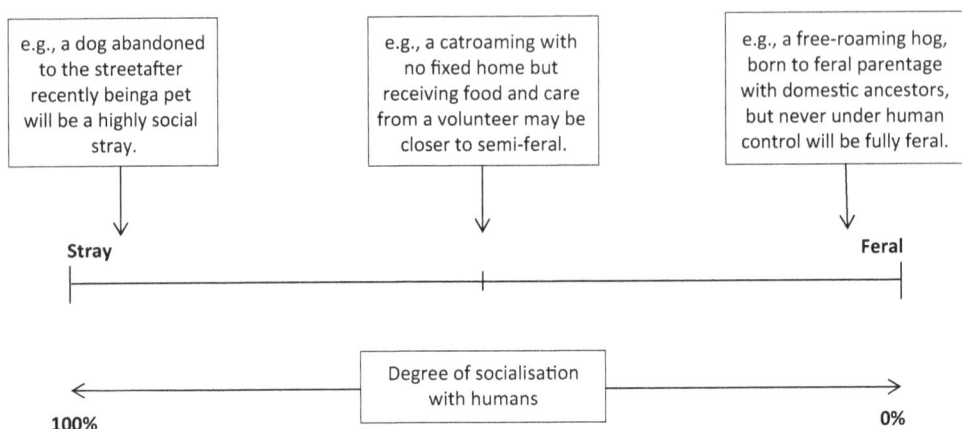

*Figure 16.1* A spectrum for stray domestic animals.

*Source*: Derived from Slater (2007).

a socially constructed message of an animal that disrupts the human ideal. During the last 170 years, especially due to humanity's urban expansion, stray domestic animals have been seen as a problem (Irvine, 2003). Who is to blame for this problem is contested, with people blaming other humans and the animals themselves. Cultural processes and human claim-makers have continuously problematised the actions and lifestyles of these animals (Jerolmack, 2008), thus embedding negative perceptions of them into everyday life. The human dominance of domestic animals and their meanings has resulted in human-beneficial narratives of strays having long gone unchallenged.

The actions and lifestyles of these animals are an enforced reality of their lives. In this way, they are not wild animals. They seek shelter and food in alleyways, waste grounds, disused buildings, junkyards, and public leisure spaces (Meijer, 2021). They also defecate in public, rifle through garbage, and get into violent conflicts with humans and other animals (Beckman et al., 2014). Often, they become dirty and unhealthy (Crawford, Calver, & Fleming, 2020). Their uncontrolled breeding can see them gather in large numbers (Strickland, 2015), intimidating at least to humans and potentially other animals, including valued domestic animals. The social identity of strays is reinforced through their actions because the realities of the strayed life contradict the ideal of a domestic animal. Nagy and Johnson (2013) state that this lifestyle leads society to think of and treat these animals with the same mentality as trash. Thus, people have negative perceptions of a nuisance animal to be avoided and dealt with by the authorities. Such a judgement ignores human responsibilities to these animals who are sentient beings with needs akin to their domesticated relatives. This way of treating and perceiving strays signifies a physical and social marginalisation of them from the human world. If animals lack power in a human-centric world, then strays are even further disempowered.

Whilst recognising the marginalised position of strays, it is important to identify that this is neither a uniform position nor one from which there is no escape. The work of a myriad of organisations, arguably led by the Society for the Prevention of Cruelty to Animals (SPCA) in its various national guises, to rehome strays and reintegrate them into society, is an example of a positive escape route. Alternatively, we increasingly see people caring for stray animals, feeding and medicating them while leaving them outside the human home. Yet such a position remains inherently precarious. The cared-for stray is still a stray, potentially better fed but still in as much danger, for example, of being run over by a vehicle. In the case of the SPCA, whether a stray may be rehomed or must be killed is dependent on human definitions of whether they are a pest. In this context, a stray dog may be rehomed, depending on breed, but a stray cat, if defined as an introduced species and killer of native fauna, may have to be destroyed even if found by an organisation dedicated to the care of animals. The thread throughout this discussion of strays is a narrative controlled by humans.

## Stray domestic animals and tourism

The enforced lifestyle of strays and the associated narratives have affected their position in the holiday space. Tourists' perceptions of strays mirror the negative judgements found throughout society. Tourists complain about stray animals' noise disturbances, dirt, threatening acts, inappropriate behaviour, and unhealthy appearances (Beckman et al., 2014). Consequently, the presence of stray domestic animals can hurt destination economies by deterring visitors (Webster, 2013). Strategies to avoid this involve controlling the presence of strays in tourism spaces (Edensor, 2000), epitomised by the eradication practices highlighted in the introduction. Other strategies aim to appear animal friendly by assisting strays with food, shelter, and/or reintegration into human society (Browne, 2019). This latter strategy is becoming more common as destination stakeholders risk the potential anger of a consumer base containing increased animal advocacy.

Ethical consumerism and an advocacy mindset have grown thanks to increases in discussions of poor animal treatment and the rights of animals. The animal protection movement has shifted from fringe demonstrations into politically and economically important topics, with significant contributions from politicians, legislators, non-profit organisations, industries, veterinarians, celebrities, scientists, and the general public (DeMello, 2012). This movement is likely to continue growing as knowledge and information communication about animal rights and welfare expands, thus intensifying it from a predominantly Western cause into a global one (Broom, 2014). It is a movement with ramifications for tourism, as tourists (or potential ones) are becoming more critical of destinations that mistreat animals. The advancement of social media is seeing global public engagement in the sector's animal issues (Shaheer & Carr, 2021), such as with campaigns like #boycottkerala, which targeted the Indian state's tourism sector due to its treatment of its stray dog population (BBC, 2015). Never before have ethical issues involving stray domestic animals in destinations been so prevalent in societal discourse.

Yet this growing voice for ethical animal treatment has developed in parallel with an ever-increasing demand for animal attractions and sanitised destinations. As such, animals, in substantial numbers, are used to meet tourism demands (Carr & Broom, 2018). What is more, destination stakeholders are continually looking at inventive ways to use the appeal of animals, which has implications for how the stray domestic animals' position in the holiday space is evolving. A growing development is for these animals to be new trendy tourist attractions. For instance, in Asia the presence of stray cats has been used to create an island destination experience (Burton-Bradley, 2018). In Africa, tourists can purchase a volunteer experience at stray dog sanctuaries (Volunteer World, n.d.). In Europe, feral horses have enabled the creation of photo-safari tour businesses (Sito-Sucic, 2021), whilst a swim with a feral pig experience is available in the Caribbean (Ross, 2017). These types of attractions remain, currently, a minority. In repurposing these animals, questions need to be asked about their status as strays. Furthermore, we need to ask whether their vulnerable status as strays opens them up to abuse as tourist attractions in ways that other animals are not exposed to.

Despite the emerging use of certain stray domestic animals as tourist attractions, the vast majority remain maltreated in destinations due to human denigration. Nonetheless, as Usui (2021) highlights, destination stakeholders can see potential economic value in stray domestic animals. If these animals were to be seen as economically valuable, then strategies for improving their lifestyle may be pursued, as has previously happened with the development of whale watching tourist attractions. It is important to recognise that seeing an economic resource is distinctly different from seeing a sentient animal. Viewing an animal as dollar signs risks ignoring their needs and wants in the same way as seeing them as trash. Improvements for the lives of stray domestic animals should be sought not for the benefit of tourism but to help them as sentient beings.

## A future guided by animal sentience

The future for all animals in tourism should be guided by their sentience. Learning and teaching about animal welfare has been central to changing animals' position in society during the past 60 years (Figure 16.2). Animal welfare relates to the states experienced by an animal when coping in its environment (Broom, 2014). As Buller et al. (2018) comment, developments in the study of welfare have aided humans' understanding of positive and negative animal experiences, allowing the adoption of practices for improved treatment. However, a central issue with the study of animal welfare is that it has been incorrectly implemented in society as human-centric. It sees people as continuing to dominate animals providing they meet (or, in reality, at least display) a human-acceptable, human-determined standard of care. This is evident in a tourism context where animal attractions focus on

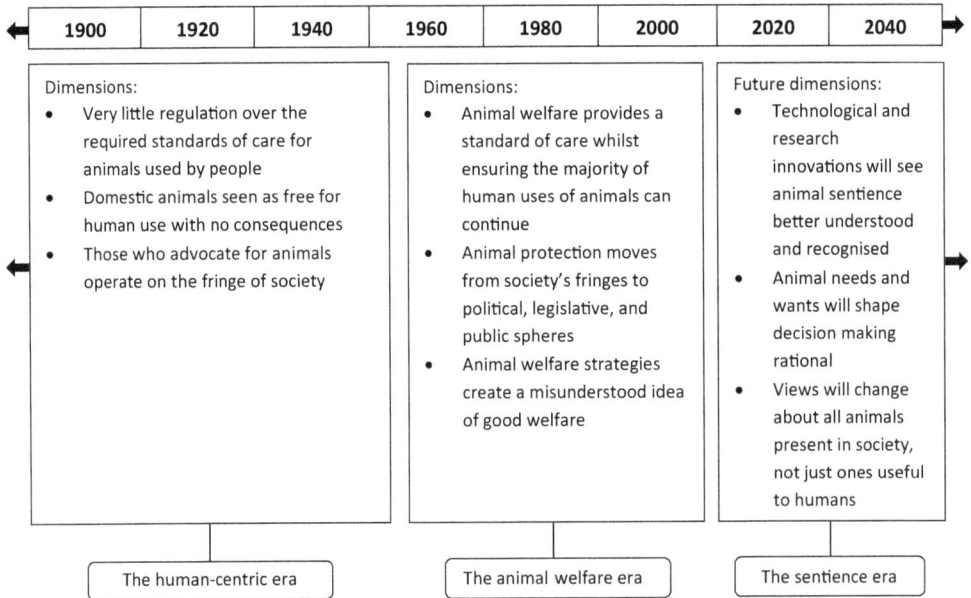

*Figure 16.2* Eras of societal animal treatment.

promotion through a simple healthy and happy story which masks more complex welfare needs (Tully & Carr, 2021). Bekoff and Pierce (2017) see that the failure to take advantage of knowledge of animal sentience has led to unsatisfactory welfare practices across society. Furthermore, implemented welfare practices have focused on those animals being labelled in a good way (such as "pet" or "farm") over those labelled as pests, including strays (Carr & Broom, 2018). This era of animal welfare has improved animals' experienced states but has not maximised welfare or been comprehensive enough. Over the coming decades, this is likely to change as the gap between societal animal welfare strategies and scientific animal sentient knowledge closes.

The closing of this divide is underway. Scientific discoveries of animal sentience are pushing the understanding of complex life experiences to the forefront of conversations (Broom, 2014). As such, it is reasoned that more just animal treatment means accepting them as thinking beings with the capability to know their own interests (Meijer & Bovenkerk, 2021). To maximise an animal's welfare requires careful listening and responding to what it actually needs and wants, as opposed to accepting human assumptions about levels of care. The knowledge discoveries of animals' emotional and cognitive capabilities provide a platform to move into an animal-centric future that recognises humans and non-humans as animals. This is a place that will identify the intrinsic value of all animals, meaning that decisions made about their treatment are done from a position of maximum animal benefits (Carr & Broom, 2018). In line with this thinking, the future of stray domestic animals in the holiday space should be based on the recognition that human activity does not have the right to take precedence. This includes the treatment of an animal in ways that view them as either trash or a new trendy attraction. All holiday spaces are shared by beings, human and non-human, with unique needs and wants. The future challenge in tourism, and society generally, is to find ways for these interests to co-exist without bias. Doing so has the potential to benefit tourism destinations by ensuring the animal welfare related concerns of tourists are aligned with the treatment of animals. Considering the welfare of all animals, including strays, in tourism aligns with the UN's sustainable development goals (SDGs), particularly 14 and 15, although animal welfare is not explicitly mentioned (Keeling et al., 2019).

Animal-centric ideas will continue to transfer into the general public's attitudes towards animal treatment. Knowledge creation was the catalyst for the development of the welfare era in society (Dwyer et al., 2021), a model the sentient era will likely follow. The continued growth of ethical consumerism and animal advocacy movements, discussed earlier, will be propelled by emerging evidence of animals' unique needs and wants. As animal-centric thinking becomes more prevalent in society, it will clash with the previously unchallenged animal ideas, such as the narrative of the stray. Yet changing narratives embedded over hundreds of years will not be easy, and many will remain sceptical about following animal-centric concepts over human-centrism (Carr & Broom, 2018). The ability of humanity to co-exist with the interests of stray domestic animals will be challenged by these competing human perspectives.

The coming decade is likely to witness an increasing presence of stray domestic animals in holiday spaces. The biggest driver of this will be humans continuing allure to pet-keeping and the subsequent abandonment of these animals (DeMello, 2016). Simultaneously, whilst humans continue to abandon their pets, these animals' uncontrolled breeding will keep populations increasing. Thus, pressure-cooker-style environments are likely in holiday spaces, as destination stakeholders battle to satisfy increasing numbers of animal advocates, to present a successful tourism product to a continued majority with misinformed imaginations, and to handle a growing animal population justly.

In these battles, tourism has the potential, as shown in wildlife encounters (Carr & Broom, 2018), to be an educational vehicle for change over the coming decades. This is due to its ability to influence, foster cooperation, and connect humanity to be a social force in changes for a more just society (Higgins-Desbiolles, 2006), one that incorporates animals as well as humans. Hence, tourism could reform its own actions towards stray domestic animals and those of the wider society. Dwyer et al. (2021) explain the importance of education in improving animal treatment and the necessity of engaging the scientific study of animal lives at all societal levels. Given the shared responsibility for treating animals better, it is education for everyone that is needed to change decision-making, attitudes, and behaviour (Dwyer et al., 2021). For tourism to realise its potential requires a coordinated effort. As tourism scholars become more engaged with animal sentient research, they will have an opportunity to inform all actors in the sector of the needs and wants of animals. If all tourism stakeholders come together and teach each other, the sector can approach stray domestic animals with animal-centric ways of thinking at all levels. This will benefit these animals' life experiences. The result can be a changing of the communications of the future from one of eradication to one of co-existence. The whole of tourism has a chance to shape a more just society for stray domestic animals. To achieve this will mean it is being guided by conceptualisations of sentience at all levels.

## Moving forward

The future for stray domestic animals in tourism depends on humans continuing to close the divide between what is thought, known, and done. As such, researchers of the sector need to consider stray domestic animals as sentient beings, and they need to work with animal scientists to show what an animal needs and wants. In addition, those invested in providing a tourism product need to adopt new approaches to stray domestic animals. Destination stakeholders should not simply kill these animals or make decisions based on the negative influence of the dominant social narrative. They should also be wary of turning them into an economic resource. Instead, new approaches should come from what is best for the animal. In liaising with researchers of tourism and animal sentience, industry stakeholders can join the growing trend of animal-centric thinking to identify between what animal welfare in society currently offers and what animals need, want, and deserve. Doing so would align with the underlying ethos of the UN's SDGs. If animal sentience is embraced, new policies and ways of treating stray domestic animals could start to be implemented, offering encouragement in attempts to meet tourists' desires and change other tourists' perceptions. This implementation can benefit destination stakeholders by appeasing a growing animal advocacy movement and educating those who are still misguided by the outdated and unhelpful narrative of the stray. This requires buy-in from tourists, who must approach tourism's future with a willingness to learn and follow the evidence. The science must be allowed to shift the narrative into one of animal sentience. In this learning process across the tourism sector, teachings can challenge the dominant attitudes and beliefs and, thus, begin to change human actions. In doing so, the quality of life experienced by stray domestic animals can be improved within tourism and beyond. The fundamental requirement is for industry stakeholders to embrace the notion of stray domestic animals as sentient beings with a right to life. This will align them with a view held by a growing proportion of the global society, enabling them to care for rather than eradicate or manipulate these animals to the satisfaction of potential visitors and the wellbeing of the animals themselves. Doing so requires working with, rather than in opposition to, animal rights and welfare organisations. The work between ABTA and the Born Free Foundation since 2004 to improve animal welfare in tourism demonstrates the potential of this (Turner, 2018). Twenty years ago, the tourism industry cared little for the welfare needs of wild animals, and people taking their pet dogs on holiday with them were a poorly catered to niche market. Both issues are central to the industry and the experiences it offers (Carr, 2014; Carr & Broom, 2018). The same issues of animal welfare concerns and the emotional attachments between animals and humans that drove these changes are now being applied to the issue of stray domestic animals. Tourism industry stakeholders over the next 10 to 20 years need to begin to apply the same thinking we have seen developed around relationships between wild animals and pets to stray domestic animals. Failure to do so will negatively impact the lives of these animals and the sustainability of an industry faced with a consumer body increasingly concerned for animal welfare. It can help to lead this change through education of the general public or potentially be ostracised if it refuses to change.

## References

Atkinson, T. (2018). *Practical feline behaviour: understanding cat behaviour and welfare*. CABI. doi: 10.1079/9781780647838.0000

BBC News. (2015, August 9). How far will India's dog-lovers go to save strays? Retrieved 1 November 2021, from https://www.bbc.com/news/blogs-trending-33822511

Beckman, M., Hill, K. E., Farnworth, M. J., Bolwell, C. F., Bridges, J., & Acke, E. (2014). Tourists' perceptions of the free-roaming dog population in Samoa. *Animals*, 4(4), pp. 599–611. doi: 10.3390/ani4040599

Bekoff, M., & Pierce, J. (2017). *The animals' agenda: Freedom, compassion and coexistence in the human age*. Boston: Beacon Press.

Broom, D. M. (2014). *Sentience and animal welfare*. CABI. doi: 10.1079/9781780644035.0000

Browne, B. (2019, April 15). Hotels catering to animal-loving tourists | Cyprus Mail. Retrieved 1 November 2021, from https://cyprus-mail.com/2019/04/15/hotels-catering-to-animal-loving-tourists/

Buller, H., Blokhuis, H., Jensen, P., & Keeling, L. (2018). Towards farm animal welfare and sustainability. *Animals*, 8(6), 81. doi: 10.3390/ani8060081

Burton-Bradley, R. (2018, November 2). Cat tourism: Small towns in Asia look to stray cats to lure in tourists and rescue dying communities. *ABC News*. Retrieved 18 October 2021, from https://www.abc.net.au/news/2018-11-03/cat-tourism-can-a-bunch-of-kitties-save-dying-towns-in-asia/10458128

Carr, N., & Broom, D. M. (2018). *Tourism and animal welfare*. CABI. doi: 10.1079/9781786391858.0000

Carr, N. (2014). *Dogs in the leisure experience*. Wallingford, UK: CABI. doi: 10.1079/9781780643182.0000

Crawford, H. M., Calver, M. C., & Fleming, P. A. (2020). Subsidised by junk foods: factors influencing body condition in stray cats (Felis catus). *Journal of Urban Ecology*, 6(1). doi: 10.1093/jue/juaa004

DeMello, M. (2012). *Animals and society: An introduction to human-animal studies*. New York: Columbia University Press.

DeMello, M. (2016). Rabbits multiplying like rabbits: The rise in the worldwide popularity of rabbits as pets. In M. P. Pregowski (ed.), *Companion animals in everyday life* (pp. 91–107). Palgrave Macmillan. doi: 10.1057/978-1-137-59572-0_7

Dwyer, C., Bacon, H., Coombs, T., & Langford, F. (2021). Educating the animal welfare practitioners of the future. In R. Sommerville (ed.), *Changing human behaviour to enhance animal welfare* (pp. 65–80). CABI. doi: 10.1079/9781789247237.0005

Edensor, T. (2000). Staging tourism: Tourists as performers. *Annals of Tourism Research*, 27(2), pp. 322–344. doi: 10.1016/S0160-7383(99)00082-1

Facebook. (n.d.). Kerala dog culling #boycottkerala. Retrieved 3 November 2021, from https://m.facebook.com/events/493782680776258/

Greek Cat Welfare Society. (2021). About Us. Retrieved 3 November 2021, from https://greekcats.org.uk/about-us/

Higgins-Desbiolles, F. (2006). More than an "industry": The forgotten power of tourism as a social force. *Tourism Management*, 27(6), pp. 1192–1208. doi: 10.1016/j.tourman.2005.05.020

Irvine, L., (2003). The problem of unwanted pets: A case study in how institutions "think" about clients' needs. *Social Problems*, 50(4), pp. 550–566. doi: 10.1525/sp.2003.50.4.550

Jarvis, P. J. (2020). Feral animals in the built environment. In I. Douglas, P. M. L. Anderson, D. Goode, M. C. Houck, D. Maddox, H. Nagendra, & T. P. Yok (eds.), *The routledge handbook of urban ecology* (pp. 463–471). Routledge. doi: 10.4324/9780429506758-39

Jerolmack, C. (2008). How pigeons became rats: The cultural-spatial logic of problem animals. *Social Problems*, 55(1), pp. 72–94. doi: 10.1525/sp.2008.55.1.72

Keeling, L., Tunón, H., Olmos Antillón, G., Berg, C., Jones, M., Stuardo, L., Swanson, J., Wallenbeck, A., Winckler, C. & Blokhuis, H. (2019). Animal welfare and the United Nations sustainable development goals. *Frontiers in Veterinary Science*, 6, 336. doi: 10.3389/fvets.2019.00336

Meijer, E., & Bovenkerk, B. (2021). Taking animal perspectives into account in animal ethics. In B. Bovenkerk & J. Keulartz (eds.), *Animals in our midst: The challenges of co-existing with animals in the anthropocene* (pp. 49–64). Springer. doi: 10.3920/978-90-8686-892-6_9

Meijer, E. (2021). Stray agency and interspecies care: The Amsterdam stray cats and their humans. In B. Bovenkerk & J. Keulartz (eds.), *Animals in our midst: The challenges of co-existing with animals in the anthropocene* (pp. 287–299). Springer. doi: 10.1007/978-3-030-63523-7_16

Nagy, K., & Johnson II, P. D. (2013). Introduction. In K. Nagy & P. D. Johnson II (eds.), *Trash animals: How we live with nature's filthy, feral, invasive, and unwanted species* (pp. 1–27). University of Minnesota Press. doi: 10.5749/minnesota/9780816680542.001.0001

Norlock, K. J. (2017). "I don't want the responsibility": the moral implications of avoiding dependency relations with companion animals. In C. Overall (ed.), *Pets and people: the ethics of our relationships with companion animals* (pp.80–94). Oxford University Press. doi: 10.1093/acprof:oso/9780190456085.003.0006

Ross, D. (2017, March 4). This is what really killed the famous swimming pigs. Retrieved 3 November 2021, from https://www.nationalgeographic.com/animals/article/swimming-pigs-bahamas-death

Shaheer, I., & Carr, N. (2021). Rallying support for animal welfare on Twitter: A tale of four destination boycotts. *Tourism Recreation Research*. doi: 10.1080/02508281.2021.1936411

Sito-Sucic, D. (2021, September 15). Bosnia's wild horses: Promising tourist attraction, or farmers' pest? | Reuters. Retrieved 1 November 2021, from https://www.reuters.com/article/uk-bosnia-wild-horses-idUKKBN2GB16I

Slater, M. R. (2007). The welfare of feral cats. In I. Rochlitz (ed.), *The welfare of cats* (pp. 141–175). Springer. doi: 10.1007/1-4020-3227-7_6

Stanton, J. (2015, June 18). Shocking hidden camera footage reveals how stray dogs are rounded up, tortured and killed so you can enjoy an unspoiled holiday in Mauritius… And even the puppies are not spared | Daily Mail Online. Retrieved 1 November 2021, from https://www.dailymail.co.uk/news/article-3128120/Shocking-hidden-camera-footage-reveals-stray-dogs-rounded-tortured-killed-enjoy-unspoiled-holiday-Mauritius-puppies-not-spared.html

Strickland, P. C. (2015). It's a dog's life: International tourists' perceptions of the stray dog population of Bhutan. *Journal of Arts and Humanities*, 4(12), pp. 1–11. doi: 10.18533/journal.v4i12.856

Tully, P. A. G., & Carr, N. (2021). Farm animals' participation in tourism experiences: A time for proper respect. In J. M. Rickly & C. Kline (eds.), *Exploring non-human work in tourism* (pp. 83–100). De Gruyter. doi: 10.1515/9783110664058-006

Turner, D. (2018). Managing tourism's animal footprint. In N. Carr, & D. Broom, *Tourism and animal welfare* (pp. 106–111). CABI.

Twitter. (2014). Stray Cat on Twitter - *@rainyhorizon*. Retrieved 3 November 2021, from https://twitter.com/rainyhorizon/status/476327488545886209

Usui, R. (2021). Feral animals as a tourism attraction: Characterizing tourists' experiences with rabbits on Ōkunoshima Island in Hiroshima, Japan. *Current Issues in Tourism*, pp. 1–16. doi: 10.1080/13683500.2021.1978950

Volunteer World. (n.d.). Domestic animal rescue & rehabilitation | Volunteer in South Africa 2021. Retrieved 1 November 2021, from https://www.volunteerworld.com/en/volunteer-program/domestic-animal-rescue-rehabilitation-in-south-africa-cape-town

Webster, D. (2013). The economic impact of stray cats and dogs at tourist destinations on the tourism industry. Report CANDI International. Retrieved 25 October 2021, from https://faunalytics.org/the-impact-of-stray-cats-and-dogs-on-tourism/

# Part III

# Management

In Chapter 17, **Prokopis Christou, Alexis Saveriades**, and **Maria Rigou** discuss *Well-being and tourism employees: The important role of "freedom" in the future workplace*. This chapter delivers understandings of tourism employees' wellbeing while emphasising the significant role of the notion at a personal, organisational, industrial, and societal level. The chapter proceeds by discussing current issues and future perspectives related to the notion of wellbeing within the tourism and hospitality workplace context. Practical and theoretical implications are also provided. A conceptual diagram that summarises the interrelated elements shaping the wellbeing of individuals is proposed for academic and tourism stakeholder consideration. The diagram presents the rather ignored aspect of the "freedom" sense that should be acknowledged while researching, measuring, and endeavouring to promote the wellbeing of individuals in the future.

*Resilient leadership in the tourism industry* is reviewed in Chapter 18 by **Adela Balderas-Cujudo, Marta Buenechea-Elberdin, Josune Baniandrés**, and **George W. Leeson**. Companies in the tourism industry should have resilience built into their DNA. Recovering in the face of adversity and adapting to change and trends should be intrinsic to the hospitality industry, given its changing and dynamic nature. While it is critical to build organisational resilience led by resilient leaders, research on such leaders and mechanisms for stimulating organisational resilience is still in its infancy. Consequently, the goal of this chapter is threefold: (1) to provide a deeper understanding of this resilience in the tourism industry; (2) to understand and recognise the value of fostering resilient leaders, capable of coping with disruptive changes, navigating complexity, and cultivating resilience in themselves and their teams; and (3) to provide some broader perspectives on the term's applicability relating to leadership in the industry.

**Nellie (Magdalena Petronella) Swart, Vanessa S. Bernauer**, and **Kailasam Thirumaran** cover *Women's education in tourism entrepreneurship: Trends and issues emerging from Africa* in Chapter 19. Over the past decade, international organisations have highlighted the important role of women as drivers of tourism entrepreneurship. The challenges and opportunities are documented in the United Nations Sustainable Development Goals, the Global Report on Women in Tourism, and the World Economic Forum Global Gender Gap. In addition to these systemic challenges, African women in tourism have limited education, particularly in areas related to entrepreneurship. Although peer-reviewed publications on women's entrepreneurship in tourism are widespread, this topic remains underrepresented in Africa, with studies evident from South Africa, Tanzania, and Burkina Faso. Although the need for education is a consistent recommendation in most of these publications, it is only a mainstream suggestion that does not address the specific educational

DOI: 10.4324/9781003291763-19

needs of women entrepreneurs in tourism. The need for entrepreneurial education has been exacerbated by the recent pandemic, as many women in tourism have lost their jobs and have been forced by circumstances to start a new business. The emerging trends highlight the need to profile women entrepreneurs in tourism by developing flexible and self-directed training programmes based on their current level of education and entrepreneurial needs.

*An American labour revolution* is the topic for Chapter 20 by **Sotiris Hji-Avgoustis** and **Alan Yen**. The tourism industry has never faced an economic crisis as bad as that during the COVID-19 pandemic. All segments of the industry, from hotels to restaurants to travel by sea or air, suffered great losses. It is now clear the industry has not been as resilient as many were predicting prior to this crisis. Many expected a quick recovery, especially after the lightning-fast introduction of a few COVID vaccines. However, the rebound is not proceeding smoothly as the revenue lost during the pandemic can never be recovered and the industry cannot find enough workers post-COVID. Add to that rising costs and an unpredictable supply system, at a time when consumer demand is skyrocketing, and it becomes difficult to avoid customer dissatisfaction. With the global economy rapidly recovering, why is there still a worker shortage? Some blame the generous unemployment payments and stimulus checks for making people less likely to take the traditionally low-paying hospitality jobs again. Others counter that companies could raise pay if they really wanted workers back quickly. This chapter discusses the reasons behind the worker shortage and expands on the great reassessment that is going on in the US economy to explain why so many workers are hesitant to return to work. Economists describe this phenomenon as reallocation friction. The pandemic caused some jobs in the economy to change, and workers are taking their time to figure out what new jobs are out there, what new jobs they want, and what skills they need for these different careers.

Chapter 21 by **Thi Hong Hai Nguyen, Diep Ngoc Su**, and **Hanh My Thi Huynh** reviews *The resilience of the tourism and hospitality workforce*. Resilience, which is understood as the capacity of a system to absorb disturbance and reorganise while undergoing change, has increasingly become an important topic in professional practices and academic research. In tourism and hospitality, this is an essential ability for destinations, organisations, as well as the workforce to adapt and thrive during competitive and challenging times. This chapter focuses on the workforce, aiming to provide a comprehensive understanding of resilience from the organisational perspective and the individual career perspective. It provides a thorough theoretical background of organisational resilience, employee resilience, and career resilience. Additionally, empirical data from Vietnam in the context of a major health crisis, that is, the COVID-19 pandemic, illustrates and broadens the knowledge of these issues. Two empirical studies are included. The first study, with a qualitative approach and from a managerial perspective, demonstrates human resources (HR) practices during the pandemic to ensure the resilience of the organisation and the workforce. A framework of HR practices to enhance resilience during a crisis is proposed. The second study, with a quantitative approach and from an employee perspective, explores the factors which influence career resilience in the hospitality and tourism workforce, using the foundation of the Protection Motivation Theory. The link of resilience between these two perspectives is then drawn, together with practical and theoretical implications.

*Hospitality workforce trends* is the topic of Chapter 22 by **Adesola Osinaike** and **Lorna Thomas**. The hospitality workforce crisis has been an ongoing discussion for academics, researchers, and industry practitioners. Globally, the hospitality sector is considered one of the fastest growing, but it is experiencing severe workforce challenges. To address this issue, it is necessary to consider the factors affecting employees and discouraging prospective staff. This chapter examines factors influencing the workforce shortage and provides suggestions for addressing the issues to make the industry attractive and sustainable.

**Suk Ha Grace Chan, Yue Yvonne He**, and **Binglin Martin Tang** address the question *Do customers really care about corporate social responsibility reputation in the decision of hotel patronage?* in Chapter 23. In recent decades, corporate social responsibility (CSR) has become a powerful marketing tool in the hospitality industry. Many hotels commit to participating in CSR to gain awareness of customer buying decisions. Young people tend to follow their favourite brands and use social media- and video-based platforms when making patronage decisions. Electronic word-of-mouth (eWOM) has become one of the common communication practices for young customers. Brand and ethics are their main concerns when making purchase decisions. Given that brands often reflect their values, they prefer brands that they perceive as environmentally responsible and socially ethical. Results indicate that CSR will not directly affect young customers' patronage intentions. Moreover, eWOM directly affects brand equity, which presents a mediating factor between eWOM and patronage intention. This chapter also highlights the opportunities and challenges in the hotel industry which has had to confront CSR. Predictions for the next three decades in sustainability development are discussed. Recommendations to industry practitioners and practical suggestions are delineated.

Chapter 24 is on *Post-pandemic Chinese outbound tourism: Three trends to look out for* by **Marine L'Hostis**. Pre-COVID-19 pandemic, the Chinese travel market was one of the most dynamic in the world. Since its official beginnings, China's outbound tourism has been seen as an El Dorado promising a plethora of visitors spending huge amounts of money and taking over former traditional outbound tourism markets. This should not eclipse the fact that Chinese tourism has always been closely controlled by the government, and the travel restrictions implemented in response to the COVID-19 outbreak are a reminder of this logic. We may wonder what the future of Chinese tourism will be, and if it will ever go back to its pre-pandemic levels. This chapter draws stakeholders' attention to three trends that should be looked out for in the long run: the rivalry with the United States, the isolationist move taken by Xi Jinping, and the ageing of the Chinese population. It underlines the challenges and opportunities brought by these trends, as well as some strategies to deal with them.

**Lemonia (Lenia) Papadopoulou-Kelidou** and **Andreas Papatheodorou** in Chapter 25 focus on *All-inclusive holiday packages in the post-COVID-19 era*. For decades, the all-inclusive holiday package offered by tour operators constituted a very popular tourism product, structurally intertwined with mass tourism and destination development. Large European tour operators, like TUI and Der Touristik, have engaged in horizontal and vertical integration practices with several tour operators, travel agencies, hotels, and airlines to provide all-inclusive holiday packages serving millions of travellers and recording billions of euros in annual sales. The demise of Thomas Cook in 2019 though led several analysts to conclude that the tour operations model is unsuitable for addressing the flexibility required by modern tourists; inclusive holidays do offer a one-stop shop reducing transaction costs and allowing budget control for holidaymakers. In addition to its overall adverse consequences for tourism, the COVID-19 pandemic severely impacted tourist preferences and their perception with respect to quality, safety, and social distancing. The purpose of this chapter is to elaborate on this change in tourist preferences regarding types of holidays, emphasising the role of all-inclusive holiday packages in the post-COVID-19 era.

# 17 Wellbeing and tourism employees

## The important role of "freedom" in the future workplace

*Prokopis Christou, Alexis Saveriades and Maria Rigou*

### Introduction

In an insightful and current article, *The New York Times* questioned the role of hospitality amidst the pandemic with the onus being put on the server. The dysfunction at the heart of the industry was emphasised with employees having to put up with enormous physical and psychological pressure (Rao, 2021). The wellbeing of tourism and hospitality employees was once more put on the frontline. Despite the overabundance of studies examining the notion of wellbeing within the overall context of tourism, less attention has been given by the tourism academic community regarding the wellbeing and tourism employee nexus. This does not imply that researchers have not examined the wellbeing notion from an employee/entrepreneur perspective (Agarwal, 2021; Christou et al., 2020a). Despite the fact that certain studies have investigated employees' wellbeing from specific standpoints, such as through a human resource theoretical lens (Agarwal, 2021; Teo et al., 2020), researchers have called for further insights within the tourism context (Vada et al., 2020), though there still exists a gap of conceptualising wellbeing within the tourism and hospitality employee context.

Significant social changes, such as the pandemic and its dreadful impacts upon the global tourism industry (Smart et al., 2021) and work-related practices, have altered realities as we know them, as well as perceptions of the "wellbeing" notion. Despite the increased attention given to wellbeing by philosophers, psychologists, academics, and practitioners, and its acknowledged importance at a personal, organisational, and national level (DiMaria et al., 2020), there remain gaps in our comprehension of the notion (Salas-Vallina et al., 2021). Furthermore, there are calls for further research particularly within the employee wellbeing context (Wang et al., 2020) and questions that remain unanswered regarding how our theoretical and practical knowledge of the notion will be shaped in years to come. This chapter delivers important information on wellbeing that has value for an individual (Crisp, 2017) and society, while postulating future issues linked to the notion within the context of the tourism and hospitality workplace.

### Wellbeing as a philosophical and psychosocial concept

Wellbeing has been a personal and philosophical concern since ancient times. This reflects more or less the desirability and quest of people for happiness and to live a good life. Ancient philosophers tried to understand and interpret the notion that in ancient Greek was referred to as "*ev zin*" (εὖ ζῆν), which may be vaguely translated as "good life/living". Based on Stroll (2014), much of the philosophy underpinning considerations of the concept has its origins from classical philosophers, such as Socrates and Plato.

DOI: 10.4324/9781003291763-20

The notion has been given a more psychological interpretation with distinctions existing in subjective and psychological wellbeing. In more detail, subjective wellbeing (SWB) refers to a person's cognitive as well as affective evaluations of his or her life, whereas psychological wellbeing consist of six components – autonomy, self-acceptance, sense of continued development, personal growth, purpose in life, and finally positive relationships (Agarwal, 2021). In some cases there are various types of wellbeing distinguished, such as the case of economic or emotional wellbeing, with these forms often being interrelated. Psychologists and academics within the central subject of positive psychology have equipped us with useful tools to measure wellbeing. For instance, Seligman (2018) referred to the acronym PERMA, representing certain measurable elements making up wellbeing and more specifically subjective wellbeing: positive emotion, engagement, relationships, meaning, and accomplishment. In their study, Carreno et al. (2020) used another tool, the Ryff's scales of psychological wellbeing, which through its questionnaire measures the wellbeing of individuals.

Various studies have suggested differing factors that may impact favourably upon the wellbeing of individuals, such as family harmony and self-strengthening (Wang et al., 2021), and others affecting it negatively, such as the case of the perceived threat of COVID-19 (Paredes et al., 2021). Differing lifetime and personal circumstances, such as the case of travel, may affect the wellbeing of individuals. Either way, the state of wellbeing is dynamic. It fluctuates within periods of time, while conceptions of it change with age (Li & Chan, 2020).

### The importance of wellbeing within the context of tourism

The significance of wellbeing is stressed at a societal level, despite its importance at a personal level. Amongst the UN Sustainable Development Goals, wellbeing fostering for all people holds a central place (WHO, 2021). In recent years, the notion has received increased attention by tourism academics, particularly from a consumer-tourist and/or destination community perspective (Farkić et al., 2020; Hanna et al., 2019; Suess et al., 2018). This probably reflects the fact that tourism has often been praised as a contributor to the wellbeing of people (such as locals and entrepreneurs), as well as tourism/leisure activity contributing to the wellbeing of tourists. Travel and tourism may also offer increased opportunities for people to rejuvenate their bodies and souls, indulge in gastronomic delights, relax, socialise, learn, or even offer their services to hosting communities (the case of volunteerism). Holidays may provide escape opportunities for individuals, they may trigger positive emotional states, and perhaps most importantly they may contribute to the wellbeing of disadvantaged groups (Vada et al., 2020). Tourism has been praised as a significant contributor to local communities by creating job opportunities and providing direct and supplementary income to entrepreneurs and those involved in the delivery of services. These direct economic benefits may then be translated into increased opportunities for people to engage in certain pleasurable travel and socio-cultural activities, as well as a means for them to foster positive relationships with others (i.e., friends and locals). From this perspective, tourism may contribute to personal and immediate family wellbeing. Hosts and service providers may benefit from a personal pleasure derived from the offering of hospitality, care, and attention to others (Christou, 2018).

Even so, tourism has been criticised for bringing in various negative impacts on communities and individuals (i.e., locals), who may experience a rather negative impact on their wellbeing mainly due to uncontrolled negative touristic activity, unruly visitor behaviour (Christou, 2021), and/or demanding and exhausting job-related challenges that are involved in the offering of visitor services. All these may have a negative effect on the wellbeing of individuals as well as their motivation to perform to their best in their working environments. From a workplace perspective, current research that has centred around wellbeing within the tourism/hospitality context has established the close and direct association between the wellbeing of employees and their performance (Wang et al., 2020). The wellbeing of employees is of utmost importance for organisations, while researchers call for organisations to be mindful of their employees' wellbeing (Salas-Vallina et al., 2021).

## Major contemporary issues linked to employee wellbeing within the context of tourism

The previous section addressed the importance of the wellbeing notion while delivering some understanding regarding how tourism may impact, whether favourably or negatively, upon the wellbeing of individuals. This section proceeds by presenting some major contemporary issues linked to employee wellbeing within the context of tourism. Such major issues in the contemporary employee wellbeing tourism and hospitality scene may be grouped into personal, exogenous/environmental, and endogenous/organisational factors.

The quest of people for improved wellbeing is undeniably not a current phenomenon nor one that will cease to exist in the future. People continue to travel to specific destinations while engaging in certain (e.g., spiritual) experiences that may favourably contribute towards their wellbeing (Buzinde, 2020). Hence the upsurge of demand for bodily-linked, wellness, as well as spiritual experiences. Whether hospitality and tourism employment opportunities are sought by people as a means to improve their overall wellbeing remains a highly debatable issue, and in need of further research. Just like any other job, a tourism-linked job may contribute at least to some extent to someone's wellbeing, and more specifically in the form of securing personal financial stability and/or providing additional income to spend on leisure and other personal rewarding activities and experiences. The hospitality industry has been credited for allowing a relatively easy access to people seeking employment without much prior working experience being needed (at least for certain posts/positions). The working conditions, particularly within a demanding and busy working hospitality environment, may result in physiological and emotional fatigue, as well as workplace burnout (Ayachit & Chitta, 2022). Hence, employees working in the tourism sector are constantly seeking improved work conditions that may contribute to their overall wellbeing.

The frangibility of the tourism industry towards various external impacts (e.g., wars and terrorism attacks) often result in high unemployment levels, as well as negative impacts on tourism/hospitality employees' wellbeing. The recent pandemic has taught us that the wellbeing of those working in the tourism and hospitality industry is not only affected by company policies, legislation, and organisational behaviour. The pandemic intensified the burnout phenomenon in the hospitality industry (Ayachit & Chitta, 2022). The study of Christou and Savva (2021) revealed that medical face masks had a deeper role to play than merely shielding service providers' health. Amongst other impacts, they caused difficulties in breathing and other social impacts such as miscommunication and misunderstandings with customers, resulting in anxiety and psychological pressure. The study took place in a country (i.e., Cyprus) that depends heavily on international tourist flows. Though the island's hospitality industry turned to alternative means to address the pandemic impacts, many employees revealed that they remained unemployed for large periods of time. This caused a negative impact on their wellbeing, while they made increased references to feelings of "disappointment", "despair", "insecurity", and "fear". The study findings revealed that wellbeing was perceived by employees as an amalgam of a "good" physiological, psychological, and mental state, a "good" and stable job, a stress-free life and working environment, a "good" social life, economic prosperity, free time to spend with themselves or others (e.g., friends), as well as a *sense of freedom*. Interviewees referred to aspects of liberty deprivation (to act, visit, and travel) and loss of freedom due to continuous lockdowns and other travel-linked restrictions. Informants commented on the fact that they had more free time to spend with their friends and families especially during weekends and holidays, such as Christmas and Easter. As they communicated, the nature of hospitality employment requires for them to work night shifts, for long hours, and during holidays. Parallel to these findings, the study of Chen (2021) revealed hospitality workers being financially strained, depressed, and socially isolated, all of these leading to impaired wellbeing. Despite this, other exogenous impacts in the form of technological advancement may also affect the wellbeing of employees. Even so, the manner and the extent to which technology affects employees' wellbeing may be argued. There is the automation of procedures in the workplace, advanced technology, and flexible work arrangements through technological means (Johnson et al., 2020). There rests the fear resulting from automated means and robots replacing humans in the service setting, hence depriving humans of potential job opportunities (Christou et al., 2020b).

There are also certain endogenous/organisational factors that may affect the wellbeing of employees in the current tourism scene. The contributory role of tourism in fostering the wellbeing of individuals specifically within the workplace context may be argued. On the one hand, the tourism and hospitality work setting may be characterised by rewarding personal opportunities falling beyond the financial factor. That is, tourism and hospitality entrepreneurs and employees may derive satisfaction and personal pleasure through their service to a guest or a person in need (Christou et al., 2019). On the other hand, workplace conditions may have a negative effect on someone's wellbeing. It is one thing taking vacations and another thing delivering (service) vacations. Conceivably workplace conditions have an effect on employees' overall psychology, their motivation to work, and their performance. This is why organisations focus on specific and targeted human resource management practices that may foster the wellbeing of their employees (Agarwal, 2021). Researchers (e.g., Salas-Vallina et al., 2021) call for organisations to establish a series of practices that enhance their employees' wellbeing, such as enabling them to perceive that they are being valued, organisations investing in training in which employees may feel more skilled and capable, enhancing two-way communications, and establishing a high quality leader–employee relationship. Others (e.g., Wang et al., 2020) reveal that within the stressful environment of the hospitality workplace setting, if employees perceive that the organisation accepts error occurrence then they may find that the organisation supports and values them. This eventually promotes their psychological wellbeing. Unethical behaviour, such as taking credit for another's work, may have an adverse impact on an employee's wellbeing (Sarwar et al., 2020).

Despite the aforementioned practices and issues that affect the wellbeing of employees, unfortunately the hospitality industry continues to be characterised by high demands of emotional labour, long hours, and a heavy workload, leaving employees unhappy with their workplace conditions (Wang et al., 2020; Mansour & Tremblay, 2016). Additionally, there are recent reports of sexual harassment experiences by female tour guides that have caused adverse impacts on their wellbeing (Alrawadieh et al., 2021). Also, the burnout phenomenon stubbornly remains an important issue (Ayachit & Chitta, 2022).

### Future perspectives on the nexus of wellbeing and tourism/hospitality employees

The previous sections have provided some insights into the concept of wellbeing, its importance, and how the notion is shaped and affected by endogenous/organisational and exogenous contemporary factors. It is hard, if not impossible, to predict how people's wellbeing will be shaped by 2050 or even earlier. This depends on the individual and the organisational, geographical, and situational context in which wellbeing is placed, as well as the medical, economic, technological, and environmental factors that will continue to shape the wellbeing of people in years to come. Even so, what is certain is that the wellbeing of people will remain over the coming decades of the utmost importance at a personal, organisational, and societal level.

In order to offer more specific perspectives regarding the wellbeing of tourism and hospitality employees one needs to acknowledge the current issues that the notion is facing, as well as to examine the past trends and issues linked to wellbeing. In the last few decades tourism and hospitality organisations have embraced employee legislation and human resource practices to secure and foster the wellbeing of their employees. This does not imply that all hospitality/tourism organisations have purposely and/or successfully done so or will be willing to do so in the future. History has taught us that the wellbeing of people has always been impacted by various factors, such as economic prosperity/recession, wars, serious catastrophes, and other crises. Such crises have provided us with opportunities to investigate this rather perplexed notion. It may be argued that "you must lose something to truly appreciate its value". There are various personal, organisational, and external factors that impact on the wellbeing of employees. In all likelihood, these will continue to shape such wellbeing, whether favourably or negatively. Resting on future predictions of tourism evolution (Christou, 2022), forthcoming wars and other crises will continue to trouble the global tourism scene. As a result, hospitality employees will be constantly called to address feelings of fear and insecurity. The role of national and regional governments will be extremely important since they must ensure that tourism-linked employees are provided with the necessary financial support during such periods. National stakeholders should strongly encourage and invest in domestic tourism, so that the employees of those destinations that rely heavily on international arrivals are not left jobless at times of international turmoil. Tourism and hospitality stakeholders must also look beyond egocentric, unethical, strictly profit-oriented tactics, and over-dependence on automated/robotic means to carry out tasks. Tourism employees must feel secure, as well as appreciated and valued/loved (Christou, 2018) in their future workplaces. Towards this end, future academic institutions should embrace as much as possible in their future curricula development modules that are linked to human resource management, topics that deal specifically with the wellbeing of employees. Another important aspect that has emerged particularly as a result of the recent pandemic and a careful examination of recent issues linked to the wellbeing of employees in the tourism field is the construct of "freedom". This construct would not have been brought up by employees prior to the pandemic. As discussed in the previous sections, due to regular and strict lockdowns and movement restrictions, employees perceive that they have been deprived of the freedom to travel, engage in pleasurable activities, and visit friends; and to keep a

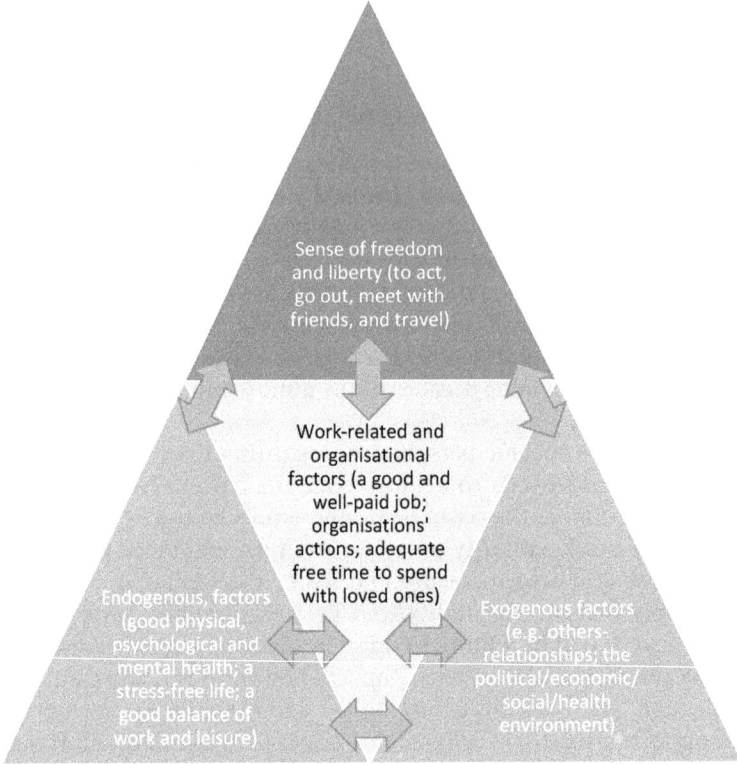

*Figure 17.1* The pyramid of wellbeing and the emerging construct of "sense of freedom".

greater physical distance from others and be obliged to wear face masks during service provision for long hours. Eventually, this has affected their overall wellbeing.

Figure 17.1 summarises the key aspects covered in this chapter and which are related to the wellbeing of tourism and hospitality employees. This pyramid-shaped diagram provides some direction for future tourism stakeholders and academics in regard to the wellbeing of those working in the tourism and hospitality industry. The diagram presents in a summarised format all interrelated factors that are crucial for the wellbeing of tourism employees. The diagram may be used beyond the sphere/context of tourism since it presents a neglected or at least highly marginalised element – that of freedom. Tourism stakeholders and more specifically managers and leaders in the future workplace will have to address this issue (i.e., of freedom/liberty) for securing and promoting their employees' wellbeing. In more practical terms this may mean further allocation of a freedom sense, such as for instance the distribution of autonomy, a sense of liberty in delivering views, complaints, and concerns. Another suggestion is for future leaders to allocate as much free time as possible for employees and even the provision of travel rewards and hotel stay vouchers. It is also well acknowledged that the tourism and hospitality industry often requires its members to work for many hours, particularly during holiday seasons, who thus are deprived of opportunities to spend important time with their loved ones. Travel or hotel vouchers that may come as a form of reward may provide increased opportunities for employees to rejuvenate and recharge their batteries. Despite the severe negative impacts caused by the recent pandemic, tourism employees were given the opportunity to spend more time with those whom they value the most. They were able to feel free to engage in and experience certain pleasurable and social activities with their friends. Hence, while allowing employees some flexibility in their work shifts and possibly a more structured/scheduled programme, managers in the future may aid employees in organising and spending more quality time with their loved ones.

On a more theoretical level, theories and future research should consider the dimension of freedom and liberty when examining the wellbeing of individuals. Quantitative studies are proposed to create scales of measuring this dimension that have emerged from this study. It is recommended that the UN Sustainable Development Goals of the WHO assess the likelihood of embracing the construct of freedom (i.e., freedom deprivation) while investigating or promoting the wellbeing of individuals in the forthcoming decades. Additional research is needed to investigate further the nexus of wellbeing and hospitality/tourism employees. More research is encouraged in regard to the relationship and the possibly causal effect of technology, artificial intelligence, and robots, over the overall wellbeing of employees. Another future research stream may investigate the relationship of social media and people's (in this case employees') wellbeing. A complexity characterises the connection of wellbeing and social media (Kross et al., 2021). Social media has certainly revolutionised how we interact with others and it will be interesting to investigate how employees share their workplace experiences and emotional states caused by workplace conditions and whether this sharing of psychological states impacts on their and others' wellbeing.

## Conclusion

This chapter has introduced useful information regarding the notion of wellbeing while highlighting its significance at an individual, industrial, and societal level. The relationship of the notion with tourism was also covered. The chapter proceeded by covering major contemporary issues linked to employee wellbeing within the context of tourism and delivered future perspectives of the nexus of wellbeing and tourism/hospitality employees. A useful conceptual diagram (Figure 17.1) was presented. This summarises the key interrelated factors contributing towards and shaping the wellbeing of individuals. The construct of "sense of freedom" is given particular emphasis, not necessarily because of its dominance over other important factors, but due to its marginalisation by the research and professional community. Certain research avenues are proposed, as well as managerial implications, that may be of great interest for tourism stakeholders and future leaders in securing the wellbeing of their workforce. Socrates (*c*.470–399 BC) once stated that the secret of happiness is not found in seeking more but developing the capacity to enjoy less. We embrace this quote, yet argue that, within the work environment, employees are not to be expected to seek or enjoy less. Their wellbeing is and will continue to be jeopardised by various factors, including unemployment risks, unethical management, egocentric management/visitor attitudes, inconsiderate guests, and physical and emotional burnout. Hence, active support, care, value, empathy, respect, appreciation, and a "sense of freedom" will be in high demand. Besides, we all have our share in securing each other's wellbeing.

## References

Agarwal, P. (2021). Shattered but smiling: Human resource management and the wellbeing of hotel employees during COVID-19. *International Journal of Hospitality Management, 93*, 102765.

Alrawadieh, Z., Demirdelen Alrawadieh, D., Olya, H. G., Erkol Bayram, G., & Kahraman, O. C. (2021). Sexual harassment, psychological wellbeing, and job satisfaction of female tour guides: the effects of social and organizational support. *Journal of Sustainable Tourism, 30*(7), 1–19.

Ayachit, M., & Chitta, S. (2022). A systematic review of burnout studies from the hospitality literature. *Journal of Hospitality Marketing & Management, 31*(2), 125–144.

Buzinde, C. N. (2020). Theoretical linkages between wellbeing and tourism: The case of self-determination theory and spiritual tourism. *Annals of Tourism Research, 83*, 102920.

Carreno, D. F., Eisenbeck, N., Cangas, A. J., García-Montes, J. M., Del Vas, L. G., & María, A. T. (2020). Spanish adaptation of the Personal Meaning Profile-Brief: Meaning in life, psychological wellbeing, and distress. *International Journal of Clinical and Health Psychology, 20*(2), 151–162.

Chen, M. H. (2021). Wellbeing and career change intention: COVID-19's impact on unemployed and furloughed hospitality workers. *International Journal of Contemporary Hospitality Management, 33*(8), 2500–2520.

Christou, P. A. (2022). *The History and Evolution of Tourism.* UK: CABI.

Christou, P. A. (2021). *Philosophies of Hospitality and Tourism.* UK: Channel View Publications.

Christou, P. A. (2018). Exploring agape: Tourists on the island of love. *Tourism Management, 68*, 13–22.

Christou, P., & Savva, R. (2021). Impacts of the pandemic: the role of 'face masks' in hospitality and tourism service provision. *Current Issues in Tourism, 25*(23), 3747–3760.

Christou, P. A., & Rigou, M. (2021). The pandemic and wellbeing: Views from tourism employees. *APacCHRIE Conference*, Singapore.

Christou, P., Hadjielias, E., & Farmaki, A. (2019). Reconnaissance of philanthropy. *Annals of Tourism Research, 78*, 102749.

Christou, P., Hadjielias, E., & Farmaki, A. (2020a). Silence, sounds and the wellbeing of tourism entrepreneurs in noisy tourism workplaces. *Current Issues in Tourism, 24*(18), 2658–2670.

Christou, P., Simillidou, A., & Stylianou, M. C. (2020b). Tourists' perceptions regarding the use of anthropomorphic robots in tourism and hospitality. *International Journal of Contemporary Hospitality Management, 32*(11), 3665–3683.

Crisp, R. (2017). *Wellbeing. The Stanford Encyclopedia of Philosophy. Metaphysics Research Lab, Stanford University.* Retrieved from: https://plato.stanford.edu/entries/wellbeing/. Accessed: 20/03/2021.

DiMaria, C. H., Peroni, C., & Sarracino, F. (2020). Happiness matters: Productivity gains from subjective wellbeing. *Journal of Happiness Studies, 21*(1), 139–160.

Farkić, J., Filep, S., & Taylor, S. (2020). Shaping tourists' wellbeing through guided slow adventures. *Journal of Sustainable Tourism, 28*(12), 2064–2080.

Hanna, P., Wijesinghe, S., Paliatsos, I., Walker, C., Adams, M., & Kimbu, A. (2019). Active engagement with nature: Outdoor adventure tourism, sustainability and wellbeing. *Journal of Sustainable Tourism, 27*(9), 1355–1373.

Johnson, A., Dey, S., Nguyen, H., Groth, M., Joyce, S., Tan, L., Glozier, N. & Harvey, S. B. (2020). A review and agenda for examining how technology-driven changes at work will impact workplace mental health and employee wellbeing. *Australian Journal of Management, 45*(3), 402–424.

Kross, E., Verduyn, P., Sheppes, G., Costello, C. K., Jonides, J., & Ybarra, O. (2021). Social media and wellbeing: Pitfalls, progress, and next steps. *Trends in Cognitive Sciences, 25*(1), 55–66.

Li, T. E., & Chan, E. T. H. (2020). Diaspora tourism and wellbeing over life-courses. *Annals of Tourism Research, 82*, 102917.

Mansour, S., & Tremblay, D. G. (2016). Workload, generic and work–family specific social supports and job stress: Mediating role of work–family and family–work conflict. *International Journal of Contemporary Hospitality Management, 28*(8), 1778–1804.

Paredes, M. R., Apaolaza, V., Fernandez-Robin, C., Hartmann, P., & Yañez-Martinez, D. (2021). The impact of the COVID-19 pandemic on subjective mental wellbeing: The interplay of perceived threat, future anxiety and resilience. *Personality and Individual Differences, 170*, 110455.

Rao, T. (2021). What is Hospitality? The current answer doesn't work. *The New York Times*. Retrieved from: https://www.nytimes.com/2021/04/13/dining/restaurant-hospitality.html?smid=em-share. Accessed: 05/2021.

Salas-Vallina, A., Alegre, J., & López-Cabrales, Á. (2021). The challenge of increasing employees' well-being and performance: How human resource management practices and engaging leadership work together toward reaching this goal. *Human Resource Management*, *60*(3), 333–347.

Sarwar, H., Ishaq, M. I., Amin, A., & Ahmed, R. (2020). Ethical leadership, work engagement, employees' wellbeing, and performance: a cross-cultural comparison. *Journal of Sustainable Tourism*, *28*(12), 2008–2026.

Seligman, M. (2018). PERMA and the building blocks of wellbeing. *The Journal of Positive Psychology*, *13*(4), 333–335.

Smart, K., Ma, E., Qu, H., & Ding, L. (2021). COVID-19 impacts, coping strategies, and management reflection: A lodging industry case. *International Journal of Hospitality Management*, *94*, 102859.

Stoll, L. (2014). A short history of wellbeing research. In *Wellbeing: A Complete Reference Guide*. New York: Wiley, 1–19.

Suess, C., Baloglu, S., & Busser, J. A. (2018). Perceived impacts of medical tourism development on community wellbeing. *Tourism Management*, *69*, 232–245.

Teo, S. T., Bentley, T., & Nguyen, D. (2020). Psychosocial work environment, work engagement, and employee commitment: A moderated, mediation model. *International Journal of Hospitality Management*, *88*, 102415.

Vada, S., Prentice, C., Scott, N., & Hsiao, A. (2020). Positive psychology and tourist wellbeing: A systematic literature review. *Tourism Management Perspectives*, *33*, 100631.

Wang, T., Jia, Y., You, X. Q., & Huang, X. T. (2021). Exploring wellbeing among individuals with different life purposes in a Chinese context. *The Journal of Positive Psychology*, *16*(1), 60–72.

Wang, X., Guchait, P., & Paşamehmetoğlu, A. (2020). Why should errors be tolerated? Perceived organizational support, organization-based self-esteem and psychological wellbeing. *International Journal of Contemporary Hospitality Management*, *32* (5), 1987–2006.

WHO. (2021). Sustainable Development Goals. Retrieved from: https://www.who.int/health-topics/sustainable-development-goals#tab=tab_2. Accessed: 20/03/2021.

# 18 Resilient leadership in the tourism industry

*Adela Balderas-Cejudo, Marta Buenechea-Elberdin, Josune Baniandrés and George W. Leeson*

## Introduction

Resilience is gaining popularity and attention across a wide range of academic disciplines and corporate sectors (Jones & Comfort, 2020; Sardá & Pogutz, 2018). Resilient industries, businesses, cities, and people can adapt, deal with adversity, and move on, allowing for rethinking, reorganisation, and even planning for future challenges. It is crucial to build resilience inside organisations – not simply in a reactive way but in a proactive one – led by resilient leaders. Although some corporate leaders are already implementing transformational changes to make companies more resilience-oriented and to match their missions and purposes with the Sustainable Development Goals (SDGs), research on resilient leaders and mechanisms for stimulating organisational resilience is still in its early stages (Lombardi et al., 2021). Following Bahadur et al. (2015), resilience is addressed directly and implicitly in a variety of suggested SDG targets to be met by 2030. Objectives 4.4, 4.7, and 8.2 allude to ensuring economic growth and adequate employment through technological advancements and innovation, as well as having a varied set of knowledge and skills for adequate employment and sustainable living. The better the environment for its residents and the greater the preparedness employees have in the face of adversity, the more resilient an economy and a person are. As suggested by Luthe and Wyss (2014), resilience requires more attention in tourism research; it has also become a popular issue in the tourism and hospitality literature (Prayag et al., 2018).

Recovering in the face of adversity and adapting to change should be intrinsic to the hospitality industry, given its changing and dynamic nature. Climate change, disruptive technology, consumer patterns, transformational changes, and the impact of COVID-19 management all appear to have pushed the concept of resilience into closer focus and perspective within the tourism industry (Jones & Comfort, 2020). Twinning-Ward et al. (2017) stated in their review of "The Global Conference on Jobs and Inclusive Growth: Partnerships for Sustainable Tourism" held in Jamaica in 2017 that

> the key word for the conference was resilience: not only how to build back better, but also how to build resilience into the everyday management of tourism, how to be better prepared, how to manage a crisis, and how to ensure greater shared economic and social benefits from tourism in the region.

Individuals, teams, and organisations all benefit from resilience and there is a growing interest in understanding its construct across all fields of psychology and the broader domain of organisational science (Britt et al., 2016). The University of Cambridge Institute for

DOI: 10.4324/9781003291763-21

Sustainability Leadership published a report in 2020 claiming: "People matter. People under-pin systems – resilient organisations cannot exist without resilient people, so we need to understand what it takes to enhance physical, emotional and psychological resilience and invest in the wellbeing of individuals" (p. 5). Being better prepared as a leader, knowing how to react, and managing a team in a resilient manner appears to be a prerequisite. Resilient leadership is a type of leadership that encourages individuals and the organisation as a whole to overcome challenges and become changemakers. Is this a style that would work in tourism industry?

The aim of this chapter is three-fold: (1) to provide a deeper understanding of this thriv-ing and growing field of resilience in the tourism industry; (2) to show the need to compre-hend and acknowledge the importance of fostering resilient leaders, who are able to cope with disruptive changes, to navigate complexity, and to cultivate resilience in themselves and in their teams; and (3) to provide some broader views on the term's applicability to leadership in the industry.

**Resilience and its role in the tourism industry**

Business innovation is creating quick and transformative changes in technology, consump-tion patterns, and lifestyle goals across the world. Simultaneously, societies are looking at businesses to drive change in response to pressing health, social, and environmental issues. These issues pose serious threats to society's stability and wellbeing. Nevertheless, they also present opportunities for adaptation. This shifting scenario is causing major upheavals across entire industries, economies, and sectors, necessitating not only new legislative and governance frameworks, but also more business accountability and resilient management and leadership (CISL, 2020).

In addition, tourism industry trends (Figure 18.1) point to a significant digital transfor-mation that will support customers' increased power as well as their intentions to promote sustainability. Customers will demand living experiences and personalised offerings, and they will have access to all of the technological tools required to meet those demands. Sus-tainability will be prioritised in this industry, including not only environmental but also social and economic sustainability. As stated at the UNWTO Executive Council meeting, the industry should look "ahead to a more sustainable, inclusive and resilient future" (UNWTO, 2022).

There are three major lessons about resilience that emerge from the research, according to Fletcher and Sarkar (2013). To begin with, resilience is made up of a number of variables that boost personal assets and protect people from the negative effects of stressors. Second, recovery and coping should be thought of as separate concepts from resilience. Finally, it has an impact on the stress response at several stages to perceived emotions, as well as on coping strategy selection. Although there is some debate on how to define "resilience", there has been an increasing interest in learning more about it. According to Britt et al. (2016), the use of the term "resilience" – to refer to a broad variety of positive attributes and processes that all have something to do with how people respond to stressful situations but have nothing else in common – is the biggest hindrance to development in the resilience literature.

The word "resilience" comes from the Latin verb *resilire*, or "to leap back", and is defined in the *Oxford Dictionary of English* as being "able to withstand or recover quickly from difficult conditions" (Soanes & Stevenson, 2006, p. 1498). Following Fletcher and Sarkar (2013), the understanding of human functioning in stressful situations has advanced

- Demand for **living experiences** and feeling immersed in local culture.
- **Self-booking** trips.
- Travel agencies' role depending on their ability to offer **unique experiences**.
- Demand for **personalising offerings** and to provide easy-to-use applications.
- Consequently, **empowered customers**.

**CUSTOMER**

**SUSTAINABILITY**

- Special care for the **environment**.
- Promoting **employment**.
- Boosting **inclusion**.

**TECHNOLOGY**

- **Technology advancements** (virtual reality, facial recognition, 3D printing, augmented reality, among others).
- **Transportation advancements** (efficient planes, fast trains and updated transportation options).

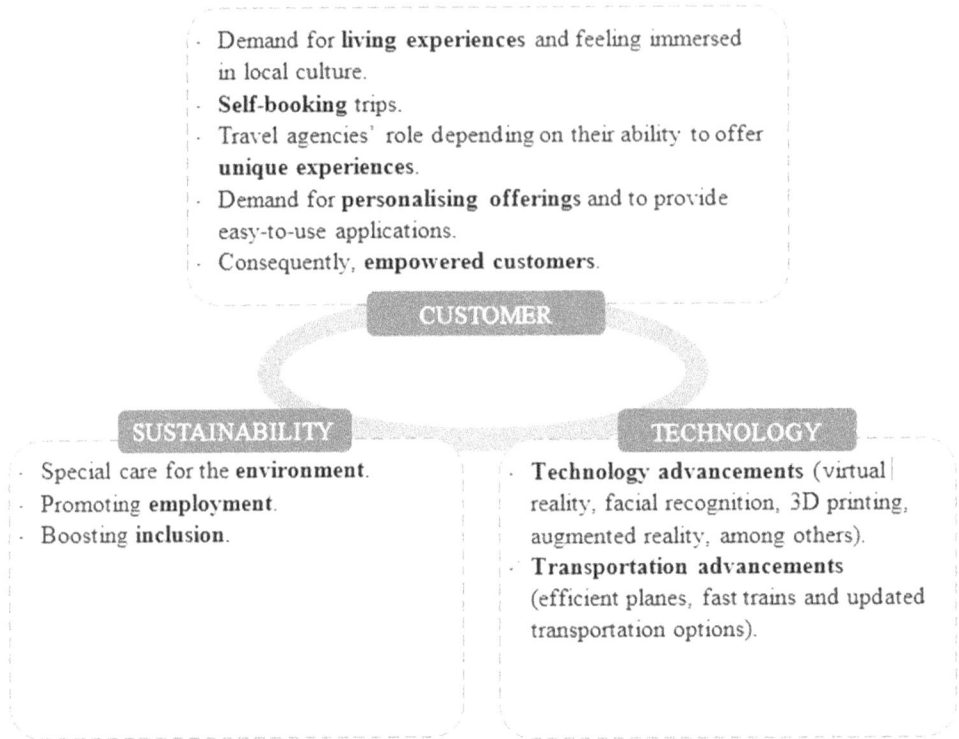

*Figure 18.1* Tourism industry trends.

*Source*: Author elaboration based on Hammond (2019), UNWTO (2022), and White (2021).

dramatically in recent decades, with resilience being studied in a variety of contexts, including businesses, education, sports performance, the military, and communities. Lengnick-Hall et al. (2011) defined "resilience" as the ability of an organisation to face a crisis and capitalise on the situation, potentially avoiding bad repercussions, expanding, and becoming stronger as a result. Giustiniano et al. (2018) define "resilience" as the ability to absorb adversity, trauma, external shocks, or any significant kind of stress, while learning from, preparing for, and adapting to changes, through a complex network of variables. These different understandings of resilience differ in whether they "require positive growth" or "simply require successful adaptation" (Britt et al., 2016, p. 380). In their article "The Art and Science of Resilience", PwC (2015) emphasised their belief that organisational resilience is the most significant. Enterprise resilience is defined as "an organisation's capacity to anticipate and react to change, not only to survive, but also to evolve" (PwC, 2015). In spite of "the construct being operationalized in a variety of ways, most definitions are based around two core concepts: adversity and positive adaptation" (Fletcher & Sarkar, 2013, p. 5).

Concurring with Eliot (2020), a study involving 223 physicians in the German healthcare system revealed a significant association between physician resilience and professional engagement. The relationship between work engagement and 11 different personal and organisational resources was analysed and the authors concluded that "resilient people show higher scores of work engagement" (Mache et al., 2014, p. 496) and tend to prepare for hardships and minimise the impact of stressful events on themselves proactively by using their psychological resources effectively (Fredrickson et al., 2008). Employee commitment, engagement, and change initiatives are major predictors of transformation-related organisational outcomes, and employee commitment is defined as "a force (mind-set) that binds an individual to a course of action deemed necessary for the successful implementation of a change initiative" (Herscovitch & Meyer, 2002, p. 475). In a highly competitive market, employees in the travel and tourism industry are a critical component of a resilient organisation and a crucial source of value and difference. In times of uncertainty and technological upheaval, great talent is crucial, and employee engagement and resilience are even more important in delivering a positive customer experience (Balderas-Cejudo & Leeson, 2022). Studies have continuously established a direct correlation between resilience and pleasant emotions in difficult conditions (Fredrickson et al., 2008), which may directly contribute to customer experience and organisational recovery.

Given the importance of resilience in organisations in general and the tourism industry in particular, which has a direct impact on service quality, recognising and strengthening resilience to current and future challenges will be critical when confronted with future major crises or adversities. Following Bethune et al. (2022), resilience in a complex, dynamic, and interconnected environment is a function of an organisation's situation awareness, diagnosis, and analysis of keystone vulnerabilities and adaptive capacity (McManus, 2008). As suggested by Luthe and Wyss (2014), the idea of resilience has much explanatory power and deserves further consideration in tourism research.

The tourism industry should demonstrate its resilience, which is urgently needed. Resilience at all value chain levels has the potential to transform the tourism industry into a new global economic system characterised by sustainability, climate change action, societal wellbeing, and community participation (Sharma et al., 2021).

## Leadership: Towards a resilient leader in resilient organisations

Leadership is a driving force in the development of employee attitudes and values, being an enhancer of the traits that identify the type of person that works for a certain company. The literature on leadership contains a variety of definitions and styles. Senge et al. (1999, p. 90) describe leadership as "the capacity of a human community to share its future, and specifically to sustain the significant processes of change required to do so". According to Prewitt et al. (2011, p. 14), "leaders create change and set the direction. A leader can take people and an organisation in a new direction with their leadership abilities".

Within the wide range of definitions of leadership that the literature accumulates, there are three basic components that different interpretations of the notion share: leadership means taking the initiative, involving others, and directing resources and behaviours toward certain objectives (Dartey-Baah, 2015).

Following Giustiniano and colleagues (2020, p. 971), "when the winds of change blow, some people build walls and others build windmills". Protecting against the wind but also figuring out how to harness its power is the hallmark of resilient leadership (Giustiniano et al., 2020). "Scholars agree that resilience is key to leadership" because leaders must be able to use all of their resources to overcome difficult situations (King & Rothstein, 2010, as cited in Lombardi et al., 2021, p. 3). Leadership is critical for resilience since leaders have a large impact on their teams, making them potential resilience infusers (Eliot, 2020).

Being a resilient leader is very demanding as such people establish a work environment that allows organisations and employees to develop and adapt, and at the same time, they should learn from, overcome problems, and improvise the company model (Lombardi et al., 2021). A resilient leader must hold opposing characteristics such as learning from mistakes and preparing for future problems, readiness and improvisation, unambiguous direction-setting and flexibility (Giustiniano et al., 2020; Lombardi et al., 2021). In addition, resilient leaders should have the ability to instil resilience in their followers (Eliot, 2020); nevertheless, resilience is not automatically spread (Lombardi et al., 2021).

Resilience, like any other ability, can be developed. Hillmann (2021) compiled a number of studies on how to enhance resilience in the workplace and found that building positive relationships and treating employees with respect were crucial. Masten (2001, as quoted in Eliot, 2020) devised ways for gaining a better understanding of how people respond to risks and improve their ability to cope with them. Resilient individuals are characterised by their improvisation skills (Lombardi et al., 2021) and their "problem solving abilities, favourable perceptions, positive reinforcement, and strong faith" (Lengnick-Hall et al., 2011, p. 245). Training is believed to be a resilience facilitator (Altshuler et al., 2021).

Some studies have highlighted a link between leadership styles and resilience development. According to Eliot (2020), servant leaders who focus on meeting followers' psychological needs can help them become more resilient. Employee resilience is necessary for organisational resilience and, in fact, employees can only make a company as resilient as they are themselves (Hillmann, 2021; Lengnick-Hall et al., 2011). A resilient business is one that continues to operate while successfully innovating and responding to rapidly changing circumstances (Robb, 2000, as cited in Dartey-Baah, 2015). Leaders have a critical role in improving employee resilience and, as a result, the organisation's ability to be resilient. Lombardi et al. (2021) suggest, "effective leaders are therefore strategic to the diffusion of resilience because they enable a process of 'learning to unlearn and learn' (Giustiniano et al., 2020), struggling to find a balance between reaction and adaptation, and transforming stressors and shocks into new energy".

Following Dartey-Baah (2015), resilient leadership mixes transformational and transactional leadership styles. What is evident is that having resilient leaders who can instil resilience in their followers, and therefore throughout the organisation, offers numerous benefits that have been linked to various leadership styles. The question is whether current people management trends are geared toward promoting organisational resilience.

## Future trends of people management

Every year, a number of reports covering a wide range of people management subjects are released. We have constructed a framework with eight trends (Figure 18.2), based on an analysis of studies published between 2017 and 2020.

The first set of trends concerns workforce composition and job design. The fourth industrial revolution has altered the work environment. A top priority in all sectors, including tourism, is revising workforce composition to meet new challenges and redesigning jobs to reflect the new technology-driven environment (KPMG, 2020; Trends, 2019). The tourist industry has begun to develop broader and more flexible job descriptions that encompass more combinations of duties and responsibilities in order to cope with decreasing demand and make workers more employable (Lytle, 2020), which in turn makes them more resilient.

Alternative workforce and talent mobility tend to be frequent among the sources of incoming staff processes (KPMG, 2020). Employers in the tourism business are currently dealing with high levels of voluntary turnover and an abundance of job applications making it difficult to identify the ideal candidate (Lytle, 2020). Additionally, downsizing tactics

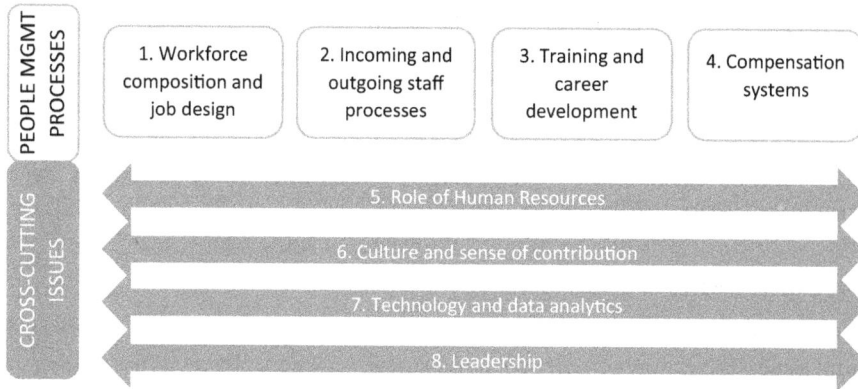

*Figure 18.2* People management trends.

are a topic of discussion for tourism leaders (Lytle, 2020). These changes raise the question of whether resilience can be built.

The fourth industrial revolution, as well as the present necessity for businesses to have a beneficial impact on their employees and society, compels firms to seek out new abilities such as analytical skills and employee value proposition management (KPMG, 2019, 2020; McKinsey Global Institute, 2017). "Learning from disaster planning and fighting the drive to turn away from failures experienced" as well as cross-training to have more adaptable staff are crucial in the tourism business (Riviera, 2020, as cited in Sharma et al., 2021, p. 5; Lytle, 2020). Training is becoming more personal and lifelong, and this long-term training is considered as assisting in the development of workforce resilience (Trends, 2019, 2020). Employee career development is critical because talent retention is a major concern, as it also is in the tourist sector (KPMG, 2020; JWU, 2021).

Another set of trends involves creating compensation systems that reflect human principles and are aligned with agile performance management models and employees' needs (Trends, 2019, 2020). Many roles in the tourism business pay poor wages, which contributes to the loss of talent (Lytle, 2020).

The fifth category includes leveraging technology and technology-related skills to boost the strategic value of the personnel department (KPMG, 2019). Trends also point to a shift in culture that welcomes failure in the quest of innovation (KPMG, 2019, 2020). The tourism industry has chosen to balance the trend of remote work with a collaborative and close engagement culture (Lytle, 2020). In the NH Hotel group, an engagement culture is fostered through communication, through talking and listening about professional and personal matters that may influence a person.

Businesses must develop a technological strategy for human resources (HR) as a result of the fourth industrial revolution (Trends, 2019). This is especially true in the tourism industry, where robots could replace people and the use of technology has risen (Lytle, 2020; Sharma et al., 2021). Which is the best combination of digital tools for enhancing the human contribution? Unfortunately, firms devote few resources to answering this issue (KPMG, 2019). Data analytics include assessing and thoroughly understanding people in order to design techniques that strengthen their sense of belonging and guide difficult employee decisions (KPMG, 2019, 2020; Trends, 2020). Because tourism is a service industry, it employs a large number of people, making the opportunity to learn more about them a significant one.

Leadership-related data indicate a disparity between words and deeds with respect to fostering an innovation-driven culture (KPMG, 2020). Leadership should be approached in a way that considers the current environment and essential skills, such as leading through change, ambiguity, uncertainty, and understanding new technology (Trends, 2019), emphasising the importance of resilient leadership. This is especially true in the tourism industry, which has recently experienced significant downturns and whose managers have been trained on how to lead in difficult situations (Lytle, 2020).

### Building resilient leaders in global tourism

From social pressures and political upheavals to technological shifts and health and environmental issues, the power of dynamic context is everywhere. To revitalise and resuscitate the tourism business, transformations such as restarting, reorganising, developing, and adapting it to the current norms and rules are required (Lew et al., 2020). Fostering resilient leadership requires working in parallel with two fields, that is, public institutions and private companies.

In terms of public institutions, they can help to professionalise the sector by promoting policies that protect employees, train them, and reduce talent drain. Resilient leadership will not work without employees with the required talent to cope with the ever-changing tourism industry environment. As a result, training current and potential employees, as well as enacting policies that reduce talent scarcity, are critical.

Leaders in private companies must understand the significance of contribution to organisational performance and how to deal with it when faced with failure or difficult circumstances. The organisation's regeneration and future prosperity are dependent on leaders' capacity to rebound from setbacks (Lombardi et al., 2021). Fear of change is one of the most common anxieties that leaders and organisations must confront and overcome (Southwick et al., 2017). It is revealing that resilient leadership entails a combination of paradoxical behaviours that necessitate a paradoxical capability: the ability to maintain regular functioning in the face of adversity. Because resilient leadership is so challenging, HR managers could use a performance evaluation process to assess leaders' abilities and then train them to develop those that are lacking. Sharma et al. (2021) and Lytle (2020) identify a few resilient skills, which are important in the tourism industry and could serve as a guideline, and emphasise the importance of mastering them.

HR professionals hold a critical role in building organisational resilience. Lengnick-Hall et al. (2011) hold that businesses could create an HR system that promotes organisational resilience and allows leaders to practise their style. Furthermore, every leader requires followers, so leadership is also dependent on having people to guide. A leader's high resilience extends to his or her followers (Eliot, 2020), yet it is important to note that resilience spread is not automatic; rather, it should be worked on, as in the perfect combination of the leader as gardener and leading while learning (Lombardi et al., 2021). The activity of resilient leaders is severely hampered by the lack of an organisational environment that has resilience and followers with appropriate skills. Therefore, HR departments' efforts are essential as they have the ability to promote working environments conducive to having versatile employees capable of assuming different roles through the creation of flexible job descriptions in combination with intensive training. These departments are also responsible for identifying and exploiting different recruiting sources that could counteract personnel scarcity, which is nowadays one of the main problems of the tourism industry.

HR managers could make effective use of the available technology and data analytics resources to understand employees' needs and desires in terms of compensation for their work and design compensation plans accordingly. As proved by Jung et al. (2021, p. 7), it is important to create environments in which employees "can express consideration and attention through mutual communication". Listening to people's voices and opinions is essential for fostering employee engagement and, as a result, employee resilience. Conversations play a vital role, as in the case of NH Hotel Group, one of the 25 largest chains in the world that operates across Europe, America, and Africa (https://www.nh-hotels.com/corporate/about-nh), which is very focused on stabilising engagement through communication. The HR department places a focus on communication skills and emphasises these interactions where everyone is free to express their opinions. This enables individuals to truly feel they are in the spotlight; to believe that they matter greatly; and as a result, to be confident in themselves and give their all despite context changes. In the same way, exit interviews are essential for communicating with people and learning about their perspectives on what works well and what needs improvement. The aim is to make the person feel significant, which is their true goal, and view the first day of incorporation as being equally significant as the final.

To summarise, the role of HR departments in enhancing resilience is to foster HR policies that support employee development while also promoting an environment where people feel cared for and the talent shortage is reduced. As Diego García de Vinuesa (NH Hotel Group Senior Director of People – Southern Europe and USA) highlights: "It is very important to listen to the voice of our people".

## Conclusions

Leaders whether in politics, industry, or civil society, must act in the context of a complex system of global forces and trends, and the challenge of leadership is to turn these dangers into opportunities (CISL, 2020). This scenario needs resilient organisations and resilient leaders. Managers of the tourism industry must support and reinforce the ecosystem by providing effective, flexible, adaptable leadership that will enable timely, accurate decisions to be made even in the face of uncertainty (Bethune et al., 2022). Managers should also work on HR policies that promote the development of a resilient environment and people with the necessary skills for dealing with turbulent changes. Healthy relations with employees and treating them with dignity are critical for increasing resilience (Hillmann, 2021).

Concurring with Jones and Comfort (2020), the concept of resilience is illuminating academic research on sustainability and influencing the development of new tourism planning strategies, as well as providing some broader reflections on the concept's use within the sector.

The goal of this chapter was to comprehend the critical need for fostering resilient leaders in the tourism industry, as well as to provide some broader perspectives on the term's applicability to leadership in the industry. We propose that resilient leaders and organisations have a two-way relationship in which leaders may inspire resilience in their followers and the environment and culture can encourage resilience in the leader.

Resilience is a concept closely related to the SDGs, which offer a roadmap to a better future and may provide a realistic approach to navigate societies (Van Zanten & Van Tulder, 2020).

## References

Altshuler, A., & Schmidt, J. (2021). Why does resilience matter? Global implications for the tourism industry in the context of COVID-19. *Worldwide Hospitality and Tourism Themes, 13*(3), 431–436.

Bahadur, A., Lovell, E., Wilkinson, E., & Tanner, T. (2015). *Resilience in the SDGs.* https://www.researchgate.net/profile/Aditya-Bahadur/publication/281031689_Resilience_in_the_SDGs_Developing_an_indicator_for_Target_15_that_is_fit_for_purpose/links/55d1d29708aee5504f68f12f/Resilience-in-the-SDGs-Developing-an-indicator-for-Target-15-that-is-fit-for-purpose.pdf

Balderas-Cejudo, A., & Leeson, G. W. (2022). Resilience of tourism employees. In Buhalis, D. (Ed.), *Encyclopaedia of tourism management and marketing.* Cheltenham: Edward Elgar Publishing.

Bethune, E., Buhalis, D., & Miles, L. (2022). Real time response (RTR): Conceptualizing a smart systems approach to destination resilience. *Journal of Destination Marketing & Management, 23,* 100687.

Britt, T. W., Shen, W., Sinclair, R. R., Grossman, M. R., & Klieger, D. M. (2016). How much do we really know about employee resilience? *Industrial and Organizational Psychology, 9*(2), 378–404.

Cambridge Institute for Sustainability Leadership (CISL). (2020). *The Implications of COVID-19 for Leadership on Sustainability.* https://www.cisl.cam.ac.uk/system/files/documents/the-implications-of-COVID19-for-leadership-on.pdf

Dartey-Baah, K. (2015). Resilient leadership: A transformational-transactional leadership mix. *Journal of Global Responsibility, 6*(1), 99–112.

Eliot, J. L. (2020). Resilient leadership: the impact of a servant leader on the resilience of their followers. *Advances in Developing Human Resources, 22*(4), 404–418.

Fletcher, D., & Sarkar, M. (2013). Psychological resilience: A review and critique of definitions, concepts, and theory. *European Psychologist, 18*(1), 12.

Fredrickson, B. L., Cohn, M. A., Coffey, K. A., Pek, J., & Finkel, S. M. (2008). Open hearts build lives: positive emotions, induced through loving-kindness meditation, build consequential personal resources. *Journal of Personality and Social Psychology, 95*(5), 1045–1062.

Giustiniano, L., Clegg, S. R., Cunha, M. P., & Rego, A. (Eds.). (2018). *Elgar introduction to theories of organizational resilience.* Cheltenham: Edward Elgar Publishing.

Giustiniano, L., Cunha, M. P., Simpson, A. V., Rego, A., & Clegg, S. (2020). Resilient leadership as paradox work: notes from COVID-19. *Management and Organization Review, 16*(5), 971–975.

Hammond, R. (2019). *The World in 2040. The Future Travel Experience.* https://www.rayhammond.com/wp-content/uploads/The-Future-Travel-Experience_The-World-in-2040-Series.pdf

Herscovitch, L., & Meyer, J. P. (2002). Commitment to organizational change: extension of a three-component model. *Journal of Applied Psychology, 87*(3), 474.

Hillmann, J. (2021). Disciplines of organizational resilience: contributions, critiques, and future research avenues. *Review of Managerial Science, 15*(4), 879–936.

Jones, P., & Comfort, D. (2020). The role of resilience in research and planning in the tourism industry. *Athens Journal of Tourism, 7*(1), 1–16.

Jung, H. S., Jung, Y. S., & Yoon, H. H. (2021). COVID-19: The effects of job insecurity on the job engagement and turnover intent of deluxe hotel employees and the moderating role of generational characteristics. *International Journal of Hospitality Management, 92,* 102703.

JWU. (15 November 2021) *10 ways to reduce hospitality industry employee turnover.* Johnson & Wales University College of Professional Studies. https://online.jwu.edu/blog/10-ways-reduce-hospitality-industry-employee-turnover

KPMG. (2019). *The Future of HR 2019: In the Know or in the No.* https://assets.kpmg/content/dam/kpmg/xx/pdf/2018/11/future-of-hr-survey.pdf

KPMG. (2020). *Future of HR 2020: Which Path Are You Taking?* https://assets.kpmg/content/dam/kpmg/xx/pdf/2019/11/future-of-hr-2020.p

Lengnick-Hall, C. A., Beck, T. E., & Lengnick-Hall, M. L. (2011). Developing a capacity for organizational resilience through strategic human-resource management. *Human Resource Management Review, 21*(3), 243–255.

Lew, A. A., Cheer, J. M., Haywood, M., Brouder, P., & Salazar, N. B. (2020). Visions of travel and tourism after the global COVID-19 transformation of 2020. *Tourism Geographies, 22*(3), 455–466.

Lombardi, S., Cunha, M. P., & Giustiniano, L. (2021). Improvising resilience: The unfolding of resilient leadership in COVID-19 times. *International Journal of Hospitality Management, 95*, 102904.

Luthe, T., & Wyss. R. (2014) Assessing and planning resilience in tourism. *Tourism Management, 44*, 161–163.

Lytle, T. (2020, December 1). Top HR challenges in the hospitality industry. *Society for Human Resource Management.* https://www.shrm.org/hr-today/news/hr-magazine/winter2020/pages/top-hr-challenges-in-the-hospitality-industry.aspx

Mache, S., Vitzthum, K., Wanke, E., Groneberg, D., Klapp, B., & Danzer, G. (2014). Exploring the impact of resilience, self-efficacy, optimism and organizational resources on work engagement. *Work, 47*(4), 491–500.

McKinsey Global Institute. (2017). *Jobs lost, jobs gained: Workforce transitions in a time of automation.* https://www.mckinsey.com/~/media/mckinsey/industries/public%20and%20social%20sector/our%20insights/what%20the%20future%20of%20work%20will%20mean%20for%20jobs%20skills%20and%20wages/mgi-jobs-lost-jobs-gained-executive-summary-december-6-2017.pdf

McManus, S. T. (2008). Organisational resilience in New Zealand. University of Canterbury. https://ir.canterbury.ac.nz/handle/10092/1574

Prayag, G., Orchiston, C., & Pennington-Gray, L. (2018). Tourism Management Perspective—Special Issue: Resilience of the Tourism and Hospitality Industry.

Prewitt, J., Weil, R., & McClure, A. (2011). Developing leadership in global and multi-cultural organizations. *International Journal of Business and Social Science, 2*(13), 13–20.

PwC. (2015). The art and science of resilience. https://www.pwc.nl/nl/assets/documents/pwc-enterprise-resilience-the-emerging-capability-every-business-needs.pdf

Sardá, R., & Pogutz, S. (2018). *Corporate sustainability in the 21st century: Increasing the resilience of social-ecological systems.* London: Routledge.

Senge, P., Kleiner, A., Roberts, C., Ross, R., Rother, G., & Smith, B. (1999). *The dance of change.* New York: Doubleday.

Sharma, G. D., Thomas, A., & Paul, J. (2021). Reviving tourism industry post-COVID-19: A resilience-based framework. *Tourism Management Perspectives, 37*, 100786.

Soanes, C., & Stevenson, A. (2006). *Oxford dictionary of English* (2nd ed.). Oxford: Oxford University Press.

Southwick, F. S., Martini, B. L., Charney, D. S., & Southwick, S. M. (2017). Leadership and resilience. In *Leadership today* (pp. 315–333). Springer, Cham.

Trends, D. G. H. C. (2019). *Leading the Social Enterprise: Reinvent with a Human Focus.* https://www2.deloitte.com/content/dam/Deloitte/cz/Documents/human-capital/cz-hc-trends-reinvent-with-human-focus.pdf

Trends, D. G. H. C. (2020). *The Social Enterprise at Work: Paradox as a Path Forward.* https://www2.deloitte.com/content/dam/Deloitte/cn/Documents/human-capital/deloitte-cn-hc-trend-2020-en-200519.pdf

Twinning-Ward, L., Perrottet, J. & Niang, C. (2017, December 7). Resilience, sustainability and inclusive growth for Tourism in the Caribbean. *World Bank Group.* https://blogs.worldbank.org/psd/resilience-sustainability-and-inclusive-growth-tourism-caribbean

UNWTO. (8 June 2022). *Turning point for tourism: UNWTO executive council looks beyond recovery.* World Tourism Organization. https://www.unwto.org/news/turning-point-for-tourism-unwto-executive-council-looks-beyond-recovery

Van Zanten, J. A., & Van Tulder, R. (2020). Beyond COVID-19: Applying "SDG logics" for resilient transformations. *Journal of International Business Policy, 3*(4), 451–464.

White, O. (27 December 2021). The travel and tourism industry by 2030. *Forbes.* https://www.forbes.com/sites/forbesbusinesscouncil/2021/12/27/the-travel-and-tourism-industry-by-2030/?sh=5171a22a402d

# 19 Women's education in tourism entrepreneurship

## Trends and issues emerging from Africa

*Nellie (Magdalena Petronella) Swart, Vanessa S. Bernauer and Kailasam Thirumaran*

### Introduction

An estimated 54% of the global tourism workforce are women, who earn on average 14.7% less than their male counterparts (UNWTO, 2019). Although research on women's entrepreneurship is well documented, some of the projects and processes followed to capture the empowerment of women entrepreneurs in tourism through education are limited in the African context (Kimbu & Ngoasong, 2016; Maliva, Bulkens, Peters, & Van Der Duim, 2018). Furthermore, little is known about how the dynamics of women's entrepreneurship affect knowledge, skills, and development initiatives, or how women's expectations influence their entrepreneurial learning experiences (Ertac & Tanova, 2020; Morgan & Winkler, 2020). In Africa specifically, research in this field is limited to a few countries including Botswana (Moswete & Lacey, 2015), Burkina Faso (Song-Naba, 2020), Cameroon (Kimbu & Ngoasong, 2016; Kimbu, Ngoasong, Adeola, & Afenyo-Agbe, 2019), Ghana (Ali, 2018), Rwanda (Iwu, Nsengimana, & Robertson, 2010), Senegal (Gomez-Perez & Jourde, 2021), South Africa (Chux Gervase & Zinzi, 2015; Hikido, 2021; Hlanyane & Acheampong, 2017; Kwaramba, Lovett, Louw, & Chipumuro, 2012; Tshabalala & Ezeuduji, 2016), and Tanzania (Maliva et al., 2018). There is a need to understand women entrepreneurs in tourism from the perspective of education, training, and skills development to achieve the United Nations (UN) Sustainable Development Goals (SDGs) in relation to women's education and empowerment, especially from an African perspective.

The UN has proposed 17 SDGs, of which five support the education of women entrepreneurs in tourism (UN, 2015). More specifically, these goals relate to poverty eradication (1), quality education (4), gender equality (5), decent work and economic growth (8), industry, innovation, and infrastructure (9), as well as partnerships to achieve these goals (17). In Africa, these SDGs are supported by the African Union Agenda 2063, which highlights seven aspirations of which a prosperous Africa based on inclusive growth and sustainable development (1) are relevant to this chapter. By addressing these goals and aspirations, the gender and economic inequalities identified in the World Economic Forum Global Gender Gap (WEF, 2019) can be addressed. To this end, it is important to identify: (i) the key trends and issues that have an impact on African women's tourism entrepreneurship education; (ii) what motivates women to become tourism entrepreneurs; (iii) what kinds of studies have been conducted on women entrepreneurs in tourism and where; (iv) what the challenges and opportunities for women entrepreneurs are; and (v) what the strategies for implementation for the next 30 years are, especially in Africa. Although this chapter focuses on Africa, it is also relevant to other developing countries, such as India (Sachs, Kroll, Lafortune, Fuller, & Wohle, 2021).

DOI: 10.4324/9781003291763-22

To support our statements, we conducted a systematic literature review of women's entrepreneurship in tourism and hospitality between 2010 and 2020 based on Briner and Denyer's (2012) five core principles. Five databases with Boolean operators yielded 70 relevant articles, with search criteria focused on (Wom*n OR female OR gender) AND (entrepreneur*) AND (hospitality OR touris*). The term "women in tourism" is used throughout this chapter, but this includes studies or references to the hospitality industry. This study focuses on African women entrepreneurs in the tourism and hospitality sectors and takes a social-business milieu approach to provide a broad overview of trends while considering a listing of future research directions.

## Major trends and issues impacting education, skills development, and training of women entrepreneurs in tourism

Over the past ten years, studies related to women entrepreneurs in tourism became a topic of choice amongst many scholars such as Kimbu and Ngoasong (2016) and Hillman and Radel (2021). Therefore, we provide an overview of prominent themes, discussions, and relationships that emerge from research on tourism entrepreneurship education and its impact on women (Figure 19.1).

In most studies on entrepreneurship in tourism, women are still underrepresented as business owners due to their limited qualifications, managerial skills, and business experience (Zapalska & Brozik, 2017). It is also known that many women entrepreneurs in tourism have insufficient tourism-specific skills and knowledge to start a sustainable business (Jaafar, Rasoolimanesh, & Lonik, 2015). Education, skills development, and training of women entrepreneurs are influenced by four factors, as shown in Figure 19.1. Continuous professional development (Perez & Bui, 2010) is enhanced by business-specific knowledge (Hikido, 2018) and language skills (Song-Naba, 2020; Zhang, Kimbu, Lin, & Ngoasong, 2020). Zapalska and Brozik (2017) and Wilson-Youlden and Bosworth (2019) highlight the need to develop policies to support female entrepreneurship in tourism. Women are encouraged to join support groups where they feel a sense of belonging based on their cultural values. These engagements in the industry can further empower women entrepreneurs financially (Chux Gervase & Zinzi, 2015). Gender stereotypes continue to be a challenge for many women in developing countries (Hillman & Radel, 2021).

Education, training, and skills development are critical to empowering women entrepreneurs in tourism. Often, gender norms hinder the development of such women, as day-to-day demands prevent them from taking advantage of available training and skills development opportunities. Studies, therefore, call for a change in the social and cultural stereotyping of gender identities (Morgan & Winkler, 2020). Yoopetch (2020) postulates that education and training can enhance women's entrepreneurial intentions and success in tourism. This also leads to the welfare of their family and general society (Ashrafi & Hadi, 2019). Family welfare is improved through higher disposable household income, women's stronger belief in themselves (Ertac & Tanova, 2020; Lapan, Morais, Wallace, & Barbieri, 2016; Zapalska & Brozik, 2017), and their financial independence (Ashrafi & Hadi, 2019; Morgan & Winkler, 2020). Furthermore, social welfare (Ertac & Tanova, 2020) is improved when more women are employed in tourism and hospitality (Hillman & Radel, 2021). Cultural principles give women entrepreneurs in tourism a sense of pride, which further encourages the development of innovative tourism products (Costa, Shah, & Korgaonkar, 2014). In terms of poverty alleviation, Xu, Wang, and Wu (2018) distinguish five dimensions: poverty alleviation related to mental

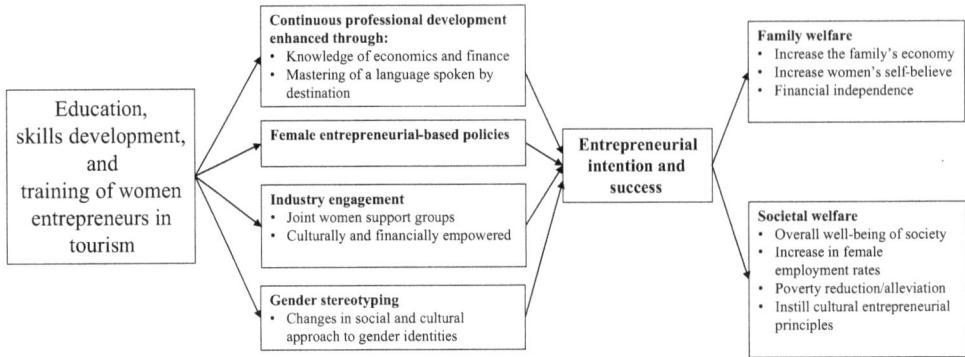

*Figure 19.1* Themes on tourism entrepreneurship education and its impact on women.

and physical health, cultural literacy, participation in public affairs, living environment, and self-esteem, highlights the importance of education, skills development, and training.

Specifically in Africa, Ahwireng-Obeng and van Loggerenberg (2011) identified in an earlier study that poverty in the sub-Saharan region weakens opportunities for women entrepreneurs who want to offer their services in medical tourism by providing recreational medical care at a holiday destination in South Africa. Nwachukwu and Auselime (2018) proposed that gender-balanced eco-entrepreneurship initiatives can alleviate poverty if women entrepreneurs have access to training and funding. For example, in the sub-Saharan belt, advocacy groups have resonated with women who are willing to venture into micro-enterprises but are still hampered by skill levels and access to credit (Gichuki, Mulu-Mutuku, & Kinuthia, 2018). In a later study by Nzama and Ezeuduji (2021) in KwaZulu Natal, South Africa, men and women found that it was not their gender differences that contributed to the level of performance in running a business, but the difference between their tertiary and non-tertiary qualifications. To alleviate the situation, the researchers suggest a wider network of mentoring and training programmes. Therefore, African scholars (Kimbu & Ngoasong, 2016; Kimbu et al., 2019; Ngoasong & Kimbu, 2019; Nwachukwu & Auselime, 2018; Zhang et al., 2020) appeal to Africa's two largest economies and assert that policy makers and planners should promote gender equality, especially in rural areas, to empower women through skills training, knowledge, and education.

## Factors motivating women to become tourism entrepreneurs

Women entrepreneurs tend to be perceived through the prism of struggles and inequalities compared to men. However, when we ask about the successes and failures of entrepreneurship, studies have shown that women are highly motivated and excel in their attributes. For example, in a study by Sibusiso and Ikechukwu (2019) conducted in the South African municipality of Mtubatuba, women are hardworking and eager to learn by attending training activities to make their work better. In an interesting methodological paper, Njaramba, Whitehouse, and Lee-Ross (2018) studied African migrant women in Townsville and Cairns, Australia. The results show that women are motivated to become entrepreneurs due to a lack of employment opportunities and a desire to make it in life for their families. These results are further echoed in several other studies.

Women in tourism are motivated to become entrepreneurs for the following reasons and as described in these studies:

1. *Creating opportunities for family members* (Hillman, Lamont, Scherrer, & Kennelly, 2021; Morgan & Winkler, 2020; Perez & Bui, 2010).
2. *Creating a sustainable business for future generations (legacy)* (Yoopetch, 2020).
3. *Using knowledge and networks to start a business* (Kimbu & Ngoasong, 2016; Maliva et al., 2018; Wilson-Youlden & Bosworth, 2019).
4. *Entrepreneurial skills are informed by value and quality services* (Çiçek, Zencir, & Kozak, 2017; Lapan et al., 2016).
5. *Personal aspirations (self-actualisation)* (Morgan & Winkler, 2020).
6. *Education and training opportunities* (Chux Gervase & Zinzi, 2015; Ertac & Tanova, 2020; Zapalska & Brozik, 2017).

Women entrepreneurs often do not have access to certain resources they need to survive in the tourism business. This observation was made in particular in the case of the Western Cape (South Africa). Nxopo and Iwu (2015) found that the various government agencies such as the Small Enterprise Development Agency (SEDA) and the Small Enterprises Finance Agency (SEFA) seem to favour men more than women entrepreneurs. Despite the success of men compared to women in the ratio of 8.0 to 4.8, women are extremely hard-working, apart from the educational challenge they face. The study also showed that this limited success is particularly pronounced in the accommodation sector of the hospitality industry. Iwu et al.'s (2010) findings indicate that, especially in the case of Kigali, Rwanda, socio-cultural pressure on women entrepreneurs is not the main reason for their lower participation in the tourism business, which was supported by Nxopo and Iwu (2015). These authors stated that women entrepreneurs need more support in the form of access to credit, lower taxes, and training programmes to compete in the tourism industry. None of this will be possible through the support of governments, general society, the tourism industry, technology, and educational institutions (Figure 19.2). Furthermore, the African Union's

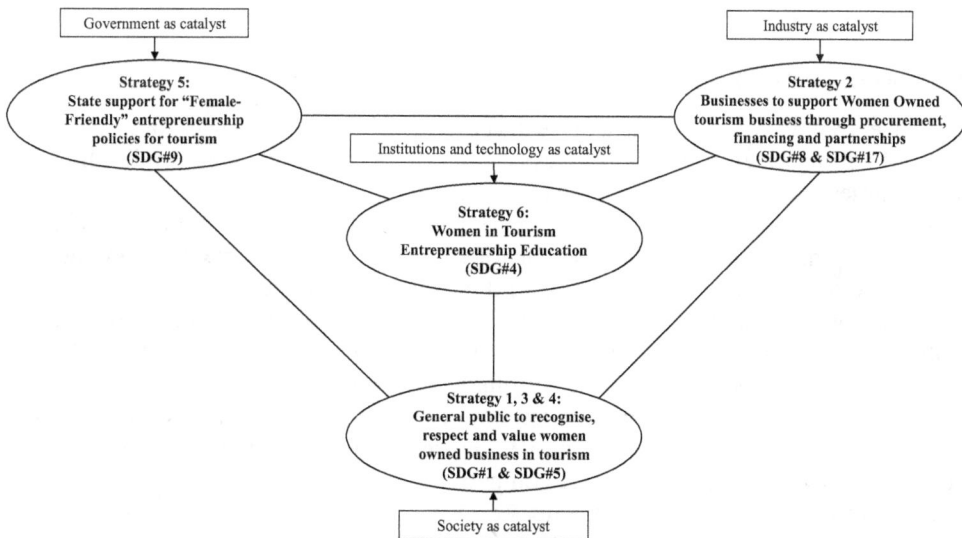

*Figure 19.2* Women in tourism entrepreneurship education in Africa.

set Agenda 2063 calls for the transformation of the continent to a powerhouse, not only for democracy but also as a land of innovation. To accomplish these the above studies suggest the need for institutional support as much as a nexus to men inclined to accept women as equals in business. These primary stakeholders, including non-governmental organisations (NGOs) and individuals in society wielding community powers, have a role to play in nurturing and further empowering aspiring women entrepreneurs (Leal Filho, Marisa Azul, Brandli, Lange Salvia, & Wall, 2021).

## Countries studied regarding entrepreneurship education for women in tourism

Since each continent and country differs in terms of its support system and opportunities for the education of women entrepreneurs, we have carefully sorted the number of studies by continent to show how those on Africa compare to those conducted worldwide. A number of studies have been conducted in Australasia (25), followed by Europe (21), Africa and the Middle East (19), and the Americas (5) on women's entrepreneurship related to education, knowledge, and skills development between 2010 and 2020. These data show that the number of publications on women entrepreneurs in Africa is not too far from those in Australasia and Europe, though these publications are limited to certain countries, namely Botswana, Burkina Faso, Cameroon, Ghana, Rwanda, Senegal, South Africa, Tanzania, and Uganda.

Of the 70 articles selected, 65 focused on one destination, 3 were conducted in multiple destinations, and 2 did not specify the country in which the study was conducted. Of the relevant studies selected for this chapter, 12 relate exclusively to the hospitality industry, 16 to hospitality and tourism, while most (38) were conducted only in tourism. A study by Garrigos, Haddaji, Segovia, and Signes (2018) covered Spain, France, and the USA, while another study by Kimbu, Ngoasong, Adeola, and Afenyo-Agbe (2019) focused on Cameroon, Ghana, and Nigeria. The training of women in tourism-related agribusiness has been examined in Haiti, Brazil, Lesotho, South Africa, Timor Leste, Indonesia, Ethiopia, and India (Favre, 2017). Studies on African women entrepreneurs in tourism were conducted in nine countries, but none of these studies suggested the development of flexible and self-directed training programmes based on women's current education levels and entrepreneurial needs.

## Opportunities and challenges for women entrepreneurs in tourism

Women entrepreneurs in tourism have faced a number of challenges over the years. Table 19.1 provides an overview of these challenges. In addition, some studies highlight opportunities that can arise from these challenges.

Women entrepreneurs in tourism are severely marginalised in education, training, and skills development. The opportunities identified provide a benchmark for how they can take charge of their education, training, and personal development to achieve the SDGs. Women can work to eradicate poverty (1) by increasing their employment rates through the establishment of cooperatives in tourism, thus having a positive impact on GDP. Quality education (4) can be supported by sensitising women on how to transform their existing skills and talents into a first (e.g., a guesthouse), second (e.g., handicrafts), or third (e.g., agricultural products) tier tourism product. Gender equality (5) can be improved through the creation of additional jobs that are compatible with women's domestic responsibilities, as well as through the psychological empowerment of them in tourism. The establishment of tourism business incubators for women entrepreneurs can support the necessary innovation and provide the infrastructure (9) to foster entrepreneurial ideas.

*Table 19.1* Challenges and opportunities related to education, training, and skills development for women entrepreneurs in tourism

| Challenges | Opportunities |
| --- | --- |
| 1. Limited access to capital and other resources<br>2. Lack of post-secondary education or no access to education<br>3. Lack of knowledge about the commercialisation of the business (especially in rural areas)<br>4. Lack of management, marketing, and accounting skills<br>5. Women are confined to the "domestic boundaries" of the business<br>6. Balancing household and work responsibilities<br>7. Women's social status is determined by marital status, education level, and religion (often patriarchal)<br>8. Although highly skilled, women who move to the countryside and become mothers stop working due to work and family demands<br>9. Inadequate business experience<br>10. Political systems that hinder education opportunities for women in tourism to support their entrepreneurial development<br>11. Finding employment due to ethnic discrimination as a result of intersections of race and gender<br>12. Gender inequity and inequality<br>13. Priority is given to children's education over their personal development | 1. Peer learning through women entrepreneurship groups<br>2. Increasing women's employment rates through the establishment of cooperatives<br>3. Positive impact on gross domestic product (GDP)<br>4. Psychologically empowering women to contribute to a flourishing society<br>5. Sensitising women on how to transform their existing skills and talents into a first (e.g., a guest house), second (e.g., handicrafts), or third (e.g., agricultural products) tier tourism product<br>6. Create complementary jobs that are compatible with their domestic responsibilities (e.g., handicraft)<br>7. Create incubators for women entrepreneurs to take up ideas in tourism<br>8. Entrepreneurial role models and mentoring |

*Sources:* Informed by Ashrafi & Hadi (2019); Chux Gervase & Zinzi, 2015; Çiçek et al., 2017; Costa et al., 2014; del Mar Alonso-Almeida, 2012; Ertac & Tanova, 2020; Hikido, 2018; Hillman et al., 2021; Kimbu & Ngoasong, 2016; Lapan et al., 2016; Maliva et al., 2018; Möller, 2012; Perez & Bui, 2010; Selvi, 2019; Surangi, 2018; Wilson-Youlden & Bosworth, 2019; Yoopetch, 2020; Zapalska & Brozik, 2017; Zhang et al., 2020.

Firstly, the challenges delineated in Table 19.1 indicate a need for lifting the many social relationships that women are trapped in in a cultural milieu. Political systems, patriarchal social values, family demands, and gender roles that rest on inequalities often create a stupendous hindrance to women taking the first step into a business venture on their own initiative.

Secondly, a means and structure, if put in place to enable and empower women, might result in a sea change in women's entrepreneurship endeavours. Education, training, creating incubators, access to mentoring programmes, and the establishment of cooperatives are some of the many ways pathways can be cleared for women entrepreneurs to emerge.

## Strategies for women's entrepreneurship education in Africa

The future education, training, and skills development of women entrepreneurs in tourism is dependent on an array of external factors to achieve the five identified SDGs. To reach these goals, we propose six strategies to be implemented over the next 20+ years (2030 until 2050) through the support of different catalysts, as illustrated in Figure 19.2.

As depicted in Figure 19.2, the six strategies align with the SDGs, as discussed below.

**Strategy 1: Cultural empowerment of women entrepreneurs in tourism for gender equality**

By establishing cultural entrepreneurship based on certain cultural values and products, such as among African tribes, women can be psychologically empowered and become people who excel in business (Ertac & Tanova, 2020). They can use their spiritual abilities, nurtured by their cultural beliefs, for the overall empowerment of their communities. Their spiritual wellbeing helps them to preserve their cultural heritage and create sustainable entrepreneurial opportunities (Zapalska & Brozik, 2017). This offers communities the opportunity to package their way of life as a unique tourism entrepreneurial product where tourists can feel that they are also contributing to the wellbeing of a community (corporate social transformation). Tourism entrepreneurs need to embrace their cultural values and customs to convey harmony, unity, and wisdom in a complex world. This strategy supports SDG 1 and SDG 5 and needs to be enabled by the general public as a catalyst.

**Strategy 2: Social independence for decent work and economic growth**

Financially independent women are not only able to support their families, but they are also more likely to be socially independent, and less likely to suffer from domestic oppression and violence (Maliva et al., 2018). Education and training in a specific tourism skill can equip women with the knowledge that enables them to become socially independent by starting their tourism businesses. Therefore, the industry needs to support women-owned tourism through partnerships to achieve SDG 8 and SDG 17.

**Strategy 3: Financial empowerment to eradicate poverty**

Governments and NGOs need to develop microfinance models to support women entrepreneurs in tourism (Ertac & Tanova, 2020; Hillman & Radel, 2021). The insurmountable difficulty of obtaining financial credit is real. Although the credit facility is available, there is a cultural barrier where women often have to compete with men who have a social advantage when it comes to obtaining the resources needed to sustain their businesses. Following on from an earlier study, Nzama and Ezeuduji (2021) found that women in KwaZulu-Natal, South Africa were unable to sustain their businesses compared to men, despite having easy access to government financial credit. In the latter study, researchers argue that training programmes should educate women in business management in trade. The recognition of women-owned tourism businesses by general society can embrace the outcomes of SDG 1 and SDG 5.

**Strategy 4: Changing social and cultural gender identities for gender equality**

In society at large, it becomes increasingly important to understand the gender lens (Costa, Breda, Bakas, Durão, & Pinho, 2016). Individual awareness and gender-specific regulations limit women's development which translates into gender inequality (Morgan & Winkler, 2020), especially for women entrepreneurs in tourism. Therefore, women need to take control of their education and training by rewriting their social and cultural directions on how they are perceived as individuals, as for example, in Nepal, where women have been encouraged to educate themselves for their future emancipation (Hillman & Radel, 2021). Furthermore, local gender structures need to be examined and understood to ensure that women are empowered and not subjected to persistent inequality due to ill-conceived gender models or gender stereotypes. This strategy supports strategies 1 and 3 to achieve SDG

1 and SDG 5 through the acknowledgment of women-owned businesses in tourism by the general public.

### Strategy 5:  *Entrepreneurial role models for innovation*

As a society, we need to reshape the ideologies of women entrepreneurs in tourism by overcoming any perceived boundaries and limitations (Hillman & Radel, 2021) in terms of education and training, especially in Africa. Such women need entrepreneurial role models to guide them through the process of risk-taking and to fully mentor them in their initiatives (Yoopetch, 2020). Role models can empower African women in tourism to grow their businesses by providing skills and mentorship. This strategy also supports SDG 9, highlighting the important role of governments to support women entrepreneurs in tourism through "female-friendly" entrepreneurship policies.

### Strategy 6:  *Educational empowerment through quality education*

Ultimately, none of the strategies will be successful if this is not supported by education, training, and skills development. There is a need to develop more specialised short learning programmes to equip women entrepreneurs in Africa with just-in-time skills and knowledge through online learning management systems. Iwu and Zinzi (2015) suggest that it may be helpful to provide knowledge and business guidance or mentorship to budding women entrepreneurs in the tourism sector to enhance their skills. Additionally, given the nature of the tourism business, women entrepreneurs also need legal and business incentives to get on track for growth. This means that women must have access to training programmes whenever and wherever they need them. This knowledge and education will strengthen women's self-esteem and confidence. However, these training programmes must cover all levels of education and be accredited by relevant organisations and government institutions to reflect women's skill levels and ensure professionalism (del Mar Alonso-Almeida, 2012). In addition, the content must be suitable for use in an African context and aligned with the pedagogy of Africanisation so that women can identify with it. These emerging trends and strategies highlight the need to profile women entrepreneurs in tourism by developing flexible and self-directed training programmes based on their current level of education and entrepreneurial needs. Further studies are needed to explore the specific training needs in the various tourism-related sectors, such as accommodation, culinary skills, transport, handicrafts, and other support services (Jaafar et al., 2015), and how this can support women's entrepreneurship. In achieving SDG 4, the challenges of access to basic services and technology are met through the various training institutions.

### Conclusion

This chapter has investigated the general principles related to training, education, and skills development for women entrepreneurs in Africa. The review of the literature has shown very clearly women entrepreneurs' challenges and successes, which have informed the proposed strategies to mitigate the challenges in reaching the SDGs. By implementing our proposed strategies for women's entrepreneurship education, African countries could contribute to the SDGs, for example, by not only having the chance to recover from the recent rise in unemployment due to the COVID-19 pandemic but also to reduce poverty and strengthen inclusive growth and sustainable development in the long run. Through education, African

*Table 19.2* Timeline of women's entrepreneurship education in tourism in Africa

| Timeline | SDGs | Strategy |
| --- | --- | --- |
| 2032 | SDG 1 and SDG 5 | Strategies 1, 3, and 4 |
| 2042 | SDG 4, SDG 8, and SDG 17 | Strategies 2 and 6 |
| 2052 | SDG 9 | Strategy 5 |

women would benefit from greater business knowledge as well as emotional, financial, and societal support for their entrepreneurship. Technological development in emerging countries will also offer new opportunities for them. In addition, women interested in or dependent on entrepreneurship in other countries that do not meet the SDGs and suffer from poverty that is similar to many African countries, such as India (Sachs et al., 2021), could also benefit from the implementation of our identified education strategies. Following SDG 17, women entrepreneurs in developed countries could work together with African women and build global partnerships and cooperation or support their educational system. Therefore, we propose the achievement of the SDGs and strategies as indicated in this staggered timeline as illustrated in Table 19.2.

In the medium term of 10–20 years (2032 and 2041), more women will emerge stronger in their societies in two streams of entrepreneurship. The first group of women would be those who are already in various industries and who gain knowledge and experience. Often a small percentage of these women when they gain a degree of confidence and the right mix of environment will go on their own to chart a new entrepreneurship career path.

The second stream of women to emerge into the landscape is the educated group. Though Africa lags all the other major continents in investing in education, there is indeed a steady rise of 3.4% in 2019 and 3.9% in 2020 in secondary and primary education investments (Gandhi, 2020). What this means is that there is potential for growth in women entrepreneurs driven by education that can liberate their minds and engage them fruitfully in society.

In the long term, of about 25–50 years (2052), another form of transformation may take place in women's empowerment, leading to an even keel in women entrepreneurs in the field of tourism and hospitality. As more women are given education opportunities in Africa and beyond, there is a high chance that their standing in both existing societies whether male dominant cultures or not, may see a shift. This optimistic view is based on globalisation and international interactive forces to which Africa as a continent is central too. Already, NGOs, governments, and educational institutions are an emerging nexus of the issues faced by women entrepreneurs, as the literature indicates. Of course, these centrifugal forces may not overturn all societies in Africa to become enablers of more women entrepreneurs, but with the SDG goals and Agenda 2063, when implemented with full commitment, we are likely to witness positive outcomes for these women in decades to come.

## References

Ahwireng-Obeng, F., & Van Loggerenberg, C. (2011). Africa's middle-class women bring entrepreneurial opportunities in breast care medical tourism to South Africa. *International Journal of Health Planning and Management*, 26(1), 39–55. doi: 10.1002/hpm.1034
Ali, R. S. (2018). Determinants of female entrepreneurs growth intentions: A case of female-owned small businesses in Ghana's tourism sector. *Journal of Small Business and Enterprise Development*, 25(3), 387–404. doi: 10.1108/jsbed-02-2017-0057

Ashrafi, A., & Hadi, F. (2019). The impact of tourism on developing Shiraz rural women entrepreneurship. *Revista Universidad Y Sociedad, 11*(4), 72–76.

Briner, R. B., & Denyer, D. (2012). Systematic review and evidence synthesis as a practice and scholarship tool. In: D. M. Rousseau (Ed.), *Handbook of evidence-based management: Companies, classrooms and research*, pp. 112–129. Oxford: Oxford University Press.

Chux Gervase, I., & Zinzi, N. (2015). Determining the specific support services required by female entrepreneurs in the South African tourism industry. *African Journal of Hospitality, Tourism and Leisure, 4*(2).

Çiçek, D., Zencir, E., & Kozak, N. (2017). Women in Turkish tourism. *Journal of Hospitality and Tourism Management, 31*, 228–234. doi: 10.1016/j.jhtm.2017.03.006

Costa, C., Breda, Z., Bakas, F. E., Durão, M., & Pinho, I. (2016). Through the gender looking-glass: Brazilian tourism entrepreneurs. *International Journal of Gender and Entrepreneurship, 8*(3), 282–306. doi: 10.1108/ijge-07-2015-0023

Costa, J. C., Shah, H., & Korgaonkar, K. (2014). From grassroots to success: A case study of a successful goan woman entrepreneur. *Prabandhan: Indian Journal of Management, 7*(2), 40.

del Mar Alonso-Almeida, M. (2012). Water and waste management in the Moroccan tourism industry: The case of three women entrepreneurs. *Women's Studies International Forum, 35*(5), 343–353. doi: 10.1016/j.wsif.2012.06.002

Ertac, M., & Tanova, C. (2020). Flourishing women through sustainable tourism entrepreneurship. *Sustainability (Switzerland), 12*(14), 17. doi: 10.3390/su12145643

Favre, C. C. (2017). The Small2Mighty tourism academy. Growing business to grow women as a transformative strategy for emerging destinations. *Worldwide Hospitality and Tourism Themes, 9*(5), 555–563. doi: 10.1108/whatt-07-2017-0034

Gandhi, D. (2020). *Figures of the Week: Public Spending on Education in Africa. Africa in Focus.* Retrieved on 2 Aug 2022 from https://www.brookings.edu/blog/africa-in-focus/2020/02/13/figures-of-the-week-public-spending-on-education-in-africa/

Garrigos, J. A., Haddaji, M., Segovia, P. G., & Signes, A. P. (2018). Gender differences in the evolution of haute cuisine chef's career. *Journal of Culinary Science & Technology, 18*(6), 439–468.

Gichuki, C. N., Mulu-Mutuku, M., & Kinuthia, L. N. (2018). Gender differences in the evolution of haute cuisine chef's career. *Journal of Culinary Science & Technology, 18*(6), 439–468.

Gomez-Perez, M., & Jourde, C. (2021). Islamic entrepreneurship in senegal: Women's trajectories in Organizing the Hajj. *Africa today, 67*(2), 104–126.

Hikido, A. (2018). Entrepreneurship in South African township tourism: The impact of interracial social capital. *Ethnic and Racial Studies, 41*(14), 2580–2598. doi: 10.1080/01419870.2017.1392026

Hikido, A. (2021). Making South Africa safe: The gendered production of black place on the global stage. *Qualitative Sociology, 44*(2), 293–312. doi: 10.1007/s11133-021-09478-z

Hillman, P., Lamont, M., Scherrer, P., & Kennelly, M. (2021). Reframing mass participation events as active leisure: Implications for tourism and leisure research. *Tourism Management Perspectives, 39.* doi: 10.1016/j.tmp.2021.100865

Hillman, W., & Radel, K. (2021). The social, cultural, economic, and political strategies extending women's territory by encroaching on patriarchal embeddedness in tourism in Nepal. *Journal of Sustainable Tourism, 30*(7), 1754–1775.

Hlanyane, T. M., & Acheampong, K. O. (2017). Tourism entrepreneurship: The contours of challenges faced by female-owned BnBs and Guesthouses in Mthatha, South Africa. *African Journal of Hospitality, Tourism and Leisure, 6*(4), 1–17.

Iwu, C. G., Nsengimana, S., & Robertson, T. K. (2010). The factors contributing to the low numbers of women entrepreneurs in Kigali. *Acta Universitatis Danubius Œconomica, 15*(6), 98–114.

Iwu, C. G., & Nxopo, Z. (2015). Determining the specific support services required by female entrepreneurs in the South African tourism industry. *African Journal of Hospitality, Tourism and Leisure, 4*(2), 13.

Jaafar, M., Rasoolimanesh, S. M., & Lonik, K. A. T. (2015). Tourism growth and entrepreneurship: Empirical analysis of the development of rural highlands. *Tourism Management Perspectives, 14*, 17–24. doi: 10.1016/j.tmp.2015.02.001

Kimbu, A. N., & Ngoasong, M. Z. (2016). Women as vectors of social entrepreneurship. *Annals of Tourism Research, 60*, 63–79.

Kimbu, A. N., Ngoasong, M. Z., Adeola, O., & Afenyo-Agbe, E. (2019). Collaborative networks for sustainable human capital management in women's tourism entrepreneurship: The role of tourism policy. *Tourism Planning and Development, 16*(2), 161–178. doi: 10.1080/21568316.2018.1556329

Kwaramba, H. M., Lovett, J. C., Louw, L., & Chipumuro, J. (2012). Emotional confidence levels and success of tourism development for poverty reduction: The South African Kwam eMakana home-stay project. *Tourism Management, 33*(4), 885–894. doi: 10.1016/j.tourman.2011.09.010

Lapan, C., Morais, D. B., Wallace, T., & Barbieri, C. (2016). Women's self-determination in cooperative tourism microenterprises. *Tourism Review International, 20*(1), 41–55. doi: 10.3727/15442721 6x14581596799022

Leal Filho, W., Marisa Azul, A., Brandli, L., Lange Salvia, A., & Wall, T. (Eds.). (2021). *Partnerships for the Goals.* Cham: Springer International Publishing.

Maliva, N., Bulkens, M., Peters, K., & Van Der Duim, R. (2018). Female tourism entrepreneurs in Zanzibar: An enactment perspective. *Tourism, Culture, and Communication, 18*(1), 9–20. doi: 10.372 7/109830418x15180180585149

Möller, C. (2012). Gendered entrepreneurship in Rural Latvia: Exploring femininities, work, and livelihood within rural tourism. *Journal of Baltic Studies, 43*(1), 75–94. doi: 10.1080/01629778. 2011.634103

Morgan, M. S., & Winkler, R. L. (2020). The third shift? Gender and empowerment in a women's ecotourism cooperative. *Rural Sociology, 85*(1), 137–164. doi: 10.1111/ruso.12275

Moswete, N., & Lacey, G. (2015). "Women cannot lead": empowering women through cultural tourism in Botswana. *Journal of Sustainable Tourism, 23*(4), 600–617.

Ngoasong, M. Z., & Kimbu, A. N. (2019). Why hurry? The slow process of high growth in women-owned businesses in a resource-scarce context. *Journal of Small Business Management, 57*(1), 40–58.

Njaramba, J., Whitehouse, H., & Lee-Ross, D. (2018). Approach towards female African migrant entrepreneurship research. *Entrepreneurship and Sustainability Issues, 5*(4), 1043–1053. doi: 10.9770/jesi.2018.5.4(24)

Nwachukwu, P. T. T., & Auselime, R. A. (2018). Shaping capabilities for gender-balanced participation in eco-entrepreneurship for sustainable development in Africa. *Gender & Behaviour, 16*(2), 11308–11323.

Nxopo, Z., & Iwu, C. G. (2015). *The Unique Obstacles of Female Entrepreneurship in the Tourism Industry in Western Cape* (Vol. 13). Pretoria: Unisa Press.

Nzama, N., & Ezeuduji, I. O. (2021). Nuanced gender perceptions: Tourism business capabilities in Kwazulu-Natal, South Africa. *Geojournal of Tourism and Geosites, 35*(2), 372–380. doi: 10.30892/ gtg.35214-661

Perez, K. T., & Bui, T. L. H. (2010). Closing doors and opening windows: opportunities for entrepreneurship in an emerging Asian Country for a seasoned woman professional. *International Journal of Entrepreneurship, 14*, 45.

Sachs, J. Kroll, C., Lafortune, G., Fuller, G., & Wohle, F. (2021). *The Decade of Action for the Sustainable Development Goals: Sustainable Development Report 2021.* Cambridge University Press. Available at: www.sustainable.development.report/reports/sustainable-development-report–2021/

Selvi, C. (2019). Women empowerment & skill development in tourism sector. *The Management Accountant, 53*(3), 71.

Sibusiso, D. N., & Ikechukwu Onyekwere, E. (2019). The attributes of successful tourism-related entrepreneurs: A case from South Africa. *EuroEconomica, 38*(2), 296–313.

Song-Naba, F. (2020). Entrepreneurial strategies of immigrant women in the restaurant industry in Burkina Faso, West Africa. *Journal of Developmental Entrepreneurship, 25*(3), 1–16.

Surangi, H. A. K. N. S. (2018). What influences the networking behaviours of female entrepreneurs?: A case for the small business tourism sector in Sri Lanka. *International Journal of Gender and Entrepreneurship, 10*(2), 116–133. doi: 10.1108/ijge-08-2017-0049

Tshabalala, S. P., & Ezeuduji, I. O. (2016). Women tourism entrepreneurs in KwaZulu-Natal, South Africa: Any way forward? *Acta Universitatis Danubius: Oeconomica*, *12*(5), 19–32.

UN. (2015). Transforming our world: the 2030 Agenda for Sustainable Development (Resolution adopted by the General Assembly on 25 September 2015 ed.): United Nations.

UNWTO. (2019). *Global Report on Women in Tourism* (2nd ed.). Madrid: UNTWO.

WEF. (2019). *Global Gender Gap Report 2020*. Geneva, Switzerland: WEF.

Wilson-Youlden, L., & Bosworth, G. (2019). Women tourism entrepreneurs and the survival of family farms in northeast England. *Journal of Rural and Community Development*, *14*(3), 125–145.

Xu, H., Wang, C., Wu, J., Liang, Y., & Nazneen, S. (2018). Human poverty alleviation through rural women's tourism entrepreneurship. *Journal of China Tourism Research*, *14*(4), 445–460.

Yoopetch, C. (2020). Women empowerment, attitude toward risk-taking, and entrepreneurial intention in the hospitality industry. *International Journal of Culture, Tourism, and Hospitality Research*, *15*(1). doi: 10.1108/ijcthr-01-2020-0016

Zapalska, A., & Brozik, D. (2017). Māori female entrepreneurship in the tourism industry. *Tourism*, *65*(2), 156–172.

Zhang, C. X., Kimbu, A. N., Lin, P., & Ngoasong, M. Z. (2020). Guanxi influences on women's intrapreneurship. *Tourism Management*, *81*. doi: 10.1016/j.tourman.2020.104137

# 20 An American labour revolution

*Sotiris Hji-Avgoustis and Alan Yen*

## Introduction

The tourism industry has never faced an economic crisis as severe as during the COVID-19 pandemic. All industry segments, from hotels to restaurants to travel by sea or air, suffered significant losses. The industry has not been as resilient as many predicted before this crisis. Many expected a quick recovery, especially after the lighting fast introduction of a few COVID vaccines. However, the rebound is not proceeding so well as the revenue lost during the pandemic is unrecoverable, and the industry cannot find enough workers post-COVID. Add to that rising costs and an unpredictable supply system when consumer demand skyrockets, and it becomes difficult to avoid customer dissatisfaction.

With the global economy rapidly recovering, why is there still a worker shortage? Some have blamed the generous unemployment payments and stimulus checks for making people less likely to retake low-paying hospitality jobs. Others have countered that companies could raise pay if they wanted workers back quickly. However, the labour issue is more complex as various social, economic, political, and cultural factors have led to the challenge we face now. Economists describe this phenomenon as reallocation friction. The pandemic caused some jobs in the economy to change in content and availability. Workers are taking their time to evaluate new job opportunities in the market, whether these opportunities appeal to them, and whether they have the knowledge, skills, and abilities for these different career paths (Long, 2021).

While each country has its unique situation associated with the labour issue, this chapter will discuss the reasons behind the worker shortage in the USA. The chapter explores the significant reassessment that is going on in the US economy to explain why so many workers are hesitant to return to work. The discussion may serve as a foundation to examine this phenomenon globally.

## Background

The U.S. Bureau of Labor Statistics regards the leisure and hospitality industry as part of the service-providing industries supersector group. It consists of two sectors: (a) the arts, entertainment, and recreation, and (b) accommodation and food services. In addition, the supersector includes various fields within the service industry, such as lodging, food services, event planning, theme parks, transportation, and other tourism-oriented products and services.

Monthly employment data collection began in January 1939, with just under two million workers employed in leisure and hospitality (U.S. Bureau of Labor Statistics, 2021a).

DOI: 10.4324/9781003291763-23

The supersector has been growing steadily, reaching 16,915,000 jobs in February 2020. However, by April 2020, the number of leisure and hospitality jobs declined by almost half (46.12%) to 8,691,000, triple the number of the next-hardest-hit industry, the government. According to the "Employment Situation Summary", published monthly by the U.S. Bureau of Labor Statistics (2021b), employment in leisure and hospitality was still down by 1.7 million, or 10%, in August 2021.

Wage data for August 2021 shared by the U.S. Bureau of Labor Statistics (2021c) shows that the average hourly rate in the hospitality sector was up around $1 compared to the pre-pandemic hourly rate. That brought the average wage in leisure and hospitality to $18.82, just about half of what all workers earn hourly. For example, in the case of the retail trade, wages continued to climb during the pandemic, reaching $22.12 by August 2021, $3.30 more per hour than leisure and hospitality.

Within the supersector, some segments struggle more than others, depending on the country's economic conditions. The hotel industry, for example, experienced unprecedented layoffs and furloughs in 2020 as properties scrambled to keep their doors open amid the COVID-19 pandemic. As a result, hotel employment was down a quarter by 2020. Moreover, the American Hotel and Lodging Association (2021) report predicts that the industry will not bounce back to 2019 employment levels until 2023.

According to a story in the trade publication *QSR* (Klein, 2021), 40% of restaurants interviewed stated they remain understaffed, and 80% said they kept at least one hiring roll always posted. Back-of-the-house positions are even more challenging to hire for, as curbside and to-go orders have grown and increased demand for those positions. In addition, since the pandemic started, restaurants have raised back-of-the-house wages by 18% on average. The events industry has also seen its share of labour challenges. Mandatory in-person meetings, team-building retreats, and conventions came to a halt in 2020, resulting in a massive layoff. Cancellations had far-reaching impacts across the supersector, including restaurants, hotels, and airlines. As business meetings slowly return, many events and conference organisers are scaling down the size of their in-person events. They re-evaluate the feasibility of hosting virtual-only or part in-person virtual meetings. While the government and businesses have relaxed some restrictions on indoor gatherings, the industry has not seen the return of group business reaching the pre-pandemic level as expected (e.g., Smith Travel Research, 2022).

**Immigration policies**

Much blame is put directly on the pandemic and its impacts on labour shortages across all industries. However, a closer examination of the labour landscape in the country reveals that labour trouble was brewing long before the COVID-19 pandemic. A sharp decline in the number of immigrant job applicants and employees put pressure on the industry's ability to meet its staffing needs.

Hotel and restaurant guests noticed the impact on everything from a shortage of front-desk employees to fewer food and beverage offerings. As a result, hospitality industry executives were demanding action from the government. Following President Trump's remarks during the State of the Union Address in February 2020, the American Hotel and Lodging Association (2020) called for immigration reform because the US labour force could not meet the increased labour demand. The opposite was true two decades before that (Campbell, 2019). As recently as 2018, immigrants entering the country fell by 70%, mainly because federal policies limited the number of available visas. According to a "Travel and

Hospitality Industry Outlook" (Deloitte, 2019), immigrants represent just under a third of the workforce but only 13% of the US population. Following the Trump administration's policy in 2018 to deploy troops to the border to stop the flow of Central American immigrants, the U.S. Chamber of Commerce president declared that the country "is out of people" (Higgins, 2018).

Even before that, the U.S. Bureau of Labor Statistics (2015) estimated that the need for immigrant workers was significant. The number of US residents working in the supersector was around 15 million, making it the fifth-largest industry in the country. But that number was way below what the actual labour needs at that time were. In the past, politicians and business leaders alike recognised the importance of legal immigration to meet the growing labour needs of the US economy. Temporary workers were allowed to enter the country as part of initiatives such as the H-2B visa programme for non-agricultural workers, created in 1986 under President Reagan. The Trump administration capped the supply of such visas to 66,000 a year. Another initiative is the short-term J-1 visa programme aimed at students or educators who enter the USA to pursue their educational interests. Restaurants and hotels often rely on the programme to fill summer jobs. In the summer of 2021, many popular US destinations, from commercial beach resorts to national parks, struggled to keep their facilities open due to labour shortages. Any employer needing to hire workers on these visas must show that they cannot fill the position with a US resident. These employers can quickly meet a requirement in the tight labour market.

The Deferred Action for Childhood Arrivals (DACA) programme's uncertainty is a pressing issue impacting the labour market. The so-called Dreamers are undocumented immigrants brought to the USA by their parents as young children under protection from the DACA programme. The programme has offered protection to this group since 2012, with ongoing discussions on how to give them a viable path to citizenship. However, by 2017, the Trump administration called for the programme's termination. Such calls sparked a backlash and multiple lawsuits in states around the country.

Nevertheless, by 2019, there was an estimated 1.3 million eligible DACA immigrants. About half of them work in the supersector of hospitality and leisure, retail trade, and construction. As a result, many trade organisations, including the American Hotel and Lodging Association, continued to call this issue an immediate immigration priority (American Hotel & Lodging Association, 2018).

In 2007, during the Obama administration, the Department of Homeland Security decided to crack down on illegal immigration by using "no-match letters" to enforce existing immigration laws. These letters informed employers that some of their employees' tax form data did not match their records. Several court battles succeeded in stopping the practice. However, a decade later, the U.S. Immigration and Customs Enforcement agency resumed the course during the Trump administration. Faced with possible criminal charges, employers, many of them in the restaurant industry, were forced to fire many of these workers. In addition, there is an expectation for employers to use E-Verify, the voluntary federal programme for checking the immigration status of new hires, to verify their employees' work eligibility. The result is a shrinking labour pool to choose from.

**The shift in US labour force demographics**

Higher pay may not be enough to retain low-wage workers in the new labour landscape. Hospitality and leisure workers continue to leave their jobs, even as some employers try to lure in workers with one-time bonuses and other perks. Moreover, economists, politicians, and business leaders are unsure of the driving force behind the work shortage. Reasons given include fear of COVID-19, childcare needs at home, mismatch of skills between the worker and the job, changing career interests, or early retirement. One overlooked trend that may also explain the shrinking labour pool and that deserves attention is the changing demographic composition of the labor force.

The trend of fewer people entering the workforce is not a new phenomenon. One only needs to casually look at the labour force participation rate over the last two decades to understand some of the reasons behind a changing labour landscape. The U.S. Bureau of Labor Statistics (2021d) reports an ageing workforce over the last few decades, and the average participation rate has consistently declined. This trend has been going on since the turn of the 21st century. In January 2000, for example, the labour participation rate was 67.3 and continued to decline, reaching a low of 62.4 in September 2015. By February 2020, immediately before the pandemic, the rate rose to 63.3. The COVID-19 pandemic dropped the speed back to 60.2 in April 2020. A quick, slight recovery followed, and by August, the rate rose to 61.7. But all demographic indicators point to rates remaining in the low 60s range in the foreseeable future. That is a significant decline from the high 60s at the turn of the 21st century.

Participation rate by age also points to similar trends. In January 1999, 84.6% of the labour force were workers in the 25–54 age group. That percentage has been declining since then, reaching 81.8 in August 2021. Moreover, unemployment rates were falling even before the COVID-19 pandemic. For example, in February 2020, the unemployment rate was 3.5% for men and 3.4% for women. However, by August 2021, the rate for both groups still hovered in the low 5% range, even during a recovering economy, an indication that the labour market cannot keep up with the needs of a growing US and global economy (U.S. Bureau of Labor Statistics, 2021b).

**Compensation and work–life balance concerns**

As the average hourly wage of leisure and hospitality workers increased slower than in other sectors, hospitality and tourism companies have offered significantly higher hourly rates to entice individuals to work in the industry – but to no avail. Many workers still find it difficult to sustain a good quality of life with their current earning levels. Empirical studies support the argument that their base earnings continue to be lower than those of their peers in different sectors of the economy (e.g., Dogru, McGinley, Line, & Szende, 2019; Jolly, McDowell, Dawson, & Abbott, 2021; Sturman, 2001; Sturman, Ukhov, & Park, 2017). In addition, younger workers face high financial burdens, ranging from student loans to other financial obligations (Friedman, 2020; Hanson, 2021). The stress from making ends meet while repaying the debt often drives leisure and hospitality workers to look elsewhere for employment.

While a competitive financial compensation package remains a critical issue in keeping individuals in the industry, other aspects of employment, such as long hours, anti-social work schedules, and physically and mentally demanding work environments, are forcing people to consider leaving the industry entirely (Cleveland et al., 2007; Magnini, 2009; Lee, Magnini, & Kim, 2011; Yen, Cooper, & Murrmann, 2013; Zhao, 2016; Zhao & Ghiselli, 2016). These studies have also raised the issue of maintaining a sustainable work–life balance. Individuals often find it challenging to balance work and personal life, especially younger generations and those with young children. The pandemic also revealed that the unemployment impact is widespread and not among different ethnic groups, as previously suggested (Parker, Igielnik, & Kochhar, 2021). As the labour force structure continues to change, younger workers are reassessing the role of work in their lives and are inclined to change jobs or careers to pursue employment opportunities that better suit their views and values.

### Transition into a new labour market

Job mobility describes the ability of workers to move quickly from one job to another. It is often linked with economic opportunity because workers change jobs for higher compensation, better working conditions, and opportunities for advancement. According to data reported monthly by the Current Population Survey (CPS), administered monthly by the U.S. Bureau of the Census (2021e) to over 65,000 households, leading up to the COVID-19 pandemic, job mobility rates have been falling, especially among younger workers between 16 and 24 years of age. The rates for workers aged 25 years and over has remained pretty much the same for the past 20 years. This, of course, is nothing new; job-to-job transitions occur more frequently earlier in life, especially under the age of 24, as younger workers are exploring career options. However, the rates stabilise over the rest of a person's work life.

CPS data paints a different picture beginning in the second half of 2020. Job mobility rates began to rise again, especially for workers with no college education. Workers with less than high school education lead all other categories with job-to-job transition rates of over 2.25%. In 2020, the leisure and hospitality supersector had the lowest median age for workers at 31.8 and the highest number of workers 24 years old or younger (U.S. Bureau of Labor Statistics, 2021f). The interpretation of the increase in job mobility among the young depends on the underlying explanations for this trend. It may be a case of more experimentation with different jobs. Another concern is that technological changes discussed below are forcing younger workers to re-evaluate the benefits of staying on a career path that these changes may disrupt. As a result, they elect to pursue careers with higher perceived job security and advancement opportunities.

### Developments in labour automation

Labour automation substitutes technology for human labour to perform specific tasks or jobs. Aaronson and Phelan (2020) addressed the trend of replacing low-wage cognitively routine and manually routine jobs with labour automation technology. Their findings point to increased automation rates for low-skill jobs, often located in rural areas. However, such advances disproportionately impact less-educated workers who are often young, male, and from minority groups.

The trend of labour automation gained steam following the 2007–2008 global financial crisis. Frey and Osborne (2013) classified occupations by their susceptibility to automation. They concluded that 47% of US workers will be at risk in the next 20 years. The leisure and hospitality supersector consists of many low-skilled jobs easily outsourced using partial or complete automation technology. Examples include automated check-ins using apps on guests' phones, Alexa-enabled systems in rooms that bypass human interactions, and food delivery apps that bypass a hotel's room service offerings.

The trend toward automation predates the pandemic, but it has accelerated at a critical moment. According to Sedik and Yoo (2021), automation gained popularity at the height of the pandemic because it offered an alternative to exposing workers to health hazards.

The increased reliance on automation displaced some low-skilled workers who had to look for jobs outside the industry. Many never returned to their old jobs.

**Quality of life issues and considerations**

Many employees, faced with losing their jobs to automation and unhappy about the working conditions, including lower pay compared to other service jobs, began searching for new career opportunities before the COVID-19 pandemic. But unfortunately, the pandemic made matters worse. According to a recent study (Chen & Chen, 2021), the layoffs and furloughs hospitality employees had to endure due to the pandemic caused them financial strain, depression, and social isolation, which resulted in an increased intention to leave the industry altogether. The choice to leave was solid among women and younger workers.

Nearly 7% of workers in the leisure and hospitality supersector quit their job in August 2021 alone, outpacing the 2.9% rate for the rest of the sectors of the US economy (U.S. Bureau of Labor Statistics, 2021b). That represents the highest rate on record since the Bureau began tracking quit numbers back in December 2000. The findings may suggest that these workers are leaving because they are finding better jobs, as the high quit rate may be interpreted as a measure of workers' confidence in their ability to land jobs elsewhere. Another explanation may be that they are leaving because of traditional low pay, a lack of benefits, long hours, and potential exposure to COVID-19. Finally, when the economy is terrible, workers often leave jobs to further their education. All these factors may be also at play and contribute to the massive labour shortages the industry is experiencing.

**Future labour trends**

Looking into the next decade, the U.S. Bureau of Labor Statistics (2021d) projects an increase of 11.9 million jobs by the end of this decade. Employment in the leisure and hospitality supersector is forecast to increase the fastest among all sectors, representing 7 of the 20 fastest growing supersectors. This growth is primarily driven by recovery from the pandemic as the public resumes recreational and in-person activities. As a result, the demand for restaurants, hotels, and arts, cultural, and recreational-related establishments is rising. Fortunately, the agency does not expect automation advances to impact the supersector negatively. However, there are indications that certain professions, such as office and administrative support, sales, and production occupations, will experience declines in overall employment due to technological changes.

Remote work for many occupations is here to stay. An online survey of human capital leaders conducted by the Conference Board (2021) points to four main challenges human capital professionals face. First, findings suggest a need to adjust to a world where a large share of employees work and prefer to continue to work remotely. Second, there are also challenges in recruiting and retaining qualified workers, especially service workers who make up a vast share of the leisure and hospitality supersector. Third, employees expressed concerns regarding poor working conditions and their impact on employee wellbeing. Fourth, challenges in managing the return to the physical workplace.

Employers benefit from switching to remote work, hiring workers in cheaper labour markets, and saving on labour costs (Upwork, 2021). In 2018, Silicon Valley posted less than 40% of tech company job advertisements outside that area. Now it is about two-thirds, primarily because of a shift to remote work. Much of the work could eventually be replaced by technology. At the same time, changes in immigration policy may lead to the employment of more new legal immigrants, which may drive labour costs down.

The leisure and hospitality supersector traditionally benefited older US residents with high disposable income. However, the pandemic and subsequent social distancing requirements contributed to their reductions in spending on in-person services during this period, much more than younger people. With a return to normalcy, a demand surge and stagnant labour supply will lead to more labour shortages and pressure businesses to remain open.

## Global labour woes

Pandemics, politics, technological advances, and wars can have unpredictable labour disruptions on national economies. For example, the pandemic prompted many foreign workers to return home to be closer to their families. Brexit forced many East Europeans to do the same. The UK's new immigration rules took away preferential treatment for European Union nationals by switching to a points-based system designed to attract mainly skilled workers. As a result, many unskilled Eastern European workers, the backbone of the UK hospitality industry, accounting for 15% of the total hospitality workforce, returned home (Food and Drink Federation, 2021). The war in Ukraine also kept many young Eastern Europeans from filling thousands of hospitality and leisure jobs across Europe and the USA. The massive layoffs during the pandemic forced hospitality workers to search for more stable careers in other fields, including manufacturing, healthcare, finance, and construction, with higher earning potential (Holzer, 2022).

Finally, as new technologies create new industries and professions, the need for skilled and unskilled employees to fill these positions from shrinking national labour pools presents hospitality employees with new career opportunities. They leave their jobs searching for more money, flexibility, and happiness. Remote work during the pandemic changed the way employees think about time and space (Neeley, 2021). It offered them an opportunity to consider what was best for them and their families.

## Closing remarks

This chapter has offered multiple factors contributing to the existing labour shortage, not just pandemic unemployment benefits. A report by Yale economists Altonji et al. (2020) found no evidence that the enhanced jobless benefits Congress authorised in March 2020 in response to the COVID-19 pandemic reduced employment participation rates. Their findings suggest that the expanded benefits neither encouraged layoffs during the beginning of the pandemic nor discouraged people from returning to work once businesses began reopening. Other factors contributing to the labour shortage include childcare issues, career changes, early retirement, and fear of COVID-19 in the workplace, especially for leisure and hospitality supersector workers.

The supersector must reinvent itself to overcome its labour challenges. A Joblist report (2021) found that more than half of hospitality workers prefer not to return to their pre-pandemic jobs. Over a third would instead leave the supersector, searching for employment elsewhere. The industry must consider ways to make careers in the sector attractive once again. Pre-pandemic research showed that about 70% of all hospitality graduates leave the industry (Blomme et al., 2009). The pandemic just made retention efforts so much worse. It is time for the industry to partner with higher education institutions to innovate new business models to attract a new generation of hospitality professionals.

Overcoming labour challenges is also one of the United Nations 2030 Agenda for Sustainable Development priorities (United Nations, 2020). The plan outlines how companies can create up to 380 million jobs by integrating one or more of its 17 goals into their strategies. Several of these goals address the issue of job creation, including in tourism and hospitality, one of the world's fastest-growing industries. For example, Goal 8.9 calls for the design and implementation of policies to create tourism-related jobs that promote local culture and products.

## References

Aaronson, D., & Phelan, B. J. (2020). The evolution of technological substitution in low-wage labor markets, Economic Studies at Brookings. Retrieved from https://www.brookings.edu/wp-content/uploads/2020/01/Phelan-Aaronson_Full-Report-Tables.pdf

Altonji, J., Contractor, Z., Finamor, L., Haygood, R., Lindenlaub, I, Meghir, C., O'Dea, D. S., Wang, L., & Washington, E. (2020). Employment effects of unemployment insurance generosity during the pandemic. Retrieved from https://tobin.yale.edu/sites/default/files/files/C-19%20Articles/CARES-UI_identification_vF(1).pdf

American Hotel & Lodging Association (2018). AHLA statement on DACA. Retrieved from https://www.ahla.com/press-release/ahla-statement-daca

American Hotel & Lodging Association (2020). AHLA statement on president trump's state of the union address. Retrieved from https://www.ahla.com/press-release/ahla-statement-president-trumps-state-union-address-0

American Hotel & Lodging Association (2021). Reports: nearly 500,000 hotel jobs won't return by year's end, room revenue will be down $44 billion from 2019. Retrieved from https://www.ahla.com/press-release/reports-nearly-500000-hotel-jobs-wont-return-years-end-room-revenue-will-be-down-44

Blomme, R. J., Van Rheede, A., & Tromp, D. M. (2009). The hospitality industry: An attractive employer? An exploration of students' and industry workers' perceptions of hospitality as a career field. *Journal of Hospitality & Tourism Education, 21*(2), 6–14.

Campbell, A. (2019). The U.S. is experiencing a widespread worker shortage. Here's why. Retrieved from https://www.vox.com/2019/3/18/18270916/labor-shortage-workers-us

Chen, C.-C., & Chen, M.-H. (2021). Wellbeing and career change intention: COVID-19's impact on unemployed and furloughed hospitality workers. *International Journal of Contemporary Hospitality Management, 33*(8), 2500–2520.

Cleveland, J. N., O'Neill, J. W., Himelright, J. L., Harrison, M. M., Crouter, A. C., & Drago, R. (2007). Work and family issues in the hospitality industry: Perspectives of entrants, managers, and spouses. *Journal of Hospitality & Tourism Research, 31*(3), 275–298.

Deloitte. (2019). 2019 travel and hospitality industry outlook. Retrieved from https://www2.deloitte.com/us/en/pages/consumer-business/articles/travel-hospitality-industry-outlook.html

Dogru, T., McGinley, S., Line, N., & Szende, P. (2019). Employee earnings growth in the leisure and hospitality industry. *Tourism Management, 74*, 1–11.

Food and Drink Federation. (2021). *Establishing the labor availability issues of the UK Food and Drink Sector*. Retrieved from https://www.fdf.org.uk/globalassets/resources/publications/reports/establishing-the-labour-availability-issues-of-the-uk-food-and-drink-sector.pdf

Frey, C. B., & Osborne, M. A. (2013). *The Future of Employment: How Susceptible are Jobs to Computerization*. Oxford Martin Program on Technology and Employment. https://www.oxfordmartin.ox.ac.uk/downloads/academic/future-of-employment.pdf

Friedman, Z. (2020). Student loan debt statistics in 2020: A record $1.6 trillion. Retrieved from https://www.forbes.com/sites/zackfriedman/2020/02/03/student-loan-debt-statistics/?sh=277321a9281f

Hanson, M. (2021). Average student loan debt. Retrieved from https://educationdata.org/average-student-loan-debt

Higgins, S. (2018). Chamber president warns U.S. is 'out of people,' needs more immigration. Retrieved from https://www.washingtonexaminer.com/policy/economy/chamber-president-warns-us-is-out-of-people-needs-more-immigration

Holzer, H. (2022). *Do Sectoral Training Programs Work? What the Evidence on Project Quest and Year Up Really Shows*. Brookings Institution. Retrieved from https://www.brookings.edu/research/do-sectoral-training-programs-work-what-the-evidence-on-project-quest-and-year-up-really-shows/

Joblist. (2021). Q2 2021 United States Job Market Report. Retrieved from https://www.joblist.com/jobs-reports/q2-2021-united-states-job-market-report

Jolly, P. M., McDowell, C., Dawson, M., & Abbott, J. (2021). Pay and benefit satisfaction, perceived organizational support, and turnover intentions: The moderating role of job variety. *International Journal of Hospitality Management*, *95*, 102921.

Klein, D. (2021). Is the restaurant industry facing a hiring crisis? Retrieved from https://www.qsrmagazine.com/employee-management/restaurant-industry-facing-hiring-crisis

Lee, G., Magnini, V. P., & Kim, B. P. (2011). Employee satisfaction with schedule flexibility: Psychological antecedents and consequences within the workplace. *International Journal of Hospitality Management*, *30*(1), 22–30.

Long, H. (2021). It's not a 'labor shortage'. It's a great reassessment of work in America. *Washington Post*. https://www.washingtonpost.com/business/2021/05/07/jobs-report-labor-shortage-analysis/.

Magnini, V. P. (2009). Understanding and reducing work-family conflict in the hospitality industry. *Journal of Human Resources in Hospitality & Tourism*, *8*(2), 119–136.

Neeley, T. (2021). *Remote Work Revolution: Succeeding from Anywhere*. New York: HarperCollins Publishers.

Parker, K., Igielnik, R., & Kochhar, R. (2021). Unemployed Americans are feeling the emotional strain of job loss; most have considered changing occupations. Retrieved from https://www.pewresearch.org/fact-tank/2021/02/10/unemployed-americans-are-feeling-the-emotional-strain-of-job-loss-most-have-considered-changing-occupations/

Sedik, T. S., & Yoo, J. (2021). Pandemics and automation: Will the lost jobs come back? *International Monetary Fund*. Retrieved from https://www.imf.org/en/Publications/WP/Issues/2021/01/15/Pandemics-and-Automation-Will-the-Lost-Jobs-Come-Back-50000

Smith Travel Research. (2022). *COVID-19: Hotel Industry Impact and Recovery*. Retrieved from https://str.com/data-insights-blog/coronavirus-hotel-industry-data-news

Sturman, M. C. (2001). The compensation conundrum: Does the hospitality industry shortchange its employees—and itself?. *The Cornell Hotel and Restaurant Administration Quarterly*, *42*(4), 70–76.

Sturman, M. C., Ukhov, A. D., & Park, S. (2017). The effect of cost of living on employee wages in the hospitality industry. *Cornell Hospitality Quarterly*, *58*(2), 179–189.

The Conference Board. (2021). The reimagined workplace a year later: Human capital responses to the COVID-19 pandemic. Retrieved from https://www.conference-board.org/pdfdownload.cfm?masterProductID=27396

United Nations. (2020). *The Sustainable Development Agenda*. Retrieved from https://www.un.org/sustainabledevelopment/development-agenda/

Upwork. (2021). Potential savings of remote workers. Retrieved from https://www.upwork.com/info graphics/working-from-home-saves-company-money

U.S. Bureau of Labor Statistics. (2015). Industry employment and output projections to 2024. Retrieved from https://www.bls.gov/opub/mlr/2015/article/industry-employment-and-output-projections-to-2024.htm

U.S. Bureau of Labor Statistics. (2021a). All employees, leisure and hospitality. Retrieved from https://fred.stlouisfed.org/series/USLAH

U.S. Bureau of Labor Statistics. (2021b). Current employment statistics – CES (National). Retrieved from https://www.bls.gov/ces/

U.S. Bureau of Labor Statistics. (2021c). Average hourly earnings of all employees. Retrieved from https://fred.stlouisfed.org/series/CES4200000003

U.S. Bureau of Labor Statistics. (2021d). Employment projections 2020–2030. Retrieved from https://www.bls.gov/news.release/pdf/ecopro.pdf

U.S. Bureau of Labor Statistics. (2021e). Labor force statistics from the current population survey. Retrieved from https://www.bls.gov/cps/

U.S. Bureau of Labor Statistics. (2021f). Household data annual averages: 18b. Employed persons by detailed industry and age. Retrieved from https://www.bls.gov/cps/cpsaat18b.htm

Yen, C., Cooper, C., & Murrmann, S. (2013). Exploring culinary graduates' career decisions and expectations. *Journal of Human Resources in Hospitality & Tourism, 12*(2), 109–125.

Zhao, X. R. (2016). Work-family studies in the tourism and hospitality contexts. *International Journal of Contemporary Hospitality Management, 28*(11), 2422–2445.

Zhao, X. R., & Ghiselli, R. (2016). Why do you feel stressed in a "smile factory"? Hospitality job characteristics influence work–family conflict and job stress. *International Journal of Contemporary Hospitality Management, 28*(2), 305–326.

# 21 The resilience of the tourism and hospitality workforce

*Thi Hong Hai Nguyen, Diep Ngoc Su and Hanh My Thi Huynh*

## Introduction

Resilience, which is understood as the capacity of a system to absorb disturbance and reorganise while undergoing changes (Walker et al., 2004), has increasingly become an important topic in professional practice and academic research. The original definition of resilience was given by Holling (1973) in the field of ecological sciences. The concept, while originating in engineering and materials science, has been adapted and applied in a wide range of fields, from the environmental and medical to social sciences and management studies. Nonetheless, there is no consensus on the concept of resilience (Hall et al., 2017). The widely known definition of resilience is "the capacity of a system to absorb disturbance and reorganise while undergoing change so as to still retain essentially the same function, structure, identity, and feedbacks" (Walker et al., 2004, p. 1).

The tourism industry is argued to be one of the most vulnerable industries to crisis (Santana, 2004). In recent years, numerous crises, including nature, health, terrorism, and economics, have created severe disturbances in the hospitality and tourism industry. Crisis management is essential for tourism destinations, organisations, and workforces to develop resilience to adapt and thrive during these challenging times (Hall, Prayag, & Amore, 2017) and to prepare for future crises and ensure sustainable development.

Resilience is an essential element of the UN Sustainable Development Goals (SDGs), where it is mentioned multiple times (UN, 2022). It is best identified in SDG 11 "Make cities and human settlements inclusive, safe, resilient and sustainable" and SDG 13 "Strengthen resilience and adaptive capacity to climate-related hazards and natural disasters in all countries" (UN, 2022). Resilience is also believed to be "a fundamental prerequisite for sustainable development and achieving the SDGs" (Global Resilience Partnership, 2022). Building resilience for the tourism and hospitality industry will facilitate the sustainable development of tourism cities, their businesses, and residents, and thus contribute to the achievement of the SDGs.

With the focus on the hospitality and tourism workforce, this chapter provides a comprehensive understanding of the multi-level resilience at the individual and organisational levels. We firstly present a thorough theoretical background of resilience, including various conceptualisations of individual resilience and organisational resilience, and the extant literature of resilience concerning resilience-building mechanisms. Then, based on empirical data from Vietnam in the context of a major crisis, two frameworks of workforce resilience during a crisis, from the organisational and individual perspectives, are illustrated. These frameworks provide insights into the resilience of the hospitality and tourism workforce and organisations when facing a global health-related crisis. Strategies are provided for

DOI: 10.4324/9781003291763-24

various stakeholders to enhance their resilience capacity and to help them be more sustainable in an uncertain world.

## Resilience and human resource management

Originating from the study of children who remained mentally stable despite having been exposed to threats of risk or adversity (Garmezy, 1974), psychologists and psychiatrists started to pay attention to the phenomenon of resilience. From the individual perspective, resilience can be understood as a capacity of a person to persist, adapt, and flourish in the face of disruption and changing events (Holling, 1973). Resilience is described as a personal trait that operates in the aftermath of a single short-term traumatic event (Bonanno, 2004, Klohnen, 1996). According to Robertson et al. (2015), resilience is also characterised as an individual trait that ensures adaptation to adversity and which can be assessed using trait variables. Rutter (1985) elaborated on the multifaceted existence of resilience, pointing out that it was influenced not just by internal personal characteristics but also by external influences. Resilience has been described as a dynamic construct that evolves over time (Fletcher & Sarkar, 2013). Luthar et al. (2000) also viewed resilience as a dynamic process rather than as a personal trait. They argued that if resilience is only viewed as a personal characteristic, it can lead to the inference that certain people do not have the necessary traits to overcome adversity. For example, an adolescent who has faced adversity but has shown solid academic adaptation may face emotional difficulties. As a result, some people who have resilience in some domains might experience issues in other domains.

The resilience of individuals in the context of the workplace is termed "employee resilience". It is defined as "a suite of adaptive, learning, and network-leveraging behaviors that signal resource availability and the individual's motivation and capacity to utilise these resources" (Kuntz, Malinen, & Näswall, 2017, p. 225). This aspect of resilience does not only refer to the individual ability to recover from adversity, but it also reflects the capability to utilise personal and workplace resources to develop that ability. To build employee resilience, Kuntz et al. (2017) identified four areas of initiatives: valuing employees, fostering learning and collaboration, human-capital development, and support for challenges at work.

Also in the working environment, along with employee resilience, career resilience is developed and studied. This concept is used to describe the resilience structures in the career domain. Career resilience is understood as a process of career development by persisting, adapting, and flourishing through overcoming challenges and disruptions (Mishra & McDonald, 2017). The research interest in career resilience has been growing, with research papers published in different fields referring to career resilience in health, education, business, banking, and psychology (Table 21.1). However, in the field of hospitality and tourism, limited research has been conducted to understand the resilience at the individual level until the world faced a global crisis event (COVID-19) starting from the end of 2019.

The focus of the research on individual, employee, and career resilience in the workplace before COVID-19 was to highlight the antecedents and outcomes in resilience promotion (Mishra & McDonald, 2017; Hartmann & Apaolaza-Ibáñez, 2012). Accordingly, individual resilience has been found as either a mediator or moderator in the extant literature of resilience and human resource management. First, resilience meditated relationships between groups of variables including personal traits (e.g., personal resources, personal

*Table 21.1* Previous studies on career resilience from an individual perspective

| Author (Year) | Topic | Field |
|---|---|---|
| Wyllie et al. (2020) | An evaluation of early-career academic nurses' perceptions of a support programme designed to build career resilience. | Health |
| Kutsyuruba et al. (2019) | Developing resilience and promoting wellbeing in early career teaching: advice from the Canadian beginning teachers. | Education |
| Cooke et al. (2019) | Can a supportive workplace impact employee resilience in a high-pressure performance environment? An investigation by the Chinese banking industry. | Banking |
| Salisu et al. (2019) | Social capital and entrepreneurial career resilience: the role of entrepreneurial career commitment. | Business |
| Papatraianou et al. (2018) | Beginning teacher resilience in remote Australia: a place-based perspective. | Education |
| Salisu et al. (2017) | Mediating effect of entrepreneurial career. Resilience between entrepreneurial careers. Commitment and entrepreneurial career success. | Business |
| Bowles and Arnup (2016) | Early career teachers' resilience and positive adaptive change capabilities. | Education |
| Kolar et al. (2017) | Resilience in early-career psychologists: investigating challenges, strategies, facilitators, and the training pathway. | Psychology |
| Mansfield et al. (2014) | "I'm coming back again!" The resilience process of early career teachers. | Education |
| Coogle et al. (2007) | Job satisfaction and career commitment among nursing assistants providing Alzheimer's care. | Health |

perception/attitudes, personal emotions, and work resources/demands) and groups of outcome variables (e.g., performance, mental and physical health, work-related attitudes, and change-related attitudes). Second, individual resilience also played the moderator role in some such relationships (Hartmann and Apaolaza-Ibáñez, 2012). The review of prior studies indicated a lack of theory-driven research on resilience in the workplace. As a result, there has been a call for applying foundation theories to understand resilience mechanisms. In addition, the multi-level analysis of resilience needs to be advanced not only at the individual level but also at the organisational level.

At a higher level, resilience is used to address the ability of organisations to survive and thrive in times of crisis (Lee et al., 2013). In recent studies, the term "resilience" has gathered new momentum in order to deepen knowledge of how the organisation survives and thrives during turbulence (Hillmann & Guenther, 2021). Besides examining the antecedents of the failure of organisations during disruption (Choo, 2008; Reason, 2000), the idea of successful organisations against disruption has attracted the interest of numerous scholars by highlighting the role of organisational resilience. Weick (1993) defined resilience as the acceptance of any change or uncertainty, the persistence to continue, and the capability of organisations to turn challenges into opportunities when facing shocks. To do so, he emphasises not only the adaptation but oriented, creative, and proactive approaches to propose solutions based on four potential sources of resilience that organisations can apply to limit severe outcomes from turbulence, namely improvisation and bricolage, virtual role systems, an attitude of wisdom, and respectful interaction (Weick, 1993). In another study concerning the World Trade Organization, Kendra and Wachtendorf (2003) viewed resilience as an ability to deal with unexpected events without any severe disruption as well as the capability of an organisation to adapt and recover after the turbulence. They proved that emergency responses such as policies, procedures, practices, and tools can fail in disaster situations and that organisational resilience should be an instrument for enhancing effective response efforts (Kendra & Wachtendorf, 2003).

The resilience of organisations in the tourism sector is an emergent research issue. From the resource-based view, organisational resilience is built upon fundamental resources including finance, human, social capital, and other core values (Biggs et al., 2015) and which can be explored using a capital-based approach with six types of capital: economic, social, physical, human, natural, and cultural (Brown et al., 2018). Moreover, capital building organisational resilience can be grouped into three critical categories: people, processes, and partnership (Hall et al., 2017). Human capital represents the core for creating interrelationships with other factors; human resource management (HRM) contributes its strategic value in developing organisational resilience (Al-Ayed, 2019, Lengnick-Hall et al., 2011).

By analysing three important pillars of resilient organisations, including specific cognitive abilities, behavioural characteristics, and contextual conditions, Lengick-Hall et al. (2017) emphasised the contribution of the individual level of knowledge, skills, abilities, and other attributes in achieving them. The resilience of individuals within the organisations, which includes employee and career resilience, also significantly contributes to the resilience of organisations. Therefore, it is suggested that HR policies and practices in HRM systems are the foundation of the resilience capacity of an organisation due to their impact on the attitudes and behaviours of employees through the process of employee experience (i.e., attraction–selection–attrition).

Facing crises and recession, a set of HRM practices must be conducted in order to enhance the resilience of organisations. Concretely, measures of cost reduction in response to unexpected events such as layoffs, payroll cuts, freezes in new recruitment, reductions in training, rigorous performance management, or downsizing are considered rather than maintaining the whole system (Santana et al., 2017; Teague & Roche, 2014). Among these measures, layoffs seem to have a negative influence on employees' morale and commitment and create a rupture in the psychological contract between employers and employees. Therefore, complementary approaches such as the high-commitment model in HR practices of the workforce should be used to develop resilience (Roche et al., 2011). This is because these behavioural HR practices facilitate employee engagement and motivation as well as accelerate social capital for organisations (Parzefall & Kuppelwieser, 2012). Social capital contributes to easing mental health problems, assisting staff and organisations in staying strong during crises, as well as enhancing the affective commitment of employees (Christodoulou & Christodoulou, 2013).

## Frameworks for workforce resilience during a major crisis

Using several studies of the Vietnamese tourism and hospitality industry during a major health crisis, that is, the COVID-19 pandemic, we established two workforce resilience frameworks from (1) the individual perspective, including employee and career resilience, and (2) the organisational perspective, that is, organisational resilience.

### *Career resilience through a major crisis: An individual perspective*

According to Mishra and McDonald (2017), prior studies have examined a wide range of personal factors influencing career resilience, including traits or individual characteristics, attitudes, skills, behaviours and career history, and contextual factors linked to career resilience, including a supportive workplace, job characteristics, and a supportive family. However, this prior literature only concentrates on examining the influential external factors from the workplace, family, and jobs of the staff. Contextual factors related to employee safety and workplace risks have not yet been taken into account, which can suggest many research ideas in the field of career resilience. Considering the context of a major health crisis, that is, the COVID-19 pandemic, the health-related risks became an issue for concern of academics and practitioners in HRM. It is crucial to understand how the workforce demonstrates its resilience to such kinds of risks. Indeed, the tourism and hospitality industry started bouncing back around the world after the long period of the global pandemic (from the end of 2019 to 2022). Travellers began going out again, with the increasing demand for domestic travel requiring the return of tourism workers. Although the tourism industry has continued to face many challenges in the recovery phase, the management of employees' health and safety is considered the greatest challenge for the industry (Hamouche, 2021; Su et al., 2021).

When returning to work in the hospitality and tourism sectors, such as hotels, restaurants, airports, or entertainment centres, the labour force faces health-related risks by making daily interactions with crowds of new people that might be a source of virus infection. Therefore, tourism and hospitality workers need to understand the types of risks in their work and make their behavioural responses to prevent infection. This psychological behaviour process can be a foundation to explain how tourism and hospitality workers develop their resilience to cope with health risks in their jobs. In order to explain this process, the protection motivation theory (PMT), derived from the discipline of health psychology, was applied to examine the direct and indirect influences of five factors: perceived vulnerability, perceived severity, self-efficacy, response efficacy, and health preventative behaviour (Figure 21.1). Perceived vulnerability and perceived severity refer to the individual's subjective evaluation of threats caused by a crisis event that is a health-related event (Janz & Becker, 1984). Self-efficacy and response efficacy were conceptualised as the individual's

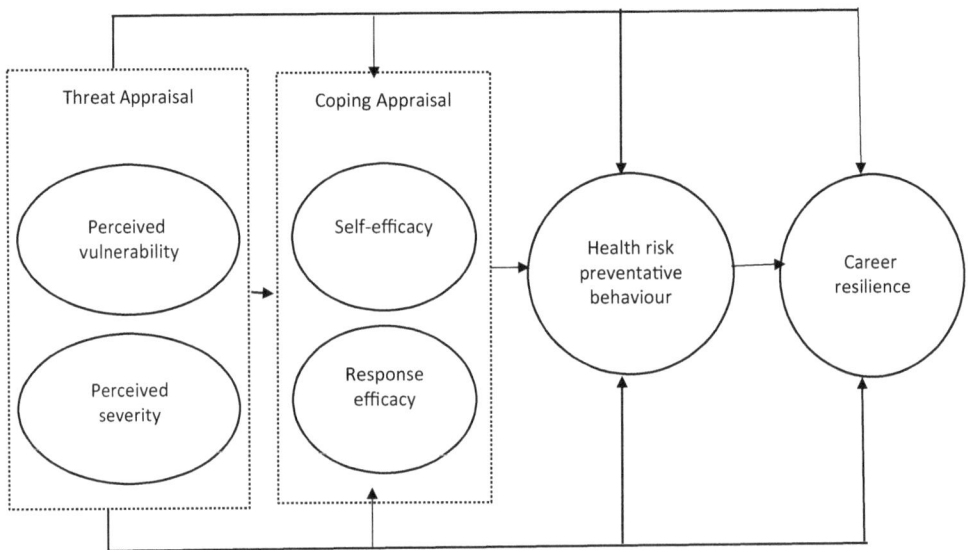

*Figure 21.1* The framework of career resilience at the individual level.

own belief and coping measures at reducing perceived threats (Lin & Bautista, 2016; Rogers, 1983). Based on PMT, career resilience was proposed as a process in which individuals develop their career competencies and resilience to manage health-related risks they perceive in a major crisis.

The model was tested using a dataset consisting of 495 hospitality and tourism workers from various sectors, including accommodation, food and beverage, travel agencies, tourist attractions, and destination management organisations. The Partial Least Squares-based Structural Equation Modelling (PLS-SEM) analysis technique tested the relationships between the five proposed constructs of PMT and career resilience. This framework of career resilience responded to a call for applying a foundation theory in the research of resilience and provides empirical support for PMT (Floyd et al., 2000) to explain the role of psychological behaviour constructs in forming career resilience in tourism and hospitality. Interestingly, out of the five predictors, response efficacy and health risk preventative behaviour were the most important determinants of career resilience. It can be concluded that tourism employees build their career resilience in a major health-related crisis by first developing response efficacy and then engaging in health-risk preventative behaviour. Therefore, managers of hospitality and tourism business organisations need to have measures to ensure the health and safety of employees in working, for example, at promoting guidelines, providing adequate personal protective equipment and facilities, providing training sessions, and establishing risk management processes and contingency plans. Such measures will help employees build their confidence and resilience in their career.

### *Resilience-enhancing HR practices during a major crisis: An organizational perspective*

From an organisational perspective, we propose the framework of HR practices during a crisis as presented in Figure 21.2. The framework applied the four Rs of crisis management within the crisis timeframe, that is, before, during, and after the crisis's main event. This framework is the outcome of a qualitative study which utilised in-depth interviews of hospitality and tourism managers (Su et al., 2021).

The framework of HR practices during a crisis includes three phases: (1) before the crisis's main event, (2) during the crisis's main event, and (3) after the crisis's main event. Prior to the main event, it is indicated that not only the physical health of employees was considered through health and safety measures, but their mental health was also emphasised. Employee wellbeing, including physical and mental, is a central pillar of organisational resilience (Hall et al., 2017). Various health and safety measures, including hand sanitisers, medical masks, protective gloves, as well as the work-from-home practice, were offered. During a crisis, job resources and demands are unstable, leading to job stress, which results in negative psychological strains such as anxiety and depression (Gazzaniga et al., 2010). A wide range of action was initiated to ensure the positive psychology of the workforce. The frequent communication and interactions between managers and employees were found to be the key element to ensuring the smooth operation of the business, that is, organisational resilience, and the high-quality performance of the employees, that is, employee resilience. Specifically, the conversations between managers and employees were often carried out to understand employees' psychological states. A bottom-up approach was used where employees were also empowered to suggest and implement alterations in operations to effectively deal with the changes due to the crisis. The support of managers and the cooperation between managers and employees are essential factors to build organisational as well as employee resilience.

| Stages of the Covid-19 crisis | Before the lockdown | During the lockdown | After the lockdown |

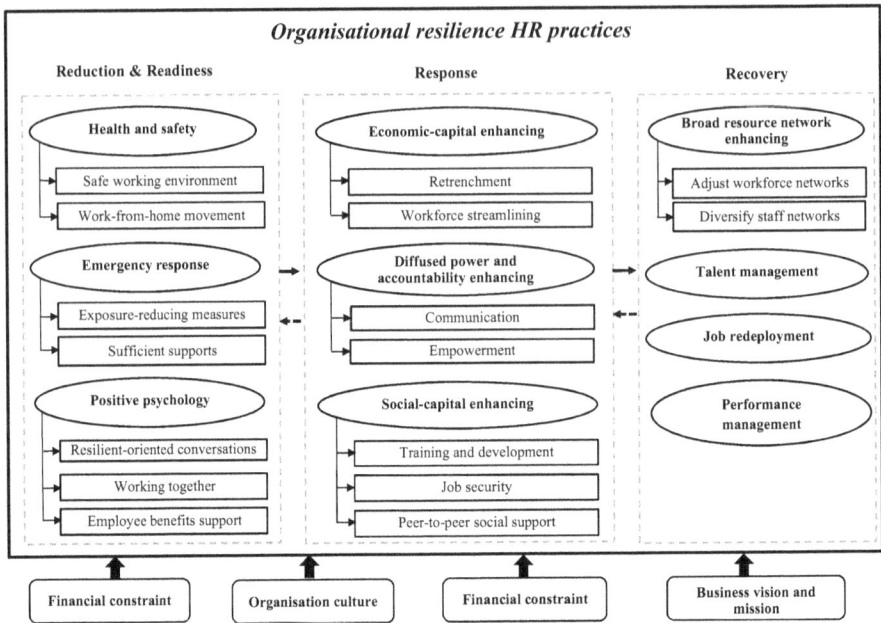

*Figure 21.2* The framework of HR practices for organisational and employee resilience during a crisis.

*Source*: Adapted from Su et al. (2021).

During the peak time of crisis when tourism and hospitality operations were disrupted, varied cost-cutting initiatives were the first actions to be executed to mitigate financial stress. These actions included cutting payroll, mandating paid or unpaid leave, and cutting training expenses. While layoffs might be the easiest practice to reduce the financial burden, it is recognised that this practice has a detrimental impact on employee morale and commitment, and therefore it was only considered as a last resort. Secondly, in addition to the above initiatives to enhance economic capital, social capital was also maintained. Social capital is indicated to positively affect employees' mental health as well as their commitment (Parzefall & Kuppelwieser, 2012). This is commonly created through the interactions between employees by training and working together (Leana & Van Buren, 1999). Therefore, even during the crisis, training and development activities were still maintained, although they were mostly done internally to reduce costs. Internal training, including introducing new skill sets such as stress management, was carried out to enhance adaptive capacity, which is an essential element of resilience. Ensuring job security during the crisis was another initiative to sustain the psychological contract and hence improve social capital. Overall, a supportive culture was strengthened during the crisis by maintaining the spirit of "together, sharing the good and bad times". Thirdly, other HR practices including empowerment and effective communication were also enhanced. These soft practices are believed to diffuse power and accountability among employees, encourage their engagement and commitment, and thus enhance resilience (Lengnick-Hall et al., 2011).

After the crisis's main event, the tourism and hospitality industry entered the recovery phase. In this stage, a series of recovery initiatives were carried out, including broadened resource networks, job deployment, talent management, and performance management.

To quickly regain supply resources, diversifying the supply network while strengthening strategic partners were concurrently applied. In terms of post-crisis human resources, restructuring was advocated, based on redeployment, relocation, and retention (Cascio, 2010). Accordingly, only a smaller number of employees who were highly skilled and capable of multi-tasking were selected. To thrive post-crisis, which is an important element of resilience, tourism and hospitality organisations also focused on long-term development and management of staff, and thus talent management and performance management were given special attention. Attracting and retaining highly qualified, talented, and committed employees were at the core of HR practices. The quality, instead of quantity and cost, of the employees was considered. Additionally, a more rigorous performance appraisal system was put in place to ensure the highest level of service quality.

In addition to the above HR practices to enhance the resilience of the organisations and employees, the framework also reveals four factors that have significant impacts on those practices: financial constraints, organisational culture, leadership, and business vision and mission. Financial resources were the major challenge for the tourism and hospitality organisations. It forced them to employ the last resort solution, which was laying off their staff and/or cutting working hours, as well as limiting their support for employee wellbeing. This significantly and negatively influenced the efforts to enhance the resilience of the organisations and their employees. Organisational culture and leadership were indicated to be the catalyst for the improvement of resilience. A people-oriented culture, where individuals are the central focus and employees are valued, listened to, and cared for, is an essential element in increasing employee commitment, performance, and resilience. Similarly, employee-oriented leadership, where the welfare of the employees is the first concern of managers, is an important determinant of resilience HR practices. Finally, organisations with clear long-term visions and missions tend to apply better resilience-enhancing practices. Indeed, resilience is not only an outcome but should be considered as a developmental process to prepare for turbulence in the future and maintain the continuous growth of the organisations (Hall et al., 2017).

## Conclusion

The two proposed frameworks of resilience in the workplace context from the individual (micro) and organisational (macro) levels provide a better understanding of how employees and their organisations develop career resilience in the face of a major crisis. While the first framework explains the formation of career resilience from the employee perspective (micro-level) of protection motivation theory, the second framework illustrates the process of HRM practices through a crisis to ensure organisational and employee resilience. Undeniably, resilience at the macro-level might not be achieved if resilience was not developed at the micro-level. Therefore, the resilience of the workforce from the individual perspective should not be overlooked. Workers and their organisations need to demonstrate their ability to survive and thrive in times of crisis.

The adaptive and flexible HRM practices of organisations through phases (e.g., before, during, and after) of a crisis are needed to help their employees to reduce the impacts of the crisis. For example, as the health crisis in 2019 was a career shock, the strategies of health and safety management, emergency response, and the positive psychology of organisations were effective for workers to gain a perception on the pandemic and then make an evaluation of their self-efficacy. Such measures also facilitate the coping strategies of employees that lead them to engage in protective behaviour such as health risk preventative behaviour and then help them to develop career resilience in a crisis. Organisation resilience practices enter in the cognitive process of the career resilience of individuals as facilitators that provide them with better insights of how they perceive the effects of crisis and how they cope with the crisis to demonstrate their resilience in career terms.

These studies provide practical implications regarding resilience-enhancing practices for various stakeholders from the individual and organizational perspectives. From an individual perspective, hospitality and tourism workers can improve their resilience by frequently updating crisis-related information from reliable resources, as well as adopting adequate preventative measures to reduce the severity of the crisis. The feelings of safety, security, and confidence are determinants of an individual's resilience during the crisis. Seeking support from social groups such as relatives/friends is also a measure of employees enhancing their resilience. From an organisational perspective, at the level of managing human resources, managers should adopt preparedness, response, and recovery plans to develop resilient human resources through each stage of a crisis. Policies and procedures for prompt reaction, regular communication, training sessions, and sufficient support are considered effective for employees to be proactive and resilient when facing the risks of a crisis. After the crisis's main event, managers should quickly conduct recovery plans through various human resources practices such as rapid job redeployment, adjusting a flexible performance management system in order to adapt to the new circumstances, permitting the return of ex-employees, and diversifying the candidate pool with lean and agile recruitment and selection strategies. Finally, organisation founders and managers should build people-oriented organisational cultures and leadership to effectively enhance the resilience of their employees as well as their organisations.

According to the "Future of Jobs Report 2020" by the World Economic Forum (2020), resilience is an emerging, yet must-have, skill for obtaining work in the future. For individuals, developing resilience enhances their employability and the chance to be successful in their careers. For organisations, to ensure their survival and sustainable development, resilience should be embedded in their talent management practices. Organisations should not only recruit talented people who possess an aptitude for resilience, but also pay attention to reskilling their current talents. Building resilience from individual and organisational perspectives in the workplace contributes to the resilience of people and the economy, thus assisting with sustainable development and the achievement of SDGs 11 and 13. In this regard, the two proposed frameworks above provide meaningful implications for building a more sustainable workforce within the hospitality and tourism industry.

## References

Al-Ayed, S. I. (2019). The impact of strategic human resource management on organizational resilience: An empirical study on hospitals. *Verslas: Teorija ir Praktika*, 20, 179–186.

Biggs, D., Hicks, C. C., Cinner, J. E. & Hall, C. M. (2015). Marine tourism in the face of global change: The resilience of enterprises to crises in Thailand and Australia. *Ocean & Coastal Management*, 105, 65–74.

Bonanno, G. (2004). Loss, trauma, and human resilience: Have we underestimated the human capacity to thrive after extremely aversive events? *The American Psychologist*, 59, 20–28.

Bowles, T. & Arnup, J. L. (2016). Early career teachers' resilience and positive adaptive change capabilities. *The Australian Educational Researcher*, 43, 147–164.

Brown, N. A., Orchiston, C., Rovins, J. E., Feldmann-Jensen, S. & Johnston, D. (2018). An integrative framework for investigating disaster resilience within the hotel sector. *Journal of Hospitality and Tourism Management*, 36, 67–75.

Cascio, W. F. (2010). Employment downsizing: Causes, costs, and consequences. In *More than bricks in the wall: Organizational perspectives for sustainable success*. Cham, Switzerland: Springer.

Choo, C. W. (2008). Organizational disasters: Why they happen and how they may be prevented. *Management Decision*, 46, 32–45.

Christodoulou, N. G. & Christodoulou, G. N. (2013). Financial crises: Impact on mental health and suggested responses. *Psychotherapy and Psychosomatics*, 82, 279–284.

Coogle, C. L., Parham, I. A. & Young, K. A. (2007). Job satisfaction and career commitment among nursing assistants providing Alzheimer's care. *American Journal of Alzheimer's Disease & Other Dementias®*, 22, 251–260.

Cooke, F. L., Wang, J. & Bartram, T. (2019). Can a supportive workplace impact employee resilience in a high pressure performance environment? An investigation of the Chinese banking industry. *Applied Psychology*, 68, 695–718.

Fletcher, D. & Sarkar, M. (2013). Psychological resilience. *European Psychologist*.

Floyd, D. L., Prentice-Dunn, S. & Rogers, R. W. (2000). A meta-analysis of research on protection motivation theory. *Journal of Applied Social Psychology*, 30, 407–429.

Garmezy, N. (1974). The study of competence in children at risk for severe psychopathology. In E. J. Anthony & C. Koupernik (Eds.), *The child in his family: Children at psychiatric risk*. Oxford, England: John Wiley & Sons.

Gazzaniga, M. S., Heatherton, T. F. & Halpern, D. F. (2010). *Psychological science*. New York: WW Norton New York.

Global Resilience Partnership. (2022). Why Building Resilience is Critical for the SDGs. Retrieved from https://www.peacewomen.org/sites/default/files/Why%20Building%20Resilience%20is%20Critical%20for%20the%20SDGs.pdf

Hall, C. M., Prayag, G. & Amore, A. (2017). *Tourism and resilience: Individual, organisational and destination perspectives*. Bristol, UK: Channel View Publications.

Hamouche, S. (2021). Human resource management and the COVID-19 crisis: Implications, challenges, opportunities, and future organizational directions. *Journal of Management & Organization*, 1–16.

Hartmann, P. & Apaolaza-Ibáñez, V. (2012). Consumer attitude and purchase intention toward green energy brands: The roles of psychological benefits and environmental concern. *Journal of Business Research*, 65, 1254–1263.

Hillmann, J. & Guenther, E. (2021). Organizational resilience: A valuable construct for management research? *International Journal of Management Reviews*, 23, 7–44.

Holling, C. S. (1973). Resilience and stability of ecological systems. *Annual Review of Ecology and Systematics*, 4, 1–23.

Janz, N. K. & Becker, M. H. (1984). The health belief model: A decade later. *Health Education Quarterly*, 11, 1–47.

Kendra, J. M. & Wachtendorf, T. (2003). Elements of resilience after the world trade center disaster: Reconstituting New York City's Emergency Operations Centre. *Disasters*, 27, 37–53.

Klohnen, E. (1996). Conceptual analysis and measurement of the construct of ego-resiliency. *Journal of Personality and Social Psychology*, 70(5), 1067–1079.

Kolar, C., Treuer, K. V. & Koh, C. (2017). Resilience in early-career psychologists: Investigating challenges, strategies, facilitators, and the training pathway. *Australian Psychologist*, 52, 198–208.

Kuntz, J. R., Malinen, S. & Näswall, K. (2017). Employee resilience: Directions for resilience development. *Consulting Psychology Journal: Practice and Research*, 69, 223.

Kutsyuruba, B., Walker, K. D., Stasel, R. S. & Al Makhamreh, M. (2019). Developing resilience and promoting wellbeing in early career teaching: Advice from the Canadian beginning teachers. *Canadian Journal of Education*, 42, 285–321.

Leana III, C. R. & Van Buren, H. J. (1999). Organizational social capital and employment practices. *Academy of Management Review*, 24, 538–555.

Lee, A. V., Vargo, J. & Seville, E. (2013). Developing a tool to measure and compare organizations' resilience. *Natural Hazards Review*, 14, 29–41.

Lengnick-Hall, C. A., Beck, T. E. & Lengnick-Hall, M. L. (2011). Developing a capacity for organizational resilience through strategic human resource management. *Human Resource Management Review*, 21(3), 243–255.

Lin, T. T. & Bautista, J. R. (2016). Predicting intention to take protective measures during haze: The roles of efficacy, threat, media trust, and affective attitude. *Journal of Health Communication*, 21, 790–799.

Luthar, S. S., Cicchetti, D. & Becker, B. (2000). The construct of resilience: A critical evaluation and guidelines for future work. *Child Development*, 71, 543–562.

Mansfield, C., Beltman, S. & Price, A. (2014). 'I'm coming back again!' The resilience process of early career teachers. *Teachers and Teaching*, 20, 547–567.

Mishra, P. & McDonald, K. (2017). Career resilience: An integrated review of the empirical literature. *Human Resource Development Review*, 16, 207–234.

Papatraianou, L. H., Strangeways, A., Beltman, S. & Schuberg Barnes, E. (2018). Beginning teacher resilience in remote Australia: A place-based perspective. *Teachers and Teaching*, 24, 893–914.

Parzefall, M.-R. & Kuppelwieser, V. G. (2012). Understanding the antecedents, the outcomes and the mediating role of social capital: An employee perspective. *Human Relations*, 65, 447–472.

Reason, J. (2000). Human error: Models and management. *BMJ*, 320, 768–770.

Robertson, I. T., Cooper, C. L., Sarkar, M. & Curran, T. (2015). Resilience training in the workplace from 2003 to 2014: A systematic review. *Journal of Occupational and Organizational Psychology*, 88, 533–562.

Roche, W. K., Teague, P., Coughlan, A. & Fahy, M. (2011). Human resources in the recession: Managing and representing people at work in Ireland. *Final Report presented to the Labour Relation Commission*, 338.

Rogers, R. W. (1983). *Cognitive and physiological processes in fear appeals and attitude change: A recised theory of protection motivation*. New York: Guilford Press.

Rutter, M. (1985). Family and school influences on cognitive development. *Journal of Child Psychology and Psychiatry*, 26, 683–704.

Salisu, I., Hashim, N. & Galadanchi, A. (2019). Social capital and entrepreneurial career resilience: The role of entrepreneurial career commitment. *Management Science Letters*, 9, 139–154.

Salisu, I., Hashim, N., Ismail, K. & Isa, F. (2017). Mediating effect of entrepreneurial career resilience between entrepreneurial career commitment and entrepreneurial career success. *International Journal of Economic Research*, 14, 231–251.

Santana, G. (2004). Crisis management and tourism. *Journal of Travel & Tourism Marketing*, 1(4), 299–321.

Santana, M., Valle, R. & Galan, J.-L. (2017). Turnaround strategies for companies in crisis: Watch out the causes of decline before firing people. *BRQ Business Research Quarterly*, 20, 206–211.

Su, D. N., Tra, D. L., Huynh, H. M. T., Nguyen, H. H. T. & O'Mahony, B. (2021). Enhancing resilience in the COVID-19 crisis: Lessons from human resource management practices in Vietnam. *Current Issues in Tourism*, 24(22), 3189–3205.

Teague, P. & Roche, W. K. (2014). Recessionary bundles: HR practices in the I rish economic crisis. *Human Resource Management Journal*, 24, 176–192.

Walker, B., Holling, C. S., Carpenter, S. R. & Kinzig, A. (2004). Resilience, adaptability and transformability in social–ecological systems. *Ecology and Society*, 9.

Weick, K. E. (1993). The collapse of sensemaking in organizations: The Mann Gulch disaster. *Administrative Science Quarterly*, 38(4), 628–652.

Wyllie, A., Levett-Jones, T., Digiacomo, M. & Davidson, P. M. (2020). An evaluation of early career academic nurses' perceptions of a support program designed to build career-resilience. *Nurse Education in Practice*, 48, 102883.

United Nations. (2022). Sustainable Development Goals. Retrieved from https://sdgs.un.org/goals

# 22 Hospitality workforce trends

*Adesola Osinaike and Lorna Thomas*

## Introduction

Tourism, of which the hospitality sector is a major component, is regarded as the world's fastest-growing business and one of the world's top generators of foreign currency. The hospitality business has made significant and diversified contributions to this overall growth in living standards, supplying needed products and services, recreational services, large-scale employment, and wealth development. According to the World Travel & Tourism Council (WTTC), the tourism industry provides one out of every ten jobs worldwide (WTTC, 2017). In 2017, the overall value of tourism in the UK was predicted to be £130 billion (UK Hospitality, 2018), generating £39 billion in tax revenue for the Treasury. However, shortages of skills and employees are the most pressing issues confronting the industry.

The hospitality sector's business environment is difficult to forecast since so much change depends on trends and the social and economic situations that affect the industry. To understand the industry's workforce and the most successful ways to manage the individuals in that workforce, it is important, first, to think about the context and environment in which the sector operates. This business environment changes as it constantly adapts, reacts, and lurches ahead in the face of strong and often unanticipated external forces. Factors such as rising nationalism mean that countries are adopting nationalist and right-wing policies, affecting migrant labour issues. There is a need to rethink how people work and interact with gender and harassment issues, as well as issues such as sustainability and changing food and drink habits. The advances in artificial intelligence (AI) and other technologies have influenced the industry's workforce to varying degrees. The sector's global reach adds to the complexity, reflecting regional, national, and cultural differences.

Many restaurants and cafes are seeing competition from new areas, such as the rise of upmarket gastro-pubs which offer fine dining and a few luxury rooms. Among other pressures in the UK, from April 2022, there has been an increase in the national minimum wage (NMW) and the national living wage (NLW) for those over 25 years of age. This is in addition to increases in business rates for premises, all of which will need to be met. Also, the economic uncertainty surrounding Brexit and the effects on staffing in terms of a shortfall in the recruitment of EU migrant workers are posing challenges to hospitality businesses (Chibili et al., 2019; Boella & Goss-Turner, 2019).

A UK inbound commissioned report, *A perfect storm? The end of free movement and its impact on the UK tourism workforce* (Thomas et al., 2019), researched and discussed the potential issues that the end of free movement would bring to the UK hospitality industry. Although it centres on Brexit, and reports on the reliance on the EU countries to staff the hospitality industry in the last few decades, it also highlights several other issues facing the

DOI: 10.4324/9781003291763-25

industry. For example, it discusses the long hours and nature of the work, particularly among kitchen staff and the relatively low-income bracket of most hospitality staff. Perceptions in society are discussed, such as the contention that, particularly for younger people, hospitality is not perceived as a deliberate career choice. The report also mentions the general skills gap in the UK labour market in terms of language and service skills, as well as how different geographical locations across the country may be suffering to a greater or lesser degree from the recruitment crisis. Varga (2021) claims that large hospitality chains have lost their focus on service delivery and caring for their staff and now mainly focus on the bottom line. She further describes how hospitality staff generally feel unappreciated and undervalued.

## Factors influencing the workforce crisis

It is important to note the many positive aspects of working in the hospitality industry. It can be good fun and interesting, and there is the chance to interact with a variety of the public and meet people from all walks of life. It can be very satisfying to make customers happy. It also offers a high degree of mobility. For example, as hospitality is a global industry, there are chances for employment nearly everywhere, and such employment can facilitate funding for travel and adventure. A wide variety of skills can be acquired, such as cocktail making, service quality, and people management skills, the latter two being highly transferrable. It is certainly not dull employment as the work can range from being employed at international hotels, restaurants, conferences, festivals, and large-scale events to working in local bars and small eateries. One main attraction of working in the industry is that it is not a routine 9 to 5 job for the most part and thus can fit around an individual's lifestyle. A lifelong, highly successful career can be cultivated in the hospitality industry. For the purposes of illustration, one just needs to look at Fred Sirieix, a French maître d'hôtel, who is best known for appearing on Channel 4's *First Dates* as the front-of-house manager. Numerous celebrity chefs and cooks all make a very good living. It is a vibrant and, at times, fast-paced team-based industry, in which employees can make lifelong friends.

Despite such positive employment attributes, the hospitality industry faces a recruitment crisis in the UK and further afield. It is easy to blame Brexit and the COVID-19 pandemic, and no doubt there will still be knock-on effects from both for some time. The pandemic has facilitated the phenomenon which has come to be known as the great resignation, a malaise which began in 2021 and has continued into 2022, where there has been a record number of resignations, seemingly across most businesses. The pandemic has led some people to evaluate the nature of their work and their work–life balance. Brexit has made UK employment for workers from the EU much harder, especially in the lower pay brackets. Such workers previously provided much of the UK hospitality workforce (Thomas et al., 2019). These two factors will continue to be influential, as the recovery ebbs and flows from the pandemic, while employment and immigration legislation is still evolving because of Brexit.

However, many more deep-rooted issues are impacting employment in the industry. The industry is dominated by Generation Z employees and there is a decline in the talent contributed by the older workforce (Goh & Okumus, 2020). Generations X and Z generally perceive the industry as not offering good prospects or providing meaningful careers. The nature of the work, such as the long hours, the belief that this kind of employment is temporary, student work, and a possibly challenging working environment (which might

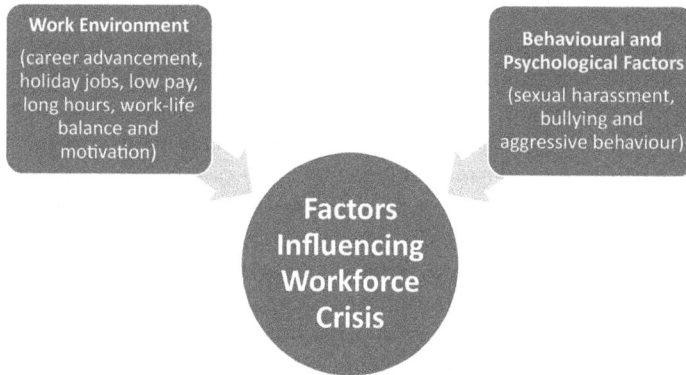

*Figure 22.1* Factors influencing the workforce crisis.

include bullying and sexual harassment), all influence people's lack of desire to enter the industry as a deliberate career path. Figure 22.1 summarises those factors influencing the workforce crisis under two main classifications, specifically the work environment itself and behavioural and psychological factors.

**The hospitality work environment**

Globally, the hospitality industry has a reputation for being low-skilled and low-paid. This impression influences the industry's shortage of employees, especially at the higher professional and managerial levels (Boella & Goss-Turner, 2019). A high turnover rate, along with a lack of employee motivation, widespread dissatisfaction, sexual harassment, and bullying have all contributed to the extant workforce crisis. The hospitality industry has long had a high labour turnover rate, with many temporary employees, including university students and backpackers, working in the industry while travelling around the region. Many young people view hospitality work as a temporary holiday job and, once the holiday is over, they seek other employment or else return to college or university.

A study conducted by Graves (2020) among UK hospitality workers revealed that the most disliked features of the sector were its unsociable, inconsistent, and long working hours. Long shifts were the most despised (by 39%), inconsistent shift patterns irritated a further 33%, and unsociable working hours annoyed 31% (Hospitality News, 2020). The zero-hour contracts that businesses use to avoid providing their employees with consistent hours are a major disadvantage of working within the hospitality industry. Personnel receive less work during off-peak hours, resulting in lower pay, yet will be overworked during peak hours. The hospitality culture encourages long working hours, meaning that time for social activities, sports, and family must be sacrificed. Priority is given to the establishment and this has been passed down through generations of hoteliers/restaurateurs to become ingrained in the unofficial law of hospitality management.

Hospitality firms naturally work different hours from most other industries. The busiest periods are when most of the working population can visit such establishments and so employees will work holidays, late nights, and weekends. The hours are also longer, while taking breaks is not always simple. This can lead to exhausted and stressed workers, which is one of the main reasons for the high employee turnover, or churn rate, within the hospitality business. Even the most loyal employees may find this lifestyle unappealing after a while and eventually seek a career that allows them to maintain a better work–life balance.

Working hours and job security are also important factors for employees' satisfaction. So, the sector needs to devise strategies and implement plans to ensure that the wellbeing of potential employees, hospitality students, and graduates is taken seriously, thereby leading to effective performance at work. Johnson (2020) points out that skilled employees want to advance and thus the industry must improve remuneration, career advancement, and employee–manager relationships to guarantee that students have favourable work-study experiences and are motivated. Unless the sector can also improve employment opportunities for graduating students, it may continue to lose highly talented workers. The hospitality business has historically had limited internal career prospects, stalling career progression and influencing staff exit behaviour.

In most hospitality establishments, the lower-level employees work harder in what is often, by its very nature, physically demanding work, for less pay than upper-level employees (Chibili et al., 2019). Employees can be irritated with management because of the salary disparity. More recently, management in the hospitality business has been accused of mostly being concerned with reaching profit targets and generating money rather than keeping their employees happy and content, leading to a lack of motivation for employees at the forefront of service to customers. Poor working relationships between managers and employees and inadequate work role clarity are factors in the workforce turnover.

Also, poor work environments play a significant role in the workforce shortage. Over the years, there have been concerns about unsuitable or uneducated management attitudes and practices that have contributed to or aggravated many of the industry's human resourcing issues (Baum, 2015). The majority of potential employees are less likely to be individually and vocationally committed until methods are developed to harness what some researchers suggest is a natural desire to work. According to Goh and Lee (2018), Generation Z is ready to work hard, but they expect to move up quickly in their career; however, this seems unachievable. Moreover, poor recruitment and selection processes can result in inefficient use of management time, high labour turnover, absenteeism, low employee morale, and ineffective management and supervision. Shiells-Jones (2021) emphasises that "until hospitality puts the same energy into looking after its staff as it does to look after its guests, the cycle of poor staff retention will continue".

Companies and industry bodies such as UK Hospitality and the Institute of Hospitality are now promoting careers in the industry. They emphasise some positive aspects of working in the sector, such as the people interaction, variety, contacts, and the fulfilment of working in a successful and dedicated team. Staff involvement in hospitality work is critical, as are good management mentors, developing employees who share the industry's passion and potential for fast-track development and promotion with secure future job prospects (MacLeod & Clarke, 2010).

### Sexual harassment, bullying, and aggressive behaviour

There are various behavioural issues in such customer-focused interactions and stressed and stressful situations regularly occur. Workplace bullying can occur in every profession, but in the hospitality industry it is frequently overlooked. Additionally, there is often the risk of sexual harassment in the workplace, an increased risk because industry personnel are constantly interacting with customers, potentially making employees feel uncomfortable and unsafe.

Sexual harassment is increasingly a daily news topic. It seems that every day there are new names added to the list of well-known people accused of sexual harassment and who are fired, and it is still prevalent in the hospitality business (Madera, Guchait, & Dawson, 2018). Despite the regulations and legislation designed to prohibit such behaviour, 10–20% of workers report workplace bullying, violence, and sexual harassment (Ram, 2018). One reason for hospitality harassment is that employees have to interact with guests, friends, family, co-workers, and others. Unfortunately, some restaurant cultures promote sexualised work conditions where sexual harassment is commonplace (Poulston, 2009).

The terms "harassment" and "bullying" are often used interchangeably (Milczarek, 2010; Ram, 2018). Sometimes hospitality jobs are portrayed as problematic, and often employees are viewed as victims of questionable attitudes or problematic employees and/or supervisors. Milczarek's (2010) survey of European working conditions highlights that 30% of hospitality workers reported at least one incident of violence, bullying, or sexual harassment, compared to 35% in the health/education sector, where 25% reported violence and threats of violence, and only 10% reported bullying and harassment. The hospitality industry was the leading sector in terms of employees' reports of bullying and sexual harassment: 8% of workers reported bullying, and 4% reported sexual harassment. In other industries, bullying was at 5% and sexual harassment at 2%. The tourism industry ranked fourth for direct violence (8%) and threats of violence (10%). Other sectors' numbers were higher than 6%.

Given that violence and intimated threats of violence are not as frequent as bullying and harassment, it may be argued that hospitality workers face greater workplace abuse than most other workers, either from customers or guests. Sexual harassment interferes with an individual's work performance, intrapersonal wellbeing, and interpersonal relations. It can also be argued that sexual harassment is harmful mostly to women's careers.

The hospitality industry is characterised by vulnerable workers who rely on their job income and are subject to abuse from supervisors and managers. These unbalanced power relations quickly translate to offensive supervisory behaviours (Ariza-Montes et al., 2017). Thus, these occurrences reveal a lack of managerial skills. The ineffectiveness of management to design and enforce suitable service policies explains a high prevalence of visitor aggressiveness and deviant behaviour towards employees (Aslan & Kozak, 2012). Previous research on bullying, violence, and sexual harassment indicate that industry norms and beliefs are important causes of aggressive behaviour and its widespread acceptance by employers and employees. Aggressive behaviour is commonly believed to be an acceptable part of the tourism and hospitality professions. Since employees are expected to comply with guests' demands (an echo of the "customer is always right" norm), abusive customer behaviour is perceived to be acceptable (Harris & Reynolds, 2004: Kensbock et al., 2015). Worsfold and McCann (2000) also suggest that those individuals attracted to a career in the hotel industry exhibit personality characteristics which result in poor harassment sensitivity. Interestingly, Poulston (2009) argues that those attracted to the industry may be more tolerant of anti-social behaviour, as it is considered "part of the job".

Toxic, violent attitudes and behaviours behind the scenes in hospitality are often justified by the hierarchy. In kitchens, for example, constant pressure and demand for perfection are used to justify violence, including shouting and bullying (Alexander et al., 2012). This aggressive kitchen culture is socialised throughout training when instructor's "ruled with iron fists and used insults to disgrace students" (Bloisi & Hoel, 2008). Thus, violence is accepted as part of the profession in hospitality, even if its explanation varies by industry.

*Figure 22.2* The hospitality workforce crisis: A way forward.

**The way forward**

It is important to determine the future of the hospitality business by contributing to the discussion on how to best manage the labour shortage in the sector. This section will consider the following aspects: talent management, institutional strategies, and policies (Figure 22.2).

If the industry is looking to plug its recruitment gap and attract skilled staff, then as noted by Hilton (2022), "if you are looking for an exceptional career, then we're looking for exceptional people". It follows that there needs to be some change toward forward-thinking strategies. Varga (2021) suggests that the sector needs to start investing in its workforce to encourage a new generation of hospitality professionals and further advocates that the sector needs to work with colleges and universities to help make this happen.

Understanding the characteristics of the future workforce is pivotal to providing solutions to existing labour issues in the hospitality sector. It has been highlighted that the Generation Z workforce is different from Baby Boomers and Generations X and Y. Generation Z demonstrates a different attitude in the workplace as they are not motivated by salary but are rather interested in job satisfaction and career prospects. It has been established that they value a sense of pride and fulfilment from their work and enjoy integrating technology into business processes. Generation Z, digitally literate for their entire lives, have seen many technological advances and believe that management should integrate technology into all areas of customer and operational processes.

This shows that for the hospitality sector to be attractive to Generation Z, who represent the future workforce, the industry should ensure that it aligns its values with future trends. There should be a continuous discussion of career progression and training, such as a management leadership training scheme within the organisation, which will give them a sense of belonging. Also, the sector should embrace technology by using management software and innovative mobile tools that will help with real-time information for guests and staff.

Some parts of the sector are now looking to drive that change and are taking it upon themselves to help change perceptions of a career in hospitality. For example, Pret a Manger now offers sophisticated apprenticeship schemes and "in-depth training, career progression and qualification ... up to a fully-funded BA Business Management undergraduate degree" (Pret a Manger, 2022). Pret encourages and will consider any employee over the age of 16 who meets the entry requirements to apply for the scheme. Some companies are enticing potential employees by using technology. For example, the Marriott hotel group has used an online game, facilitated through social media, called "My Marriott", where future recruits can virtually create their own hotel and kitchens in a deliberate attempt to position the industry as fun (Goh & Okumus, 2020).

As part of the United Nations Sustainable Development Goals to eradicate poverty and inequality and reduce climate change by 2030, the eighth goal hinges on decent work and employment growth, suggesting that if this is effectively considered within the hospitality sector, organisations will easily address the issue of a poor work environment and the behavioural and psychological factors examined in this chapter. Hospitality organisations will need to move beyond cost reduction and develop a strategic approach that focuses on employees' wellbeing and satisfaction to recruit and retain talents needed in the sector. Baum et al. (2016) also highlighted that conscious consideration and inclusion of workforce issues in sustainable hospitality conversation would pave the way for addressing the issue. Many hospitality employers are substantially investing in recruitment, training, and development, and there are some wonderful instances of organisations with highly successful recruitment and training programmes across the sector. These not only help their firms to thrive, but they also help people move up the social ladder because of the low barriers to entry for various jobs and the development opportunities available. Despite the significant expenditure, however, labour volatility undermines training and development. There are many reasons for seeking to improve the acquisition and retention of talent.

### *Talent management*

Talent management (TM) has been described as a strategy to mitigate the staff shortages facing the hospitality sector. Mensah (2019) highlights that TM is a key tool for communicating an organisation's goal to employees at all levels. The term "talent management" refers to the process of anticipating and planning for the firm's human resource needs. It is the science of employing human resource strategic planning to boost corporate value and to be an intrinsic part of an organisation's and corporation's efforts to achieve their objectives. TM makes employee recruitment, training, and personal and professional growth easier. It places a high priority on improving employee abilities and aids in assessing individuals in various settings, allowing for better problem identification at various levels (Horner, 2017).

However, as highlighted in this chapter, more enlightened managers need to see the need to reduce or eliminate these issues through more responsive management action and more professional human resource management (HRM) policies and practices. Split shift working, variable working hours, staff reliance on tips, ignorance of ways of calculating pay and distributing service charges, and management's reluctance to include employees in problems that influence their working lives are some of the industry's typical issues. Managers need to be educated on managing and leading a changing workforce.

Every company requires a well-balanced workforce to meet short and long-term goals, and TM is a fantastic way to achieve this. The TM process is critical within the organisation because it enables the company to provide safety to consumers, confirm the overall quality of the guest experience, and protect the company's brand. The organisation's talent operational plan focuses on employee engagement and motivation, which helps companies to optimise their procedures while also giving opportunities for staff to develop their creative skills.

Organisations must attract, retain, and develop talent to make it sustainable (Baum, 2018). If assigned to the correct position of work at the right moment, TM is considered the strategic organisation of high-calibre people in effective response to different industry needs. TM techniques may help the sector signal opportunities for career and development advancement, improve talent retention and attractiveness, and help to expand talent in economies that rely heavily on the industry's income but lack talent supply (Ladkin & Kichuk, 2017; Horner, 2017). Identifying a person's intrinsic qualities, attributes, and personality, as well as adjusting to an acceptable job profile, are all part of TM: each person has a unique skill that fits into a challenging work profile, where all other jobs are less attractive to that person (Dey et al., 2013).

As a result, TM is a critical component of HRM. Morale, training, and staff retention are vital in the tourism business since career impressions can be unfavourable (Murillo Othón, 2019). All client-facing firms such as the hospitality sector must retain and hire well-trained and contented employees. The basic goal of TM is to retain, recruit, and reward key employees. Some companies focus on reskilling and upskilling current employees through specialised training, programmes, mentoring, and on-the-job experiences. In contrast, others invest in new technologies to uncover skills and connect employees with possibilities.

According to Watson et al. (2018), older hospitality workers are more likely to demonstrate higher commitment, though the industry is dominated by a younger workforce (Goh & Okumus 2020). Generally, hospitality recruitment strategies tend to aim at the younger generations, for example school or university leavers. The industry should look to develop an action plan to attract older generations; as well as possessing a wealth of experiences and skills, this should also help younger employees to picture themselves in the industry longer and view hospitality as a career. There is potentially a wealth of untapped talent in other demographics. The industry should look to develop an action plan to attract the over 50s, who were the hardest hit in terms of job losses during the lockdowns of the pandemic (UK Hospitality, May 2022). As well as plugging a recruitment gap, it could help younger employees to view themselves in the industry as a long career choice. Such an action plan should help the hospitality industry work towards UN-SDG 8.6, by reducing the proportion of youth not in employment, education, or training.

There is also potentially a wealth of untapped talent in other demographics, and the action plan should target other groups, such as people with disabilities and ex-prisoners, whose wealth of skills and experiences could prove invaluable to the industry.

*Institutional strategies and policies*

The industry is, by nature, fast-paced and responsive to market changes; this means the talent and willingness are already present to be able to react and react quickly where necessary. The industry must focus on enticing and successfully managing its future workforce. This section notes the main areas where current hospitality leaders and managers can respond to the current crisis.

In February 2021, the Institute of Hospitality reported on a meeting entitled "How can universities and the wider industry work together to maintain the hospitality stars of the future?" (Institute of Hospitality, 2022). At this meeting there was a discussion on the importance of working with educational institutions, such as colleges and universities, to help position careers in hospitality as valid and valuable. Universities and colleges can help provide high-level qualifications, such as degrees, degree apprenticeships, and postgraduate qualifications. Robert Richardson FIH (2022), IoH CEO, commented: "we must encourage more young people to choose hospitality as a career so we can combat the ongoing staff shortages challenging our fantastic industry" (IoH, 2022). Working with educational institutions may help the industry to achieve this. Additionally, there is a need to bring hospitality ambassadors and mentors into schools to talk to pupils and career advisers.

The industry needs to unite locally, regionally, and nationally to discuss and write strategies and policies on how best to proceed and provide a visual career pathway. Such a tactic can help to plug the skills gap within the hospitality UK workforce (Thomas et al., 2019).

Staff need to feel valued and cared for to ensure their longevity and to advance leadership skills within the industry. The service quality ethos that most hospitality establishments are keen to express to their customers should always be extended to staff. Richard Branson, the founder of the Virgin Group, famously said on Twitter in 2015: "If you look after your staff well, they will look after your customers". A lifelong-learning policy can be adopted by hospitality associations and organisations, one that equally values new recruits and the older workforce and their accumulated work skills and experiences. Policies that facilitate a clear visual career path with prospects should be visible to all. Closely working with educational institutions such as colleges and universities will help to ensure that the industry recruits appropriately qualified and trained employees.

A change in culture in some hospitality spaces, such as kitchens, is much needed. The great resignation of 2021–2022 has made many hospitality workers such as chefs re-evaluate their working life and their need to tolerate a bullying and toxic culture. Although changing organisational cultures is often challenging, human resource policies can be brought up to date relatively quickly to kick-start a culture change. In parallel, training and mentoring sessions for staff on knowing the law, their rights, and how to deal with bullying and all types of harassment from colleagues and the wider public are really important to equip the workforce with the right tools and confidence to deal with any issues. It is not acceptable in any industry that a member of staff should routinely experience bullying and sexual harassment. Companies and associations, such as the Institute of Hospitality, need to encourage a culture of whistleblowing, where staff feel safe to report any issues.

Brexit and recovering from the COVID-19 pandemic cannot be ignored when looking toward and planning for the future. In many ways, both can be used as opportunities to review the status of recruitment within the industry. A study by Gebbels (2020) outlines a strategy for supporting the hospitality industry pre/post-Brexit. Some of the post-Brexit recommendations are to create financial incentives for partner higher education institutions and hospitality businesses, aiming to help with growth prospects and cash injections to the sector. Additionally, as the great resignation of 2021–2022 continues, the hospitality industry can use this as an opportunity to entice skilled and experienced workers from other industries who have recently resigned. The positive attributes of working in the industry should be immediately advertised to help plug the current recruitment crisis, at least in the short term.

**Conclusion**

Every period has its unique environment for change, and the greatest managers and firms are those that respond, adapt, and change most effectively in response to the new difficulties they face. To make a more lasting change, owners, managers, and practitioners must aim to modify how the hospitality business operates. It should be seen as an industry for those who enjoy taking on new challenges daily, as well as a location where teamwork and learning and development are greatly supported and encouraged. Managers will have to figure out what makes individuals want to work in this industry to solve the recruitment dilemma. Focusing on internal promotion to reduce attrition and alleviate the talent shortage is a prominent trend, especially for specialised professions. More regular performance assessments, training, and wellness initiatives will not be enough to solve the recruitment dilemma. It will require managers to delve deeper and discover what motivates staff to work in this business. Of course, a wage raise would tempt individuals back in the short term, but potential employees currently have a poor perception of the industry. Different scholars and industry practitioners have proposed long-term plans to reduce bullying, guest violence, and sexual harassment. This could be considered a step towards modifying industry norms that encourage violent behaviour towards employees. The need to rethink and question the "customer is always right" mentality should also be addressed; a pessimistic outlook seems prevalent with customers (and other workers) behaving as they want rather than should.

This chapter contributes to the existing discussion on the hospitality workforce and advocates for organisations within this sector to understand their employees' characteristics, especially Generation Z's entrance into the workforce. For businesses to anticipate and create workplaces suitable for Generation Z, which will affect organisational performance, it is important to identify factors that may affect the success of their recruitment and retention efforts by paying attention to what makes this generation take interest in work and make a contribution.

A future discussion should embrace a significant interest in sustainability and a sustainability agenda. As highlighted above, the eighth sustainable development goal emphasises decent work; the hospitality industry bodies and organisations must reconsider the industry's work culture, image, and perception. Management should create a good working environment to enhance employees' wellbeing and satisfaction. Transparency, independence, adaptability, and personal freedom are key elements of Generation Z's work ethics and failing to consider them could lead to demotivation, decreased productivity, and a lack of employee engagement. Employers should start developing open, inclusive rules that promote transparency. Overall, the hospitality business must reform from its core foundation, thus preserving sustainability by continuing to innovate and improve.

## References

Alexander, M., MacLaren, A., O'Gorman, K., and Taheri, B. (2012). "He just didn't seem to understand the banter": Bullying or simply establishing social cohesion? *Tourism Management*, 33(5), pp. 1245–1255.

Ariza-Montes, A., Arjona-Fuentes, J. M., Law, R., and Han, H. (2017). Incidence of workplace bullying among hospitality employees. *International Journal of Contemporary Hospitality Management*, 29(4), pp. 1116–1132.

Aslan, A., and Kozak, M. (2012). Customer deviance in resort hotels: The case of Turkey. *Journal of Hospitality Marketing and Management*, 21(6), pp. 679–701.

Baum, T. (2015). Human resources in tourism: Still waiting for change?–A 2015 reprise. *Tourism Management*, 50, pp. 204–212.

Baum, T. (2018). Sustainable human resource management as a driver in tourism policy and planning: A serious sin of omission? *Journal of Sustainable Tourism*, 26(6), pp. 873–889.

Baum, T., Cheung, C., Kong, H., Kralj, A., Mooney, S., Nguyễn Thị Thanh, H., Ramachandran, S., Dropulić Ružić, M., and Siow, M. L. (2016). Sustainability and the tourism and hospitality workforce: A thematic analysis. *Sustainability*, 8(8), p. 809.

Bloisi, W., and Hoel, H. (2008). Abusive work practices and bullying among chefs: A review of the literature. *International Journal of Hospitality Management*, 27(4), pp. 649–656.

Boella, M., and Goss-Turner, S. (2019). *Human Resource Management in the Hospitality Industry*. 10th Edition. Oxon: Taylor and Francis.

Chibili, M., Bruyn, S., Benhadda, L., Lashley, C., Penninga, S., and Rowson, B. (2019). *Modern Hotel Operations Management*. Netherlands: Taylor and Francis.

Dey, M., Tripathy, P. K., and Dey, M. B. (2013). Talent management – A theoretical perspective. In S. Jasanoff, G. Markle, J. Peterson, and T. Pinch (Eds), *Handbook of Management. Technology and Social Sciences*. Thousand Oaks, CA: Sage, pp. 591–603.

Gebbels, D. M. (2020). What the Government Can Do to Help the Hospitality Industry before and after Brexit? Available at: https://gala.gre.ac.uk/id/eprint/23451/ (Accessed 01/02/2022).

Goh, E., and Lee, C. (2018). A workforce to be reckoned with: The emerging pivotal Generation Z hospitality workforce. *International Journal of Hospitality Management*, 73, pp. 20–28.

Goh, E., and Okumus, F. (2020). Avoiding the hospitality workforce bubble: Strategies to attract and retain generation Z talent in the hospitality workforce. *Tourism Management Perspectives*, 33, p. 100603.

Graves, S. (2020). Revealed: The Hospitality Sector's Biggest Pet Peeves. Available at: https://opsbase.com/hospitality-biggest-pet-peeves/ (Accessed 20/01/2022).

Harris, L. C., and Reynolds, K. L. (2004). Jaycustomer behavior: An exploration of types and motives in the hospitality industry. *Journal of Services Marketing*, 18(5), pp. 339–357.

Hilton. (2022). Available at: https://jobs.hilton.com/emea/en (Accessed 13/01/2022).

Horner, S. (Ed.). (2017). *Talent Management in Hospitality and Tourism*. Oxford: Goodfellow Publishers Ltd.

Hospitality News. (2020). Zero Hours Contracts and Long Shift Patterns among Hospitality Workers' Biggest Pet Peeves. Available at: https://catererlicensee.com/zero-hours-contracts-and-long-shift-patterns-among-hospitality-workers-biggest-pet-peeves/ (Accessed 20/01/2022).

Institute of Hospitality. (2022). Available at: https://www.instituteofhospitality.org/ (Accessed 01/02/2022).

Johnson, A. G. (2020). We are not yet done exploring the hospitality workforce. *International Journal of Hospitality Management*, 86, p. 102402.

Kensbock, S., Bailey, J., Jennings, G., and Patiar, A. (2015). Sexual harassment of women working as room attendants within 5-star hotels. *Gender, Work and Organization*, 22(1), pp. 36–50.

Ladkin, A., and Kichuk, A. (2017). Career progression in hospitality and tourism settings. *Talent Management in Hospitality and Tourism*, 1, p. 69.

MacLeod, D., and Clarke, N. (2010). Leadership and employee engagement: Passing fad or a new way of doing business? *International Journal of Leadership in Public Services*, 6(4), pp. 26–30.

Madera, J. M., Guchait, P., and Dawson, M. (2018). Managers' reactions to customer vs co-worker sexual harassment. *International Journal of Contemporary Hospitality Management*, 30(2), pp. 1211–1227.

Mensah, J. K. (2019). Talent management and employee outcomes: A psychological contract fulfilment perspective. *Public Organization Review*, 19(3), pp. 325–344.

Milczarek, M. (2010). Workplace violence and harassment: A European picture. Report, EU-OSHA, Belgium.

Murillo Othón, E. (2019). Why do employees respond to hospitality talent management: An examination of a Latin American restaurant chain. *International Journal of Contemporary Hospitality Management*, 31(10), 4021–4042.

Poulston, M. J. (2009). Working conditions in hospitality: Employees' views of the dissatisfactory hygiene factors. *Journal of Quality Assurance in Hospitality and Tourism*, 10, pp. 23–43.

Pret a Manger (2022). Available at: https://www.pret.co.uk/en-GB/pret-jobs (Accessed 13/02/2022).

Ram, Y. (2018). Hostility or hospitality? A review on violence, bullying and sexual harassment in the tourism and hospitality industry. *Current Issues in Tourism*, 21(7), pp. 760–774.

Richardson, R. (2022). The Institute of Hospitality. Available at: https://www.instituteofhospitality.org/ (Accessed 01/02/2022).

Shiells-Jones, M. (2021). Is Hospitality to Blame for its Own Staffing Crisis? Available at: https://www.hospitalityandcateringnews.com/2021/06/is-hospitality-to-blame-for-its-own-staffing-crisis/ (Accessed 27/02/2022).

Thomas, K., Scott, J., Butcher, J., O'Donoghue, D., and Thomas, L. (2019) *A Perfect Storm? The end of Free Movement and its Impact on the UK Tourism Workforce*. UK: UK Inbound.

UK Hospitality. Available at: https://www.ukhospitality.org.uk/ (Accessed 28/05/2022).

Varga, J. R. (2021). Is there a Solution to the Hospitality Staff Crisis? Available at: https://www.hospitalitynet.org/opinion/4107067.html (Accessed 13/01/2022).

Watson, A. W., Taheri, B., Glasgow, S., and O'Gorman, K. D. (2018). Branded restaurants employees' personal motivation, flow and commitment: The role of age, gender and length of service. *International Journal of Contemporary Hospitality Management*, 30(3), pp. 1845–1862.

Worsfold, P., and McCann, C. (2000). Supervised work experience and sexual harassment. *International Journal of Contemporary Hospitality Management*, 12(4), pp. 249–255.

# 23 Do customers really care about the corporate social responsibility reputation in the decision of hotel patronage?

*Suk Ha Grace Chan, Yue Yvonne He and Binglin Martin Tang*

## Introduction

In the service industry, ethical consumption is crucial since it represents customers' cultural aspects and ethical behaviour (Davies & Gutsche, 2016). Preference for solo travel and awareness of carbon footprints presents a chance for travel operators and organisations to link tourists with locals and the environment (Bourne et al., 2020). It is important to have a better understanding of customers' buying decisions in relation to the value of corporate social responsibility (CSR) in hotel patronage. Given the overwhelming development of technology in the past decade, e-communication has become vital for customers, particularity the younger ones. Therefore, young customers' perceived brand values may provide a major source of positive word-of-mouth. This ultimately leads to positive buying behaviour in the younger demographic, of whom previous studies have mostly understated their capacity to generate profit for the hotel and tourism sector.

With the increasing importance of CSR initiatives in the hospitality industry, along with the Internet's role as a new media platform for tourism development, electronic word-of-mouth (eWOM) has become a space for young customers to share their daily life experiences. Implementation of CSR initiatives may potentially lead to eWOM and purchasing decisions, although admittedly there is a lack of thorough discussion on the impacts of young tourists in groups. This chapter attempts to better understand how young customers are affected by CSR initiatives; specifically, how brand imaging and eWOM affect the process of choosing a hotel. This chapter aims to widen the current stream of literature regarding this field and show the attitude of young customers towards the tourism industry at large.

## Marketing trends and issues

Over the last two decades, new trends have emerged with the increasing importance of CSR initiatives in the hospitality industry and with the Internet and new media platform development. Hence, eWOM has become an important platform for customers to share their experiences. Online reviews, for example, which feature product experience write-ups on platforms such as Taobao, Tmall, or RED, as well as the opinions of friends and family on social media, remain effective drivers of customer influence in China. Most customers prefer e-commerce platforms due to their greater transparency and quality assurance. Arica et al. (2022) highlighted how tourists' interactivity and identity are essential factors of increased travel-experience sharing on social media platforms (Okazaki, Andreu, & Campo,

DOI: 10.4324/9781003291763-26

2017). The implementation of CSR initiatives may increase brand building, positive eWOM, and buying decisions. However, further exploration and discussion of young tourists remain limited. Factors that affect young customers' personal patronage behaviours include their specific tastes, access to high-quality service, information collection, and ego. Liang, Choi, and Joppe (2018) highlighted that the social value and organisation of a certain brand may also contribute to final purchasing decisions. eWOM has become an indispensable tool in enhancing customers' selection and confidence in a brand by providing benefits and thus guaranteeing customer loyalty. This shows the investments that hospitality organizations must make in achieving wider external marketing and more effective branding strategies.

In today's highly competitive industry environment, guests increase their involvement when looking for lodgings, therefore pushing hospitality organisations to translate a hotel's subjective perceptions and service quality into tangible services (Su et al., 2016). Studies such as Martínez and Nishiyama's (2019) highlighted how brand equity shares many elements with CSR. Organisations play an active role in society, given their stakeholders and customers. CSR can increase the level of hospitality and brand credibility while stakeholders can readily identify themselves with certain brands' CSR initiatives. Since CSR reputation and brand equity affect an organisation's reputation, marketing management must be aware of brand communication and how to effectively target the right population segments (Hur et al., 2014).

UNWTO included tourism as part of the 2030 agenda, which is built on the historic Millennium Development Goals (MDGs) for the better establishment and protection of our planet. In the 17 Sustainable Development Goals (SDGs) listed, goals 8, 12, and 14 are tightly linked with the tourism industry. These are inclusive and sustainable economic growth, sustainable consumption and production (SCP), and the sustainable use of oceans and marine resources (UN, 2015).

Since tourism provides employment opportunities for many, promoting history and culture can bring about significant economic benefits to a certain region. Given its significance in the 17 SDGs, development tourism has become the trend for many in the industry. Its importance is emphasised in areas such as the Macao Special Administrative Region (MSAR), where tourism is the leading industry.

### Objectives and expected outcomes

In this chapter we aim to achieve six goals: (1) to examine the extent to which CSR affects eWOM, (2) to evaluate the relationship between hotel patronage intention and the brand equity of young tourists, (3) to examine the cognitive process behind hotel patronage intention, (4) to investigate the eWOM of tourists, (5) to identify ways to improve the brand building strategy of hospitality operators, and lastly (6) to produce relevant marketing strategies for potential tourists.

The expected outcomes provide implications for marketers to better understand CSR and brand equity, anticipating possible and future business ramifications. We also aim to expand the existing scope and gain a complete understanding of young Chinese tourists' motives and influences.

From a marketing standpoint, young customers carve out a niche segment of business opportunities. Over 80% of young customers choose to obtain information from people they personally know. Their eWOMs effectively amplify the noise of updates on future products, discounts, and promotions. With the updating of product information, marketers can utilise their CSR marketing opportunities to create their hospitality profile and gain more opportunities to improve and predict product sales. Through this, we aim to disclose essential and comprehensive suggestions to the tourism industry.

## Theoretical background

### *CSR theory and the tourism industry*

Carroll (1979) suggested four definitions of CSR – financial, legal, moral, and discretionary responsibilities – which are embedded in a conceptual framework of Corporate Social Performance (CSP). Businesses that participate in CSR can be perceived as good corporate citizens, ethical, follow the law, and earn revenue simultaneously. Given that financial and legal obligations are largely anticipated in businesses, philanthropic and ethical responsibilities following the model of Carroll (1991) are usually considered business CSR. Many CSR studies are related to hospitality and focus on financial returns (Franco et al., 2020), reporting (Holcomb et al., 2007; De Grosbois, 2012), and enhancing customer and employee expectations and perceptions (Kim et al., 2017; Park & Levy, 2014; Gonzalez-Rodriguez et al., 2018). The tourism sector uses CSR activities to improve customer awareness and build a reputation. Scholars highlight the relationship between CSR and customers' brand responses. Their reaction to CSR initiatives undoubtedly influences their brand loyalty (Rashid et al., 2021), especially their willingness to pay for fairness pricing and attribution blame during brand crises (Klein & Dawar, 2004). Thus, businesses, including hotels, must determine whether a relationship exists between CSR initiatives and brand awareness, especially the kind of CSR initiatives that significantly affect positive brand image development. Furthermore, given the popularity of social media platforms, those in the industry must determine the relationship between eWOM, CSR activities, and brand equity. Such knowledge helps practitioners produce new products and services fit for the needs of young customers and better cater for these needs.

The theoretical concepts of sense giving, sense making, and attribution are often used in describing CSR communication. Individuals tend to connect a person's behaviour with a brand's CSR activities and its causes. These interpretations influence their responses to behaviours (Ng et al., 2018). Perceived causation in these studies is recognised as attribution theory, which denotes the perception or inference of cause. General models of attribution were proposed earlier by scholars Kelley and Michela (1980) among others. Attribution theories illustrate why things happen and explain the communicative and mental processes involved in using information to judge, comprehend and act on social and individual events in daily life (Manusov & Spitzberg, 2008).

Business ethics theory was first proposed by Carroll (1999) and was developed into three main theories: deontological, utilitarianism, and norm theories. Business ethics theory refers to the moral principles implemented by a company to ensure employees in the company behave ideally. In previous studies, business ethics have always been considered a part of CSR and work to constrain a company in terms of ethical responsibility, although some scholars regard these two constructs as interchangeable.

Researchers of CSR communication have proposed that customers will try to understand companies' CSR initiatives due to companies' intrinsic or extrinsic motives when carrying out CSR activities (Bogan & Sarusik, 2020). Businesses are recognised as inherently driven to carry out CSR when these activities are viewed as genuinely intended for public service. In contrast, businesses are also recognised as superficially driven when perceived as performing CSR for public relations only so as to increase revenue

(Schaltegger & Burritt, 2018). De Jong and Van der Meer (2017) proposed that customers process CSR attribution via cognitive elaboration, and that CSR initiatives are attributed according to the available CSR information. Stakeholders that view the company as having inherent motivations will suppose its CSR activities as favourable to its identity and thus react positively (Panagopoulos et al., 2016).

### How customer-based perceptions of CSR affect eWOM

The spread of content in online communication includes positive and negative eWOM. Social media platforms have shown a new dynamic to this relationship because they provide multiple communication channels such as the dialogue between the organisations and stakeholders (Xu & Saxton, 2019). Given the access of customers to updated information, they have become more responsive to ethical and sustainability issues (Ballew et al., 2020) and information, which can significantly impact customer engagement (Uzunoğlu et al., 2017). Martínez et al. (2014) highlighted that a negative perception of CSR may potentially be shared and spread through eWOM. Carroll (1979) suggested that the different aspects of CSR, namely ethical, philanthropic, and economic, affect customer evaluations of the organisation for embedding in a conceptual framework of CSP. Business performance is expected to align with society's philanthropic anticipations and provide economic and human resource opportunities to advance the community's quality of life (Carroll, 1991).

Climate change is crucial to the community since many marketers adopted CSRs when considering SDGs so as to achieve higher levels of reputation. Many hotels focus on sustainable food systems and collaborate with green suppliers to achieve goodwill. Hotels transform their technical, policy, and capacity enhancement in finance and are more likely to reduce their ecological impacts on stakeholders in the community (Campbell et al., 2018). Therefore, SDGs often come up as a selling point for the hotel especially when they are identified as geared towards community service.

### eWOM and brand equity

Brand equity and eWOM are important influences for companies in terms of customer satisfaction and loyalty (Febrian & Fadly, 2021). eWOM considerably builds strong brand equity, which not only increases customer trust in service and quality, but also affects the future profitability of the company (Murtiasih et al., 2014). Strong brand equity allows tourists to derive added value from products and services which they cannot obtain from other products. Brand equity and eWOM elements positively impact tourist satisfaction (Hendrata et al., 2021).

### Brand equity and patronage intention

Buzdar et al. (2016) pointed out that the customer-based brand equity dimension has four factors: brand awareness, brand loyalty, brand image, and service quality (Ishaq et al., 2014). Yoong and Lian (2019) alternatively defined these four dimensions as customer patronage intentions, namely surveillance, social interaction, sharing of information, and attraction.

A significant positive relationship exists between the brand equity dimension and patronage intention, thereby highlighting the importance of brand equity in terms of service competitiveness in the hotel industry (Seric et al., 2018). Brand equity is the main predictor of patronage intention, and the causality between brand attitude and patronage intention, introduced as a mediating factor, is stronger and more pronounced (Liu et al., 2017). The data analysis from studies such as Chakraborty's (2019) revealed that brand awareness and perceived value partially mediated the effect between credible online reviews and purchase intentions.

### eWOM and patronage intention

Purchase intention is a complex process in which customers consider various conditions and factors in making a purchase decision (Pradhan, 2018). Zhang et al. (2015) argued that the relevance, timeliness, and comprehensiveness of information affect customers' perceptions of the amount of information available. In sectors such as e-commerce, hotel, and tourism, greater informativeness, which helps users to better compare products, is what enables customers to make better purchasing decisions, and therefore has a direct relationship with their perceptions of usefulness (Filieri et al., 2018). Persuasive arguments can also significantly influence customer responses (Chang et al., 2015). Gunawan and Huarng (2015) also found that the persuasive strength of a message on social media is important in shaping visitors' attitudes towards the message and its usefulness. eWOM, therefore, substantially contributes to brand image, impacts visitors' willingness to purchase, and is an important basic element of any brand's online marketing.

### Hotel CSR in Macao

To identify the importance of CSR and test the extent to which it can influence eWOM on the decision of hotel patronage, we launched a survey focused on Macao, where hospitality serves as the leading industry. The judgemental sampling method was used to ensure that all respondents were identified as young customers. To eliminate concerns and biases, the beginning of the questionnaire explained the purpose of data collection and allowed interviewees to respond anonymously. The questionnaire was distributed to students from four local universities in Macao through social-media-sent links for ease and convenience. From February to April 2020, 520 valid questionnaires were collected over three months. The questionnaires were initially designed in English. To avoid unnecessary misunderstandings and potential data errors, the questionnaire was then professionally translated into Chinese to ensure that there were no ambiguities between the two versions.

CSR activities which included philanthropic, ethical, and economic responsibilities positively affected eWOM, indicating a strong relationship with one another. This conclusion paralleled studies such as Martínez et al. (2014) which found that a negative CSR will potentially affect information sharing and spreading. Meanwhile, eWOM also significantly affected brand equity, which was consistent with studies such as Murtiasih et al. (2014) which posited that eWOM played a prominent role in building brand equity. Brand equity also positively influenced young customers' patronage intentions, proving the opinion of Bian and Liu (2011) that brand equity is an important impetus for competitiveness in the hotel industry.

Surprisingly, however, eWOM did not show a significant influence on patronage intentions. This finding may be explained by how young customers use their own judgement regarding their decisions and are therefore less susceptible to influence from outside information. Although Gunawan and Huarng (2015) found that the persuasive strength of a message on social media is important in shaping visitors' attitudes towards the message and its usefulness, the same is not true however for young customers.

The research also uncovered that eWOM did not show the mediating effect between CSR and patronage intention directly, further illustrating young customers' independence in decision-making. Compared to eWOM, it was brand equity that is truly significant for young customers' patronage intentions. Young customers are more concerned about things themselves rather than its over-packaging or marketing. However, eWOM directly affected brand equity, which in turn presented a mediating factor between eWOM and patronage intentions. Although CSR does not affect young customers' patronage intentions directly, its effects still seep through via eWOM and brand equity.

### Making a CSR effort by hotels in the adoption of sustainability

Sustainability is an important factor in urban and corporate development. In Macao, this is evident in many hotels and tourism establishments. Businesses such as Wynn Hotel in Macao, a one-stop luxury resort, harbours a high reputation in the local hotel industry. Its success does not come from remaining stagnant, but in its commitment to sustainability. From its Wynn Macau 2020 environmental, social, and governance (ESG) report, Wynn Hotel has promoted a sustainable plan called the "Goldleaf Sustainability Program" that has achieved a good response since its inception a few years back (Wynn, 2020). The programme's core principle is "Care for our Guest and Our Planet", which fully echoes the UNWTO 17 SDGs (UN, 2019a). As a representative of the larger Macanese hotel industry in philanthropy and sustainability, Wynn Hotel gave more than USD15,000 worth in donations and sponsorship, recycled a total of more than 8,500 lb of soap in partnership with Clean The World, and donated 14,000 volunteer hours in 2020. Over 50% of local employees are female, and local employees reach 74%, which guarantees the rights of local Macanese and women in the workforce. A well-managed hotel organisation can benefit local people by providing jobs, promoting the development of green industries by recycling resources, and helping society through donations and volunteer time (Wynn, 2020). Wynn Hotel has set an example in sustainable practices, which hopefully will be followed by others. The benefits provide awareness to stakeholders, such as arriving visitors and delivery suppliers, in pursuit of novel sustainable approaches.

### Opportunities and challenges in Macao's hotel industry

Our research in this chapter found that building eWOM, alongside its positive impacts, must improve brand equity, and not just entice customers to stay. Given the excess of information available the authenticity of some online reviews is open to scrutiny. If a customer chooses a hotel because of attractive eWOM but the experience is subpar, it will negatively impact brand equity. Therefore, the establishment of eWOM should start from its CSR activities, which in turn enhance brand equity. Strong brand equity will become the core competency of the hotel brand to attract more customers, with CSR being the key to success. Even though CSR is not the ultimate determinant in the decision of hotel patronage, it serves an invaluable role as a business's foundation, and it goes without saying that only a firm foundation can support its subsequent weight.

Ultimately, this reveals that marketers must refine their existing marketing strategies. Although eWOM can be a powerful tool in delivering brand equity and organisation image, marketers must make more effort by utilising eWOM to increase customers' confidence and loyalty. Accordingly, this encourages young customers to choose hotel patronage. Practitioners from the hospitality industry can carefully monitor eWOM and reduce any harm to the brand image and reputation. Given that young customers perceive their own values and standards differently from other earlier generations, marketers must develop their corporate image to match their needs.

## Prediction for the next three decades

### SDG predictions up to 2030

Tsalis et al. (2019) highlighted that business organisations need to incorporate the SDGs in their own strategies and meet the current needs of protecting, sustaining, and enhancing the human and natural resources necessary for future use. For future years up to 2030, trends show and project that the SDGs may demonstrate more diversification. Higher living standards often allow more people to be informed about larger sustainable issues. Overall, education, knowledge, and experience of sustainability should be well delivered. Given this trend of societal consciousness on sustainability, consumers are more likely to consider sustainable products and practices as motivating their purchasing decisions. Marketers are more likely to consider sustainable development and build more effective marketing plans highlighting the SDGs. Hotels are pushed to form business partnerships and strengthen their implementation plans for achieving sustainable development as well.

### SDG predictions up to 2040

By 2040, some developing countries are predicted to gain mid-class level development. In other Western countries, the maturity of technology, economy, governance, infrastructure, mobility, and all other social factors may shape the trajectory of sustainable development. However, economic slumps, population booms, and slower birth rates may lead to reluctance in changing buying behaviour. Many customers may perceive scarcity and the collapse of ecosystems as a normal situation (Broo et al., 2021).

### SDG predictions up to 2050

The increased population of young people may lead to higher demand for open and green spaces which will uplift green leisure tourism in cities with many countries that aim to have a stable workforce supporting these green infrastructure projects. This may also lead to more comfortable work with high technology and the utilisation of virtual reality (VR) and artificial intelligence (AI) technologies (Churkina et al., 2020). More consumers will support sustainable products which will be aimed at affordability and innovation. Technology support may result in time-effective and cost-effective products and services. More autonomous transportation may lead to decarbonisation and efficiency. Many in the hospitality industry may adopt green products which use renewable sources.

## Contributions

This chapter has provided theoretical and practical implications. From a theoretical perspective, it has expanded the research on young customers and enriched the existing CSR literature. Academics can further examine young customers' perception of brand equity, which leads to their buying decisions. Young customers differ from other generations since they are not easily swayed by external factors in their decision-making, contrasting previous findings and providing new research ideas for future studies. Academics can further examine the psychological behaviour of young customers to explore their buying behaviour such as their motivation. Future studies can focus on young customers' values and social standards as new perspectives for further discussion.

Hospitality marketers should prioritise fulfilling their CSR regarding philanthropic, ethical, and economic values to obtain market strength and compete with other companies. Although it is ideal to increase customer patronage intentions through marketing techniques, establishing brand equity in marketing management delivers better results, given that brand equity affects eWOM, whereas young customers' hotel patronage may not be affected by eWOM.

Marketers must uplift their brand building to execute the right marketing strategies, aiming at the right perspectives to help businesses grow. Managers should pay more attention to continue investing in CSR initiative programmes. Hotel managers must consider young customers and their perceived brand value and establish brand equity based on their perspectives. Brand communication may explain the benefits of socially responsible activities to make young customers further understand the benefits. Young customers' perceptions must be enhanced through different channels, such as social media platform usage, organisation websites, and public relations in promoting CSR activities.

Lastly, hotels must not only provide a quality service but should also develop a novel CSR programme to enhance young customers' involvement and participation. Hotel organisations should commit to fulfilling the goals of the UNWTO SDGs by taking measures that include but are not limited to reducing the waste of plastic packaging by putting shampoo and shower gel in extruded containers, calling for customers to bring their own daily necessities, and installing other reusable materials to replace plastic. Ultimately, the hybrid measures of optimising e-marketing strategies and actively creating an excellent brand image through CSR can lead to greater hotel patronage.

## References

Arica, R., Cobanoglu, C., Cakir, O., Hsu, M.-J., & Della Corte, V. (2022). Travel experience sharing on social media: Effects of the importance attached to content sharing and what factors inhibit and facilitate it. *International Journal of Contemporary Hospitality Management, 34*(4), 1566–1586.

Ballew, M. T., Pearson, A. R., Goldberg, M. H., Rosenthal, S. A., & Leiserowitz, A. (2020). Does socioeconomic status moderate the political divide on climate change? The roles of education, income, and individualism. *Global Environmental Change, 60*, 102024.

Bian, J., & Liu, C. (2011). Relation between brand equity and purchase intention in hotel industry. *International Journal of Services and Standards, 7*(1), 18–34.

Bogan, E., & Sarusik, M. (2020). Organization-related determinants of employees' CSR motive attributions and affective commitment in hospitality companies. *Journal of Hospitality and Tourism Management, 45*, 58–66.

Bourne, J. E., Cooper, A. R., Kelly, P., Kinnear, F. J., England, C., Leary, S., & Page, A. (2020). The impact of e-cycling on travel behaviour: A scoping review. *Journal of Transport & Health, 19*, 100910.

Broo, D. G., Lamb, K., Ehwi, R. J., Pärn, E., Koronaki, A., Makri, C., & Zomer, T. (2021). Built environment of Britain in 2040: Scenarios and strategies. *Sustainable Cities and Society, 65*, 102645.

Buzdar, M., Janjua, S., & Khurshid, M. (2016). Customer-based brand equity and firms' performance in the telecom industry. *International Journal of Services and Operations Management, 25*(3), 334–346.

Campbell, D., Shotter, J., & Draper, R. (2018). *The socially constructed organization.* London: Routledge.

Carroll, A. B. (1979). A three-dimensional conceptual model of corporate performance. *Academy of Management Review*, *4*(4), 497–505.

Carroll, A. B. (1991). The pyramid of corporate social responsibility: Toward the moral management of organizational stakeholders. *Business Horizons*, *34*(4), 39–48.

Carroll, A. B. (1999). Corporate social responsibility: Evolution of a definitional construct. *Business & Society*, *38*(3), 268–295.

Chakraborty, U. (2019). The impact of source credible online reviews on purchase intention: The mediating roles of brand equity dimensions. *Journal of Research in Interactive Marketing*, *13*, 142–161.

Chang, Y. T., Yu, H., & Lu, H. P. (2015). Persuasive messages, popularity cohesion and message diffusion in social media marketing. *Journal of Business Research*, *68*(4), 777–782.

Churkina, G., Organschi, A., Reyer, C., Ruff, A., Vinke, K., Liu, Z., Reck, B., Graedel, T., & Schellnhuber (2020) Buildings as a global carbon sink. *Nature Sustainability*, *10*, 1038.

Davies, I., & Gutsche, S. (2016). Consumer motivations for mainstream "ethical" consumption. *European Journal of Marketing*, *50*(7/8), 1326–1347.

De Grosbois, D. (2012). Corporate social responsibility reporting by the global hotel industry: Commitment, initiatives and performance. *International Journal of Hospitality Management*, *31*(3), 896–905.

De Jong, M., & van der Meer, M. (2017). How does it fit? Exploring the congruence between organizations and their corporate social responsibility. *Journal of Business Ethics*, *143*, 71–83.

Febrian, A., & Fadly, M. (2021). The impact of customer satisfaction with eWOM and brand equity on e-commerce purchase intention in Indonesia moderated by culture. *Binus Business Review*, *12*(1), 41–51.

Filieri, R., Mcleay, F., Tsui, B., & Lin, Z. (2018). Consumer perceptions of information helpfulness and determinants of purchase intention in online consumer reviews of services. *Information & Management*, *55*, 956–970.

Franco, S., Caroli, M., Cappa, F., & Del Chiappa, G. (2020). Are you good enough? CSR, quality management and corporate financial performance in the hospitality industry. *International Journal of Hospitality Management*, *88*, 102395.

Gonzalez-Rodriguez, M., Martin-Samper, R., Ali Koseoglu, M., & Okumus, F. (2018). Hotels' corporate social responsibility practices, organizational culture, firm reputation and performance. *Journal of Sustainable Tourism*, *27*(3), 398–419.

Gunawan, D. D., & Huarng, K.-H. (2015). Viral effects of social network and media on consumers' purchase intention. *Journal of Business Research*, *68*(11), 2237–2241.

Hendrata, A. A., Tinaprilla, N., & Safari, A. (2021). The effect of brand equity and electronic word of mouth (E-WOM) on customer satisfaction and loyalty in e-commerce marketplace. *International Journal of Research & Review*, *8*, 308–315.

Holcomb, J. L., Upchurch, R. S., & Okumus, F. (2007). Corporate social responsibility: What are top hotel companies reporting? *International Journal of Contemporary Hospitality Management*, *19*(6), 461–475.

Hur, W. M., Kim, H., & Woo, J. (2014). How CSR leads to corporate brand equity: Mediating mechanisms of corporate brand credibility and reputation. *Journal of Business Ethics*, *125*(1), 75–86.

Ishaq, M. I., Hussain, N., Asim, A. I., & Cheema, L. J. (2014). Brand equity in the Pakistani hotel industry. *Revista de Administração de Empresas*, *54*, 284–295.

Kelley, H. H., & Michela, J. L. (1980). Attribution theory and research. *Annual Review of Psychology*, *31*(1), 457–501.

Kim, H., Woo, E., Uysal, M., & Kwon, N. (2017). The effects of corporate social responsibility (CSR) on employee wellbeing in the hospitality. *International Journal of Contemporary Hospitality Management*, *30*(3), pp 1584–1600.

Klein, J., & Dawar, N. (2004). Corporate social responsibility and consumers' attributions and brand evaluations in a product–harm crisis. *International Journal of Research in Marketing*, *21*(3), 203–217.

Liang, L. J., Choi, H. C., & Joppe, M. (2018). Understanding repurchase intention of Airbnb consumers: Perceived authenticity, electronic word-of-mouth, and price sensitivity. *Journal of Travel & Tourism Marketing, 35*(1), 73–89.

Liu, M. T., Wong, I. A., Tseng, T.-H., Chang, A. W.-Y., & Phau, I. (2017). Applying consumer-based brand equity in luxury hotel branding. *Journal of Business Research, 81*, 192–202.

Manusov, V., & Spitzberg, B. H. (2008). Attributes of attribution theory: Finding good cause in the search for theory. In D. O. Braithwaite & L. A. Baxter (Eds.), *Engaging Theories in Interpersonal Communication* (pp. 37–49). Thousand Oaks, CA: SAGE.

Martínez, P., & Nishiyama, N. (2019). Enhancing customer-based brand equity through CSR in the hospitality sector. *International Journal of Hospitality & Tourism Administration, 20*(3), 329–353.

Martínez, P., Pérez, A., & Del Bosque, I. R. (2014). CSR influence on hotel brand image and loyalty. *Academia Revista Latinoamericana de Administración, 27*(2), 267–283.

Murtiasih, S., Sucherly, S., & Siringoringo, H. (2014). Impact of country of origin and word of mouth on brand equity. *Marketing Intelligence & Planning, 32*(5), 616–629.

Ng, T., Yam, K. C., & Aguinis, H. (2018). Employee perceptions of corporate social responsibility: Effects on pride, embeddedness, and turnover. *Personnel Psychology, 72*, 107–137.

Okazaki, S., Andreu, L., & Campo, S. (2017). Knowledge sharing among tourists via social media: A comparison between Facebook and TripAdvisor. *International Journal of Tourism Research, 19*(1), 107–119.

Panagopoulos, N., Rapp, A., & Vlachos, P. (2016). I think they think we are good citizens: Meta-perceptions as antecedents of employees' reactions to corporate social responsibility. *Journal of Business Research, 69*(8), 2781–2790.

Park, S. Y., & Levy, S. E. (2014). Corporate social responsibility: Perspectives of hotel frontline employees. *International Journal of Contemporary Hospitality Management, 26*(3), 332–348.

Pradhan, S. (2018). Role of CSR in the consumer decision making process – The case of India. *Social Responsibility Journal, 14*(1), 138–158.

Rashid, N., Nika, F., & Thomas, G. (2021). Impact of service encounter on experiential value and customer loyalty: An empirical investigation in the coffee shop context. *Sage Open, 11*(4), 1–12.

Schaltegger, S., & Burritt, R. (2018). Business cases and corporate engagement with sustainability: Differentiating ethical motivations. *Journal of Business Ethics, 147*, 241–259.

Seric, M., Mikulic, J., & Gil-Saura, I. (2018). Exploring relationships between customer-based brand equity and its drivers and consequences in the hotel context. An impact –asymmetry assessment. *Current Issues in Tourism, 21*(14), 1621–1643.

Su, L., Swanson, S., & Chen, X. (2016). The effects of perceived service quality on repurchase intentions and subjective wellbeing of Chinese tourists: The mediating role of relationship quality. *Tourism Management, 52*, 82–95.

Tsalis, T., Malamateniou, K., Koulouriotis, D., & Nikolaou, L. (2019). New challenges for corporate sustainability reporting: United nations 2030 Agenda for sustainable development and the sustainable development goals. *Corporate Social Responsibility Environmental Management, 27*, 1617–1629.

UN. (2015). The 17 Goals, https://sdgs.un.org/goals

Uzunoğlu, E., Türkel, S., & Akyar, B. Y. (2017). Engaging consumers through corporate social responsibility messages on social media: An experimental study. *Public Relations Review, 43*(5), 989–997.

Wynn Resorts 2020 Environmental, Social and Governance Report. (2020).

Xu, W., & Saxton, G. D. (2019). Does stakeholder engagement pay off on social media? A social capital perspective. *Nonprofit and Voluntary Sector Quarterly, 48*(1), 28–49.

Yoong, L. C., & Lian, S. B. (2019). Customer engagement in social media and purchase intentions in the hotel industry. *International Journal of Academic Research in Business and Social Sciences, 9*(1), 54–68.

Zhang, K. Z., Benyoucef, M., & Zhao, S. J. (2015). Consumer participation and gender differences on companies' microblogs: A brand attachment process perspective. *Computers in Human Behavior, 44*, 357–368.

# 24 Post-pandemic Chinese outbound tourism

*Marine L'Hostis*

## Introduction

On 27 January 2020, 20 days after confirming the discovery of a new coronavirus, Chinese authorities required travel agents to suspend the sale of outbound group and package travel, as part of a set of measures meant to stop the spread of the virus inside China's borders. Individual travel was made very difficult as well, because of strict tests, quarantine, and travel policies, in addition to passport issuance restrictions (even confiscation).

These restrictions, in addition to official recommendations against non-essential travel abroad, have been very dissuasive to potential Chinese tourists planning to cross China's border for leisure, and they have caused a dramatic decrease of Chinese outbound tourism flows, as shown in Figure 24.1.

According to the Chinese Outbound Tourism Research Institute (COTRI), border crossings in 2021 reached 8.5 million, which brings them below the 2000 level (Arlt, 2022). At the time, Chinese outbound tourism was still in its beginnings, after China signed a new series of bilateral agreements (known as Approved Destination Status, hereafter "ADS agreements") allowing its nationals to travel for tourism to certain destinations. Since then, progressively, but pretty steadily, China attained in 2012 first place in the world in the tourism source market in terms of spending (UNWTO, 2013), and in 2018 it was still in first place (UNWTO, n.d.). Even before then, Chinese tourists were already a very coveted market, and appeared as the "goose that laid the golden eggs" to the tourism industry's professionals. As a consequence, and quite understandably, the restrictions still weighing on Chinese outbound tourists, two years after they were implemented, now raise the question of their return. Not only the date of China's borders reopening, but also the potential before/after effect of the pandemic on Chinese tourists' practices need to be addressed in order to anticipate possible evolutions and adapt to them.

The potential effects of the COVID-19 pandemic on Chinese tourism have already been addressed by a few authors. Some of these papers focus on the media coverage of Chinese tourism during the time of the pandemic (Chen et al., 2022a; Hasenzahl & Cantoni, 2021; Zheng et al., 2020). Other authors address the pandemic consequences for the tourism industry in China (Wang et al., 2022; Zhong et al., 2022). For the largest part, publications study the impact of the pandemic on Chinese tourists' choices and practices. Some focus on the immediate outcome (Huang et al., 2021; Nazneen et al., 2022) and others anticipate post-pandemic Chinese international tourism through different aspects, for example, arrival predictions and destination recovery (Polyzos et al., 2021), tourists' practices and behaviour (Jin et al., 2022; Wang et al., 2021; Wen et al., 2021a; Wen, Wang, et al., 2021b), and host/guest relations (Armutlu et al., 2021).

DOI: 10.4324/9781003291763-27

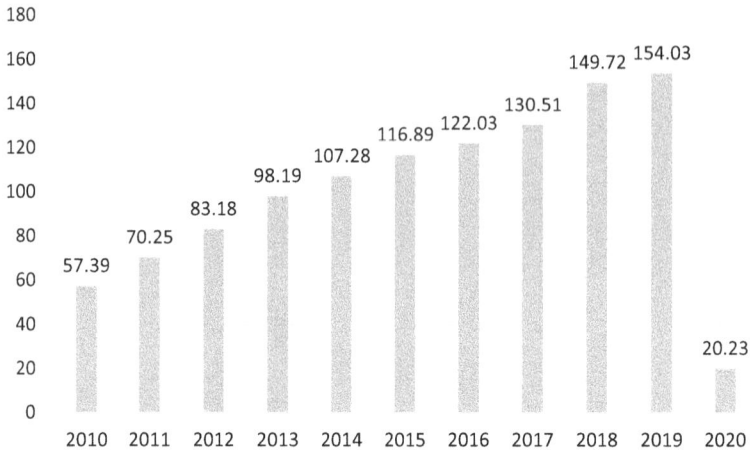

*Figure 24.1* Number of outbound tourists departing from China with an estimate for 2020.

*Sources*: Statista and Ministry of Culture and Tourism (China) 2022.

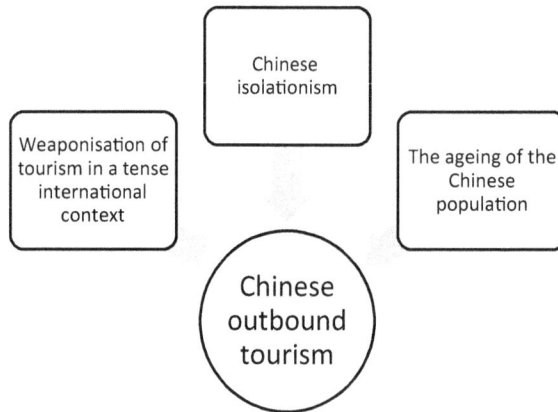

*Figure 24.2* Chinese outbound tourism: Future challenges.

No publication directly addresses the broader geopolitical context in which Chinese outbound tourism will resume, and the ways this context could possibly affect its recovery. Nor do authors address the isolationist turn taken by the Chinese authorities and the demographic ageing of the population. Yet, we will see that these three factors (Figure 24.2) are likely to have an important impact on Chinese outbound flows in the next decades. Using the case of France, we will establish a picture of pre-COVID Chinese outbound tourism trends and of the stakeholders involved in the market development. We will describe how the three aforementioned trends could influence the growth of outbound tourism in the years to come and finally address the implications for stakeholders and possible strategies for facing future challenges.

## Pre-COVID-19 Chinese outbound tourism trends: The case of France

France officially started receiving Chinese tourists in 2004, after the European Union signed an ADS agreement with China. However, prior to 2004, Chinese were already visiting France for tourism, but only for business trips or official delegations (Taunay, 2013). Some would also come via Germany, since this country signed an ADS agreement before the rest of the EU and would issue Schengen visas allowing Chinese tourists to enter other European countries (L'Hostis, 2020).

### Chinese outbound tourism: A key market for France

Since these official beginnings, France has been easing entry conditions for Chinese tourists, granting them tourist visas in 48 hours as of 2014. This gesture shows the interest of France in this market, and the will to remain competitive with other European popular destinations such as Italy and Germany. Yet, France's biggest source markets for non-resident tourists are still neighbouring countries like Germany, Belgium, or the UK. In comparison, China was the ninth largest tourist market for France in 2018 with 2.2 million visitors (DGE, 2019).

In order to better understand the French policy, we need to consider how inbound tourism to France has been evolving.

Figure 24.3 shows that, although Europeans markets are still largely outnumbering non-European markets in terms of tourist arrivals, the latter follow a steadier increase. In 2016 (after the 2015 terrorist attacks), they compensated for the drop in European markets, for example.

If we look in detail (Figure 24.4), we can see that traditional tourist source markets for France (such as Germany, Belgium, the Netherlands, or the UK) are stagnating or decreasing. In comparison, non-European markets such as China and the USA are growing. As a consequence, the development of the Chinese market appears as an opportunity to make up for the downturn of traditional markets. This role of China as a growth driver is also reflected in the income this market generates. In 2017, China represented the fifth largest revenue source among all foreign tourist markets in France, with almost €4 billion of spending (DGE, 2018). A significant part of these expenditures was on shopping. In Paris for example, 50% of Chinese tourists practice shopping, which represents 24% of their budgets (CRT Paris Île-de-France, 2020).

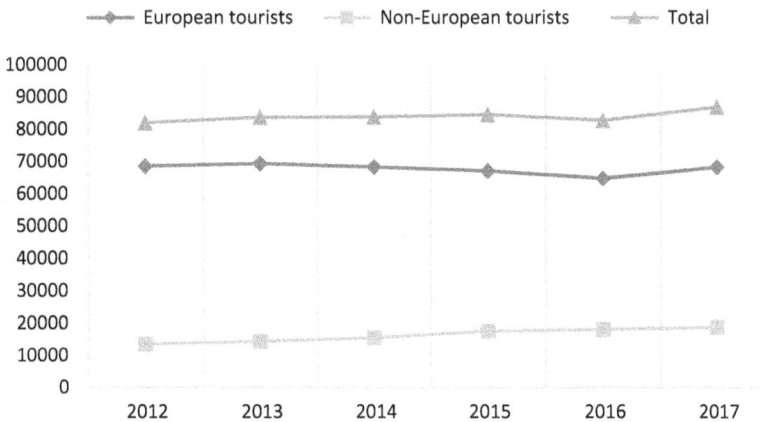

*Figure 24.3* European and non-European tourist arrivals in France, 2012–2017.
*Source*: DGE (2018).

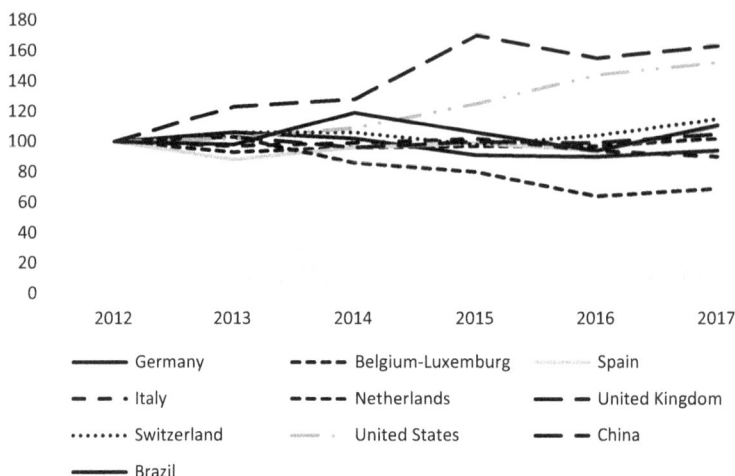

*Figure 24.4* Evolution of tourist arrivals in France by country of origin, 2012–2017.

*Notes*: This graphic is based on the ten major countries of origin in 2017; 2012 = 100.

*Source*: DGE (2018).

While China represents a key market for France, France's position as a destination for Chinese tourists should be put into perspective. When Chinese tourism was officially authorised in France, the local media were very enthusiastic about the opening of this huge market, overlooking the fact that Chinese outbound tourists, for the biggest part, primarily travel to close destinations such as Hong Kong, Macao, Taiwan, South Korea, Thailand, and Japan. In 2018, Asia held an 89.03% market share of outbound Chinese tourism, against 3.83% for Europe (Dragon Trail, 2019). The lack of harmonised data makes it difficult to determine how France compares to its European competitors in terms of Chinese tourist reception. According to statistics issued by the European Commission (on the website Eurostat), France ranks second, between Italy and Germany, in terms of arrivals at tourist accommodation establishments (before Brexit, in 2016, the UK was in seventh position).

### A dynamic, but challenging market

In terms of trends, pre-COVID, Chinese tourists in France were getting more adventurous and independent, exploring regions outside Paris and travelling more on their own. This is the result of an evolution that has taken place over the last two decades.

Originally, most Chinese first-time tourists would visit several European countries with a tour group and Paris would be their only stop-off in France (Taunay, 2013). Over the years, other French regions progressively attracted more Chinese visitors, especially along the routes connecting France to Switzerland, and Paris to the region of Provence-Alpes-Côte d'Azur (L'Hostis, 2020). Paris is still unquestionably a must-see, but as Chinese tourists gain experience and familiarity, they tend to deepen their exploration of France. Nice (located on the French Riviera) is among the cities enjoying a rise of Chinese tourist flows. During the fieldwork we conducted in this city in 2017, we observed that the main drivers of this diffusion were repeaters, coming to Nice after visiting Paris during a previous trip to France. These visitors preferred individual tourism because they saw it as more flexible than travelling with a group. Some of them were willing to "go off the beaten track" and were renting cars to do a road-trip around Provence. We also observed a curiosity for the local way of life and the will to get immersed in it. For instance, some tourists chose Nice because they wanted to see "where French people spent their holidays" or were staying in Airbnbs in order to experience "living like a local". Unlike the stereotype of gregarious and consumerist tourists travelling in groups, Chinese tourists are actually getting more autonomous, emancipating themselves from organised tours, opting for flexibility and freedom, and individualising their itineraries and practices according to their needs and aspirations.

During our research, we also noticed many tourists aged 20–45 and who were travelling with their parents and/or grandparents. This phenomenon reflects the mutations Chinese society has been going through since the 1980s and the economic reforms initiated by Deng Xiaoping. The generations who grew up prior to these reforms and under the Maoist regime were not allowed to travel for leisure and therefore had little experience of going abroad and dealing with a foreign context. Many of these people aged 45 and above need help to travel abroad and will either choose a tour group with a professional guide or their own children. As the urban lifestyle spreads in China and tourism becomes a legitimate activity, more of them use their money and time to discover destinations that were forbidden in their youth. Some authors actually think these senior visitors could be the third wave of Chinese outbound tourists (Bao et al., 2019).

We also studied the stakeholders interested in the Chinese market in France. We conducted interviews with representatives from tourism boards, local councils, and with destination management company (DMC) owners. These stakeholders are interested in the Chinese market as a source of income, given the exceptional size of the Chinese population and the purchasing power attributed to Chinese tourists. However, professionals from the French tourism industry are faced with different kinds of obstacles when they deal with China. The public sector (tourism offices, local and regional institutions) suffers from a lack of means to keep a count of Chinese visitors and identify their practices, which makes their vision of these tourists incomplete and biased. In the private sector, DMC managers dealing with the Chinese industry complain about unfair competition and unethical practices altering the quality of service, sparking the dissatisfaction of tourists, and ultimately degrading the destination image and the working conditions of local professionals (L'Hostis, 2020). Finally, in 2020 and 2021, the French press was echoing the concerns of hotels, tourist shops, and Parisians department stores concerning the prolonged absence of Chinese tourists because of travel restrictions. The resulting loss of income is barely compensated by customers from other nationalities (e.g., Garnier, 2021).

After enjoying the liberalisation of Chinese outbound tourism, these stakeholders are now hoping for an impending resumption of these tourist flows. However, their return will probably be progressive and be submitted to tight control by the Chinese state. As a result, it will take a while for frequencies to reach their pre-pandemic levels.

## Future trends and issues affecting the Chinese tourist market: Geopolitical context and demographic crisis

After outlining the Chinese outbound tourism context up until 2022, I will now present three trends that stakeholders interested in the Chinese market should look out for in the years to come. My intention is not to be exhaustive about the future of Chinese international tourism, but to draw readers' attention to certain trends and issues that have not been addressed by researchers in tourism so far, and that could interfere with the evolution of Chinese leisure mobilities in the future.

### The rivalry with the USA and the Taiwanese issue

Over the past decade, China has been more assertive in its ambition to attain world leadership and to export its model of governance. This ambition is best illustrated by the Belt and Road Initiative, an infrastructure project connecting China to Europe by land and by sea. This project relies on cooperation with countries in Asia, Central Asia, Europe, Africa, and even stretches to South America. The objective is to sustain China's economic growth by developing new export markets, but also to ensure its food and energy sufficiency, and to create safe provision routes. Politically, China aims at protecting its territorial integrity (especially stabilising Xinjiang) and to act as a pacifying power counterbalancing the Russian influence in Central Asia. Ultimately, this New Belt and Road Initiative is the spearhead of the restoration of China as the world's top superpower, an ambition that was set as an objective to be reached by 2049 in Xi Jinping's opening speech of the 19th Chinese Communist Party Congress in 2017.

The USA is not passive about facing this competitor. Under the Obama administration, the USA started focusing its attention on the Indo-Pacific region, but the rivalry with China really started during Trump's mandate and took an aggressive turn. Both countries are competing in several sectors (the economy, technology, military capacity), but the epicentre of their conflict is Taiwan because of Chinese ambition to take possession of it. In response to this threat, the USA sent several signs of support to Taipei, implying they might take military action in the case of invasion. Generally speaking, the Biden administration keeps following this harsh line towards China, but unlike Trump Biden opts for a multilateral approach and builds alliances such as AUKUS, a security pact composed of the USA, Australia, and the UK, officially meant to secure and stabilise the Indo-Pacific region.

The open competition with the USA, US multilateralism, and the Taiwanese issue will very probably impact Chinese outbound tourism in the future. In order to understand why and how, we need to remember that China tends to use commercial sanctions in order to pressure countries deemed unfriendly (as was the case when Australia requested an investigation into the origins of COVID-19 in 2020, for example). This logic applies to tourism as well, since outbound flows have been used as a diplomatic tool since they were permitted (e.g. Tse, 2011 and Zhu et al., 2021) and calls to boycott a destination are part of the Chinese commercial sanction arsenal. South Korea already suffered the consequences of this strategy in 2017, when Washington delivered an anti-missile system to Seoul, which Beijing saw as a threat to its own national security. As a result, Chinese authorities retaliated through a set of commercial sanctions, including suspending tourism to South Korea (Kim, 2020). This boycott call was not only followed by travel companies, but also by individual tourists, as we found out during our research. In 2017, we conducted an interview with two Chinese students who told us that they would no longer travel to South Korea because of diplomatic tensions with China (L'Hostis, 2020), which shows that the patriotic call resonates at the individual level too.

In the future, Chinese outbound tourism flows will certainly be affected by this complex international context, not to mention the eventuality of an actual invasion of Taiwan in the decades to come. As the Chinese Communist Party (CCP) strives to expand its influence over the world, it will be rewarding allies or punishing opponents and the access to the Chinese tourist market will most likely be favoured by the support offered by destinations to Chinese ambitions.

Before we even consider this hypothesis, the conditions in which Chinese outbound tourism will resume are yet to be seen. For the time being, there is nothing but speculation as to when and how this will happen, and the ambiguity maintained by the Chinese state around this matter could be the symptom of the isolationist move initiated by Xi Jinping since the beginning of his mandate in 2012.

### Xi Jinping's isolationist turn

According to researcher Jean-Pierre Cabestan (quoted in McLaughlin, 2022), "China's decision to turn more inward preceded the pandemic but has been intensified by it". This general move is illustrated by several policies initiated by Xi Jinping: for example, "Document no. 9", a confidential document circulating in the CCP from 2012 on, and warning against Western ideas; the "Made in China 2025 strategy" issued in 2015; or the Hong Kong national security law adopted in 2020. Cabestan states that these decisions reflect the same aspirations: "Move away from the West and attack its ideology; reduce China's dependence upon the outside; enhance its economic and technological self-reliance". In this context, strict restrictions related to COVID-19 (especially borders restrictions) appear particularly timely in order to accelerate the transition towards a self-sufficient society.

This isolationism can be attributed to a more general return to socialist fundamentals, which would involve an increased supervision of private companies by the CCP, and a tighter grip on civil society. For example, since 2021, minors have been forbidden to play online video games for more than three hours a week. This restriction is not only a way for the CCP to rein in technology companies, but it also completes other policies interfering in youth education: for instance, teaching Xi's thoughts in schools and banning after-school private tutoring companies.

In this context of isolationism, anti-Western discourse, and increasing control over society, what will be the place of tourism, and more specifically, outbound tourism in Chinese society? The development of tourism in China is the fruit of Chinese society's liberalisation and the economic reforms of the 1980s, but the government currently seems at odds with these policies. While domestic tourism is coherent with the CCP's objective to rely more heavily on the domestic market and can be used for nationalistic purposes (as is the case of red tourism, for example), outbound tourism does not fit as much the society project promoted by Xi Jinping. Unsurprisingly, the five-year tourism plan released by the Chinese government in January 2022 mainly focuses on developing and modernising domestic tourism and has only limited places for inbound and outbound tourism (4 sub-sections out of 36, including one specifically dedicated to Hong Kong, Macao, and Taiwan). The plan does not provide any resumption deadline and subjects the reopening of outbound tourism to the pandemic context. What particularly draws my attention is the intention to "strengthen the orientation and the management of outbound tourists", basically in order to make them good ambassadors of China, and the will to "strengthen cooperation with countries along the Belt and Road", which implies a use of tourism as a soft power tool to help expand China's influence (China Ministry of Culture and Tourism, 2022). This confirms the trend I presented earlier about the use of outbound tourism as a diplomatic leverage.

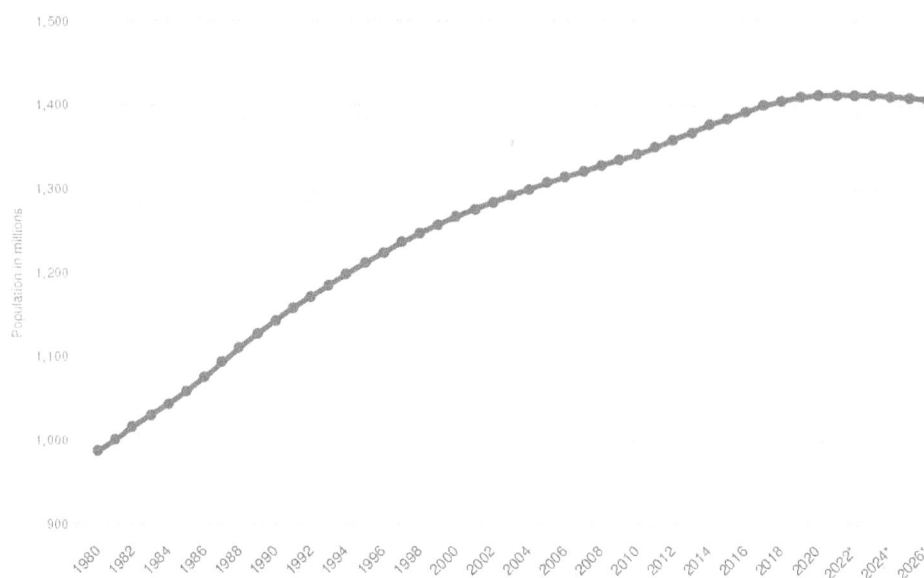

*Figure 24.5* Total population of China from 1980 to 2021 with forecasts until 2027.

*Sources*: IMF; CEIC; National Bureau of Statistics of China and Statista 2023.

### The ageing of the Chinese population

The last long-run trend I would like to address here is the ageing of the Chinese population and the impact of this demographic evolution on outbound tourism.

At the moment, with 1.4 billion inhabitants, China has the largest population in the world, but it is on the edge of a demographic decline. From 2000 to 2020, the population only increased by about 0.5% per year (Attané, 2020) and according to researcher Xiujian Peng, the fertility rate reached 1.15 births per woman in 2021, against 2.6 in the late 1980s, which means China is probably culminating at its demographic peak: "The Shanghai Academy of Social Sciences team predicts an annual average decline of 1.1% after 2021, pushing China's population down to 587 million in 2100, less than half of what it is today" (Peng, 2022).

While a smaller population is not necessarily a problem in itself, China is going to face a transformation in its population structure, which will have heavy repercussions on society and the economy. Indeed, as life expectancy is extending, the share of people older than 60 is increasing rapidly and could reach 30% by 2040 (Attané, 2020) (Figure 24.6).

This will certainly change the face of the Chinese tourist market and have implications for stakeholders from the tourism industry. Indeed, senior tourists have special needs and expectations requiring adaptation, as has been shown in the literature already (e.g., Fleischer & Pizam, 2002; Kim, 2015). Chinese senior tourists are no exception. Those we met during our fieldwork were sometimes suffering from physical conditions necessitating a slow-paced trip or special food regime. Being retired, they had more spare time than their children and were more flexible in terms of dates, but they also needed help and guidance to cope with a foreign environment. This lack of autonomy results from the fact that generations who

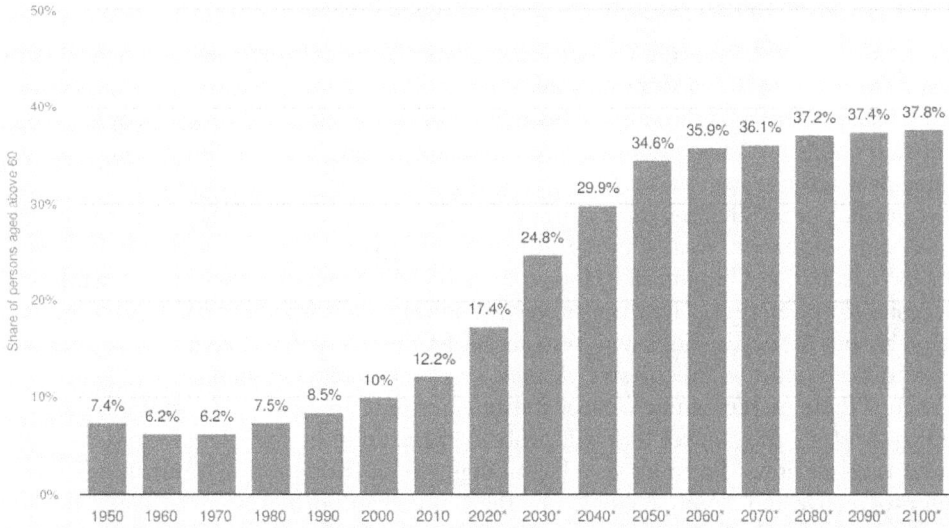

*Figure 24.6* Share of population aged 60 and older in China from 1950 to 2010 with forecasts until 2100.

*Sources*: UN DESA; National Bureau of Statistics of China and Statista 2023.

were born prior to the economic reforms had very little experience of travelling before tourism was officially authorised. The next generations of senior Chinese tourists probably will not have such difficulties, since they were born in a society where tourism was allowed and are the first generations enjoying the possibility to travel abroad for pleasure. These future senior tourists currently account for about two-thirds of the tourist market (Bao et al., 2019) and belong to the well-educated and urban middle class from first and second-tier Chinese cities. They will most likely keep travelling in the future, since it is been shown that "the better educated seniors have more money, are predisposed to recreational spending, and tend to travel farther from home if their health allows" (Zimmer, Brayley & Searle, 1995, quoted in Bao et al., 2019). However, in China's case, the question is not so much to find out if age will interfere with senior Chinese tourist mobilities, but to observe if and how the Chinese state will make it possible for its population to grow old in health and financial conditions decent enough to allow travelling abroad.

This concern is raised by the fact that the aforementioned fast ageing of the Chinese population goes hand in hand with a shrivelling workforce. According to Peng, the latter is decreasing after reaching a peak in 2014 and is projected to shrink to less than one-third of that peak by 2100, whilst the Chinese population aged 65 and above is expected to pass China's working-age population around 2080. "This means that while there are currently 100 working-age people available to support every 20 elderly people, by 2100, 100 working-age Chinese will have to support as many as 120 elderly Chinese" (Peng, 2022). Consequently, China will have to reform its economy in order to sustain the ageing of its population.

If current projections are confirmed, by 2100, the Chinese population will be older and more than half smaller than it is today. In the meantime, not only the demographics of Chinese tourists are going to evolve, but the future of Chinese outbound tourism will also depend on the country's capacity to provide its elderly with sufficient purchasing power.

**Challenges, opportunities, and strategies to develop the Chinese market**

Stakeholders interested in the Chinese market will be faced with a few challenges in the decades to come. In the short term, some analysts predict a fast recovery of tourism flows once China reopens, and they are confident about the eagerness of Chinese people to travel to distant countries. This optimism needs to be mitigated, however. Some market studies do demonstrate a strong interest in foreign destinations from the Chinese public. For example, according to a McKinsey survey, the desire for overseas travel has reached pre-pandemic levels, and South East Asia, Europe, Russia, and Japan are named as the most desired overseas destinations. Nonetheless, the same survey also points out two main constraints to the desire to travel: first, Chinese visitors will be very cautious about safety and they will preferably opt for destinations with zero cases of COVID-19 (Chen et al., 2022b). Second, quarantine on return is a dissuasive obstacle to go on holidays abroad.

Even if these market studies reflect a strong desire to travel abroad, we think they also tend to overlook the control that will probably be exerted by the Chinese state over outbound tourism flows. From its very beginnings, the liberalisation of tourism has always been progressive and closely regulated by the Chinese authorities, so once again, the government will probably reopen borders and lift restrictions gradually and cautiously. For example, priority might be given to business trips. When international tourism resumes, we will not immediately see flights full of Chinese tourists heading to Paris, New York, or Bali. Chinese authorities will more likely gradually extend the list of allowed destinations according to their health and safety, and progressively relax the restrictions. From this perspective, Chinese outbound tourism probably will not reach pre-pandemic levels before 2024 or 2025. In a more unofficial way, Chinese authorities might also influence tourists' choices according to its strategic interests. For example, we can imagine simplified procedures and strong promotion for China's partner countries such as those participating in the Belt and Road Initiative (that is already the logic at work with Thailand for example; Li, 2019). Conversely, countries bonding with Taiwan, or whose alliance with the USA would be seen as a threat, could be smeared in official state media and eventually be boycotted. Stakeholders involved in the Chinese market should be very attentive to the diplomatic relationships between China and tourist destinations. At the moment, despite the threats, it is impossible to say if China will attack Taiwan one day and when this could happen. The unpredictability of political events will necessitate flexibility in order to adapt to their consequences over tourist flows. Conversely, destinations bonding closely with China could see opportunities rising in terms of Chinese tourist reception, the signing of an ADS agreement being usually granted as part of wider commercial cooperation.

In times of uncertainty and travel restrictions, technology offers ways to stay connected with the Chinese source market and even to sell tourism products. For example, Atout France, France's national tourism development agency, has been organising destination livestreams in order to keep France in the Chinese public's imagination and to stir the desire for travel. The principle is to present a live virtual visit of a tourist place and to interact with the viewers during the tour. Pre-pandemic, livestreams were mostly aimed at selling products online, but the tourism industry is now taking advantage of this tool in order to adapt to travel bans and to stay connected with potential visitors. This might interest stakeholders from the public and private sectors, such as national or local tourism agencies, museums, tourist sites, and DMCs.

Another way to cope with travel bans is to communicate with international Chinese students. Even if this target is also sensitive to the diplomatic context, at the moment they are not subjected to Chinese travel restrictions. If we take the case of France, there were 29,731 Chinese students in 2020, which represented 8% of foreign students in total, and the second nationality after Moroccan students (Campus France, 2021). These numbers are not huge, but students represent an opportunity because, as we have observed during our research, they enjoy travelling around their host country and the neighbouring ones. They also act as ambassadors for the destination and, when possible, invite friends and family to visit them, so attracting them to a destination which may prove fruitful in the long term. Some stakeholders already have communication strategies specifically aimed at Chinese students, such as VisitScotland, which has launched a campaign on Chinese social media Weibo and WeChat (Paulis-Cook, 2022).

## Conclusion

In this chapter, we have seen that, pre-COVID, China was a dynamic outbound tourism market, evolving towards more independent and adventurous tourists. This dynamic is hindered by trends that were already at work pre-pandemic and that have intensified with the COVID-19 outbreak: the rivalry with the USA, the Taiwanese issue, and the isolationist turn. In this political context, Chinese outbound tourism will certainly be impacted by the Chinese authorities' propensity to weaponise tourist flows according to their diplomatic agenda. The climax of these tensions could be an invasion of Taiwan triggering US intervention, though it would be very risky to predict if and when this could happen. In the longer run, tourism stakeholders should also prepare to see the face of Chinese tourism evolve, with the fast ageing of the population. Senior tourists will have specific needs and expectations to be met, but more importantly, with a shrinking workforce, their ability to travel will depend on China's capacity to ensure decent living conditions for its elderly.

In these times of uncertainty, stakeholders interested in the Chinese market should be attentive to diplomatic relations between China and foreign destinations, because their evolution might bring challenges but also opportunities, depending on how these relations threaten or favour Chinese interests. Technology will bring tools for destinations to stay connected with the Chinese market and to foster the desire to travel. Finally, for the time being, targeting Chinese students already living in destinations with communication campaigns may prove profitable in the short and long run. First because they take their studies as an opportunity to travel around and second because they are ambassadors of their host country among their relatives.

Overall, in the next decades, the Chinese outbound tourism market will probably be quite challenging and will necessitate flexibility in order to adapt to sudden shifts reflecting a global instability.

## References

Arlt, W. (2022, janvier 21). Editorial: Percentages, products and prayers for China. *COTRI*. https://china-outbound.com/editorial-percentages-products-and-prayers-for-china/

Armutlu, M. E., Bakır, A. C., Sönmez, H., Zorer, E., & Alvarez, M. D. (2021). Factors affecting intended hospitable behaviour to tourists: Hosting Chinese tourists in a post-COVID-19 world. *Anatolia*, *32*(2), 218–231. https://doi.org/10.1080/13032917.2020.1855595

Attané, I. (2020). Vieillissement démographique et ralentissement économique en Chine. *Cités*, *82*(2), 87–97. Cairn.info. https://doi.org/10.3917/cite.082.0087

Bao, J., Jin, X., & Weaver, D. (2019). Profiling the elite middle-age Chinese outbound travellers: A 3rd wave? *Current Issues in Tourism*, 22(5), 561–574. https://doi.org/10.1080/13683500.2018.1449817

Campus France. (2021). *Chiffres clés de la mobilité étudiante dans le monde*. Beijing: Chine.

Chen, G., Saxon, S., Yu, J., & Zhang, C. (2022a). *China Tourism in 2022: Trends to Watch|McKinsey*. McKinsey. https://www.mckinsey.com/industries/travel-logistics-and-infrastructure/our-insights/outlook-for-china-tourism-in-2022-trends-to-watch-in-uncertain-times

Chen, H., Huang, X., & Li, Z. (2022b). A content analysis of Chinese news coverage on COVID-19 and tourism. *Current Issues in Tourism*, 25(2), 198–205. https://doi.org/10.1080/13683500.2020.1763269

China Ministry of Culture and Tourism. (2022, January). "十四五"旅游业发展规划 *14th Five-Year Tourism Plan*. http://www.gov.cn/zhengce/content/2022-01/20/content_5669468.htm

CRT Paris Île-de-France. (2020). *Repères 2020 Touristes Chinois*. Paris: Comité Régional de Tourisme Île-de-France.

DGE. (2018). *Memento du tourisme*. Paris: Direction générale des entreprises.

DGE. (2019). *More than 89 Million Foreign Tourists in France in 2018*. Paris: Direction générale des entreprises.

Dragon Trail. (2019). *CTA: Annual Report on China Outbound Tourism Development 2019—Dragon Trail International*. https://dragontrail.com/resources/blog/cta-annual-report-on-china-outbound-tourism-development-2019

Fleischer, A., & Pizam, A. (2002). Tourism constraints among Israeli seniors. *Annals of Tourism Research*, 29(1), 106–123. https://doi.org/10.1016/S0160-7383(01)00026-3

Garnier, J. (2021, October 1). A Paris, les boutiques de luxe apprennent à vivre sans les touristes chinois. *Le Monde.fr*. https://www.lemonde.fr/economie/article/2021/10/01/a-paris-les-boutiques-de-luxe-apprennent-a-vivre-sans-les-touristes-chinois_6096695_3234.html

Hasenzahl, L., & Cantoni, L. (2021). "Old" and "New" Media Discourses on Chinese Outbound Tourism to Switzerland before and during the COVID-19 Outbreak. An Exploratory Study. In W. Wörndl, C. Koo, & J. L. Stienmetz (Éds.), *Information and Communication Technologies in Tourism 2021* (pp. 530–542). Springer International Publishing. https://doi.org/10.1007/978-3-030-65785-7_50

Huang, S. (Sam), Shao, Y., Zeng, Y., Liu, X., & Li, Z. (2021). Impacts of COVID-19 on Chinese nationals' tourism preferences. *Tourism Management Perspectives*, 40, 100895. https://doi.org/10.1016/j.tmp.2021.100895

Jin, X. (Cathy), Bao, J., & Tang, C. (2022). Profiling and evaluating Chinese consumers regarding post-COVID-19 travel. *Current Issues in Tourism*, 25(5), 745–763. https://doi.org/10.1080/13683500.2021.1874313

Kim, H. L. (2015). *An Examination of Salient Dimensions of Senior Tourist Behavior: Relationships among Personal Values, Travel Constraints, Travel Motivation, and Quality of Life (QoL)*. Blacksburg, VA: Virginia Polytechnic Institute and State University.

Kim, V. (2020, novembre 19). *When China and U.S. Spar, It's South Korea that Gets Punched*. Los Angeles Times. https://www.latimes.com/world-nation/story/2020-11-19/south-korea-china-beijing-economy-thaad-missile-interceptor

L'Hostis, M. (2020). *La diffusion du tourisme chinois en France—Influence des représentations et du capital spatial*. Angers: Université d'Angers.

Li, M. (2019). *Processus d'adaptation et logiques d'acteurs face au développement du tourisme chinois en Thaïlande*. Angers: Université d'Angers.

McLaughlin, T. (2022, February 5). *Can China Ever Reopen?* The Atlantic. https://www.theatlantic.com/international/archive/2022/02/china-COVID-zero-policy-restrictions/621476/

Nazneen, S., Xu, H., Ud Din, N., & Karim, R. (2022). Perceived COVID-19 impacts and travel avoidance: Application of protection motivation theory. *Tourism Review*, 77(2), 471–483. https://doi.org/10.1108/TR-03-2021-0165

Paulis-Cook, S. (2022, May 25). VisitScotland campaign targets Chinese students in the UK - Dragon trail international. *Dragon Trail International*. https://dragontrail.com/resources/blog/scotland-is-calling-2021

Peng, X. (2022, May 29). China's population is about to shrink for the first time since the great famine struck 60 years ago. Here's what it means for the world. *The Conversation*. http://theconversation.com/chinas-population-is-about-to-shrink-for-the-first-time-since-the-great-famine-struck-60-years-ago-heres-what-it-means-for-the-world-176377

Polyzos, S., Samitas, A., & Spyridou, A. E. (2021). Tourism demand and the COVID-19 pandemic: An LSTM approach. *Tourism Recreation Research*, *46*(2), 175–187. https://doi.org/10.1080/02508281.2020.1777053

Taunay, B. (2013). The increasing mobility of Chinese repeat visitors to France. *Tourism Planning& Development*, *10*(2), 205–216.

Tse, T. S. M. (2011). China's Outbound Tourism as a Way of Ordering: 以秩序整理模式分析中国出境旅游. *Journal of China Tourism Research*, *7*(4), 490–505. https://doi.org/10.1080/19388160.2011.627031

UNWTO. (n.d.). *UNWTO Tourism Data Dashboard*. Consulté 12 mars 2022, à l'adresse https://www.unwto.org/country-profile-outbound-tourism

UNWTO. (2013). *China—The New Number One Tourism Source Market in the World\UNWTO*. https://www.unwto.org/archive/global/press-release/2013-04-04/china-new-number-one-tourism-source-market-world

Wang, C., Meng, X., Siriwardana, M., & Pham, T. (2022). The impact of COVID-19 on the Chinese tourism industry. *Tourism Economics*, *28*(1), 131–152. https://doi.org/10.1177/13548166211041209

Wang, M., Jin, Z., Fan, S., Ju, X., & Xiao, X. (2021). Chinese residents' preferences and consuming intentions for hotels after COVID-19 pandemic: A theory of planned behaviour approach. *Anatolia*, *32*(1), 132–135. https://doi.org/10.1080/13032917.2020.1795894

Wen, J., Kozak, M., Yang, S., & Liu, F. (2021a). COVID-19: Potential effects on Chinese citizens' lifestyle and travel. *Tourism Review*, *76*(1), 74–87. https://doi.org/10.1108/TR-03-2020-0110

Wen, J., Wang, C. C., & Kozak, M. (2021b). Post-COVID-19 Chinese domestic tourism market recovery: Potential influence of traditional Chinese medicine on tourist behaviour. *Anatolia*, *32*(1), 121–125. https://doi.org/10.1080/13032917.2020.1768335

Zheng, Y., Goh, E., & Wen, J. (2020). The effects of misleading media reports about COVID-19 on Chinese tourists' mental health: A perspective article. *Anatolia*, *31*(2), 337–340. https://doi.org/10.1080/13032917.2020.1747208

Zhong, L., Sun, S., Law, R., Li, X., & Yang, L. (2022). Perception, reaction, and future development of the influence of COVID-19 on the hospitality and tourism industry in China. *International Journal of Environmental Research and Public Health*, *19*(2), 991. https://doi.org/10.3390/ijerph19020991

Zhu, J. (Jason), Siriphon, A., Airey, D., & Mei-Lan, J. (2021). Chinese tourism diplomacy: A Chinese–style modernity review. *Anatolia*, 1–14. https://doi.org/10.1080/13032917.2021.1978515

# 25   All-inclusive holiday packages in the post-COVID-19 era

*Lemonia (Lenia) Papadopoulou-Kelidou and Andreas Papatheodorou*

## Introduction

The tourism market in the Mediterranean region works under oligopoly (a market shared by a small number of companies)–oligopsony (a market where there is a small number of buyers) conditions, especially in peripheral destinations, with a few mass tour operators dominating the market. TUI and Der Touristik are two of the largest mass tour operators, buying rooms in bulk, creating holiday packages, and selling them to tourists. That dual role of buying and selling, along with the limited number of mass tour operators, make them powerful over hotels since they are the intermediaries between tourists and hotels (Buhalis, 2000; Bastakis et al., 2004). Although their role is to facilitate tourist distribution and hotel performance, opportunistic behaviours have been recorded with the acquisition by mass tour operators of many hotels and airlines, that is, hotel and vertical integration, and becoming competitors rather than facilitators.

The majority of hotels in the Mediterranean region are small to medium tourism enterprises (SMTEs). They are confronted with serious economic, managerial, and marketing issues, leading to overdependence on mass tour operators and sharing their profit margins with them. Direct bookings remain at a low level compared to the bookings gained via the tour operators. Direct bookings for hotels are smaller compared to the number of bookings from distribution channels, that is, tour operators' packages including flight, accommodation, and other amenities. The above-mentioned issues of hotels prevent them from reducing their overdependence on tour operators and from increasing the number of direct bookings.

## European tourism market and tourist preferences before COVID-19

The collapse of Thomas Cook one of the largest tour operators in the European area was the first shock in the EU tourism market structure. It further shrank the number of mass tour operators and reinforced the oligopoly–oligopsony aspect of the market. Before analysing the collapse and its effect on the market it is worth looking closely at some other mass tour operators to understand their market power compared to SMTEs' market power.

Large European tour operators, count under their control a significant number of tour operators, travel agencies, hotels, airlines, and aircraft, which serve a millions of travellers per year and recording billions of annual sales. The number of passengers/tourists being served by these large tour operators, the amount of sales per annum, and the realised horizontal and vertical integration draw attention to research on the role of these operators in the tourism industry and more specifically on their relationship with hotels.

DOI: 10.4324/9781003291763-28

*Figure 25.1* Vertical integration by TUI.

The TUI Group, a German tour operator, is one of the largest and most powerful mass tour operators in the world tourism market, reporting 1,600 travel agencies and online portals; clustered tour operators like TUI Deutschland; 150 aircraft; airlines like TUI Airways, TUI Fly, and TUI Belgium; 380 hotels; and six luxury cruise vessels. The recorded sales for 2018 were €19.5 billion; and TUI Group is listed on the London and Frankfurt stock exchange markets. The TUI policyLOUNGE web portal is a gate for accessing and communicating with the TUI Group for political representatives regarding tourism policies. The horizontal (travel agencies and clustered tour operators) and vertical integration (hotels, airlines, cruise vessels) of the TUI Group (Figure 25.1) along with its access to high level political decision making processes is evidently led by the target of enlarging its tourism market share.

The Der Touristik tour operator originated in 1917 as Der Deutsches in Berlin, equally shared between Norddentsche Lloyd and Hamburg America Line. In 1980, Lufthansa bought a good share of the company. In 1983, the company was renamed Dertour. In 2006, Dertour expanded its business under the brand "Dertour deluxe" and initiated a collaboration with FCm Travel Solutions, the Australian Company, and Austrian tour operators. In 2012, Dertour entered the gay market under the tourism product "Gay Travel" and acquired the Czech tour operator Exim Tours. Counting more than 60 companies in 2014, Der Touristik acquired the large European tour operator Kuoni, counting 28,500 employees, 48 hotels, 16 tour operators, 2,400 travel agencies, one airline, and 7.1 million passengers/ customers per year. In 2017 it reported €6.5 billion, indicating that Der Touristik is a big player in the tourism industry.

Thomas Cook's success story started in 1841 when the Baptist preacher used his social awareness, regarding alcohol, to persuade the Midland Railway Company to dispose a train for a shilling per head for the transfer of temperance supporters to a temperance meeting 12 miles away. The success of that trip was followed by a three-year period of organised rail trips at almost no cost. In 1845, Thomas Cook realised his first rail trip to Liverpool at low prices (15 shillings for first-class passengers and 10 shillings for second-class passengers). That was the beginning of a profitable business providing various railway routes across

Britain that were expanded in 1855 to other European destinations, enabling Thomas Cook to offer the first holiday package for non-UK destinations. In 1863, the collaboration with Lyons and the Mediterranean Railway assured for Thomas Cook the ability to issue tickets in English and French for the Paris to Alps route. The very first working-class customers were gradually replaced by middle-class customers demanding better quality accommodation, clearing the ground for the partnership with hotels in 1874. In 1869, two steamers were hired to expand trips to North America, China, and India. In 1875, cruise trips officially took place. In 1902, Thomas Cook's newspaper, named the *Excursionist*, was changed to the magazine *Traveller's Gazette*. In 1919, the promotion of holiday air trips was the beginning of a new area for Thomas Cook. The first holiday package including flights and hotel accommodation took place in 1927 for six people travelling from New York to Chicago while the first chartered aircraft appeared in 1939. In 1981, a private view data platform enabled Thomas Cook's customers to get access to the reservation system. In 1995, the website was established, whereas in 1996 horizontal integration led to the acquisition of Sunworld and Time Off (short-haul and EU tour operators). In 1998, the Flying Colours Leisure Group was acquired and in 1999 a merger between Thomas Cook and Carlson Leisure Group successfully took place. The JMC was created in 1999 and included all the above-mentioned acquired companies and which emerged as the strongest tour operator and airline business. In 2003, Thomas Cook Airlines was a fact. In 2016, Thomas Cook began Casa Cook, one of its hotel brand names for independent travellers.

Thomas Cook was one of the largest tour operators in the tourism market, reporting sales of £9 billion in 2018, 190 own brand hotels, 93 aircraft, 20 million customers, 22,000 employees in 17 countries, 1.7 million downloads of mobile apps, and targeting profitable growth. Its economic collapse in 2018 changed the market structure in Europe, squeezing the number of tour operators, distributing its market share with them, and consequently boosting their market power. It is worth analysing the reasons for the collapse and its effect on the tourism market.

Thomas Cook collapsed after 178 years in business, raising the question of how this happened. Many factors led to this historical collapse. The beginning of the disaster was the decision to merge with MyTravel, which was reporting losses of £1.5 billion and suffering serious economic issues – it had not recorded any profits since 2001 (https://www.the guardian.com/business/2019/sep/23/thomas-cook-as-the-world-turned-the-sun-ceased-to-shine-on-venerable-tour-operator). The merger was the beginning of the collapse, since Thomas Cook absorbed the huge debt of MyTravel; but this was not the only factor that led to this result. The failure of Thomas Cook to appreciate the change in travelling and holidays was another crucial factor that accelerated the collapse. City break holidays were replacing beach holidays, and the way of booking holidays had also changed, since most people had started choosing online booking rather than visiting a travel agency and buying a holiday package. These factors were combined with two other changes in the tourism market: the rising popularity of Low Cost Carriers (LCCs) and the emergence of the sharing economy, for example, Airbnb and Uber. Independent travellers started booking online low cost flights, Airbnb apartments, or rooms in hotels at convenient prices. Holiday packages and the traditional booking process only applied to tourists aged 60 years and above, who were not familiar with technology. The failure of Thomas Cook to detect that change in tourism and to adapt itself to the new trends in the market, in combination with a bad merger, led to its collapse. A big share of the company was absorbed by the Chinese group of companies Fosun after having acquired France's Club Med and Canada's Cirque de Soleil.

Tourists' changes in preferences and priorities were recorded before Thomas Cook collapsed in 2018. This was followed by the COVID-19 outbreak in 2019 that reinforced and reshaped again the way of travelling and travellers' needs and decision making processes.

## The EU tourism market under the COVID-19 crisis

Thomas Cook's collapse in 2018 was only the beginning of the subsequent crises. The COVID-19 outbreak has been established as the biggest disease after the Spanish flu in 1918 and which was spread in three waves, recording many deaths, with India reaching 12.5 million (https://www.britannica.com/event/influenza-pandemic-of-1918-1919). The outbreak in 2019 attacked countries globally, rich and poor, led to loss of human lives, spread fear about the following day, and affected the economy, society, development, politics, trade, and of course tourism. The dominant feeling of fear for health and whether humanity could ever reshape its life back to normality, as people used to perceive normality, left no space for thoughts about travelling or tourism for a long period of time. The three waves of the disease led to sequential national lockdowns and the closing of countries' borders, resulting in destructive consequences for tourism and travelling. Airports were closed; airlines grounded all aircraft, suffering million dollar losses; hotels were closed; and people wondered what the new normality would look like if they were lucky enough to stay alive. For mass tour operators things were not better. Liquidity issues emerged as well as economic losses, since people that had already bought their holiday packages were unable and unwilling to travel, and therefore tour operators had to refund them. The powerful position of tour operators in the market was also shocked, though without changing the ultimate power imbalance between them and SMTEs. Tourism and travelling would never be the same.

After the last wave and the last lockdown in the broader EU area, measures for travelling were the next obstacle for tourists. Even though borders opened and travelling was allowed again, rapid tests, quarantines, and then vaccination certificates were only some of the measures that were applied by most countries. Policies were different from country to country, while some prohibited temporarily outbound tourism from countries with a high number of infected COVID-19 people. Tourism, politics, and international relations played a key role during that period. Having to go through rapid tests or quarantine was an obstacle for people to travel and to choose a tourism destination. Destinations with looser measures could attract more people and could restart tourism much easier than destinations with tough health measures in airports and when entering a country.

The pandemic reshaped tourists' perceptions about notions like quality, safety, wellbeing, social distancing, value of time, and experiences. The fear of death and the loss of beloved relatives and friends from COVID-19 led to the reordering of values and criteria when taking leisure holidays.

## Tourist preferences in the aftermath of COVID-19 and the European market response

Before the COVID-19 outbreak tourists' preferences and decision making used to be based mainly on financial criteria. Research had emerged that the cost of holidays was the main factor affecting tourists' choices for destinations, holiday packages, and hotels. Price offers, flash sales, and early bird discounts also used to affect tourism demand. Mass tourism clearly dominated in areas like Barcelona, Malta, and Venice, along with holiday packages created mainly by mass tour operators (Chilembwe et al., 2019). Hotels' marketing campaigns were also based on promoting early bookings, flash sales, and special offers. Holiday costs can be divided into two main parts: travelling and accommodation (Boto-Garcia et al., 2020). Holidays with cheap flights and cheap accommodation (Picazo & Moreno-Gil, 2018) affected tourists' decisions regarding leisure holidays (Pan & Truong, 2018). Wellbeing, authentic experiences, risky sports, and other factors were also determinants of tourist choices. Nevertheless, for the middle classes the price used to be the "king" factor, whereas for upper-class tourists luxury remained the main criterion.

The pandemic made the rich and poor suffer. Good health, safety, and wellbeing replaced the price factor in tourists' decision-making processes. The threat to human life changed tourists' perceptions along with the notion of normality. Poor and rich people had the same priority: health and wellbeing. Social distancing was the new term that was absorbed into the tourism industry, since tourists' needed to keep a distance from others when having holidays or avoided mass tourism services. Prices did not count that much anymore and revenge travelling after two years of lockdowns meant that people wanted their life and wellbeing back, even at high prices. The term "revenge travelling" was coined during the pandemic, and it reflects the willingness of people to travel again after two years of strict measures and to get their lives back (Wang & Xia, 2021).

Safety was another main factor that affected tourists' decision making for leisure tourism (Ghose & Johann, 2018). Being safe was redefined to mean COVID-free, or to have a very low risk of being infected when on holiday. Culture, the environment, political conditions (Lin et al., 2018; Demir, 2021), travelling measures against COVID-19 (Shubtsova et al., 2020), and the vaccination levels of a tourism destination also emerged as key determinants of tourist choices and decision making (Yin & Ni, 2021).

Independent travelling was reinforced during the crisis. National governments imposed health measures and certain health certificates and protocols by law to all accommodation units, restaurants, airlines, airports, transportation, museums, and cafes. Therefore, tourists could also be safe when travelling independently, without having the need to follow a mass tour operator's holiday package for health protection. The new travelling trend emerging before the collapse of Thomas Cook was reinforced by the COVID-19 crisis; as it added more reasons for independent travel, for online bookings, for avoiding mass tourism, and for looking for authentic experiences and wellbeing under social distancing.

The question raised concerns how well-prepared mass tour operators were so as to adjust themselves to these new trends and emerging tourist needs. Had they taken into account the reasons why Thomas Cook collapsed in 2018 and the changed normality in tourism and in travelling needs?

**The way forward**

The post-COVID-19 era is a period that aggregates the change in tourists' preferences and way of travelling that began in 2007, and the changes in tourists' preferences and criteria for decision making. This clearly covers the period 2007–2021.

Since 2007 tourists started preferring online bookings and taking advantage of the LCCs and the sharing economy, hence building up the independent traveller. At the same time alternative types of tourism emerged, like wine tourism, dark tourism, sport tourism, religious tourism, gastronomy tourism, and wellbeing tourism, indicating that tourists had started looking for new experiences and authenticity blended with local culture. Tourists do not buy traditional holiday packages so much; instead, they book online and create their own customised holiday packages. Mass tour operators' power has been challenged by changes in tourist preferences. Hotels started to have more direct online bookings and more potential to offer customised rather than mass services.

The inability of one of the largest mass EU tour operators, Thomas Cook, to detect these trends and to adjust itself to them, led to one of the most noisy and destructive collapses in the EU tourism market. The pandemic reinforced tourists' changes in preferences and ways of travelling, such as contactless and customised services, independent travel, wellbeing services, safety, and authenticity.

Perceptions of safety, quality, and value for money have also changed. Safety has been defined as being healthy, quality has been linked with isolation and customised services, while value for money has been attributed to wellbeing services that are also consistent with safety and quality (Kim et al., 2021a; Sie et al., 2021). These perceptions are subjective and cannot be defined via stereotypes (Li et al., 2021). Nostalgia for travel and tourists' intentions of revenge travelling, after sequential lockdowns, combined with wellbeing, are the new trends (Sie et al., 2021). Tourists are looking for authentic and memorable experiences. Living the unforgettable, searching for authenticity in every single detail, is the new path that guides tourists' choices in a subjective way. Prices and economic holiday packages are not anymore the main criteria for choosing destinations or types of holidays, without meaning that budget constraints are not taken into account. Tourists are looking for holidays that will give them value for money within their budget frames.

Value for time has also emerged as a key factor, and the term "time rebound effect" (TRE) reflects the value of time for busy explorers and the effect of technology in saving time for travelling and extending the length of stay (Kim et al., 2021b). Technology saves time when searching for information, when booking, and when travelling, since some processes are addressed in a shorter period of time digitally than manually (e.g., check-in processes in hotels or airports). This in turn means that people save time that can be spent on their holidays instead, which in turn means extra money spent during holidays and extra income for tourism enterprises.

Although mergers and acquisitions have taken place as a result of the economic consequences of the pandemic, the market structure has not changed overall, that is, it remain oligopsonistic–onligopolistic (Liulov et al., 2020). Some of the most important and largest mergers and acquisitions for 2021 were: Hyatt's acquisition of Apple Leisure Group for $2.7 billion, the acquisition of Apollo by the USA-based Hertz for $4.7 billion, and the acquisition of Booking Holidays by the Swedish flight-tech Etraveli Group for $1.8 billion. Regarding mass tour operators, the economic collapse of Thomas Cook reinforced the market power of the remaining EU tour operators without changing the overall balance in the EU tourism market.

The way forward could be the adaptation of the market to tourists' needs. Hotels in the European market could seize the chance to offer higher value for money for their goods and services, to customise their services, to give emphasis to wellbeing services, and to provide authenticity and memorable experiences. Tourists' preferences are moving towards direct communication with hotels, and skipping travel agencies and mass tourism services. Hotels can exploit this opportunity, by customising services, giving value for money and high quality along with authenticity.

## References

Bastakis, C., Buhalis, D., and Butler, R. (2004). The perception of small and medium sized tourism accommodation providers on the impacts of the tour operators' power in Eastern Mediterranean. *Tourism Management*, 25(2), 151–170. https://doi.org/10.1016/S0261-5177(03)00098-0

Boto-García, D., Mariel, P., Pino, J. B., & Alvarez, A. (2020). Tourists' willingness to pay for holiday trip characteristics: A discrete choice experiment. *Tourism Economics*, 28, 349–370.

Buhalis, D., (2000). Relationships in the distribution channel of tourism: Conflicts between hoteliers and tour operators in the Mediterranean region. *International Hospitality, Leisure and Tourism Administration Journal*, 1(1), 113–139. https://doi.org/10.1300/J149v01n01_07

Chilembwe, J. M., Mweiwa, V. R., & Mankhomwa, E. (2019). The Role of Tour Operators in Destination Tourism Marketing in Malawi. In M.E. Camilleri (Ed.), *Strategic Perspectives in Destination Marketing* (pp. 295–321). Hershey, PA: IGI Global.

Demir, Ş. Ş. (2021). The Effect of COVID-19 Measures in Hotels on Tourists' Perceptions of Safe Tourism Service. In M. Demir, A. Delgic, & F. Ergen (Eds.), *Handbook of Research on the Impacts and Implications of COVID-19 on the Tourism Industry* (pp. 372–392). Hershey, PA: IGI Global.

Ghose, S., & Johann, M. (2018). Measuring tourist satisfaction with destination attributes. *Journal of Management and Financial Sciences*, 34, 9–22.

Kim, J. J., Han, H., & Ariza-Montes, A. (2021a). The impact of hotel attributes, wellbeing perception, and attitudes on brand loyalty: Examining the moderating role of COVID-19 pandemic. *Journal of Retailing and Consumer Services*, 62, 102634.

Kim, S., Filimonau, V., & Dickinson, J. E. (2021b). Tourist perception of the value of time on holidays: Implications for the time use rebound effect and sustainable travel practice. *Journal of Travel Research*, 62, 362–381.

Li, Y., Song, H., & Guo, R. (2021). A study on the causal process of virtual reality tourism and its attributes in terms of their effects on subjective wellbeing during COVID-19. *International Journal of Environmental Research and Public Health*, 18(3), 1019.

Lin, L. P. L., Yu, C. Y., & Chang, F. C. (2018). Determinants of CSER practices for reducing greenhouse gas emissions: From the perspectives of administrative managers in tour operators. *Tourism Management*, 64, 1–12.

Liulov, O. V., Us, Y. O., Pimonenko, T. V., Kvilinskyi, O. S., Vasylieva, T. A., Dalevska, N., … & Boiko, V. (2020). The link between economic growth and tourism: COVID-19 impact. *Proceedings of the 36th International Business Information Management Association (IBIMA)*, Granada, Spain, 4–5 November, pp. 8070–8086.

Pan, J. Y., & Truong, D. (2018). Passengers' intentions to use low-cost carriers: An extended theory of planned behavior model. *Journal of Air Transport Management*, 69, 38–48.

Picazo, P., & Moreno-Gil, S. (2018). Tour operators' marketing strategies and their impact on prices of sun and beach package holidays. *Journal of Hospitality and Tourism Management*, 35, 17–28.

Sie, L., Pegg, S., & Phelan, K. V. (2021). Senior tourists' self-determined motivations, tour preferences, memorable experiences and subjective wellbeing: An integrative hierarchical model. *Journal of Hospitality and Tourism Management*, 47, 237–251.

Shubtsova, L. V., Kostromina, E. A., Chelyapina, O. I., Grigorieva, N. A., & Trifonov, P. V. (2020). Supporting the tourism industry in the context of the coronavirus pandemic and economic crisis: Social tourism and public-private partnership. *Journal of Environmental Management & Tourism*, 11(6), 1427–1434.

Wang, J., & Xia, L. (2021). Revenge travel: Nostalgia and desire for leisure travel post COVID-19. *Journal of Travel & Tourism Marketing*, 38(9), 935–955.

Yin, J., & Ni, Y. (2021). COVID-19 event strength, psychological safety, and avoidance coping behaviors for employees in the tourism industry. *Journal of Hospitality and Tourism Management*, 47, 431–442. https://skift.com/2021/12/27/10-largest-2021-travel-acquisitions-set-to-reconfigure-industry/ (12-03-2022)

# Part IV
# Technology

Lixian Zhou and Jing Li review *Tourist-generated short videos arousing aspirations to visit a place: Perceiving authenticity?* in Chapter 26. Tourist-generated short videos (TGSVs) have increasingly contributed to the popularity of destinations among Chinese people, with rapidly rising number of them making and watching such videos. Core characteristics of TGSVs include being short in time extent and amusement-oriented. Watching TGSVs makes audiences feel that the place and travel style in the videos are close to their life. Therefore, it is hypothesised that the videos are perceived as trustworthy and prompt aspirations to travel. The chapter explores the capacity of TGSVs to improve travel intentions through studying the authenticity perception and multi-dimensional travel motivations. Mixed methods were employed. Firstly, three authenticity-relevant themes – authentic video content, authentic producer, and authentic self – emerged from a content analysis of 186 reviews of TGSVs. Secondly, a scale measuring the perceived authenticity among video audiences was developed as a tool for further study, based on 155 valid online surveys. Thirdly, quasi-experiments involving 32 participants were conducted to verify the impacts of the short videos. Higher levels of authenticity perception, travel motivations, and intention to visit Iceland were manifested among participants watching a TGSV, compared with those watching an official destination marketing video of the same length. The findings facilitate a better understanding of the ways in which convenient channels available on a mobile phone are influencing Chinese tourist behaviours. This research is significant for the online marketing strategies of destination management organisations (DMOs).

Chapter 27 is on *Travel vlogging and its role in destination marketing* by **Maria Criselda Badilla, and Carl Francis Castro**. In recent years, the highly interactive nature of social media has given tourists a place to dream, plan, book, experience, and share travel. Travel vlogging has become one of the powerful ways for tourists to discover places, form images, and create expectations. Travel vlogs emerged as shared through social media, involving many tourism aspects such as travel activities, accommodation, food, and adventures. With the advancement of online video sites (e.g., YouTube), ordinary people are empowered to manage their self-making videos more efficiently, giving rise to the popularity of many professional and amateur vloggers. Given the enormous marketing possibilities of travel vlogs, analysing their content helps marketers understand their roles in shaping destination images and influencing travel behaviour. Through a content analysis of the narratives contained in the vlogs of foreign tourists stranded in the Philippines during the early part of the COVID-19 pandemic, five dominant themes relevant to the destination image emerged, namely incomparable sites, Filipino hospitality, unique food experience, Filipino qualities, and the Philippines as home. The themes showed how stranded foreign vloggers gained a

DOI: 10.4324/9781003291763-29

positive view of the Philippines and Filipinos despite the erstwhile negative experience of being stranded.

**Evrim Çeltek** reviews *Augmented reality and virtual reality in tourism* in Chapter 28. This chapter aims to give details about augmented reality (AR) and virtual reality (VR) application trends in the tourism industry. AR and VR technologies and types, tourism industry applications, the advantages and challenges of these technologies in different areas of tourism, and predictions for the future are described in this chapter. This chapter is an effective source for those who want to learn about tourism AR and VR applications and their advantages and challenges.

*Tourist engagement throughout the customer journey: A service ecosystem approach* is the focus for Chapter 29 by **Rodoula H. Tsiotsou** and **Ronald E. Goldsmith**. Tourists are increasingly becoming active consumers of tourism offerings, engaging throughout the tourism journey. Although engagement is an important concept in tourism marketing, the literature is replete with nomological approaches investigating its antecedents and outcomes. Thus, there is no available study in delineating how tourist engagement (TE) may co-create/destroy value at each stage of the consumption process to advance our knowledge on the topic. Therefore, this chapter aims to explain TE throughout the consumption process of tourism offerings. To achieve this goal, the chapter proposes an integrative conceptual framework that delineates TE in the three stages of the customer journey and at all levels of the service ecosystem. The chapter contributes to the literature by advancing our knowledge of tourist engagement and its role in value co-creation/destruction. Moreover, it identifies research voids and provides valuable future research recommendations.

**Vera Antunes, Gisela Gonçalves**, and **Cristina Estevão** in Chapter 30 provide *Perspectives for communication in social media: The case of thermal spas*. This chapter analyses the comments on social networks and the opportunities that may arise when looking at the future of communication. The results come from a content analysis of the social networks of Termas de Chaves and Termas de São Pedro do Sul located in Portugal, where it is identified that people are more motivated to experience thermal services and products in a fast-changing world. This situation needs rethinking, as we look to the future of thermalism and communication.

Chapter 31 discusses *Urban mobility and mobility-as-a-service (MaaS) trends* and is authored by **Xu Zhao, Claire Papaix**, and **Yufang Zhou**. The growth of tourism has increased the need for visitors to negotiate foreign languages, unknown transport systems, complex networks, asynchronous payment methods, and different currencies in tourism destinations, which has added to travel stress. This chapter explores the use of Mobility-as-a-Service (MaaS), a digital innovation designed to manage urban mobility trips, as a solution to improve tourists wellbeing. Beijing MaaS was developed in 2019 in China, alongside the city's carbon trade incentive approach, to contribute to the city's 2030 carbon-neutral target and enhance visitors' wellbeing and levels of satisfaction related to travel. This chapter presents users' attitudes towards Beijing MaaS in 2020 and 2021. It explores tourists' concerns, needs, and values towards tourism and MaaS in this particular case study, compared to other places in the world. It is concluded that MaaS has considerable potential to improve tourists' wellbeing through providing instant and accessible travel information in a seamless, convenient, and user-centric mobility service.

**Muhammet Necati Çelik** in Chapter 32 provides *Digital nomads and destination characteristics: A conceptual analysis*. Digital nomadism is a fast-growing phenomenon all over the world. The current situation and historical development of this phenomenon are discussed in order to clarify the concept. The research topics related to digital nomads are also examined by the research. Because of the digital nomadism movement, digital nomad destinations have been emerging. It is necessary to determine the characteristics of the most appropriate digital nomad destination with the purpose of contributing to the relevant literature. A wide range of information resources including 23 scientific publications, 22 web platforms and 3 reports were examined. The needs and motivations of digital nomads were determined and qualifications were identified for preferring a destination as a digital nomad. A model called the UDDN-Model was developed in order to present convenient characteristics for digital nomad destinations. The developed model is expected to be a pathfinder for destinations and a guide for digital nomads.

Chapter 33 features *Gamification in museum tourism* by **İge Pirnar, Duygu Çelebi**, and **Muruvvet Deniz Sezer**. Gamification is a newly emerged concept which has many uses in a wide variety of fields. Tourism as an industry is one of those areas in which gamification practices are gaining popularity. The purpose of gamification in tourism is to engage tourists in numerous activities and tasks to enhance their touristic experience by utilising gaming dimensions. In recent years, museum tourism has become a crucial motivational factor for touristic movement of all kinds of tourists. The rising popularity of gamification in the tourism industry is also reflected in museum tourism. The main objective of this chapter is to provide deep insight into how gamification practices can be used to create a demand for museum tourism by creating real experiences. This chapter examines existing studies on this topic in the literature in detail and compiles different applications of gamification practices in museum tourism.

In Chapter 34, **Jonatan Gómez Punzón** and **Nuria Recuero-Virto** discuss *The influence of museum user generated content to improve the experience design*. Museums have shifted to a new hybrid cultural venue format, open to be enjoyed through multiple channels and multiple platforms, spreading the voice of art all around the world. Museums are challenged to draft new ways to interact with potential visitors. Technology has built a new framework of opportunities for museums to engage with clients and art and culture lovers, allowing these consumers to be part of the museum experience from any place. This chapter seeks to evaluate through a conceptual analysis the use of #musetech or #musesocial as a tool to define the new technological adoptions of potential visitors, matching the digital needs and measuring the level and depth of adoption of these museums' disruptive technologies. An initial exploratory analysis is offered to determine the current trending topics reframing service design.

*Presence in virtual hotel experience and purchase intention: The mediating role of decision comfort* is the focus of Chapter 35 by **Sima Rahimizhian, Farzad Safaeimanesh, Mobina Beheshti**, and **Olayinka Afolabi**. In the hospitality industry, the emerging virtual reality (VR) technology is increasingly aligned with various opportunities. Although VR as a marketing tool has the potential to transport consumers to virtually experience a hotel or tourist destination, few studies have been conducted on the factors that convince customers to make an online purchase or booking using VR experience. This chapter investigates whether the perceived VR presence (PVP) and the decision comfort (DC) of users positively affect their purchase intentions for online hotel bookings by applying the stimulus-organism-response theory. The results of this research imply that the sense of presence in the VR experience aided by head-mounted display (HMD) has a positive and significant impact on online purchase intention towards the hotel highlighted in the virtual experience. PVP also had a positive influence on DC. VR is an opportunity for immersive marketing that provides value through personalised content and shapes marketing strategies to deliver more engaging experiences for consumers. The theoretical and managerial implications, as well as recommendations for future research, are further discussed.

**Nabila Norizan and Norhazliza Halim** author *Co-designing the smart tourism experience for all-inclusive hotels as a new trend in staycation experience* in Chapter 36. Emerging trends of smart tourism theories primarily focus on how information and communications technology and big data influence marketing, product, and destination development. However, an overt technical approach to formal outcomes, including product, services, and technology, diminishes the concern about social dimensions and practices of those involved. This chapter explores theoretical reviews and elaborates a few selected case studies of co-designing as an approach to evoke smart tourism experience through the all-inclusive hotel staycation experience. Through the integration of Service-Dominant Logic and the Resource-Based View Theory Model, co-designing moves from designing for people to co-designing with people. The chapter highlights the co-designing between hotel management and guests staying in the hotel. This approach is discussed in terms of stimulating a creative and "designerly mindset" in the search for latent possibilities by linking existing tourism practices in the hotel industry with technologies and big data and through new social relations between stakeholders. In smart tourism, key concepts for enhancing tourism experiences are described as value-adding experiences and value co-creation mediated by technology, smart devices, and real-time data. A central argument and contribution is made in theorising that the potential value of big data to evoke innovation is only as great as the successful interplay of sensemaking and co-designs underpinning big data practices. The expected key discussion from the reviews includes that smart tourism is collaborative and its successful creation requires the initiating and sustaining of new relations in designing new experiences for tourists.

Chapter 37 poses the question *A digital safe-zone tourism network: Are we ready to travel again?* and is by **Norhazliza Halim, Nabila Norizan**, and **Thinaranjeney Thirumoorthi**. As 90% of the global population adjusted to life under travel restrictions and others stayed home in fear of the virus itself, the travel and tourism sector came to a near-total standstill. Traveller preferences and behaviours have shifted toward the familiar, more predictable, and trusted journey. Transparent communication will be even more important to travellers in spurring demand. Health and safety are paramount in this new era. Personal experiences, advice from experts, and concerns for distancing will guide consumer behaviour in the short to mid-term. The lack of real-time determination of tourism safe areas hence calls for the need to identify tourism safe zoning through the implementation of a Safe-Zone Travelling Network. Businesses will have to collaborate even more closely with their extended value chains to ensure readiness and the implementation of like-minded protocols, such as safe travel protocols, standard operating procedures, and health certificates. In this chapter, the travel and tourism sector's quest for innovation and the integration of new technologies such as the Smart Health Travel Certificate as well as the Digital Safe-Travel Zoning Corridor Network are discussed and an integrative framework is proposed to facilitate travellers' decision making regarding travel, to enhance tourism-related business preparation, and to achieve government policy acceptance, especially for cross-border travelling. The aim is for destinations to successfully offer a virus-free environment, aid the international readiness of tourism destinations for the new norm, and transform government policy support into creating new tourism hotspots. Consequently, the integrative framework guideline functions as a catalyst of innovation and a tool to build tourism resilience.

**Lina Zhong** and **Mengyao Zhu** in Chapter 38 discuss *Reconstructing tourism development in China: The role of the Internet industry*. Information and communication technology has integrated with all industries closely and rapidly. In China, the tourism industry has experienced a high-level digitalisation exploration. With big data, cloud computing, artificial intelligence, VR, AR, and other digital tools, industrial digitalisation is powering the tourism industry toward sustainable and intelligent production. The Chinese government and Internet platform companies actively embraced global digitalisation opportunities to supply high-quality public services. Based on the value of co-creation, these companies used technical superiorities to aid governance and tourism market supervision. At the same time, this action could also help themselves transform from the Consumer Internet to Industrial Internet services. In September 2017, IDC Government Insights rated an innovative travel platform named GO-Yunnan as the best Asia-Pacific Smart City project in the "Economic, Tourism, Art, Library, Culture and Public Space". This smart travel platform enabled the connection between online and offline public services from the government to merchants, representing the current changes in China's tourism industry. Therefore, the authors choose the GO-Yunnan smart travel platform as an example for discussing the Internet industry's role in reconstructing the Chinese tourism industry. Three trends are summarised related to the industry, tourists, and government.

# 26 Tourist-generated short videos arousing aspirations to visit a place

## Perceiving authenticity?

*Lixian Zhou and Jing Li*

### Introduction

Convenient channels that millions of Chinese access via mobile phones anytime and anywhere have been competitive marketing venues for various industries, including tourism. The average time spent on making travel decisions has shrunk among the Chinese. More decisions are made based on fragmented information or just an instant impression of a place. In particular, flipping through short music videos has become a habit of most Chinese as a way of filling gaps in the day. It is not uncommon among young Chinese to go somewhere on a-spur-of-the-moment decision made when glancing at a place on short music videos. Therefore, making use of short music videos has been taken into account in destination marketing strategies by destination management organisations (DMOs).

There are two main types of short music videos that encourage audiences to visit a place, including official promotional videos and documentary videos generated by tourists for non-profit purposes. Tourists shoot and post short videos on social media to share their experience, build social identity, or just record moments in their lives. Tourist-generated short videos (TGSVs) are more likely to focus on personal perception, storytelling, and distinctive styles, such as humour, while official videos tend to demonstrate the advantages of a place in a general way. It is arguable whether TGSVs are more persuasive than official promotions, considering the short time involved in watching a video.

Audiences prefer short music videos produced by ordinary people and tend to trust independent vloggers because such user-generated content (UGC) makes them perceive a similarity between their own lives and the ones shown. Therefore, this chapter investigates TGSVs' effects on travel intentions from the perspective of authentic perception. The goal is to verify the hypothesis that TGSVs are more likely to arouse audiences' aspirations to travel to a destination than short music videos produced by official DMOs.

Young Chinese born after 1990 who make up the biggest segment of the tourist market and the short music video audience were selected. TGSVs' effects are explored by answering three questions. Firstly, do potential tourists perceive higher levels of authenticity in TGSVs than in official promotional videos? Secondly, do TGSVs exert greater impact on potential travel motivations? Thirdly, are TGSVs more powerful in producing tourist intentions to visit places than official videos?

DOI: 10.4324/9781003291763-30

## Literature review

### *Destination marketing via new media*

Digitalisation has characterised the recent development of the marketing strategies of various organisations, including tourism enterprises and destinations. Internet marketing is not only a means for spreading information, but also a contemporary way for providers to relate to (potential) customers. The openness of online media and the advancement of information and communication technologies (ICTs) allow for comprehensive brand engagement, which emphasises dynamic people-to-brand connections. The influence of key opinion leaders (KOLs) and hyper-social relationships in people's consumption behaviours were determined in an investigation of 409 followers of two Instagram accounts (Farivar, Wang, & Yuan, 2020). However, the potential of Internet marketing has not always been fully realised. In Shaltoni's (2017) research, Internet marketing was being used as a one-way communication means rather than as a networking platform by half of the investigated organisations. Such a discovery should prompt further research on Internet marketing that would facilitate improvements in such marketing strategies.

Marketing tactics incorporating social media have exerted tremendous impacts on customers' decision-making and consumption behaviours in emerging markets. Internet branding has been widely utilised to build destination brands and enhance potential tourists' engagement with the brand before on-site encounters (Barna & Semak, 2020). A diversity of content producers, multiple information sources, and interactions in virtual communities are features of marketing strategies on social media. Such comprehensive engagement effectively improves the perceived place attachment and motivates people to visit places.

Short music videos are one of the newest forms of social media that have contributed to destination marketing. A large number of places rose to fame overnight and became popular among tourists owing to their exposure in popular short videos, although little research has stressed the power of short videos in marketing and there is not yet a sound conceptual knowledge of why and how short videos influence tourist decision making (Gibbs, 2007). Case studies of successful marketing campaigns dominate the current research of short videos in the Chinese situation. Problems of short video marketing practices have also been analysed. For example, it is alleged that some superficial "live + short video" exhibitions are detrimental to the experience of cultural tourism (Guo, 2021).

Although the marketing significance of short videos has been studied from the user perspective, these studies are oriented by value creation. Improving perceived value was found to be the most effective way to promote China as a desirable travel destination on the platform of short videos (Shani et al., 2010).

The core advantage of online destination marketing lies in the capacity to improve customer engagement with the established destination brand. However, research has been dominated by fragmented case studies from the angle of business. It is time to conceptualise how watching short videos influences tourist behaviours from the perspective of the psychological experience of the audience.

## Making destinations meaningful for tourists

A wide range of studies have analysed factors influencing the attempted outcomes of destination marketing – tourist intentions to travel. The focus of successful marketing strategies are on establishing appropriate destination images and assisting the market to recognise and perceive the nature and atmosphere of the promoted destination (Enrique Bigné, Isabel Sánchez, & Sánchez, 2001; Mackay & Fesenmaier, 1997). A destination image that conveys symbolic and cultural significance can effectively build emotional connections between people and places (Blake, 2010).

The means of conveying marketing information must serve the ultimate goal of making destinations meaningful to audiences. Technological applications are considered effective marketing tools because immersive and authentic experiences build meaningful connections between individuals and places before consumption. Individuals' environmental embeddedness, immersion, and restorative perception are enhanced by the vivid experiences of virtual reality (VR) tourism (Huang, 2021; An, Choi, & Lee, 2020). Virtual communities also contribute to the meaningfulness of the marketing (Wang & Fesenmaier, 2004).

Other experience-centred marketing tactics, such as sensory marketing and story marketing, also focus on the improvement of the perception of authenticity and people-to-place interactions. Krishna, Cian, & Sokolov (2016) advocated the using of sensory marketing to build up and spread destination images that allows perceptions of authentic destinations. The perception of authenticity helps strengthen place attachment among potential tourists and that contributes to emotional connections to the destination (Ram, Björk, & Weidenfeld, 2016; Shi, Zhang, & Ma, 2021). Story marketing, especially individual narratives, works effectively in promoting destinations because it links place images to individuals' self-images (Li, 2014; Rickly-Boyd, 2009). The perceived closeness and oneness with a place pushes individuals to visit a place for self-confirmation.

Technology-based marketing means are useful in constructing the meaningfulness of a destination's image. However, the ways in which short videos – a marketing means that has made many places famous overnight – work to boost the effect of destination marketing has not been studied. Neither the roles of perception of authenticity nor motivation, important antecedents of meaningfulness, have been figured out.

### Pursuits of young Chinese tourists

Understanding tourist motivation is significant for analysing divergent responses to marketing information (McCain & Ray, 2003). Underlying pursuits play an important role in constructing the tourist experience (Sharpley, 2005). Motivations influence how tourists recognise and perceive the destination through shaping behaviours, ranging from decision-making to post-travel sharing. Therefore, there have been numerous studies of the motivations of different categories of travellers.

In particular, younger generations tend to travel for pursuits beyond what can be explained by classic motivational theories. In addition to traditional travel pursuits, namely escaping routines and achieving psychological compensation, young travellers attach more importance on factors contributing to subjective wellbeing, such as self-actualisation and sociality (Iso-Ahola, 1982; Xing, Yang, Huang, Li, & Wang, 2018; Wang & Ma, 2020). Tourists born after 1990 rank the main pursuits motivating them to travel overseas as accommodation, making friends, and improving interpersonal relationships. Specifically, interpersonal relationships and self-actualisation are named as the most important motivations for those who cycle along the Sichuan–Tibet Road (Wang & Ma, 2020). The sense of ritual shared among members of a social community accounts for the travel motivations aroused by watching short videos.

Wang (2017) categorises the motivations of young Chinese into five groups through empirically studying a large number of Chinese born after 1990. The pursuits motivating young Chinese to travel include knowledge, self-esteem, punishment minimisation, reward maximisation, and self-actualisation. This five-dimensional framework facilitates measuring the levels of travel motivation aroused by short videos in this chapter.

### *Perception of authenticity in tourism*

The concept of authenticity is widely employed to study tourist experience in situations including, but not limited to, heritage tourism. The tourism context is regarded as a convergence of people, place, and other objects. Therefore, tourists' perception of authenticity has been conceptualised from objective, constructive, and existential perspectives (Cohen, 2008; Reisinger & Steiner, 2006). Investigations of the connection between tourism and authenticity are often focused on whether tourists feel the encounters are real, genuine, and meaningful (Moore, Buchmann, Månsson, & Fisher, 2021).

The widespread use of authenticity in branding tourism products, sites, destinations, and experiences can be understood by considering the increasing variability of objects and the concealability of people in the post-modern society (Bullingham & Vasconcelos, 2013). In particular, the sculptured destination and performed experience on the stage of new media have been suspected of deviating from the real (Hogan, 2010; Bareket-Bojmel, Moran, & Shahar, 2016). The level of authenticity perceived by audiences can predict the effectiveness of online promotion (Liu & Yan, 2021). In the field of tourism, authenticity has been considered an important driver, motive, or value for tourists (Grayson & Martinec, 2004). Therefore, perception of authenticity is employed in this chapter to study the capacity of short videos in encouraging audiences to visit a destination.

There are six levels of perceived authenticity, namely origin, true, real, sincerity, originality, and life course. Authenticity as life course has been used to represent the unexpected experiences in travel (Cohen, 2008). Travelling is viewed as an essential part of an ongoing life course which is significant for being oneself. The perception of authenticity is investigated on three dimensions, namely the place in the video, the video producer, and the awareness of being self-aroused by watching the video.

## Methodology

Mixed methods were employed to investigate the capacity of tourist-generated short videos to motivate audiences to visit a destination. Firstly, 186 textual reviews of short music videos about travelling were sampled and examined by using thematic analysis. Concepts and themes that emerged from the reviews facilitated the understanding of perceptions of the authenticity of short video audiences, particularly to explore the structure of the constructs involved. Secondly, surveys were conducted online to develop measurements for the variables. Scales measuring the levels of perceived authenticity and motivation to visit the place were achieved from an analysis of 155 valid surveys. Thirdly, experiments were organised to justify the hypothesised advantage of tourist-generated short videos over official promotion short videos in motivating people to visit a place. The hypothesised moderating role of authenticity perception was also verified.

*Figure 26.1* TikTok (*Douyin*) is a popular video-oriented social media platform in China.

### Theme analysis of video reviews

The reviews were sampled from three video-oriented social media platforms that are popular with Chinese, namely Tik Tok (Figure 26.1), Bilibili, and Little Red Book. Apart from the huge number of users and published short music videos, these media were selected as appropriate information sources because their communities are functioning well. Users are active in commenting and communicating.

The most popular videos about travelling that include only one destination in one video were targeted by checking the number of reviews of and likes to the video. The researchers read the abundant reviews – published from 1 to 31 March 2021 – of the top ten popular travel videos on each platform. Valid data were sampled based on two criteria. Firstly, the destination, the producer, and the audience's own self had to be all mentioned in the comments. Secondly, the comment had to contain one or more of the pre-set keywords relevant to authenticity. These keywords were decided based on a literature review and examination of the promotional documents of authentic products, destinations, and experiences. As a result, 186 reviews containing relatively rich information were collected to facilitate a comprehensive understanding of the perceptions of authenticity of the short video audience.

The sampled reviews were subject to analysis to explore concepts and themes that constitute the short video audience's perception of authenticity. Specifically, the researchers identified concepts from original texts, then grouped concepts with the same meaning into categories, and finally made a further abstraction from the groups into theoretically meaningful themes.

### Surveys exploring construct structure

Preliminary scales were designed based on the literature review and results from a qualitative analysis of the reviews. Specifically, five items were used to represent the most important pursuits of young Chinese tourists: the combination of escaping, compensation, socialisation, knowledge, self-esteem, punishment minimisation, reward maximisation, and self-actualisation. Thirteen items were designed to measure the perceived authenticity from the perspectives of authentic destination, genuine producer, and identified true self. All items were measured by seven-point Likert scales.

There were 155 valid surveys out of the 375 returned. Exploratory and confirmatory factor analyses were administrated to screen usable items, explore the dimensions of constructs, and test the measurements' reliability and validity. The verified instruments were used in the subsequent experiments.

### Controlled experiment

We decided to use short videos about Iceland as the study material because Iceland is relatively unfamiliar to most Chinese. Two three-minute music videos were selected; one was shot by an ordinary tourist while the other was created by the Icelandic tourism authority.

Thirty-two participants were recruited through snowball sampling, incorporating 16 males and 16 females. Firstly, they were asked three questions to verify homogenisation, in terms of convergent level of understanding, interest, and perception of Iceland before watching the videos. Then participants were randomly and averagely allocated into an experimental group and a control. The experimental group were presented with the tourist-generated short music video, while the control group watched the official promotional short video. All participants were required to finish the survey after watching the videos. The levels of perception of authenticity and the aroused motivation to visit Iceland were reported on seven-point Likert scales.

An independent sample t-test was run by employing SPSS 25.0 software to statistically compare the capacities of tourist-generated and official promotional videos in producing perceptions of authenticity and travel motivations. The advantage of TGSVs would be verified if the capacity of the official videos was considered as a baseline.

### Findings

#### Perceived authenticity of video content, producers, and self

In Table 26.1, eight groups of concepts emerged from the sampled reviews and were further abstracted into three themes. The themes of authentic video content, authentic producers, and authentic self corresponded with the three theoretical approaches to authenticity (objective, constructive, and existential).

There were three groups of homogeneous concepts constituting the theme of perceived authentic video content. These concepts were identified from 48 reviews that commented

*Table 26.1* Results of grounded theory analysis

| Theme | Groups of concepts | Number of reviews |
| --- | --- | --- |
| Authentic video content | Real and attractive tangible objects and environment | 13 |
| | Touching overall atmosphere | 28 |
| | Striking resemblance to certain scenes | 7 |
| Authentic producers | Relying on the producer to construct their own knowledge of the world | 46 |
| | The producers are trustworthy | 37 |
| Authentic self | Self-expression | 14 |
| | Self-discovery | 35 |
| | Realisation of desire | 6 |

on real and attractive tangible objects and the environment (e.g., "feeling the destination presented in the video quite beautiful"), touching overall atmosphere (e.g., "it seems that I am there"), or striking resemblance to certain scenes (e.g., "this looks exactly where that film was shot").

The second perception theme of an authentic video producer emerged from 83 reviews. Concepts supporting this theme were identified from comments on the people who created and published the videos. Some of this category of reviews acknowledged the trustworthiness of the producers (e.g., "She is as kind as my neighbour girl telling her travel story"); others expressed the audience's willingness to rely on the producer to construct their own knowledge of the world (e.g., "I really enjoy observing the world through the producer's eyes").

The theme of authentic self was comprised of three groups of concepts: representing self-discovery, self-expression, and realisation of desire. Thirty-five reviews commented on the video by saying "finding or feeling oneself" or "having a better understanding of oneself". Fourteen reviews indicated that the audience was encouraged by the video to speak out about what they were thinking, such as "it makes me want to let the world know my desire". The other six reviews expressed the audience's determination to act out the travel they had imagined or planned.

The above results indicate that audiences who had never been to the destination presented in short videos really perceived authenticity more or less and became interested in the destination. It also shows some kind of effect of authenticity perceived by the audiences on the intention to travel to the destination.

### *Measurements for authenticity perception and travel motivation*

The reliability and validity of preliminary measurements were tested with the 155 valid surveys. As for the scales measuring authenticity perception, the Cronbach alpha coefficient in Table 26.2 was 0.936 and the Kaiser-Meyer-Olkin (KMO) value in Table 26.3 was above 0.8, demonstrating reliability and validity. The scales of travel motivation were confirmed as reliable and valid too, with a Cronbach coefficient in Table 26.4 of 0.917 and a KMO value in Table 26.5 of 0.838.

*Table 26.2* Cronbach reliability analysis of the scales of authenticity perception

| Name | Correction item total correlation | Item removed alpha coefficient | Cronbach α coefficient |
|------|------|------|------|
| Item 01 | 0.759 | 0.930 | |
| Item 02 | 0.699 | 0.932 | |
| Item 03 | 0.672 | 0.933 | |
| Item 04 | 0.662 | 0.933 | |
| Item 05 | 0.741 | 0.930 | |
| Item 06 | 0.783 | 0.929 | |
| Item 07 | 0.780 | 0.929 | **0.936** |
| Item 08 | 0.618 | 0.934 | |
| Item 09 | 0.760 | 0.930 | |
| Item 10 | 0.700 | 0.932 | |
| Item 11 | 0.631 | 0.935 | |
| Item 12 | 0.732 | 0.931 | |
| Item 13 | 0.655 | 0.933 | |

*Table 26.3* Validity analysis of the scales of authenticity perception

| Name | Factor load factor | | Commonality (common factor variance) |
| --- | --- | --- | --- |
| | Factor 1 | Factor 2 | |
| Item 01 | 0.668 | 0.444 | 0.644 |
| Item 02 | 0.709 | 0.311 | 0.600 |
| Item 03 | 0.810 | 0.142 | 0.677 |
| Item 04 | 0.754 | 0.202 | 0.610 |
| Item 05 | 0.699 | 0.387 | 0.638 |
| Item 06 | 0.771 | 0.355 | 0.721 |
| Item 07 | 0.774 | 0.344 | 0.718 |
| Item 08 | 0.714 | 0.393 | 0.664 |
| Item 09 | 0.635 | 0.339 | 0.518 |
| Item 10 | 0.310 | 0.697 | 0.582 |
| Item 11 | 0.329 | 0.794 | 0.739 |
| Item 12 | 0.204 | 0.854 | 0.771 |
| Item 13 | 0.381 | 0.774 | 0.744 |
| **KMO value** | **0.923** | | |

*Table 26.4* Cronbach reliability analysis of the scales of travel motivation

| Name | Correction item total correlation | Item removed alpha coefficient | Cronbach α coefficient |
| --- | --- | --- | --- |
| Item 14 | 0.838 | 0.889 | **0.917** |
| Item 15 | 0.718 | 0.913 | |
| Item 16 | 0.819 | 0.892 | |
| Item 17 | 0.790 | 0.898 | |
| Item 18 | 0.777 | 0.901 | |

*Table 26.5* Validity analysis of the scales of travel motivation

| Name | Factor load factor | | Commonality (common factor variance) |
| --- | --- | --- | --- |
| | Factor 1 | Factor 2 | |
| Item 14 | 0.900 | 0.810 | 0.900 |
| Item 15 | 0.816 | 0.666 | 0.816 |
| Item 16 | 0.891 | 0.795 | 0.891 |
| Item 17 | 0.871 | 0.759 | 0.871 |
| Item 18 | 0.859 | 0.738 | 0.859 |
| **KMO value** | **0.838** | | |

The result of the first exploratory factor analysis reminded us to remove two items of the authenticity measurement that were loaded evenly high on two factors with loading coefficients above 0.4. Therefore, the second exploratory analysis was run after deleting these two items, resulting in two principal factors. Reliability and validity were improved, with the Cronbach alpha coefficient in Table 26.6 rising to 0.93 and the KMO value in Table 26.7 reaching 0.917. Each item has only one factor load coefficient above 0.4. The dimension of authentic producer was merged into the factor of authentic content. The theme of authentic self was reserved.

*Table 26.6* Cronbach reliability analysis of the scales of authenticity perception

| Name | Correction item total correlation | Item removed alpha coefficient | Cronbach α coefficient |
|---|---|---|---|
| Item 01 | 0.685 | 0.925 | |
| Item 02 | 0.657 | 0.926 | |
| Item 03 | 0.659 | 0.926 | |
| Item 04 | 0.731 | 0.923 | |
| Item 05 | 0.790 | 0.921 | |
| Item 06 | 0.784 | 0.921 | **0.930** |
| Item 07 | 0.620 | 0.927 | |
| Item 08 | 0.761 | 0.922 | |
| Item 09 | 0.696 | 0.924 | |
| Item 10 | 0.627 | 0.928 | |
| Item 11 | 0.734 | 0.923 | |
| Item 12 | 0.658 | 0.926 | |

*Table 26.7* Validity analysis of the scales of authenticity perception

| Name | Factor load factor | | Commonality (common factor variance) |
|---|---|---|---|
| | Factor 1 | Factor 2 | |
| Item 02 | 0.695 | 0.317 | 0.583 |
| Item 03 | 0.796 | 0.147 | 0.655 |
| Item 04 | 0.761 | 0.202 | 0.620 |
| Item 05 | 0.690 | 0.390 | 0.629 |
| Item 06 | 0.782 | 0.361 | 0.742 |
| Item 07 | 0.785 | 0.348 | 0.738 |
| Item 08 | 0.720 | 0.394 | 0.674 |
| Item 09 | 0.648 | 0.338 | 0.534 |
| Item 10 | 0.316 | 0.697 | 0.586 |
| Item 11 | 0.322 | 0.797 | 0.738 |
| Item 12 | 0.197 | 0.856 | 0.772 |
| Item 13 | 0.382 | 0.775 | 0.747 |
| **KMO value** | **0.917** | | |

All of the five items measuring travel motivation converged onto one factor with each item loading above 0.4.

Consequently, a two-dimensional scale for authentic perception and a single dimensional measurement for travel motivation activated by the video were developed. Specifically, eight items comprise the dimension of perceived authentic video content: four items for the dimension of authentic self and five items for motivation. The scales were used in the following experiments.

**Advantages of short music videos generated by tourists**

Thirty-two valid surveys administered to two groups of participants in the experiment were subject to a dependent sample t-test. The results suggest advantages of TGSVs over official promotional videos in facilitating authenticity perceptions and motivating audiences to visit the destination.

*Table 26.8* Independent sample t-test

| | | Mean equality t-test | | |
| | | Two-tail | Difference 95% confidence interval | |
| | | | Lower limit | Upper limit |
|---|---|---|---|---|
| Authentic content | Assume equal variances | 0.017 | 0.082195 | 0.777180 |
| perception | Assume unequal variances | 0.018 | 0.081109 | 0.778266 |
| Authentic | Assume equal variances | 0.033 | 0.05146 | 1.13604 |
| self-perception | Assume unequal variances | 0.033 | 0.05064 | 1.13686 |
| Travel motivation | Assume equal variances | 0.004 | 0.2472 | 1.2028 |
| | Assume unequal variances | 0.004 | 0.2470 | 1.2030 |

*Table 26.9* Independent sample t-test

| | | Mean equality t-test | | |
| | | Two-tail | Difference 95% confidence interval | |
| | | | Lower limit | Upper limit |
|---|---|---|---|---|
| Item 14 | Assume equal variances | 0.139 | −1.479 | 0.229 |
| | Assume unequal variances | 0.139 | −1.480 | 0.230 |
| Item 15 | Assume equal variances | 0.003 | −1.995 | −0.505 |
| | Assume unequal variances | 0.003 | −1.996 | −0.504 |
| Item 16 | Assume equal variances | 0.149 | −1.503 | 0.253 |
| | Assume unequal variances | 0.150 | −1.505 | 0.255 |
| Item 17 | Assume equal variances | 0.017 | −1.570 | −0.180 |
| | Assume unequal variances | 0.018 | −1.576 | −0.174 |
| Item 18 | Assume equal variances | 0.008 | −1.687 | −0.313 |
| | Assume unequal variances | 0.008 | −1.695 | −0.305 |
| Travel motivation | Assume equal variances | 0.006 | −1.4578 | −0.2922 |
| | Assume unequal variances | 0.008 | −1.4745 | −0.2755 |

As for all three factors, t-test outputs in Table 26.8 report that the values of the two-tail were below 0.05, indicating statistically significant differences between the two groups within a 95% confidence interval. The average values of the experimental group were higher than that of the control group. Therefore, audiences of TGSVs perceived higher levels of authentic content and authentic self than those who watched official promotional videos. Moreover, TGSVs exerted more effect on producing travel motivation.

As for the moderating effect of authenticity perception, only the perception of self-authenticity exerted significant influence on a video's capacity to motivate travel. As demonstrated in Table 26.9, the higher the level of authentic self-perception, the stronger the effect a TGSV would exert on travel motivation. In particular, perception of authentic self partially moderates the effect by intensifying the effect of TGSVs on three items of travel motivation, including self-esteem, reward maximisation, and self-actualisation. The other items of motivation were exempt from moderation.

## Conclusions

It can be concluded from this chapter that the prevalent social media – short music videos – can improve young Chinese people's perceptions of destinations. It is easier to make an audience experience authentic regarding the destination by using short music videos created voluntarily by ordinary tourists, compared with videos produced by professionals. The similarity between audience and the video producers strengthens the producer–audience connection and resonance.

This research proves the effect of short videos with reliable experiments and valid data and therefore affirms the value of short video marketing. The findings can serve as a reference for Internet-based destination marketing strategies. It is important for DMOs to make good use of short music videos and pay attention to the role of Internet celebrities so as to create a more authentic image and attract more tourists.

This research has its limitation. The selection of experimental material – videos provided to participants – was inevitably influenced by our subjectivity. It is difficult to include multiple styles of videos due to time constraints. It is expected that future studies could exclude the influence of irrelevant factors, such as materials and researcher bias. There is still space to improve our understanding of the relationships between short videos, authenticity, and travel motivation.

## References

An, S., Choi, Y., & Lee, C.-K. (2020). Virtual travel experience and destination marketing: Effects of sense and information quality on flow and visit intention. *Journal of Destination Marketing & Management*, 19. doi:10.1016/j.jdmm.2020.100492

Bareket-Bojmel, L., Moran, S., & Shahar, G. (2016). Strategic self-presentation on Facebook: Personal motivess and audience response to online behavior. *Computers in Human Behavior*, 55(2), 788–795.

Barna, M., & Semak, B. (2020). Main trends of marketing innovations development of international tour operating. *Baltic Journal of Economic Studies*, 6(5), 33–41. doi:10.30525/2256-0742/2020-6-5-33-41

Blake, K. S. (2010). Colorado fourteeners and the nature of place identity*. *Geographical Review*, 92(2), 155–179. doi:10.1111/j.1931-0846.2002.tb00002.x

Bullingham, L., & Vasconcelos, A. C. (2013). 'The presentation of self in the online world': Goffman and the study of online identities. *Journal of Information Science*, 39(1), 101–112.

Cohen, E. (2008). *Chapter 2 Tourism and Disaster: The Tsunami Waves in Southern Thailand*. Bingley, UK: Emerald.

Enrique Bigné, J., Isabel Sánchez, M., & Sánchez, J. (2001). Tourism image, evaluation variables and after purchase behaviour: Inter-relationship. *Tourism Management*, 22(6), 607–616. doi:10.1016/S0261-5177(01)00035-8

Farivar, S., Wang, F., & Yuan, Y. (2020). Opinion leadership vs. para-social relationship: Key factors in influencer marketing. *Journal of Retailing and Consumer Services*. doi:10.1016/j.jretconser.2020.102371

Gibbs, C. (2007). Mms marketing a hit abroad, on deck in U.S. *RCR Wireless News*, 26, 12–13

Grayson, K., & Martinec, R. (2004). Consumer perceptions of iconicity and indexicality and their influence on assessments of authentic market offerings. *Journal of Consumer Research*, 31(2), 296–312. doi:10.1086/422109

Guo, J. (2021). Research on the Marketing Innovation of "Live + Short Video". In *The Culture and Tourism Industry in We Media Era. E3S Web of Conferences*, 251, 03036. doi:10.1051/e3sconf/202125103036

Hogan, B. (2010). The presentation of self in the age of social media: Distinguishing performances and exhibitions online. *Bulletin of Science Technology & Society*, 30(6), 377–386.

Huang, T. L. (2021). Restorative experiences and online tourists' willingness to pay a price premium in an augmented reality environment. *Journal of Retailing and Consumer Services*, 58, 102256. doi:10.1016/j.jretconser.2020.102256

Iso-Ahola, S. E. (1982). Toward a social psychological theory of tourism motivation: A rejoinder. *Annals of Tourism Research*, 9(2), 256–262. doi:10.1016/0160-7383(82)90049-4

Krishna, A., Cian, L., & Sokolov, T. (2016). The power of sensory marketing in advertising. *Current Opinion in Psychology*, 142–147. doi:10.1016/j.copsyc.2016.01.007

Li, Y. M. (2014). Effects of story marketing and travel involvement on tourist behavioral intention in the tourism industry. *Sustainability*, 6(12), 9387–9397. doi:10.3390/su6129387

Liu, H., & Yan, M. (2021). Influence of mobile short-form video on tourist behavioral intentions. *Tourism Tribune*, 10, 62–73. doi: 10.19765/j.cnki.1002-5006.2021.10.009

Mackay, K. J., & Fesenmaier, D. R. (1997). Pictorial element of destination in image formation. *Annals of Tourism Research*, 24(3), 537–565. doi:10.1016/S0160-7383(97)00011-X

McCain, G., & Ray, N. M. (2003). Legacy tourism: The search for personal meaning in heritage travel. *Tourism Management*, 24(6), 713–717. doi:10.1016/S0261-5177(03)00048-7

Moore, K., Buchmann, A., Månsson, M., & Fisher, D. (2021). Authenticity in tourism theory and experience. Practically indispensable and theoretically mischievous? *Annals of Tourism Research*, 89, 103208.

Ram, Y., Björk, P., & Weidenfeld, A. (2016). Authenticity and place attachment of major visitor attractions. *Tourism Management*, 52(Feb.), 110–122. doi:10.1016/j.tourman.2015.06.010

Reisinger, Y., & Steiner, C. J. (2006). Reconceptualizing object authenticity. *Annals of Tourism Research*, 33(1), 65–86. doi:10.1016/j.annals.2005.04.003

Rickly-Boyd, J. M. (2009). The tourist narrative. *Tourist Studies*, 9(3), 259–280. doi:10.1177/1468 797610382701

Shaltoni, A. M. (2017). From websites to social media: Exploring the adoption of internet marketing in emerging industrial markets. *Journal of Business & Industrial Marketing*, 32(7), pp. 1009–1019. doi:10.1108/JBIM-06-2016-0122

Shani, A., Chen, P. J., Wang, Y., & Nan, H. (2010). Testing the impact of a promotional video on destination image change: Application of China as a tourism destination. *International Journal of Tourism Research*, 12(2), 116–133. doi:10.1002/jtr.738

Sharpley, R. (2005). *Taking Tourism to the Limits*. Retrieved from https://www.researchgate.net/p ublication/284256980. doi:10.1016/B978-0-08-044644-8.50023-0

Shi, Y., Zhang, R., & Ma, C. (2021). Emotional labor and place attachment in rural tourism: The mediating role of perceived authenticity. *International Journal of Marketing Studies*, 13(3), 33. doi:10.5539/IJMS.V13N3P33

Wang, H., & Ma, Z. (2020). A study on tourist motivation of cycling tourism on the Sichuan-Tibet line and its impact on the cyclers' happiness: The mediating effect of tourist satisfaction. *Tourism Science*, 6, 53–65. doi:10.16323/j.cnki.lykx.2020.06.005.

Wang, X. (2017). An empirical study on the authenticity of tourism from the perspective of tourists. *Agriculture of Henan*, 3, 48–55. doi:10.15904/j.cnki.hnny.2017.03.022.

Wang, Y., & Fesenmaier, D. R. (2004). Towards understanding members' general participation in and active contribution to an online travel community. *Tourism Management*, 25(6), 709–722. doi:10.1016/j.tourman.2003.09.011

Xing, N., Yang, S., Huang, Y., Li, M., & Wang, J. (2018). The motivation and value pursuit of outbound tourism of post-90s. *Tourism Tribune*, 9, 58–69. doi:CNKI:SUN:LYXK.0.2018-09-012

# 27 Travel vlogging and its role in destination marketing

*Maria Criselda Badilla and Carl Francis Castro*

## Introduction

The emergence of Web 2.0 and the sophistication of technological gadgets have greatly advanced the tourism industry by enabling travellers to easily share and exchange information about their travel experiences with others (Alrawadieh, Dincer, Dincer, & Mammadova, 2018). Its online environments offer opportunities for connecting with each other and creating spaces for the exchange and development of knowledge in blogs, forums, Google docs, and wikis (Zammit, 2016).

The highly interactive nature of social media has given tourists a place to dream, plan, book, experience, and share travel. Social media makes it possible to stay in touch with friends and maintain interest in a place. Dutta, Sharma, and Goyal (2021) stated that digital media has changed from being a transmitting medium to one that allows for client involvement. Travel influencers have emerged as an effective method for many hospitality and tourism organisations to engage with their target markets and build brand recognition. The majority of tourists utilise social or digital media while travelling (Abad & Borbon, 2021). Social media platforms have increased brand exposure, customer interaction, and customer input to improve the quality of goods and services, and decision-making transparency (Surugiu, Surugiu, & Mazilescu, 2019).

According to wearesocial.com, globally there are 4.48 billion active social media users, or 56.8% of the total population, who spend an average of two hours and 24 minutes every day on the platforms. YouTube is the second most visited website, second only to Google. com according to semrush, similarweb, and alexa (wearesocial.com). Further, 25.5% of global Internet users aged 16–64 years old watch influencer videos and vlogs. These figures show the vast impact of social media and vlogging on the online community and how much people consume these platforms.

## Development of travel vlogs

Blogging was brought to a new level with the rapid evolution of technology by utilising smartphones and digital cameras to record videos, post, and share travel experiences. According to Peralta (2019), this craze has ushered in a new era of vlogging, which combines video blogging and video recording. The advent of social media platforms and the advancement of network information technology have had an effect on how people gather information and make decisions. With the development of various video media platforms and technologies, vlogging has steadily taken the lead as the most significant tool for people to document their lives and showcase their personalities. Videography and posting to

DOI: 10.4324/9781003291763-31

social media platforms have become common behaviours, particularly among the post-1990s and post-2000 generation (Chen, Guo, & Pan, 2021).

On YouTube, a vlog is similar to a blog, but in video format. Tourists begin filming themselves, emphasising the activities and experiences they had while travelling, and sharing it on social media (Birch-Jensen, 2020). Travel vlogs have emerged as the new face and phase of travel blogs which are greatly shared on social media (Peralta, 2019). Travel vlogs offer great marketing potential and provide endless possibilities for tourism marketers. Several scholars have written about how travel vlogs provide beneficial information in the decision-making and travel behaviour of tourists (Lodha & Philip, 2019).

In recent years, travel vlogging has become ultimately one of the powerful ways for tourists to discover unique places, form images, and create expectations. Social media has given rise to the sharing of travel vlogs that include a variety of tourist attractions such travel activities, lodging, food, and adventures (Peralta, 2019). The popularity of many professional and amateur vloggers has increased due to the development of Internet video platforms like YouTube, which has enabled regular people to manage their self-made films more effectively (Cheng, Wei, & Zhang, 2020). Travel vlogging has gained popularity because, despite certain overlaps and similarities, other nations offer narratives that differ in nature, character, and traditions (Chakravarty et al., 2021). In the study of Choi and Lee (2019), they went into great detail about how travel vloggers interact with viewers in the same way that celebrities do, but because they are thought of as regular people, they play the same role as a viewer's buddy or acquaintance. As a result, the contact between the vlogger and the viewer might be seen as a parasocial connection between audience and media.

According to Elliot (2012), a number of destination management organisations (DMOs) use vlogs as electronic consumer-to-consumer (C2C) distribution channels for user-generated content (UGC) – also known as user-created content (UCC) – which is a broad term that refers to a variety of media and creative content types that have been created or have been at least significantly co-created by "users" – that is, by contributors working outside of traditional professional environments. Although UGC in non-digital formats has a longer history than UGC in digital formats, the term became more widely known with the participative shift in Web design and practices that occurred in the early years of the new millennium, and it is frequently referred to as Web 2.0 emergence (Bruns, 2016).

This chapter examines travel vlogs in relation to their use as a marketing tool for DMOs. Through a literature review and thematic analysis of the vlogs of stranded foreign vloggers in the Philippines during the COVID-19 lockdown in March 2020, the advantages, impacts, opportunities, and challenges of travel vlogging are identified and elucidated throughout the chapter.

**Advantages of travel vlogs**

The promotion of tourism has been significantly impacted by travel vlogs, which can be used as a marketing tool by firms that specialise in destination marketing. Vlogs are easily accessible through social media platforms like YouTube and Facebook, and this ease of access has significantly influenced how tourists make decisions. As a result, the choice of destination is influenced by the sharing of personal stories and real-life experiences in vlogs. The following benefits of travel vlogs are audience interaction, easy access to information, customisation of experiences that lend credibility to the content, and availability of information.

*Information: Access and availability*

People are using vlogs to obtain information through the informative videos made by vloggers. People are using various search engines to search for different informative videos that spread awareness about several causes. Access to information is a crucial prerequisite for travel, given that tourism is one of the booming sectors in many parts of the world. There is a need for greater information and inspection because the tourism product is an experience made up of numerous diverse elements, including transportation, lodging, activities, attractions, cuisine, and facilities. As consumption has grown in scope and intensity over time, anyone can access vast sources of knowledge (Birch-Jensen, 2020). Social media is increasingly being used as a source of information. People use social media to research options, weigh them, and come to a choice (Paul, Roy, & Mia, 2019).

With a major emphasis on tourism and word-of-mouth marketing, digital technologies have expanded word of mouth beyond the small circle of people we know. These tourism companies have benefited from this intriguing technology by stepping up their promotion of locations and goods on social media in an effort to reach as many people as possible (Mukherjee & Nagabhushanam, 2016). Travellers view travel blogs and vlogs as trustworthy sources of information because, through social media, travel planners can get feedback, comments, and/or ratings, which raises the reliability rating factor for visitors when considering a particular tourist attraction, booking platform, airline, or hotel (Azucena, 2020). Vlogs are becoming a significant source of information and a critical component of how consumers make decisions (Mannukka, Maity, Reinikainen, & Luoma-aho 2019).

*Credibility of information*

Coursaris and Van Osch (2016) stated that, as the second-largest search engine, YouTube has emerged as a key resource for consumers for knowledge and information. Information recipients must believe the source of the information to be credible, knowledgeable, and trustworthy in order for it to be seen as dependable. Therefore, people are better able to judge the legitimacy of the vlogger's recommendations the more time they spend on and engage with the vlog, the vlogger, and other followers (Mannukka et al., 2019).

*Audience engagement*

With the capacity to connect to these sites via mobile phones, tablets, and computers, social media has become an integral part of many people's everyday lives. The usage of information that is made available to each and every one on a social networking site raises serious issues with personal privacy as well as issues regarding how this information might be utilised to attack and endanger lives. The simplicity of information sharing has made it possible for people to stay in touch with friends and family and notify them about changes in their lives, their opinions on various topics, their collaboration on projects, and much more (Wise & Shorter, 2014).

**Impacts of travel vlogs**

People make vlogs, and some of them have gained many followers. Vloggers post films on YouTube about their personal lives, product reviews they have written, or even reviews of places they have been. Moreover, vloggers provide valuable information about certain products or places. As consumption of travel vlogs caught the fancy of many potential tourists, it is noteworthy to mention some of their impacts on the tourism industry for the tourist and the tourism destination. Travel vlogs have an impact on: (1) tourists, on their perceptions, travel behaviour, and experiences; (2) on the DMO; and (3) on achieving Sustainable Development Goal 12 of sustainable consumption and production patterns.

### *Impact on tourist perception, travel behaviour, and visitor experiences*

The influence of vloggers is one of the most powerful tools nowadays to encourage people. It has an impact on people's decisions and opinions. One of the most popular tourist destinations may experience a surge in business when a vlogger comes to record and post about their visit. Increasing the browsing from followers and other viewers' travel vlogs has many positive implications. Viewers are allowed to experience something they might not be able to and can even be educated about the cultures and traditions of other nations. They may affect how a customer or traveller perceives things and makes decisions. Additionally, they are regarded as being more real and approachable (Marinas et al., 2021). Furthermore, in the study of Abdurrahim, Najib, and Djohar (2019), one's attention, interest, and desire to learn more about the content being promoted via social media has a significant impact. Additionally, one's level of interest and focus has a significant influence on travel choices. Arora and Lata (2020) stated that when travellers are deciding which place to visit, they reflect critically and assess the YouTube channel's content. Vloggers should therefore use their videos to provide people with pertinent, comprehensive, and accurate location information.

Some top travel vloggers have shared their travel experiences to millions of people. Some famous vloggers with millions of subscribers and billions of video views include Mark Weins with 8.51 million subscribers, Devin SuperTramp (Devin Graham) with 6.18 million subscribers, and Drew Brinsky with 3.08 million subscribers.

### *Impact on DMOs*

Liang (2020) discussed the benefits of vlogging as a means of disseminating tourism information as well as its rising popularity among those looking for information online. The actual visit to a destination and the actuality of the destination can change tourists' perceptions of the destination. Watching the vlogs may provide the viewers with a virtual trip to the location, even though they may not have actually been there, and pique their interest in travel. Travel vlogging affects visitor experiences because it creates a self–other distinction, mediates one's own experiences, and modifies those of others (Chen et al., 2021).

Travel vlogs are helpful for tracking visitors' perceptions of a particular location; they can gather much data that later can be used to modify the best approach to pique visitors' interest. By doing this, the location can develop a distinctive brand that inspires travellers to experience and return occasionally. The tourist industry and tourism marketers must come to see travel vlogs as a brand-new technological phenomenon and their role in destination marketing. As a result, according to Sangeetha and DineshBabu (2021), YouTube travel video channels help young people learn through the varied media provided by travel bloggers.

Llanos (2021) highlighted that travel vlogs have become important for destinations as tourists become ambassadors sharing travel experiences and encouraging others to consider visiting the destination. Travel vlogs facilitate the storytelling of tourists and have become virtual diaries that are available and accessible to everyone.

These days, we refer to this as digital word-of-mouth advertising. Vlogs are similar to virtual diaries in that they are only kept online and are always accessible. Young people are given more knowledge on tourist destinations by a variety of travel vloggers who post travel-related videos, which influences their travel plans to visit the tourist locations described by the YouTubers. Travellers now rely heavily on these movies for tourism-related information, which has an impact on their behaviour and intentions. Vloggers' use of beautiful, interactive, perceived-enjoyable, and useful travel videos has a significant positive effect on viewers' behavioural intentions to travel (Chen et al., 2021). People can share their vlog entries with their friends and family, allowing for discussion and the development of new visitor relationships. Travel vlogs have been found to play a significant function in the online world, which can aid in publicising the places.

### Impact on SDG 12

Additionally, UN Sustainable Development Goal 12 aims to promote sustainable patterns of consumption and production. Vlogs on travel offer a lasting method of advertising goods and services. The adoption of sustainable consumption and production (SCP) patterns denotes an increase in productivity and efficiency throughout the supply chain and the product life cycle, immediately and in the long run (Hoballah & Averous, 2015). Without producing brochures and other marketing materials, vlogs are used to advertise locations in a creative way to reach a wider audience and develop sustainable marketing strategies.

### Trends and issues in travel vlogs

Several studies discuss the trends and current issues in the world of vlogging. These are: (1) more content spectators than producers; (2) vloggers' attractiveness and credibility as a factor of influence; (3) perceived entertainment value, interactivity, and usefulness of the platform; and (4) advances in technology and digital applications.

### *More video-content spectators than producers*

Internet platforms serve as repositories for user-generated data. Although audiences create as much material as they consume, there are still more people watching video content than making it. Audiences now frequently consume the content that media users create and share. Additionally, media consumers frequently create content, such as comments and recommendations on social media, that helps advertisers deliver brand messaging and that may also be used as information to gauge the success of advertising campaigns (Napoli, 2015).

### *Vloggers' attractiveness and credibility as a factor on influence*

In their study, Chen et al. (2021) suggested three important characteristics that will have an impact on a user's inclination to travel. They are: perceived interactivity, perceived enjoyment, and perceived utility of vloggers. The more trustworthy the vlogger is, the more likely viewers are to follow his or her advice, and the more determined they are to travel after viewing the vlog.

Le and Hancer (2021) claim that audience desire to identify with travel vloggers was influenced by their physical appeal, social attractiveness, and credibility, with credibility having the greatest influence. The impact of the trip vlogger's social attractiveness on viewer wishful identification was noticeably stronger when the vlogger and the viewer were of different genders (Le & Hancer, 2021).

### Perceived entertainment value, interactivity, and usefulness of the platform

According to Statista.com (2019), US residents watch YouTube primarily for entertainment (25%), music (20%), people (20%), and blogs (19%). Tourists learn about destinations through social media platforms, gain experience from travel vlogs, and interact with others' travel stories through comments. A potential traveller is more likely to visit a location that the travel vlogger has recommended if they find the video entertaining (Chen et al., 2021). From the standpoint of tourism marketing, if viewers do not find the trip vlog entertaining, the strategy of converting viewers into future tourists will not be successful (Chen et al., 2021). It stands to reason that viewers are more likely to endorse vacation vlogs that make them feel amused, at ease, and pleasant (Cheng et al., 2020).

### Advances in technology and digital applications

The introduction of Internet technology, particularly social media, has altered how individuals find information, shop, connect with others online, and even plan their travels (Beham, 2015). The travel and tourism sector has changed as a result of technological innovation. Consumers of tourism are becoming more able to contact with suppliers of goods and/or services at multiple levels, even before they reach their destinations, thanks to the Internet and social media. Due to the increased accessibility of tourism customers to technologies like the Internet and social media, the providers and users of a tourism product and service have altered long before travellers reach their vacation destinations. Huang, Backman, Backman, and Chang (2015) assert that the availability of technology enables travellers to visit a variety of locations and gauge the calibre of such sites, goods, tools, and services. They are important and may even suggest whether a place is worth visiting. In this regard, these customers assess locations on social media before deciding whether or not to visit them. Visitors can view the feedback and views of other travellers who have visited these destinations on social media networks. Therefore, technology has the potential to play a significant part in helping travellers make informed judgements about the places they travel to.

YouTube videos are starting to gain more and more traction. Duffett (2020) claims that YouTube is the most popular online video platform with more than two billion users. Every day, over a billion hours of YouTube videos are watched, mostly by young people. In order to reach the profitable Generation Z cohort, first born in the late 1990s, and sway their notoriously erratic purchase decisions, YouTube has grown into a significant commercial communication channel.

Online video editing activity has significantly expanded as a result of the rising popularity of vlogging. There are numerous websites and tools available now that enable users to edit videos online directly from their browsers without having to download any additional software. These websites and applications for editing videos, such International Remix, JumpCut, Videoegg, Eyespot, Motionbox, Photobucket, or One True Media, are typically distinguished by their exceptional usability, which makes it simple for beginners to perform basic and complex editing (Gao, Tian, Huang, & Yang, 2010).

Vlogging is currently a popular Internet fad. About one in three people who use the Internet, or one billion people, according to YouTube (2015), use YouTube. Compared to those targeted by television, the majority of YouTube users are between the ages of 35 and 49, with 40% of those users accessing the site via their mobile devices.

According to Mindruta (2015), the vloggers, who primarily are young adults, make interesting films in which they appear in their home environments and depict details of a typical day in their lives. Making movies of one's daily life has become very popular among young people. In addition to edited videos, live streaming is becoming more popular. Additionally, live streaming enables vloggers to communicate in real time with their audience.

**Opportunities and challenges offered by travel vlogs**

User-generated content abounds on the Internet and vlogs offer new ways of sharing information. The literature has shown how vlogs have provided opportunities to various tourism stakeholders. These opportunities include: (1) helping tourists seeking third-party information about destinations; (2) providing insights on market research; (3) tourism marketing and promotions potential; and (4) enhancing the quality of life of the audience.

First, travel vlogs offer a wealth of trustworthy, unbiased information on various locations. Personal trip accounts and true stories are featured in travel vlogs, which are four times more popular with viewers than any other type of social media content (Think with Google, 2014). Vlogs give first-person narratives that are uninfluenced by destination marketers and present fresh approaches to storytelling (Cheng et al., 2020). Travel vlogs tend to be more emotional than cognitive because the vloggers highlight their own experiences, which causes the audience to shift their attention away from facts and rational qualities and toward their emotional connection to the location (Chen et al., 2021). The DMO materials are supplemented by the vlogs since the videos can provide experiential and narrative validation of the knowledge.

Second, since watching travel vlogs has an effect on electronic word of mouth, which can result in increased travel intent, travel vlogs can offer fantastic market insights for hospitality and tourism marketers on what can influence potential customers in their choice of travel locations (Chen et al., 2021). Insights on understanding tourist experiences, motivation, behaviour, and niche markets are just a few examples of the types of information that may be found in the vlogs that can aid in market research. Social media is a significant instrument for analysing consumer views and creating successful businesses (Surugiu et al., 2019). As the vloggers discuss their trip experiences and behaviour patterns in the videos, the vlogs turn into data that DMOs can examine. When it comes to popular activities, positive and bad experiences, and information on how to enhance visitor amenities and activities inside the destination, these become natural sources of information for DMOs.

Third, as values are shared and communicated through films, travel vlogs have the ability to promote tourism and destinations (Chen et al., 2021). Travel vlogs have a huge marketing potential for giving authentic experiences for destination marketing given the phenomenal web traffic generated, notably during the COVID-19 pandemic (Peralta, 2019). By transporting viewers to a specific location and enhancing the viewers' perspectives of particular destinations, travel tales in vlogs aid in the promotion of international tourism (Chakravarty et al., 2021). Due to the new reality of safe travel, travel vlogs provide an affordable alternative to explore virtual tourism places (Chakravarty et al., 2021). User-generated social media material that describes unique and/or memorable experiences is a potent tool for drawing tourists (Du, Liechty, Santos, & Park, 2020). The trip vlogs help DMOs expand their promotional ideas and audience. The DMO's social media platforms allow for the sharing and reposting of the vlogs. The vlog uploads and expands the destination's reach by tapping into the vlogger's online audience, which the DMO might not otherwise be able to access.

Fourth, because these films give vloggers the chance to visit uncharted locales and document their experiences on camera, travel vlogs are a type of virtual travel that improves the quality of life of their viewers. Vlogs give the vloggers and their audiences the chance to spread a brand-new, previously unheard of, form of travel awareness (Chakravarty, Chand, & Singh, 2021). When involvement and feedback increase, social media has the ability to alter the nature of one's social life at the interpersonal and communal levels (Baruah, 2012).

As the emergence of travel vlogs have given way to opportunities for the tourism industry, more companies have started strategically embedding them in their marketing communications to improve their strategic use (Surugiu et al., 2019). However, it also comes with challenges that stakeholders need to be aware of. Some challenges identified in the literature include: (1) the shift of control of content from producer to consumer; (2) credibility of vlogs; and (3) the rise of influencer marketing.

First, the shift in content production to end users has been brought about by the growth of UGC production and consumption. Due to this, the term "produser", a contraction of the words "producer" and "user", was created to describe the various ways that consumers today produce content. Producers are individuals who use or consume information from the Internet as well as produce or create content for use by others (Zammit, 2016). Marketers no longer have complete control over the image of their destination since users can now provide content and frame it in a completely new way. Vlogs, one of the most popular forms of UGC, have transformed vloggers from storytellers to endorsers, giving them the same prominence that superstars had in the 1980s (Mannukka et al., 2019). Since travel vlogs are UGC, DMOs cease to have control of the destination's image.

Second, consumers might build a favourable opinion of the brand thanks to endorsers with a high level of credibility, which eventually results in a higher level of purchase intent (Mannukka et al. 2019). Massive online audiences enable vloggers to persuade their viewers more effectively.

Lastly, the rise of influencer marketing, where DMOs pay social media influencers to vlog about their locations, may further cast doubt on the veracity of the vlogs. Due to the increase in UGC consumption on social media, travellers now have more opportunity to share their travel-related experiences with other travellers (Surugiu et al., 2019). There may come a point when the validity of the videos may be questioned due to its financial purpose as vloggers experience financial rewards in developing their online audiences. Vloggers and viewers may encounter moral conundrums when creating and watching vlogs. Tourists now place more trust in other travellers' recommendations than in marketing firms' guidance, thus these organisations are starting to work with vloggers to covertly advertise their goods.

It is essential to talk about the need for social media influencer regulation in a world where YouTube and Instagram have replaced television. Children and teenagers closely follow the advice given by social media celebrities, and aspire to be influencers rather than doctors, firefighters, or astronauts (Goanta & Ranchordas, 2019). The rigorous monitoring of these social media influencers has recently received much media attention and the interest of watchdog groups. Hidden advertising has received attention from the U.S. Federal Trade Commission and the Italian Competition Authority. The Federal Trade Commission emphasised that truth in advertising also applies to social media in 2018 by reminding

90 influencers that they should disclose any material link with marketed consumer products. As a result, modern US literature has made an effort to investigate the legal norm that applies to influencer advertising. Over the past year, more research on European legislative frameworks like the Unfair Commercial Practices Directive has been appearing (Goanta & Ranchordas, 2019).

### Future trends

Destination marketing strategists will continue to incorporate social media because of its capacity to raise awareness, engage consumers, and affect buying and post-purchase decisions (Surugiu et al., 2019). UGC will have a significant impact on how destination images are formed. Marketers must be involved and attentive to the content posted on social media.

Direct interactions between customers and producers have been made possible by the way technology has revolutionised the travel and tourism sector, particularly in terms of marketing and distribution. Vlogs may introduce a profusion of new intermediaries and co-creation of value, which will lead to the invention of new methods of doing things (Buhalis, 2020). Vloggers have the capacity to forge close bonds with their followers and build sizable online travel communities. Online travel communities, robotics and artificial intelligence, big data analytics, and the sharing economy are seen as four of the biggest trends in tourism, according to Bowen and Whalen (2017). In the years to come, vloggers' influence and position in the tourism industry will only increase as more travel experiences at all phases of the journey are documented through video blogs.

Marketers and travellers alike will create and consume more social media because it is a common source of information used during the travel decision-making and consumption process (Paul et al., 2019). The emergence of social media, according to Glucksman (2017), "has created a new avenue for brands to communicate with customers more directly and naturally". Destination companies have made the shift from text to audio and video to podcasts; in the future, we may see shorter videos on TikTok and live streaming.

Vlogging is a new form of tourism that has been shaped by the rapid advancement of technology. This area needs more research and thoughtful attention (He, Xu, & Chen, 2021). Vlogging may become a crucial component of market research, the creation of marketing strategies, and image building. The development of technology will continue to influence vlogging.

### Theoretical contributions

Katz and Paul Lararsfeld's Two-Step Flow of Communication (Littlejohn and Foss, 2009) study on the social influence of opinion leaders verifies the impact vloggers have on destination image. Opinion leaders are members of specific organisations who enlighten less active people in their sphere of influence in depth. The flow of media communications is changed as leaders add their own opinions to the actual content and/or filter the actual information for their group, enabling them to comprehend the information (www.communicationtheory.org). Vloggers are comparable to these opinion leaders who were targeted by political leaders during elections and who were earlier theorised by Katz and Lazarsfeld. The way that vloggers construct their stories affects the choices that their online communities make. Such messaging is also filtered through the mediatised gaze in Urry's (2002) tourist gaze. The tourist gaze reveals to visitors a different reality than that which the DMO intends to present. With the development of digital technology, regular travellers can now construct their own perceptions of locations based on their own experiences, independent of destination marketers. The tendency toward UGC and the co-creation of value further emphasises the destination's framing within the context of vloggers.

## Practical and social contributions

Vlogging gives rise to many practical and social contributions. For the DMO, vlogging provides countless opportunities for marketing and promotions. The destination can have exposure from unsolicited social media which can generate unexpected tourist visitation. On the downside, however, negative vlog posts can also affect the destination's image portrayal. There will be an easing of market research data, as access to data and research methodologies become available online.

As YouTube vloggers grow in popularity and influence, their personal lives have become available to the public. There is a blurring of what is public versus what is private as well as what is authentic and what is staged. The use of vlogging as a tourism marketing tool will continue to grow as more people search for travel information with ease through social media. UGC will also dominate Internet content. The vloggers and his or her audience will be in constant search of new and exciting content to produce and consume.

## Conclusion

Throughout this chapter, we have discussed the role of travel vlogging in destination marketing by explaining the development or evolution of travel vlogging and highlighting its advantages, impacts, trends and issues, opportunities, challenges, and predictions, together with its theoretical, practical, and social contributions. A brief case study on the impact of travel vlogging on tourism marketing has shown the power of vlogging in shaping the image of a destination. Inarguably, travel vlogging plays a key role in destination marketing, from image formation, influence in travel behaviour, and a new way of experiencing destinations, among others. Travel vlogs enable viewers to appreciate the featured destination and eventually influence destination choice.

## Recommendations

The benefits of vlogging as a destination marketing tool have been iterated throughout this chapter. The DMO should be able to maximise its use for marketing and image formation. Authenticity needs to be guarded in exchange for commercialism and possible financial gain motivation for the vloggers.

There has also been a dearth of literature on travel vlogs wherein most studies highlighted the practical implications of travel vlogs of destination image and tourism marketing (Bosangit, 2021). Future research could focus on the theoretical implications of travel vlogs, the impact of travel vlogging on the whole travel experience, the political economy of vlogging, the rise of vloggers as social media influencers, and the future trends of vlogging such as live streaming and use of shorter videos.

Tourists also need to be more critical of the vlogs they consume. As some popular vloggers may use their following for financial gain rather than expressing personal convictions, followers should be more discriminating of the vlogs and vloggers they follow. Vloggers, too, need to protect their online communities and be more discriminating of products to endorse and companies to collaborate with. DMOs will make use of vlogs for marketing destinations more extensively in the future as people become more dependent on them for travel information. This will not be cost free as vloggers who already have a substantial following are charging hefty fees for brand collaborations and partnerships.

## References

Abad, P. E. S., & Borbon, N. M. D. (2021). Influence of travel vlog: Inputs for destination marketing model. *International Journal of Research Studies in Management, 9*(3), 47–66. https://doi.org/10.5861/ijrsm.2021.m7729

Abdurrahim, M. S., Najib M., & Djohar, S. (2019). Development of Asia's model to see the effect of tourism destination in social media. *Journal of Applied Management, 17*(1), 133–143.

Alrawadieh, Z., Dincer, M., Dincer, F., & Mammadova, P. (2018). Understanding destination image from the perspective of Western travel bloggers, the case of Istanbul. *International Journal of Culture, Tourism and Hospitality Research, 12*(2), 198–212.

Arora, N., & Lata, S. (2020). YouTube channels influence on destination visit intentions: An empirical analysis on the base of information adoption model. *Journal of Indian Business Research, 12*(1), 23–42. https://doi.org/10.1108/JIBR-09-2019-0269

Azucena, Hillary Estelle. (2020). A Study of the Impact of Travel Bloggers and Vloggers on the tourists' Decision-Making Process and Experience when travelling to Australia.

Baruah, T. D. (2012). Effectiveness of social media as a tool of communication and its potential for technology enabled connections: A micro-level study. *International Journal of Scientific and Research Publications, 2*, 1–10.

Beham, A. (2015). Role of social media in Generation Y travellers' travel decision-making process, BA thesis in Tourism and Hospitality Management, Modul University, Vienna.

Birch-Jensen, J. (2020). Travel vloggers as a source of information about tourist destinations (Dissertation). Retrieved from http://urn.kb.se/resolve?urn=urn:nbn:se:umu:diva-171751

Bosangit, C. (2021). Travel Vlogs. In: Buhalis, D. (ed.), *Encyclopedia of Tourism Management and Marketing*. Cheltenham, UK: Edward Elgar Publishing.

Bowen, J., & Whalen, E. (2017). Trends that are changing travel and tourism. *Worldwide Hospitality and Tourism Themes, 9*(6), 592–602. https://doi.org/10.1108/WHATT-092017–0045

Bruns, A. (2016). User-generated content. *The International Encyclopedia of Communication Theory and Philosophy*. https://doi.org/10.1002/9781118766804.wbiect085

Buhalis, D. (2020). Technology in tourism-from information communication technologies to eTourism and smart tourism towards ambient intelligence tourism: A perspective article. *Tourism Review, 75*(1), 267–272

Chakravarty, U., Chand, G., & Singh, U. N. (2021). Millennial travel vlogs: Emergence of a new form of virtual tourism in the post-pandemic era? *Worldwide Hospitality and Tourism Themes, 13*(5), 666–676.

Cheng, Y., Wei, W., & Zhang, L. (2020). Seeing destinations through vlogs: Implications for leveraging customer engagement behavior to increase travel intention. *International Journal of Contemporary Hospitality Management, 30*(10), 3227–3248.

Chen, Yingying, Guo, Zhaojuan, & Pan, Qiuyue. (2021). Analysis on the characteristics of travel vlog video and its impact on users' travel intention. 10.2991/assehr.k.210519.034.

Choi, W., & Lee, Y. (2019). Effects of fashion vlogger attributes on product attitude and content sharing. *Fash Text, 6*, 6. https://doi.org/10.1186/s40691-018-0161-1

Coursaris C. K., & Van Osch, W. (2016). Exploring the Effects of Source Credibility on Information Adoption on YouTube. In: Nah, F. H., & Tan, C. H. (eds.), *HCI in Business, Government, and Organizations: eCommerce and Innovation. HCIBGO 2016*. Lecture Notes in Computer Science, vol. 9751. Cham: Springer. https://doi.org/10.1007/978-3-319-39396-4_2

Du, X., Liechty, T., Santos, C. A., & Park, J. (2020). 'I want to record and share my wonderful journey': Chinese Millennials' production and sharing of short-form travel videos on TikTok or Douyin. *Current Issues in Tourism, 25*(21), 3412–3424.

Duffett, Rodney. (2020). The YouTube marketing communication effect on cognitive, affective and behavioural attitudes among generation Z consumers. *Sustainability*, *12*, 1–25. 10.3390/su12125075.

Dutta, K., Sharma, K., & Goyal, T. (2021). Customer's digital advocacy: The impact of reviews and influencers in building trust for tourism and hospitality services. *Worldwide Hospitality and Tourism Themes*, *13*(2), 260–274. https://doi.org/10.1108/WHATT-09-2020-0123

Elliot, S. (2012). Understanding the role of social media in destination marketing. *Turismos: An International Multidisciplinary Journal of Tourism*, *7*(1), 193–211.

Gao, Wen, Tian, Yonghong, Huang, Tiejun, & Yang, Qiang. (2010). Vlogging: A survey of videoblogging technology on the web. *ACM Computing Surveys (CSUR)*, *42*, 1–57.

Glucksman, M. (2017). The rise of social media influencer marketing on lifestyle branding: A case study of lucie fink. *Elon Journal of Undergraduate Research in Communications*, *8*(2), 77–87.

Goanta, Catalina, & Ranchordas, Sofia. (September 20, 2019). The Regulation of Social Media Influencers: An Introduction. In: Goanta, C., & Ranchordas, S. (eds.), *The Regulation of Social Media Influencers* (Edward Elgar, 2020, Forthcoming), University of Groningen Faculty of Law Research Paper No. 41/2019, Available at SSRN: https://ssrn.com/abstract=3457197 or http://dx.doi.org/10.2139/ssrn.3457197

He, J., Xu, D., & Chen, T. (2021). Travel vlogging practice and its impact on tourist experiences. *Current Issues in Tourism*. 10.1080/13683500.2021.1971166

Hoballah, Arab, & Averous, Sandra. (2015). *Goal 12 - Ensuring Sustainable Consumption and Production Patterns: An Essential Requirement for Sustainable Development*. https://www.un.org/en/chronicle/article/goal-12-ensuring-sustainable-consumption-and-production-patterns-essential-requirement-sustainable

Huang, Yu, Backman, Kenneth, Backman, Sheila, & Chang, Lan-Lan. (2015). Exploring the implications of virtual reality technology in tourism marketing: An integrated research framework: The implications of virtual reality technology in tourism marketing. *International Journal of Tourism Research*, *18*. https://doi.org/10.1002/jtr.2038.

Le, L. H., & Hancer, M. (2021). Using social learning theory in examining YouTube viewers' desire to imitate travel vloggers. *Journal of Hospitality and Tourism Technology*, *12*(3), 512–532. https://doi.org/10.1108/JHTT-08-2020-0200

Liang, A. (2020). Exploring the formation and representation of destination images in travel vlogs on social media. Tour related decisions. *International Journal of Education and Research*, *5*(5), 203–208.

Littlejohn, S., & Foss, K. (2009). *Two-step Flow Model of Communication by Katz and Lazarsfeld Encyclopedia of Communication Theory*. London: Sage Publications.

Llanos, Carl Angelo. (2021). *The Impact of Travel Vlogging on Local Tourism Industry*. https://zdocs.ro/doc/chapter-1-final-36ow80qrejpl

Lodha, R., & Philip, L. (2019). Impact of travel blogs and vlog on decision-making among students of Bangalore. *International Journal of Scientific Research and Review*, *7*(3), 2279–2543.

Mannukka, J., Maity, D., Reinikainen, H, & Luoma-aho, V. (2019). Thanks for watching'. The effectiveness of YouTube vlogendorsements. *Computers in Human Behavior*, *93*, 226–234.

Marinas, Ed. D., Ungui, Nathan, Vitan, Justine, Quirimit, Mark, Barredo, Laeticia, Friginal, Sophia, Lising, Maryam, & Piedad, Yullia. (2021). Student vloggers: Altering the students pathway in academic life. *International Journal of Research Publications*. *70*. 10.47119/IJRP100701220211732.

Mindruta, R. (2015). *YouTube Strategy: Tips for Building an Audience & Working with Vloggers*. Retrieved from: https://www.brandwatch.com/2015/01/top-9-social-media-trends-2015/

Mukherjee, A., & Nagabhushanam, M. (2016). Role of social media in tourism marketing. *International Journal of Science and Research*, *5*(6). http://doi.org/10.21275/v5i6.NOV164776

Napoli, Philip. (2015). The audience as product, consumer, and producer in the contemporary media *Marketplace*. http://doi.org/10.1007/978-3-319-08515-9_15

Paul, H. S., Roy, D., & Mia, R. (2019). Influence of social media on tourist's destination selection decision. *Scholars Bulletin*. 10.36348/SB.2019.v05i11.009

Peralta, R. (2019). How vlogging promotes a destination image: A narrative analysis of popular travel vlogs about the Philippines. *Place Branding and Public Diplomacy*, *15*, 244–256. https://doi.org/10.1057/s41254-019-00134-6

Sangeetha, U. B., & DineshBabu, S. (2021). *A Study on the Influence of Travelogue Videos among the Youngsters*. https://www.annalsofrscb.ro/index.php/journal/article/view/7580

Surugiu, C., Surugiu, M., & Mazilescu, R. (2019). Evolution of management in the era of globalization. *Manager*, 29.

Think with Google. (2014). *YouTube Data*. Available at: www.thinkwithgoogle.com/consumerinsights/travel-content-takes-off-on-youtube/ (accessed 20 July 2019).

Urry, J. (2002). *The Tourist Gaze* (2nd ed.). London: Sage Publications Ltd.

Wise, E., & Shorter, J. (2014). Social networking and the exchange of information. *Issues in Information Systems*, *15*(II), 103–109. https://doi.org/10.48009/2_iis_2014_103-109.

YouTube. (2015). *Statistics*. Retrieved from: https://www.youtube.com/yt/press/statistics.html.

Zammit, Katina (2016) Collaborative Writing: Wikis and the Co-construction of Meaning. In B. Guzzetti & M. Lesley (Eds.), *Handbook of Research on the Societal Impact of Digital Media*. Texas: IGI Global.

# 28 Augmented reality and virtual reality in tourism

*Evrim Çeltek*

## Introduction

Developments in smart technologies have revealed effects that change and transform the tourism industry, tourist behaviour, and the practices of tourism enterprises. With the development of smart technologies, many different technologies have started to be used in the tourism sector. One of these technologies is augmented reality (AR) and virtual reality (VR) in tourism. Using the recognition feature of devices such as smart glasses, tablets, smartphones, and overlaying virtual objects on real images is called AR (Çeltek, 2020, tom Dieck & Jung, 2017). VR is a computer-generated 3D environment where individuals experience the feeling of being there (Yovcheva & Buhalis, 2013).

In 1957, Morton Helig, considered the leader of virtual reality technology, produced the first virtual reality glasses called the Sensorama. The first computerised VR began to be used in the late 1960s; but the name VR was not used until the 1980s when head-mounted displays and computer-attached tactile gloves became commercially available (Barnes, 2016). In 1966, US computer engineer Ivan Sutherland produced the device called a "Stereoscopic-Television Apparatus for Individual Use", which enables the visual, sensory, and olfactory senses to be used, which is very similar to today's head mounted display (HDM) technology (Carmigniani et al., 2011). VR began to appear in computer games in the early 1990s, while Ford began using it to design and manufacture its vehicles in 1999 (Barnes, 2016). In 1992, L. B. Rosenberg succeeded in developing the first AR system, which he called "Virtual Fixtures" (Carmigniani et al., 2011). In 1996, Ars Electronica Electronic Arts Museum in Linz, Austria publicly presented its CAVE VR system (Sherman & Craig, 2018). In 1998, the University of North Carolina in the USA developed the first 3D AR technology (Carmigniani et al., 2011). In 1999, Hirokazu Kato and Mark Billinghurst developed the ARToolKit device used for the development of AR applications. In 2003, Second Life, a virtual world game, was released by Linden Labs (Sherman & Craig, 2018). In 2008, the first Android-based smartphone, G1Android, was released. The phone was compatible with Wikitude AR Travel Guide, the first mobile AR app. In 2009, the Massachusetts Institute of Technology (MIT) produced a wearable AR device, which is used to reflect human movements and body language, with the SixthSense project. In 2012, Google produced Project Glass, the first AR glasses free of additional hardware. In 2015, Microsoft introduced Hololens glasses, a holographic computer that provides augmented and virtual reality to the user. In 2019, the Oculus Quest VR headset, which is used without the need for a computer and phone, was released.

In 1998 the Hellenic World Foundation, a Greek cultural heritage institution, established a VR department and presented educational VR exhibits related to the Hellenistic

DOI: 10.4324/9781003291763-32

period. Also in the same year, Disney opened the first of DisneyQuest family adventure centres with numerous VR attractions using HMD and projection-based visual screens (Guttentag, 2010). Tuscany+ is the first AR app built specifically for tourism. The application features an interactive, mobile tourist guide for the Tuscany region (Kounavis et al., 2012). Holiday Inn is the first hotel chain to make the best use of AR technology. With the application, guests are able to see virtual Olympic and Paralympic athletes at the reception, in the hall, or in their hotel rooms.

One of the first AR applications in the field of cultural heritage is the ArcheoGuide AR system. The application was developed for an ancient temple: Olympia in Greece (Yovcheva & Buhalis, 2013). AR, which started with fighter pilots in the Second World War, and VR, which started to develop with the first VR glasses "Sensorama", are used in many different areas. In the tourism industry, AR and VR are used in museums, hotels, restaurants, tourism destinations, and historical areas to give information, promote, advertise, navigate, guide, and for gamification. This chapter aims to provide details about AR and VR application trends in the tourism industry. AR and VR technologies and types, tourism industry applications, the advantages and challenges of these technologies in different areas of tourism, and predictions for the future are described in this chapter, which is an effective source for those who want to learn about tourism AR and VR applications and their advantages and challenges.

## Augmented reality and virtual reality

AR technology is the addition of digital layers on top of an existing environment, bridging the gap between the virtual and physical world (Manjunath et al., 2020). In fact, AR enables digital content such as graphics, audio, and video to be displayed on screen using a device's camera. AR is the 3D visualisation of a message provided with visual and auditory methods. According to another definition, AR is the integration of digitally created objects in a computer environment with the real world; it is a visualisation technique that synthesises real images and virtual images (Chung et al., 2015). With its feature of adding virtual information to physical objects and environments, AR can display virtual objects on real-world objects (Chung et al., 2015). AR is used by smartphones equipped with GPS technology and allows users to specify a location to regulate device orientation (Jung et al., 2015). In order to benefit from AR technology, there must be Internet access and smart devices (smartphones, tablets, smart glasses) that will define AR in the current environment, and one of the applications defined must be installed on the device to be used. Special software developed for AR can recognise and track 2D or 3D targets and display various images such as 3D models, animations, or videos on them.

VR is electronic simulation of environments experienced via head-mounted eye goggles and wired clothing enabling the end user to interact in 3D situations (Barnes, 2016). VR is an alternative world filled with computer-generated images that respond to human movements. These simulated environments are usually encountered with the aid of an expensive data suit that features stereophonic video goggles and fibre-optic data gloves (Steuer, 1992). According to another definition, VR is a computer-generated, 3D, virtual or simulated environment in which the user can navigate and interact, and feel real with their five senses (Guttentag, 2010). Special gloves and controllers have been developed for VR that allow the user to interact with virtual objects in the virtual environment. Gloves, called Data-Gloves, are designed to detect moving objects, touch, and to feel physical events. VR is a program. With the software that accompanies VR glasses, the user can watch a movie, play

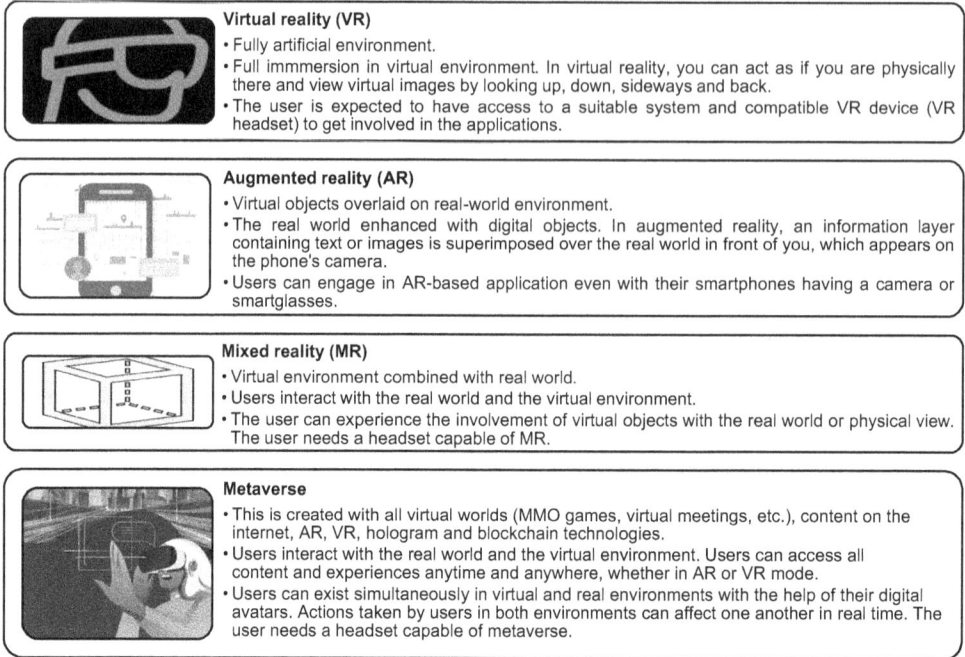

**Virtual reality (VR)**
- Fully artificial environment.
- Full immmersion in virtual environment. In virtual reality, you can act as if you are physically there and view virtual images by looking up, down, sideways and back.
- The user is expected to have access to a suitable system and compatible VR device (VR headset) to get involved in the applications.

**Augmented reality (AR)**
- Virtual objects overlaid on real-world environment.
- The real world enhanced with digital objects. In augmented reality, an information layer containing text or images is superimposed over the real world in front of you, which appears on the phone's camera.
- Users can engage in AR-based application even with their smartphones having a camera or smartglasses.

**Mixed reality (MR)**
- Virtual environment combined with real world.
- Users interact with the real world and the virtual environment.
- The user can experience the involvement of virtual objects with the real world or physical view. The user needs a headset capable of MR.

**Metaverse**
- This is created with all virtual worlds (MMO games, virtual meetings, etc.), content on the internet, AR, VR, hologram and blockchain technologies.
- Users interact with the real world and the virtual environment. Users can access all content and experiences anytime and anywhere, whether in AR or VR mode.
- Users can exist simultaneously in virtual and real environments with the help of their digital avatars. Actions taken by users in both environments can affect one another in real time. The user needs a headset capable of metaverse.

*Figure 28.1* Differences between virtual reality, augmented reality, mixed reality, and metaverse.
*Source*: Adapted from Çeltek (2021).

a game, or visit Niagara Falls in a virtual environment. Users can see the images in VR by moving their heads just as in real life. That is, the image is adjusted according to the physical position of the user's head.

## Types and technologies

AR is divided into two main groups, depending on which method the virtual and real views are superimposed: (a) marker based and (b) markerless. Markerless AR is divided into four groups. A detailed explanation of AR types follows (Chung et al., 2015; Çeltek, 2015, 2020; Sinha, 2021).

### Marker-based AR (recognition-based AR)

Marker-based AR has a special shape to trigger apps. As soon as the QR code, a unique image or code associated with the application, is recognised, the application is triggered, and digital content is placed on the shape. Usually, a mobile device or AR glasses can be used for this. This type of AR is used in tourism, in education books, advertising brochures, billboards, and tickets.

### Markerless AR

Markerless AR is more versatile than marker-based AR. The user can employ the AR application with a mobile device or AR glasses without carrying any pointers. Markerless

AR software provides the necessary information from smartphone hardware such as camera, GPS, digital compass, and accelerometer. With markerless AR, there is no need for any object tracking system thanks to the latest technological developments in cameras, sensors, and artificial intelligence algorithms. This type of AR works with digital data obtained by these sensors, which can record a physical area in real time. There are four categories of markerless AR:

- *Location-based AR* aims to reveal 3D virtual objects at the user's physical location. This technology uses a smart device's position and sensors to position the virtual object at the desired location or point of interest. The best example of this type of AR is the smartphone game Pokémon GO. This type of AR is used in tourism to provide guidance services to tourists.
- *Projection-based AR* is a little different from other types of markerless AR. No mobile device is needed to view the content. Using light instead, digital graphics are projected onto an object or surface to create an interactive experience, though it would be more accurate to call it an interactive hologram. This type of AR is used in digital art museums.
- *Superimposition-based AR* uses object recognition. The related object is recognised and superimposed on its VR image. This type of AR can be used to show what a destroyed historical statue looked like in the past. The disadvantage of this technology is the requirement for full object recognition. When scanning the object from different angles, the device's camera may not be able to detect all features of the object.
- *Outlining-based AR (contour-based AR)* apps recognise borders and lines to help in situations which the human eye cannot see. AR outlines use object recognition to understand a user's immediate surroundings. It allows for driving in low light conditions or viewing a building's structure from the outside.

VR is divided into four groups according to the immersion or interaction rate: non-immersive, semi-immersive, fully immersive, and collaborative (Halarnkar et al., 2012; Çeltek, 2020):

- *Non-immersive VR* is a type in which the user interacts with a virtual environment, usually via a computer screen, and can control certain characters or activities within the experience, though the virtual environment does not directly interact with the user.
- *Semi-immersive VR* is something between non-immersive and fully immersive. The user can navigate the virtual environment with a computer screen or VR glasses. 360 videos are a good example of semi-immersive virtual technology.
- *Fully immersive VR (CAVE VR-Computer Assisted Virtual Environment)* is the opposite of non-immersive VR. It provides a realistic virtual experience. With this VR, if possible, the user's five senses are addressed. It consists of VR sensors, speakers that broadcast sound/music from different angles, and wall and floor projection. The user feels as if they are physically in the virtual world and events that occur in the virtual world are happening to the user. This type of VR requires special equipment such as VR glasses, gloves, or body detectors equipped with sensory detectors. The data from these tools are used by the computer and the virtual world responds in real time to give users a realistic virtual experience.
- *Collaborative VR* is a type of VR or virtual universe where users from different countries come together in a virtual environment in the form of 3D avatars. Games where players' avatars come together, and interact with each other's virtual personalities in the same environment, are a good example of this VR.

AR and VR

AR — VR

**AR side:**
Types of AR — AR devices

Types of AR:
- Marker based AR
- Markerless AR
  - Location based
  - Projection based AR
  - Superimposition based AR
  - Outlining based AR

AR devices:
- Head mounted displays (HMD)
- Holographic displays
- Smart glasses
- Handheld displays (smartphones and tablets)
- Retinal displays and contact lens, neural implants (future tech.)

**VR side:**
Types of VR — VR devices — Types of VR experiences

Types of VR:
- Non-immersive VR
- Semi-immersive VR
- Fully-immersive VR
- Collaborative VR

VR devices:
- PC (personal computer)/console /smartphone
- Head mounted displays (HMD)
- Input devices (handheld controller, data gloves, eye gaze, hand tracking, full body tracking)
- Fully immersive sphere
- Projection/large screen
- Retinal displays and contact lens, neural implants (future tech.)

Types of VR experiences:
- Monoscopic, immersive spaces (360-degree panorama movies, limited interactivity)
- Stereoscopic, immersive experiences (3D graphics, high interactivity)

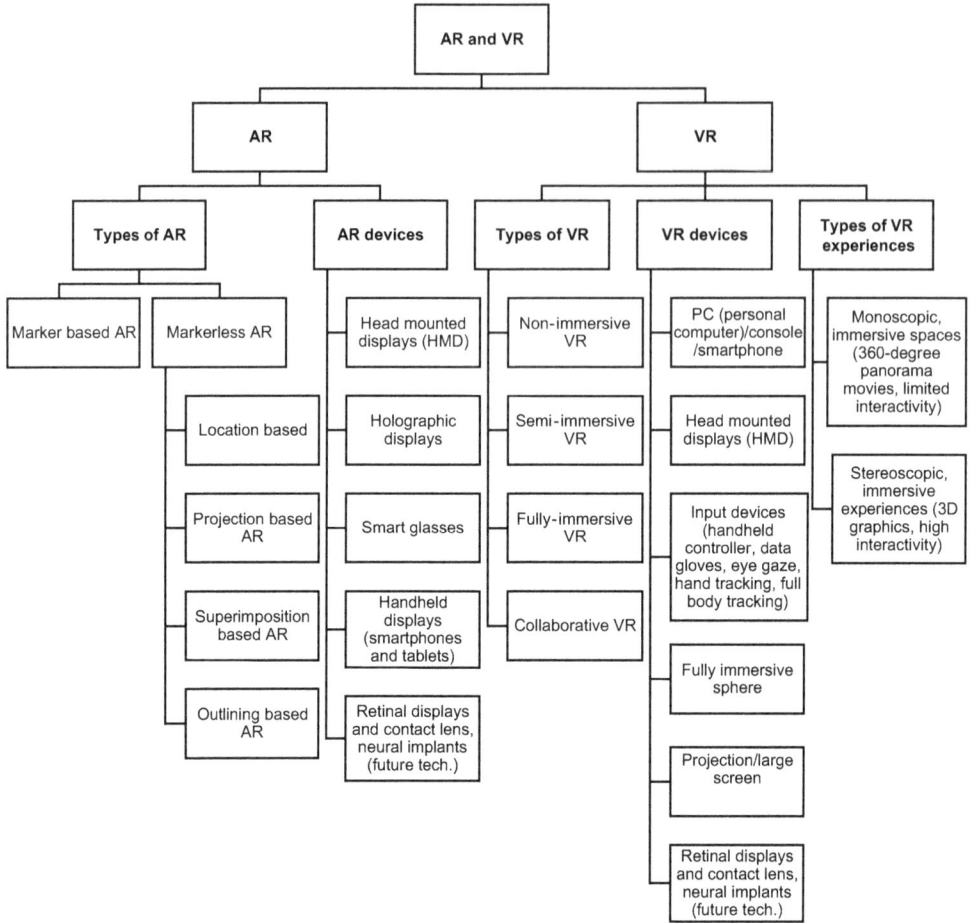

*Figure 28.2* Types of AR/VR and AR/VR devices.

In VR users are immersed in computer generated environments with HDM, gloves, 3D graphics, and body tracking. In AR, virtual images are blended with the real world images. Users see virtual images through HDM, handheld displays, and viewpoint tracking. Figure 28.2 shows the types and devices for AR and VR.

## Tourism industry applications

### Destinations

Studies show that when tourists use VR and AR experiences, their perceptions about the destination image improve and help them establish an emotional bond with the destination (Pantona & Servidio, 2001). Most tourist destinations in the world have AR-based digital guide applications. These applications can be downloaded from Apple Store and Google Play. One of the AR tourist guide examples is the Discover Hong Kong application. The app serves as a comprehensive guide to more than 100 tourist attractions, 5,500 retail stores, 2,000 restaurants, and events in Hong Kong. AR tourist guides enable the tourist to visualise points of interest (POI) located in their nearby environment in the countries or cities. POI databases include restaurants, hotels, WiFi hotspots, and subway stations (Çeltek, 2015). Departures Switzerland is one of the public transport AR apps in Switzerland. The app shows all public transport options in the user's surroundings through an AR browser. Paris Then and Now City Guide application aims to enrich the experience of tourists with AR applications while visiting historical places. With the application, tourists can see the status of more than 2,000 different historical places 100 years ago through the cameras of their tablets or smartphones (EY, 2019). Tripventure is a location-based AR game and guide which can be played in many cities in Europe. It allows tourists to experience virtual stories in real life via AR. While exploring the actual city, tourists interact with virtual people, solve mysteries, and find hidden items to reach the answers to the questions given (Tripventure, 2020). National Geographic Explore VR is an interactive VR game that allows the user to explore Antarctica and the Andes Mountains in Peru (Oculus, 2022a). In The Grand Canyon VR Experience players can take an interactive boat ride in the Grand Canyon. The player can choose between a boat tour under the hot, sweltering sun or under a cool starry sky. Wander is a VR app that enables users to travel around the world (and see the Pyramids, the Grand Canyon, the Louvre, Berlin, Mumbai, and Kyoto) with 360° photos. This application is a multiplayer, and more than one person can use it at the same time. BRINK Traveler is a VR app designed for nature lovers. It is possible to travel virtually to 12 different parts of the world, shot in full 3D with photogrammetry (Death Valley, Arches National Park, Mount Whitney, Iceland's Háifoss waterfall, and more). When We Stayed Home is a VR app where several documentarians recorded empty streets with cameras in their hands when the COVID-19 pandemic hit in 2020. The app offers VR videos of Paris, Venice, Jerusalem, and Tokyo that are quiet and devoid of tourist crowds (Dingman, 2021). Most countries and cities have VR videos on platforms such as VeeR and Steam. Airpano. com has VR videos of tourist destinations from the Northern Lights to Antarctica. On the 360cities.net site, there are 360° photos of many destinations, and these can be seen with VR glasses. In destinations, VR can be used to visit historical places and lifestyles that no longer exist, to see places where disabled people cannot go, or to visit places that are dangerous for visiting. In addition, with VR, tourists can make a purchase decision by virtually visiting the destination they want to see. With AR and VR applications, countries and cities can be marketed to tourists with more detailed and visually rich images.

*Hotels*

Holiday Inn uses an AR app for customers to see realistic virtual depictions of celebrities who have stayed at their hotels. Another example of AR applications in accommodation is the interactive wall map in each room of Premier Inn's Hub hotels, where guests can interact with their phones and check out local attractions nearby (Sharma, 2019). The Mansion at Casa Madrona uses AR-based brochures to promote its accommodation facilities to its guests. Hotel customers, who scan the interactive areas of the hotel's brochures with the Blippar application, can access much more content than the brochure in an easy and fun way. Marriott Hotels market their hotels with VR and customers can even rent VR glasses in the hotels. Virgin Holidays has developed a VR application that enables customers to experience their tours before purchasing. Virgin Holidays recorded an ambient sound to create a real and sensory experience of what the holiday would sound like, then turned it into a visual customer experience using VR glasses (EY, 2019). Amadeus VR travel search and booking experience is a project-stage application that allows travellers to visit a destination, search for flights, tour the inside of the plane to choose their seat, review different rental cars, and pay after they have decided (Vallantin, 2017). Most hotels in the world have uploaded VR videos that allow visiting their rooms by customers on YouTube or websites. Some of these hotels are Atlantis Dubai hotel, Novotel Hotels, Holiday Inn Express Sydney, and Relaxa Hotel Munich. Thomas Cook travel agency used the "Try Before You Fly" VR application to allow its customers to experience the holiday spirit and to feel themselves on a virtual holiday island, in a crowded city centre, or by the sea. In-agency VR application provides customers with a realistic experience of their future travel by taking them to a hotel or destination before making a reservation. After the VR application, Thomas Cook's sales increased by 80% (EY, 2019). AR and VR can be powerful tools for any hotel promotional campaign. By using AR in hotels, the customer experience is improved. With AR and VR, hotels can be advertised, the information needed by their customers can be provided, and hotel employees can be trained.

### Museums

In most museums, AR adds extra materials such as descriptions, pictures, maps, and videos to the artefacts, making the museum more interesting and educational (Jung & tom Dieck, 2017). In June 2021, the Muséum National d'Histoire Naturelle in Paris launched an AR application (using Microsoft Hololens). The app is called "REVIVRE" ("To Live Again"), and in it visitors can see digital versions of now-extinct animals. In 2021, The National Gallery in London displayed its collections with QR codes on the busy streets of central London. Users who scanned the codes had the opportunity to see the works in the museum. The National Museum of Singapore is running an immersive AR installation called "Story of the Forest". The exhibition uses 69 images from the William Farquhar Collection of Natural History Drawings. These images have been transformed into 3D animations that visitors can interact with. In 2017, the Smithsonian Institution (Washington DC) created an AR app for the skeletons in the Museum's Bone Hall. With the Skin and Bone AR application, 13 skeletons can be examined and users can see how skin and muscles look on bones and how animals move (Coates, 2021a). AR was used in the Mirages & Miracles exhibition, which created works by Adrien M and Claire B at the Ars Electronica digital art museum. The works in the exhibition turn into moving animations with the AR application installed on the tablet (ars.electronica.art, 2022).

Smartify is an AR application that works with image recognition and artificial intelligence to give users access to background information about objects in museums. In the app, the user holds the device and uses its camera to scan and learn about the artwork (https://smartify.org, 2022). The Dali Museum in Florida immerses its visitors in a picture of Dali with 360° VR (EY, 2019). In October 2019, the Louvre Museum in Paris used interactive design, sound, and animated images in the VR app "Mona Lisa: Beyond the Glass". The application has been used by visitors for four months. This app allowed visitors to approach the painting more closely and personally than ever before, see the brushstrokes invisible to the naked eye, and see the techniques Leonardo da Vinci used in his creation. At the National Museum of Finland in Helsinki, visitors can travel back in time to 1863 with a VR application of R. W. Ekman's painting *The Opening of the Diet 1863 by Alexander II*. With the Natural History Museum's "Hold the World" VR application, users can meet dinosaurs, and hold and resize objects in a virtual environment (Coates, 2021b). With the National Archaeological Museum: Live the Past VR application, the player lives the experience of visiting several towns in the history of Spain and learning how people lived at different times (Oculus, 2022b). AR and VR offer museums the opportunity to diversify their audience by offering new avenues of art experience. AR and VR enable visitors to obtain in-depth, rich, relevant, and easily accessible and personalised information about museums. They are used to create museum tours, make exhibits interactive, and bring scenes to life. They also help curators put works in context and show their true scale.

### Restaurants

Pizza-maker Domino's has put up 6,000 posters across Britain that look like a normal promotion poster but also serve as an AR marker when viewed through the Blippar Application. The user has the option to download deals for their nearest Domino's store, get the Domino's mobile ordering app, or view their local menu (Çeltek, 2015). Shanghai Roastery in Shanghai, China is the world's first Starbucks that offers customers an AR experience. The AR app acts as a kind of virtual tour guide for customers, explaining the different steps of the coffee making process. Customers can watch an animated version of the newly roasted beans falling into the cask with the AR application they download to their phones. In the app, customers unlock a virtual badge at every step, and when all badges are collected, they get a special roastery filter that can be shared on social media (Dahlstrom, 2017). Sublimotion is a projection-based AR application created to change the theme of the environment while serving dinner to 12 people around a table. Sublimotion, which was first used in a restaurant in Ibiza in 2014 ($2,000 a plate), is now used in the Mandarin Oriental Jumeira in Dubai. In the application, a fun atmosphere (according to the theme created) is offered to the diners with images projected on every surface from the walls to the table (Mandarinoriental, 2022). Inamo restaurant in London has applied interactive projection-based AR on the dining tables. With this application, customers can play games at the tables, learn about the local environment, and decorate their tables with pictures and graffiti according to their taste (Inamo-restaurant, 2022). Kabaq is an AR, VR, and mixed reality application that provides 3D visualisations of the food on the menus of restaurants. In the application, the food of the restaurants was scanned in 3D and digitally visualised. With the application, customers can examine the food on the digital menu in 3D on a tablet or smartphone and learn about their ingredients (Kabaq, 2022). Bareburger, a famous burger chain in the USA and Dubai, is one of the first restaurants to use Snapchat in its AR food menu (AR food menus can be customised to show information about each meal, such as 360-degree visualisation, nutrition and calorie information, ingredients used, portion sizes). Snapchat is an application that provides the opportunity to sell products through augmented reality filters created specifically for advertising brands. In the application, advertisements are placed on selfies and videos that users share with personalised augmented reality filters. When the user clicks on the "Shop Now" button, they are directed to the product page without leaving the application (Grigonis, 2018). In 2017, Honeygrow restaurant (23 businesses in nine cities) started training their employees on five topics (from food safety to good customer service) with VR. It used Google's Daydream VR headset for this. With interactive VR applications, for example, employees can learn to cook noodles and store food in a gamified way (Sharp, 2022). There is also a VR game (called The Hard Way) played with Oculus Rift, which teaches KFC's employees how to cook fried chicken. AR and VR are used in restaurants to introduce the meals on the menu with 3D visuals and to provide information about the contents of the meals. In addition, staff can be trained with AR and VR, and the waiting time for customers for orders can be made fun.

**Advantages and challenges**

In tourism AR and VR are used to show the features of destinations, museums, historical sites, package tours, or hotel rooms in the virtual and real world. In addition AR and VR are used in tourism to give information, promote, advertise, navigate, guide, and for gamification. Advantages of AR and VR applications for tourism businesses, tourists, and sustainable tourism are shown in Table 28.1 (Kounavis, et al., 2012; Sing & Pandey, 2014; Chung et al., 2015).

- advertisements, brochures and menus for tourism products
- AR provides information and facilitates activities throughout the travel experience. VR provides information before the tourism experience and helps decision making
- Marketing and presentation of cities, destinations, historical and touristic values, and hotels
- Promotion and marketing of tourism services
- Sale of tours, making reservations
- Tourist guide and city guide service
- With the help of AR, tourists can create their favourite lists, rate businesses, share their experiences and comments on social networks
- Those who cannot travel, especially the elderly, can travel virtually with VR and see places that are dangerous to go
- The impact of tourism on vulnerable destinations can be reduced with VR applications. For example, the creation of virtual twins of sensitive historical sites
- Creates unforgettable and unique experiences for tourists. Creates unique brand interaction for tourism destinations and businesses
- Virtual hotel rooms, museums and destinations and historical areas can be created with VR

The challenges of AR and VR applications for tourism businesses, tourists, and sustainable tourism are shown in Table 28.1 (Carmigniani et al., 2011; Berryman, 2012).The responsibilities of the stakeholders for the development of more effective and beneficial AR and VR applications in tourism are:

- Technology companies and tourism businesses that create AR and VR applications should research customer needs. It should be emphasised that the content is of high quality, meets the needs of tourists, and is easy to use.
- Most tourists are unaware of AR and VR applications or do not know how to use them. The tourism business should promote and provide information on this issue.
- What can be effective practices within the scope of sustainable tourism should be investigated.
- The number of interactive stereoscopic VR applications should be increased.
- AR and VR applications to be used in tourism education should be diversified. For example, AR-based tourism books.
- Interactive, fun, massively multiplayer (MMOG) tourism AR and VR games should be developed.
- Tourism businesses, destination management organisations (DMOs), sustainable tourism planners, and city managers should plan and implement AR applications, especially in destinations with high demand. For example, navigation for disabled tourists, interactive AR events, AR art exhibitions.

*Table 28.1* Advantages and challenges of AR/VR; the today and tomorrow of AR/VR

|  | *Advantages* | *Challenges* |
|---|---|---|
| **AR** | • Marketing and presentation of cities, destinations, historical and touristic values, and hotels. Sending personalised and geotargeted advertising messages. Cost effective advertising.<br>• Giving personalised, location-based tourist guide and city guide service.<br>• Providing information and facilitating activities throughout the travel experience.<br>• Creating interactive advertisements, menus, brochures, books, and catalogues for tourism products.<br>• Creating interactive digital art museums.<br>• Providing users to share their experiences and reviews on social media. | • Requires approval or participation from the user.<br>• Consumer privacy, unauthorised collection of consumer information.<br>• AR can cause accidents while using in an outdoor setting.<br>• Can cause digital fatigue, addiction to the virtual world.<br>• If the camera is not properly focused, the virtual object will not be displayed.<br>• Some apps have poor content quality.<br>• Not every AR app is compatible with every phone. Some AR apps require iOS operating system, and some require Android operating system. |
| **VR** | • Informing about the city/destination/country's attractions, tourism businesses, and environment before travelling.<br>• Marketing tourism businesses and increasing the demand for tourism. Cost effective advertising.<br>• Creating travel experience centres before travelling.<br>• Creating digital heritage and gamification.<br>• Sustainability and protection of heritage.<br>• Creating digital twins of historical sites, artefacts, and cities.<br>• Visiting difficult to reach, dangerous areas in the virtual environment.<br>• Providing unique and memorable virtual experience. | • Unknown health effects.<br>• Possible harm of VR headset to the eyes (can dry eyes).<br>• User's adaptation.<br>• In tourism, some applications are still experimental.<br>• VR tools are still expensive for the general user.<br>• User addiction to the virtual world.<br>• Prolonged use of an HMD device may cause headaches and nausea. The weight of the HMD is also another issue for a user.<br>• Users can feel slow in some VR apps.<br>• Some VR apps requires powerful hardware.<br>• Every VR content is not used with all HDM. Different VR applications are required for different HDM. |
| | **Today's Device Features** | **Tomorrow's Device Features** |
| **AR** | Billinghurst (2021) summarises the features of AR today:<br>• Handheld display technologies and lightweight head mounted displays.<br>• Location based, marker based, image based, and hybrid tracking.<br>• Screen based, simple gesture and tangible interaction. | Billinghurst (2021) summarises the features of AR for the future:<br>• Projected AR displays, retinal displays, contact lens displays, wide FOV (Field of view) see trough.<br>• Model based and environmental based tracking.<br>• Natural gesture, multimodal, intelligent interfaces, sensor based interaction. |
| **VR** | Billinghurst (2021) summarises the features of VR today:<br>• High quality VR graphics.<br>• Semi-immersive (110–150 degrees) display technologies.<br>• Interaction with handheld controller/some gestures.<br>• Limited movement.<br>• Multi-user number is limited (few users). | Billinghurst (2021) summarises the features of VR for the future:<br>• Photo realistic VR graphics.<br>• Full immersive display technologies (360 degree).<br>• Full interaction with gesture, body, and gaze.<br>• Natural movement and navigation.<br>• Millions of users can meet and interact with each other. |

*(Continued)*

*Table 28.1* (Continued)

| | Advantages | Challenges |
|---|---|---|
| | **Today's Applications** | **Tomorrow's Applications** |
| **AR** | • AR-based digital restaurant menu.<br>• AR tourist, city, museum, art, and hotel guide.<br>• Marker-based AR advertisements and books.<br>• Location based, personalised advertisements in AR guides.<br>• Location-based multiplayer city and museum AR games.<br>• AR-based interactive art exhibitions.<br>• Educational AR applications. | • Interactive walled, projection-based AR classrooms, halls, rooms in hotels.<br>• Artificial intelligence, big data, blockchain, the internet of everything and recognition technology integrated AR apps, glasses, and HDM.<br>• Tourism AR advertisements on buildings surfaces.<br>• Outlining-based AR apps for unmanned autonomous vehicles.<br>• MR and XR based glasses, retinal displays and contact lens, neural implants. Merged AR and VR apps. |
| **VR** | • Interactive city, destination, and museum VR games that can be played with hand controller.<br>• 360-degree images and 360 degree VR videos of hotels, destinations, museums, paintings (Dali, Picasso) on web sites.<br>• Semi-immersive city, museum, destination VR tours. | • Interactive city, destination, hotel, restaurant, and museum VR games that can be played with data gloves, eye gaze, hand tracking, full body tracking.<br>• Full immersive (CAVE VR) city, museum, destination VR tours.<br>• MR and XR based glasses, retinal displays and contact lens, neural implants. Merged AR and VR apps.<br>• VR tourism reservation and buying apps and systems.<br>• Full immersive VR travel experiences for the elderly.<br>• Metaverse tourism destinations in which millions of users can meet and interact with each other with avatars. |

*Source*: Adapted from Çeltek (2020).

## Predictions for 2032, 2042, and 2052

With the development and progress of AR and VR technology, its use in the field of tourism will become widespread in the future and the speed, quality, and resolution of images will increase. In addition, AR and VR devices will become cheaper and their size will be reduced. The following are future predictions about AR and VR applications in tourism.

## 2032

• The proliferation of tourist guiding applications: AR can act as a tour guide for tourists during travel, with navigation, views of notable 3D animations and videos of past events at certain locations, foreign language translation, voice signals, and dialogues.
• Gamification: Interactive city, destination, hotel, restaurant, and museum AR and VR games that can be played with data gloves, eye gaze, hand tracking, and full body tracking.
• Metaverse platforms: Users will be able to visit the metaverses of tourism businesses with their avatars and make presentations as holograms at conferences. With metaverse, hybrid conferences and meetings can be held.

- The proliferation of tourism AR and VR marketing platforms: With 3D AR and VR applications, previews of hotels, destinations, and historical sites can be created, and tourists can get information about their features. These previews are used in the field of marketing and sales in tourism. In addition, AR will be used extensively in marketing filters, printing advertisements, brochures, and menus in tourism.
- Indoor and outdoor AR navigation: With mobile AR applications, tourists can get location analysis and directions in complex indoor and outdoor areas such as airports and cities. These technologies accurately detect the location of tourists and show virtual routes and arrows that guide them.
- Widespread use of Web AR applications: Web AR applications that work only by clicking on a URL or QR code, without downloading a mobile application with a single click, will become widespread in tourism as they are easy to use and do not require a mobile application.
- The proliferation of AR and VR education applications: One of the biggest advantages of AR and VR technologies in education is that they make dangerous, difficult, or expensive environments reliable, easily accessible, and inexpensive and present them to students (Wu et al., 2013). With virtual environments created with AR and VR, students and tourism personnel can learn by doing and having fun.
- Development and cheaper pricing of AR and VR technologies: Faster, lighter, more affordable VR technologies will develop. With advances in smartphone technology (like better cameras and processors), the quality of AR apps will improve.
- Interactive user manuals: The use of interactive AR user manuals will become widespread, especially in complex self-service applications (use of coffee machines, check-in/check-out in smart hotels, MaaS applications).
- Preservation of historical and cultural sites: AR and VR technology are used in the digitisation of historical, cultural, and architecturally valuable objects, destinations, and important assets, which may deteriorate and become sensitive, fragile, or destructible, and create a 3D view. Thus, these valuable assets can be visited virtually as if they were real, and they are prevented from being destroyed by tourists (Guttentag, 2010).

## 2042

- Realistic VR tourism applications: When viewing a hotel or destination with high-resolution VR glasses, a virtual experience with breathtaking detail and reality will be experienced, and the user will be able to experience virtual travel with all five senses at the same time.
- Smaller wearable AR and VR devices: In the future, technology will begin to integrate more seamlessly with the human body. Examples of these are AR glasses that work in an integrated way with the smartphone, smart glasses, AR and VR contact lenses, and wearable robotic bots that provide the feeling of walking according to your movement in the VR glasses. Also, VR headsets will include brain–computer interfaces. This VR headset will allow users to activate actions just by thinking about them. Thus, the user who wants to visit the hotel or destination with VR will be able to direct the application with his or her thoughts.
- Realistic travel experiences: Virtual touristic trips that make you feel like you are travelling in real life will be possible through brain–computer interface devices mounted on the human body. The images and experiences created by these devices will be virtually indistinguishable from reality. Brain–computer interfaces will allow controlling avatars, various objects, and digital processes with brain signals.

**2052**

- Cities and destinations equipped with MR: Cities will be equipped with AR layers that provide information according to the geographical location of the tourists. Cities will be filled with all kinds of information that can be filtered as needed using AR apps. For example, discount coupons, nearby museums, restaurants, hotels, bus stops, the history of the city, old buildings, 3D animations of famous people in history.
- MR tourism applications: The combination of artificial intelligence, big data, the Internet of Things, blockchain, AR and VR technologies in a single lens or glasses will enable the customisation of VR and AR content as customers require. For example, a receptionist wearing mixed reality (MR) glasses will be able to greet the customer by name the first time they arrive at the hotel.
- Multiple viewing modes in tourism MR applications: There will be multiple viewing modes in tourism applications with MR glasses, similar to normal glasses, where AR and VR technology is integrated. AR mode will allow you to see holograms in historical sites and museums. MR mode will allow you to see realistic virtual objects in the real world; for example, a real human avatar tourist guide will provide information. VR mode will provide the opportunity to visit the virtual destination with very realistic images. In AR and MR modes, the user will be able to manipulate multiple apps, files, and content using brain activity.
- Real VR and metaverse tourism destinations: Realistic touch, smell, and taste will be possible with VR contact lenses. Thanks to this technology, the user will be able to feel the cold, warm weather, light winds, or the feeling of swimming in a virtual sea. Real digital twins (Photorealistic Graphics) of cities and destinations will be created in metaverse environments. These environments will be connected to real life with holograms. The avatar will make the user's actions in the real world in the metaverse.

**Conclusion**

Tourists can visit destinations without the need for guides by using AR applications and can make individual tours by the use of navigation. They can examine the artefacts in museums with AR and VR and can even see them online from their homes and get information about them without going to museums and historical areas. With the AR menus of restaurants, information about the real image and content of the food can be obtained. Tourism businesses can advertise with AR markers placed in magazines, brochures, and on billboards. With VR games, difficult to get to and dangerous areas can be visited in a fun way in the virtual environment. Even past lifestyles can be experienced with AR and VR games, and historical areas and buildings without ruins can be visited with digital twins created.

AR and VR applications have advantages in interactive communication (sharing on social media, communicating with tourism businesses and customers), providing visual and written information and measurement (who used the application, how long they stayed on the site, which options attracted more attention, categorising users) and creating up-to-date content.

In the long run, tourism businesses will widely use AR and VR applications integrated with artificial intelligence, big data, blockchain, the Internet of Things, and recognition technologies. Soon, AR advertisements of tourism businesses will begin to be used on buildings. In the future, scanned 3D virtual copies of the real world, that is, digital twins, will be created, and these twins will be visited by thousands of people at the same time in metaverse environments. In the future, unmanned autonomous vehicles will benefit from outlining-based AR infrastructures, especially in navigation systems. VR applications, where interaction is limited today, will be able to be used with facial and bodily movements in the future. Individuals who cannot even go will be able to dive underwater in the Maldives in the virtual world with their body movements or climb the most dangerous places on Mount Everest and even physically feel the climate of the mountain. With projection-based AR applications, hotel rooms and restaurants will be created in the desired theme and image.

In the future, the separation of VR and AR will disappear, and MR applications will intensify. In addition, AR and VR applications will merge on contact lenses or glasses and will turn into a tool that we carry on us without disturbing our daily life. Again, in the future, tourists will be able to get information about historical sites and tourism businesses by wearing AR and VR-based contact lenses. Also, VR technology is expected to be of much higher resolution in terms of tracking sensors. With SixthSense technology, the physical characteristics of the user will be reflected in the virtual world by detecting features such as skin pigmentation, clothes worn, and hair colour. With the same technology, the body movements of the user in the real world will be reflected in the avatar in the virtual world. In other words, whatever body movement the user is performing in the real world, the avatar in the virtual world will do the same. Thus, the virtual and real worlds will turn into an intertwined environment.

## References

ars.electronica.art. (2022). *Mirages & Miracles – Fantastical Worlds, Brought Alive.* Retrieved 01.01.2022 from https://ars.electronica.art/homedelivery/de/mirages-et-miracles/

Barnes, S. (2016). Understanding virtual reality in marketing: Nature, implications and potential. *Implications and Potential* (November 3).

Berryman, D. R. (2012). Augmented reality: A review. *Medical Reference Services Quarterly*, 31(2), 212–218.

Billinghurst, M. (2021). Grand challenges for augmented reality. *Frontiers in Virtual Reality*, 2, 578080. doi: 10.3389/frvir.2021.578080

Carmigniani, J., Furht, B., Anisetti, M., Ceravolo, P., Damiani, E., & Ivkovic, M. (2011). Augmented reality technologies, systems and applications. *Multimedia Tools and Applications*, 51(1), 341–377.

Çeltek, E. (2015). Augmented Reality Advertisements in Tourism Marketing. In N. Ö. Taşkıran, & R. Yılmaz (Eds.), *Handbook of Research on Effective Advertising Strategies in the Social Media Age* (pp. 125–146). IGI Global. http://doi:10.4018/978-1-4666-8125-5.ch007

Çeltek, E. (2020). Progress and Development of Virtual Reality and Augmented Reality Technologies in Tourism: A Review of Publications from 2000 to 2018. In E. Çeltek (Ed.), *Handbook of Research on Smart Technology Applications in the Tourism Industry* (pp. 1–23). IGI Global. https://doi.org/10.4018/978-1-7998-1989-9.ch001

Çeltek, E. (2021). Gamification: Augmented Reality, Virtual Reality Games and Tourism Marketing Applications. In F. Xu, & D. Buhalis (Eds.), *Gamification for Tourism*. Bristol: Channelview Publications.

Chung, N., Han, H., and Joun, Y. (2015). Tourists' intention to visit a destination: The role of augmented reality(AR) application for a heritage site. *Computers in Human Behavior*, 50, 588–599.

Coates, C. (2021a). *How Museums are Using Augmented Reality.* Retrieved 01.01.2022 from https://www.museumnext.com/article/how-museums-are-using-augmented-reality/

Coates, C. (2021b). *Virtual Reality is a Big Trend in Museums, but what are the Best Examples of Museums Using VR?* Retrieved 01.01.2022 from https://www.museumnext.com/article/how-museums-are-using-virtual-reality/

Dahlstrom, L. (2017). *Through the Looking Glass: Starbucks' First In-Store Augmented Reality Experience*. Retrieved 01.01.2022 from https://stories.starbucks.com/stories/2017/starbucks-first-in-store-augmented-reality-experience/

Dingman, H. (2021). *Virtual Vacation: 11 VR Apps and Films That Let You Travel the World from Home*. https://www.oculus.com/blog/virtual-vacation-11-vr-apps-and-films-that-let-you-travel-the-world-from-home/

EY. (2019). *Turizm Sektörü Dijitalleşme Yol Haritası*. Türkiye.

Grigonis, H. (2018). *You AR what you Eat—Augmented Reality Menus are Coming to Snapchat*. Retrieved 01.01.2022 from https://www.digitaltrends.com/social-media/kabaq-ar-menus-on-snapchat-bareburger/

Guttentag, D. A. (2010). Virtual reality: Applications and implications for tourism. *Tourism Management*, 31(5), 637–651. doi:10.1016/j.tourman.2009.07.003

Halarnkar, P., Shah, S., Shah, H., Shah, H., & Shah, A. (2012). A review on virtual reality. *IJCSI International Journal of Computer Science Issues*, 9(6), 325–330.

Inamo-restaurant. (2022). *Our Restaurants*. Retrieved 01.01.2022 from https://www.inamo-restaurant.com/

Jung, T., Chung, N., & Leue, M. C. (2015). The determinants of recommendations to use augmented reality technologies: The case of a Korean theme park. *Tourism Management*, 49, 75.

Jung, T. H., & tom Dieck, M. C. (2017). Augmented reality, virtual reality and 3D printing for co-creation of value for visitor experience at cultural heritage places. *Journal of Place Management and Development*. doi:10.1108/JPMD-07-2016-0045

Kabaq. (2022). *Augmented Reality Food*. Retrieved 01.01.2022 from https://www.kabaq.io/

Kounavis, C. D., Kasimati, A. E., & Zamani, E. D. (2012). Enhancing the tourism experience through mobile augmented reality: Challenges and prospects. *International Journal of Engineering Business Management*, 4, 10.

Mandarinoriental. (2022). *SUBLIMOTION*. Retrieved 01.01.2022 from https://www.mandarinoriental.com/dubai/jumeira-beach/fine-dining/magical-cuisine/sublimotion

Manjunath, S. S., Navalli, R. R., Mallikarjun, G., & Anjum, R. (2020). Augmented reality based tourism application. *International Research Journal of Innovations in Engineering and Technology*, 4(5), 70.

Oculus. (2022a). *National Geographic Explore VR*. Retrieved 01.01.2022 from https://www.oculus.com/experiences/quest/2046607608728563/

Oculus. (2022b). *National Archaeological Museum: Live the Past*. Retrieved 01.01.2022 from https://www.oculus.com/experiences/go/1144207145676653/?ranking_trace=0_1144207145676653

Pantona, E., & Servidio, R. (2001). An exploratory study of the role of pervasive environments for promotion of tourism destinations. *Journal of Hospitality and Tourism Technology*, 2(1), 50–65. doi:10.1108/17579881111112412

Sharma, D. (2019). *3 Reasons Why Hotels Should Invest in Augmented Reality*. Retrieved 01.01.2022 from https://hospitalitytech.com/3-reasons-why-hotels-should-invest-augmented-reality

Sharp, E. (2022). *How Honeygrow Uses Virtual Reality to Train New Employees*. Retrieved 01.01.2022 from https://www.fox.temple.edu/posts/2018/01/how-honeygrow-uses-virtual-reality-to-train-new-employees/

Sherman, W. R., & Craig, A. B. (2018). *Understanding Virtual Reality: Interface, Application, and Design*. San Francisco, CA: Morgan Kaufmann.

Sing, P., & Pandey, M. (2014). Augmented reality advertising: An impactful platform for new age consumer engagement. *IOSR Journal of Business and Management*, 16(2), 24–28.

Sinha, D. (2021). *An Overview: Understanding Different Types of Augmented Reality*. Retrieved 01.12.2022 from https://www.analyticsinsight.net/an-overview-understanding-different-types-of-augmented-reality/

Smartify. (2022). *Smartify*. Retrieved 01.01.2022 from https://about.smartify.org/about-us

Steuer, J. (1992). Defining virtual reality: Dimensions determining telepresence. *Journal of Communication*, 42(4), 73–93.

tom Dieck, M. C., & Jung, T. H. (2017). Value of augmented reality at cultural heritage sites: A stake-holder approach. *Journal of Destination Marketing & Management*, 6(2), 110–117. doi:10.1016/j. jdmm.2017.03.002.

Tripventure (2020). *tripventure*. Retrieved from https://tripventure.soft112.com

Vallantin, C. (2017). *An Amadeus Company, Navitaire, Unveils the World's First Virtual Reality Travel Search and Booking Experience*. Retrieved 01.01.2022 from https://amadeus.com/en/insights/press-release/an-amadeus-company-navitaire-unveils-the-worlds-first-virtual-reality-travel-search-and-booking-experience

Wu, H. K., Lee, S. W. Y., Chang, H. Y., & Liang, J. C. (2013). Current status, opportunities and challenges of augmented reality in education. *Computers & Education*, 62, 41–49.

Yovcheva, Z., & Buhalis, D. (2013). Augmented reality in tourism: 10 unique applications explained. *Digital Tourism Think Tank*, 1–12.

# 29 Tourist engagement throughout the customer journey

## A service ecosystem approach

*Rodoula H. Tsiotsou and Ronald E. Goldsmith*

## Introduction

Tourism services have moved away from passive consumption to become more interactive, customised, and technology-driven. The rise of digital technology has empowered tourists and transformed them into digital content co-producers. New technology benefited tourism businesses by developing value-added experiences for travellers and boosting service efficiency (Buhalis et al., 2019). Tourists are well-connected and informed; they participate in the development of tourism experiences and co-create value (Tsiotsou, 2019, 2022a, 2022b) before, during, and after their trips (Tsiotsou & Wirtz, 2015). Moreover, tourists demand highly customised services, engage socially and technologically, communicate dynamically via social media, co-create experiences and content, and utilise mobile devices across multiple contact points of the customer journey (Buhalis et al., 2019). The spectacular increase of user-generated content (UGC) on social media platforms, for example, gives a large quantity of information that enables a first-hand assessment of tourist consumers' experiences, thoughts, and sentiments (Tsiotsou, 2019, 2022a, 2022b). As a result, tourists are increasingly becoming active consumers of tourism offerings, engaging throughout the tourism consumption process (Tsiotsou & Wirtz, 2015).

Although consumer engagement (CE) is a pivotal concept in tourism marketing, the literature is replete with nomological approaches investigating its antecedents and outcomes. Thus, there is no available study delineating tourist engagement (TE) at all stages of the consumption process. Therefore, we aim to explain TE throughout the consumption process of tourism offerings. To achieve this goal, we propose an integrative conceptual framework that delineates TE in the three stages of the consumption process (Tsiotsou & Wirtz, 2015) and at all levels of the service ecosystem. The proposed framework considers the pre-consumption, the service encounter, and the post-encounter stage as well as the micro-, meso-, and macro-levels of the tourism ecosystem. The chapter contributes to the literature by providing a comprehensive framework of TE and advancing our knowledge on the topic. It identifies knowledge gaps and provides useful future research directions.

The proposed framework is presented by analysing TE in all stages of the customer journey; the chapter concludes with future research directions.

## A proposed conceptual framework of tourist engagement

Engagement is a multidimensional concept including cognitive, emotional, and behavioural dimensions that reflects "actors' dispositions to invest resources in their interactions

DOI: 10.4324/9781003291763-33

with other connected actors in a service system" (Brodie et al., 2019, p. 183). Engagement is highly interactive, social, collective, and dialectic in nature, including dyadic, triadic, and networked interactions in a service ecosystem among various actors such as customers, service firms, technologies, robots, employees, competitors, and advertisers (Brodie et al., 2019; Tsiotsou, 2021). Thus, a dynamic network of structures in engagement exists in service ecosystems where actors invest their resources to co-create or co-destroy value (e.g., Brodie et al., 2019; Tsiotsou, 2021).

The present chapter uses the three-stage model of service consumption (Tsiotsou & Wirtz, 2015) to delineate TE. It takes a service ecosystem approach that informs the type (dyadic, triadic, and networked) and level of engagement (micro-, meso-, and macro-level) as well as the institutions that guide TE. The proposed framework named Tourist Engagement throughout the Customer Journey (TECUJO) includes the main interactions (e.g., tourist interactions with tourism personnel, with other tourists, with local communities, and with technologies) and the form of engagement (cognitive, emotional, and behavioural) at each stage of the tourism consumption process and at each level of the tourism ecosystem, and their outcomes (Figure 29.1).

### Tourist engagement in the pre-purchase stage

Need arousal (internal or external) triggers tourists to begin searching for information (cognitive, emotional, and behavioural engagement) and assess alternatives (cognitive and emotional engagement) before they make a buying decision during the pre-purchase stage.

Tourists typically engage in an information search from multiple sources to examine and assess alternative tourism options, form performance expectations of offerings, save money, and decrease risk (Tsiotsou & Wirtz, 2015). In addition to sources such as advertising and promotions paid for and sponsored by hospitality and tourism firms, tourists engage in an information search from different types of informal sources such as the Internet, social

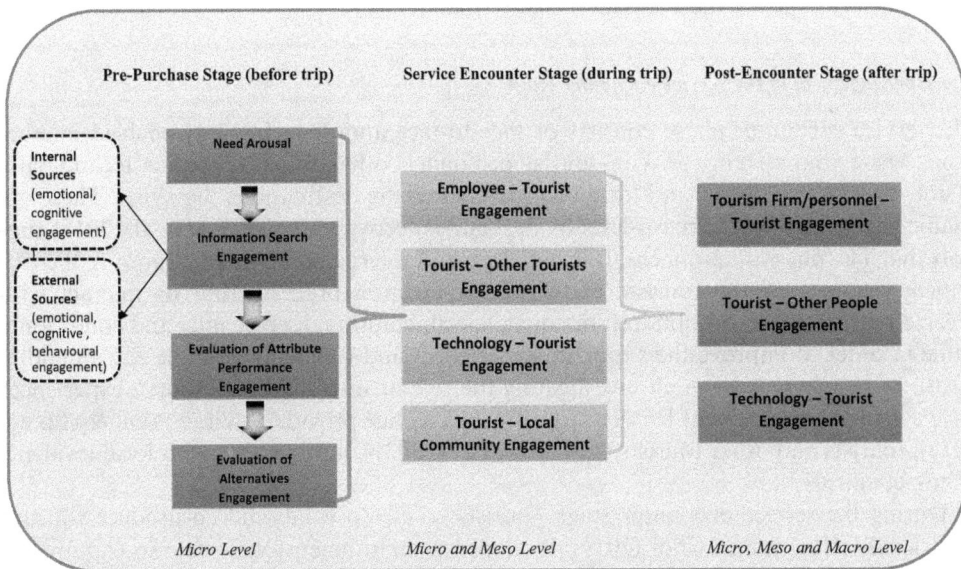

*Figure 29.1* The Tourist Engagement throughout the Customer Journey framework.

media, and online travel forums (Roozen & Raedts, 2018). Moreover, they rely heavily on personal sources of information such as family, friends, and co-workers because they consider them more trustworthy (Tsiotsou & Wirtz, 2015).

The Internet constitutes another source of information to compare tourism offerings (e.g., metasearch engines) and search for independent online reviews and ratings of destinations, hotels, restaurants, attractions, and tours (Tsiotsou & Wirtz, 2015). According to Fuel (2016), nearly half (48.4%) of North Americans use search engines when beginning to research a trip. Tourists' perceptions of tourist destination online content directly impact their visit intentions, while their satisfaction mediates this relationship (Majeed et al., 2020). The order of the search for tourism information categories varies according to the characteristics of the tourists, such as contingency factors (i.e., the purpose of the trip, the composition of the travel party, and the number of visits) and search outcomes (i.e., length of stay and number of attractions visited) (Kang et al., 2021).

In addition, social media can be used as an information source for tourists (Tsiotsou, 2022a, 2022b). As social media has grown in popularity, tourist-generated content (e.g., online reviews) has become a powerful source of information because they are read and trusted by a large number of individuals (Tsiotsou, 2022a, 2022b). It has been reported that 95% of travellers check travel reviews before booking and spend an average of 30 minutes online doing this. The goal of this online engagement is to learn about the location, activities, and experiences offered by individuals who share information about their recent tourist experiences (Bilgihan et al., 2016). In addition, some online tourist-to-tourist engagement is targeted at locating like-minded people to participate in leisure activities (Torres & Orlowski, 2017).

With their environmental embedding effects, new technologies such as augmented reality (AR) mirror increase TE, excite visitors to pay a premium price, and assist in building an effective AR tourism destination experience in the pre-purchase phase while monitoring tourists' behaviour (Huang, 2021). Moreover, new apps such as Elude match travellers with trips that fit their budgets, interests, and reward points and thus facilitate information searching and expedite the evaluation of alternatives.

*Tourist engagement in the service encounter stage*

The service encounter phase consists of face-to-face and/or technology-enabled interactions where tourists engage with tourist personnel, other tourists, service technologies (Tsiotsou & Wirtz, 2015), and local communities in the destinations they visit. These are triadic interactions at the meso-level of the tourism ecosystem. Campos et al. (2015) support that TE (physical and mental) in activities and interpersonal interactions are the two critical elements of the tourism experience co-creation process. Tourists interact with diverse tourism actors to obtain functional, social, emotional, epistemic, and conditional value in order to improve their tourism experiences and wellbeing (Randle & Zainuddin, 2020). Most previous research investigating the role of interactions in tourist experiences have examined engagement between tourists and service providers (Virabhakul & Huang, 2018), tourists and other tourists (Lin et al., 2019, 2021), and tourists with local residents (Zhou et al., 2015).

During the service encounter stage, tourists co-create value and co-produce tourism experiences. Huang and Choi (2019) developed a multidimensional scale to capture TE during service encounters consisting of four dimensions: relatedness, social interactions, interactions with employees, and activity-related TE. The interactions between tourism personnel and tourists constitute arguably a significant part of the service experience (Virabhakul & Huang, 2018), where value is created through participation (Yachin, 2018) and engagement in tourism-related activities. The considerable interactivity of the tourist experience significantly enhances engagement as tourists meet frontline personnel such as hotel personnel and tour guides. Inviting, involving, and giving room for tourists; adopting an experiential discourse; and including supporting moments designed to allow tourists to socialise with service personnel such as tour guides have been identified as facilitating practices that stimulate TE and transform tourists into active participants (Yachin, 2018). TE in this stage of the tourism experience can create emotional ties that bind the tourist to the destination through emotional and rational bonds. Tourists' interaction and engagement with personnel exhibiting emotional intelligence, especially empathy, situational sensitivity, and personnel availability, are the most significant elements in luxury tourism services (Iloranta & Komppula, 2022).

Research has also looked into tourist-to-tourist engagement and its effect on travel and tourism experiences. Tourist-to-tourist engagement may lead to friendships among visitors during a trip, substantially influencing their experiences. Social engagement among tourists in a coach tour and among backpackers has been found to play a pivotal role in their overall travel experiences (Murphy, 2001). Actually, the possibilities for social engagement during a backpacking trip determine to a great degree the choice of backpacking as a mode of transportation (Murphy, 2001). Interactions among tourists lead to relationships contributing to better cruise (Huang & Hsu, 2010) and conference experiences (Wei et al., 2017). Furthermore, interpersonal interactions among tourists impact their engagement level during the tourism experience co-creation process (Minkiewicz et al., 2014). Lin et al. (2019) demonstrated that self-disclosure increases visitors' perceived coherence and closeness, hence influencing their engagement. Lin et al. (2021) recently found that a sense of closeness and control are direct predictors of TE, whereas tourism information sharing and self-disclosure are indirect determinants.

TE with local communities is also an important factor that shapes tourism experiences. The tourist–local community engagement allows tourists to gain more reliable and honest insights about their travel experiences (Zhou et al., 2015). In this sense, tourists who want to engage with locals frequently do so since they believe such interactions enhance the memorability of their trips (Kim, 2010). Cordina, Gannon, and Croall (2019) report that sincere interactions and engagement with local fans in spectator sport tourism play a vital role in shaping authentic, memorable, and enjoyable tourist experiences. Lin et al. (2017) studied locals' engagement with visitors and discovered that economic and socio-cultural advantages, life satisfaction, and age were all positively associated with value co-creation processes.

In addition, tourist–local community engagement may lead to transformative outcomes and secure sustainability. The value co-creation interactions influence positively subjective wellbeing and residents' support for tourism growth (Chen et al., 2020). However, there is great diversity among residents' perceptions of their role in tourism creation experiences. Tourist behaviour, length of staying, changes in tourism business models, and labour market developments impact the depth of relationships between host communities and tourists, indicating the importance of not only tourists' characteristics such as origin, lifestyle, and mode of travelling but also developments in the tourism host community (Huber & Gross, 2021).

While at a destination, tourists also engage with technologies to enhance their tourism experiences. Tourists engage with a variety of technologies such as mobile apps, integrated payment methods, smart cards, and robots as well as information sources such as online travel agents, personal blogs, destination websites, tourism firm websites, social media, virtual reality (VR), and AR (Yang et al., 2020). Technology-mediated TE maximises the value of tourism service experiences (Um & Chung, 2021) and influences value co-creation (Buhalis et al., 2019). Engagement with smart tourism technologies has been linked to memorable tourism experiences and tourist happiness (Lee et al., 2018). In this phase, tourists engage mainly with city guide apps, mobile payments, Google maps, and the map locations of tourist attractions (Azis et al., 2020). Moreover, robots are increasingly being used to staff hotel front desks, assist human staffers, and interact with guests to increase TE during the hotel stay (Yang et al., 2020).

Finally, other elements of the tourism experience, such as buildings, architectural design, interiors, costumes, uniforms, weather, scenery, and ambient sounds, can positively or negatively influence customer TE (Virabhakul & Huang, 2018), thus raising satisfaction with the experience and promoting positive word-of-mouth in the post-encounter stage. These elements are of particular importance to people experiencing vulnerabilities (e.g., people with disabilities), influencing their tourism experience negatively (e.g., lack of ramps for wheelchairs result in limited access to venues) and subsequently their satisfaction and well-being (e.g., increased frustration) (Rubio-Escuderos et al., 2021).

### Tourist engagement in the post-encounter stage

Tourists can continue to be engaged with tourism actors (e.g., tourism firms, destinations, other tourists and technologies) passively and actively after they return home. Passively engaged, they can relive specific moments of the experience and form an overall evaluation of them and how well they met their needs and expectations formed during the pre-encounter stage. They can mainly dwell on interactions that enhance their enjoyment and engagement with the experience and venue (Tsiotsou, 2016). Actively engaged, tourists can describe and comment on their experience to others via social media, review sites, in person, and directly to the provider in the form of feedback. Some tourists will be much more active in this regard than others (Tsiotsou, 2019, 2022a).

Because of market saturation, customer empowerment, and greater competitiveness owing to the Internet and social media, TE is especially crucial for tourism services in creating value (Tsiotsou, 2022a). Existing research indicates that TE with tourist service providers or destination management organisations results in various positive outcomes, including value co-creation, commitment, loyalty, satisfaction, positive word-of-mouth and online reviews, and purchase intentions (Lin et al., 2019; Litvin, Goldsmith, & Pan, 2018; Tsiotsou, 2022a, 2022b).

Moreover, transformative value is created for tourists in the form of enhanced wellbeing (Lee et al., 2018) and local communities by securing their sustainability (Chen et al., 2020). TE promotes cultural engagement directly and indirectly, forming memorable tourist experiences, which positively affect intentions to return and to recommend the destination to others (Chen & Rahman, 2018). Highly engaged visitors are more likely to spend more time at destinations, spend more money on ancillary items (e.g., souvenirs, meals, extra activities), have more favourable attitudes about their travel experiences, and are more inclined to advocate or revisit tourism destinations (Gannon et al., 2017).

Social media's rapid development has empowered tourists and enabled two-way information communications in tourism services. Nowadays, tourists write many online reviews on tourism firms' and organisations' websites and/or social media as well as on online third-party sites like Tripadvisor, Google Business, Yelp, and Zomato. TE in providing third-party reviews is very valuable because ratings, testimonials, photographs, and videos increase the credibility and reputation of tourism businesses and destinations. Over the years, there has been a significant increase in online travel reviews, particularly in the hotel industry (Tsiotsou, 2022a, 2022b).

Online reviews are expressions of high-intensity TE that may create value for the reviewers, companies, and other tourists in tourism services owing to their experiential nature (Tsiotsou, 2022a, 2022b). When it comes to online reviews, tourists write them for a variety of reasons. These include gaining respect and recognition, increasing self-esteem, maintaining and/or augmenting social capital, strengthening social bonds, enjoying the online activity, altruistic motives (assisting others and preventing them from making poor decisions), and achieving enhanced cooperation (Munar & Jacobsen, 2014). Tourists' perceived behavioural control, subjective standards, and sense of belonging contribute to increased information sharing in the form of online reviews (Dixit et al., 2019; Qu & Lee, 2011). Moreover, tourism businesses derive value from online reviews through building loyalty and positive word-of-mouth, and increasing sales and market share. Other tourists also derive value through the information provided and recommendations, which reduce risk, save time and effort, and enable better purchasing decisions (Tsiotsou, 2019, 2022a). Thus, all actors involved derive value from online reviews. However, tourists can also destroy value via negative reviews, especially when their motive is revenge (Tsiotsou, 2022a, 2022b).

Culture seems to moderate the volume and valence of reviews. Tourists from collectivistic cultures with low socioeconomic status, less developed service settings, and high power distance are likely to submit more reviews, have lower service quality standards, and be more generous in their hotel ratings (Tsiotsou, 2019). When it comes to luxury hotels, Eastern and Northern Europeans are reported to be more generous in their review ratings than are Western and Southern Europeans. Furthermore, tourists from various cultures assess and interpret service quality differently. Eastern Europeans value more the physical evidence/environment of the hotel, Western Europeans prioritise the main product (room and cuisine), while Southern Europeans and Northern Europeans regard service personnel as the most important aspect of hotel services (Tsiotsou, 2022a).

Through reviews, tourism firms and destinations gain feedback, identify desirable features or flaws of their services, and gather valuable information in improving their services while enhancing their credibility and image (Zhang et al., 2016). At the macro-level, tourist evaluations may persuade governments and policymakers to improve their destinations' infrastructure and embrace new technologies to deliver better tourism experiences (e.g., smart cities) while preserving the environment (Buhalis et al., 2019; Um & Chung, 2021). At the meso-level, high review ratings positively impact the price of tourism services (Kim et al., 2015), increase sales (Kim et al., 2015), lead to higher revenue per availability (Phillips et al., 2015) and market share (Duverger, 2013), and boost online hotel bookings (Ye et al., 2011).

Tourists may engage in a dialogue with tourism services personnel after posting an online review (micro-level). When compared to no response, providing a timely and human (vs professional) response to negative reviews from the hotel elicits considerably higher favourable trust and customer concern inferences (Sparks et al., 2016). Tourism firms engage in online reviews and respond to tourists complaints for two purposes: for problem-solving (resolving customer complaints as quickly, efficiently, and discreetly as possible) and for accomplishing strategic goals such as relationship building (to engage tourists and the general public in a more ongoing relationship) and improving operational efficiency and effectiveness (Park & Allen, 2013).

## Conclusions and future research directions

The purpose of this chapter was to present a new framework, TECUJO, for understanding TE with tourism services throughout the customer journey and the service ecosystem. TECUJO proposes that tourists engage in various ways and levels before, during, and after the consumption process, and illustrates TE as a pivotal aspect of tourism services. The proposed framework suggests that hospitality and tourism managers can create a seamless series of encounters with customers to collect information to enhance engagement at each stage of the experience.

The growth of modern technologies has greatly facilitated this process by providing a variety of ways tourists can engage and communicate and co-create their experiences. Thus, TECUJO signifies that TE involves multiple actors (individual and collective, human and techno-actors) being in dyadic, triadic, and networked relationships in the tourism ecosystem. Moreover, TECUJO recognises that TE may create transformative value for tourists, tourism organisations/firms, local communities, tourism destinations, governments, and other relevant actors of the tourism ecosystem. Thus, TE does not or should not take place at the expense of tourism destinations' sustainability but it should contribute to local cultural and financial development. Furthermore, tourism services need to be inclusive and therefore a design-for-all approach to tourism premises (e.g., hotels, transportation, museums, and restaurants) is a necessary prerequisite in respecting, securing, and fulfilling human rights goals and securing sustainability.

The proposed framework provides a number of future research directions. One possible avenue of research would be to integrate studies of TE across all three stages or all levels of interaction for the same topics/subjects. This perspective would give a realistic view of the entire TE instead of the piecemeal approach of focusing on the stages independently. As tourism marketing evolves, researchers must broaden their perspective beyond the core tourism experience. Researchers have historically been preoccupied with the tourism service encounter (Yachin, 2018), disregarding to a great extent the pre- and post-consumption stages of TE, resulting in a type of myopia. Researchers, in particular, are losing opportunities to apply service theories and principles to pre-and post-consumption TE where tourism firms can differentiate and gain a competitive advantage. Thus, an integrative and comprehensive view might yield findings of theoretical and managerial significance.

TECUJO encourages future research to build models and hypotheses predicting the relative quantities of resources that tourism firms and organisations should invest in the three stages of the service consumption process and the three levels of interaction to facilitate and increase TE in a sustainable way. The initial focus should be on identifying the key drivers of TE that explain the differences in these relative amounts of investment in the three stages and levels of the tourism ecosystem. Increasing TE would provide tourism services with the opportunities to interact with tourists to increase their satisfaction with tourism experiences and learn valuable information on which to base service improvements and new tourism services development. TECUJO also signifies the importance of creating transformative value for all actors involved, including vulnerable people and, thus, supporting a transformative service research perspective in future TE research and practice (Tsiotsou & Diehl, 2022). However, cultural differences should be taken into account because these are very relevant in tourism services.

# References

Azis, N., Amin, M., Chan, S., & Aprilia, C. (2020). How smart tourism technologies affect tourist destination loyalty. *Journal of Hospitality and Tourism Technology*, *11*(4), 603–625. https://doi.org/10.1108/JHTT-01-2020-0005

Bilgihan, A., Barreda, A., Okumus, F., & Nusair, K. (2016). Consumer perception of knowledge-sharing in travel-related online social networks. *Tourism Management*, *52*, 287–296.

Brodie, R. J., Fehrer, J. A., Jaakkola, E., & Conduit, J. (2019). Actor engagement in networks: Defining the conceptual domain, *Journal of Service Research*, *22*(2), 173–188.

Buhalis, D., Harwood, T., Bogicevic, V., Viglia, G., Beldona, S., & Hofacker, C. (2019). Technological disruptions in services: Lessons from tourism and hospitality, *Journal of Service Management*, *30*(4), 484–506. https://doi.org/10.1108/JOSM-12-2018-0398

Campos, A. C., Mendes, J., Valle, P. O. D., & Scott, N. (2015). Co-creation of tourist experiences: A literature review. *Current Issues in Tourism*, *21*(4), 369–400.

Chen, H., & Rahman, I. (2018). Cultural tourism: An analysis of engagement, cultural contact, memorable tourism experience and destination loyalty. *Tourism Management Perspectives*, *26*, 153–163.

Chen, Y., Cottam, E., & Lin, Z. (2020). The effect of resident-tourist value co-creation on residents' well-being. *Journal of Hospitality and Tourism Management*, *44*, 30–37.

Cordina, R., Gannon, M. J., & Croall, R. (2019). Over and over: Local fans and spectator sport tourist engagement. *The Service Industries Journal*, *39*(7–8), 590–608.

Dixit, S., Badgaiyan, A. J., & Khare, A. (2019). An integrated model for predicting consumer's intention to write online reviews. *Journal of Retailing and Consumer Services*, *46*(C), 112–120.

Duverger, P. (2013). Curvilinear effects of user-generated content on hotels' market share: A dynamic panel-data analysis. *Journal of Travel Research*, *52*(4), 465–478.

Fuel. (2016). *Most Travelers Use Search Engines When Planning a Trip. They're Also Using Mobile, and Will Continue To*. Available at https://www.emarketer.com/Article/Most-Travelers-Use-Search-Engines-Planning-Trip/1013745

Gannon, M. J., Baxter, I. W., Collinson, E., Curran, R., Farrington, T., Glasgow, S., & Yalinay, O. (2017). Travelling for Umrah: Destination attributes, destination image, and post-travel intentions. *The Service Industries Journal*, *37*(7–8), 448–465.

Huang, J., & Hsu, C. H. (2010). The impact of customer-to-customer interaction on cruise experience and vacation satisfaction. *Journal of Travel Research*, *49*(1), 79–92.

Huang, S., & Choi, C. H.-S. (2019). Developing and validating a multidimensional tourist engagement scale (TES). *The Service Industries Journal*, *39*(4), 1–29. https://doi.org/10.1080/02642069.2019.1576641

Huang, T. L. (2021). Restorative experiences and online tourists' willingness to pay a price premium in an augmented reality environment. *Journal of Retailing and Consumer Services*, *58*, 102256. https://doi.org/10.1016/j.jretconser.2020.102256

Huber, D., & Gross, S. (2021). Local residents' contribution to tourist experiences: A community perspective from Garmisch-Partenkirchen, Germany. *Tourism Review*, ahead-of-print. https://doi.org/10.1108/TR-08-2020-0401

Iloranta, R., & Komppula, R. (2022). Service providers' perspective on the luxury tourist experience as a product. *Scandinavian Journal of Hospitality and Tourism*, *22*(1), 100568. https://doi.org/10.1080/15022250.2021.1946845

Kang, S., Kim, W. G., & Park, D. (2021). Understanding tourist information search behaviour: The power and insight of social network analysis. *Current Issues in Tourism*, *24*(3), 403–423. https://doi.org/10.1080/13683500.2020.1771290

Kim, J. H. (2010). Determining the factors affecting the memorable nature of travel experiences. *Journal of Travel & Tourism Marketing*, *27*(8), 780–796.

Kim, W. G., Lim, H., & Brymer, R. A. (2015). The effectiveness of managing social media on hotel performance. *International Journal of Hospitality Management*, *44*, 165–171.

Lee, H., Lee, J., Chung, N. & Koo, C. (2018).Tourists' happiness: Are there smart tourism technology effects? *Asia Pacific Journal of Tourism Research*, *23*(5), 486–501.

Lin, H., Zhang, M., & Gursoy, D. (2021). Effects of tourist-to-tourist interactions on experience cocreation: A self determination theory perspective. *Journal of Travel Research*, *76*(6), 153–167. https://doi.org/10.1177/00472875211019476

Lin, H., Zhang, M., Gursoy, D., & Fu, X. (2019). Impact of tourist-to-tourist interaction on tourism experience: The mediating role of cohesion and intimacy. *Annals of Tourism Research*, *76*, 153–167.

Lin, Z., Chen, Y., & Filieri, R. (2017). Resident-tourist value co-creation: The role of residents' perceived tourism impacts and life satisfaction. *Tourism Management*, *61*, 436–442.

Litvin, S. W., Goldsmith, R. E., & Pan, B. (2018). A retrospective view of electronic word-of-mouth in hospitality and tourism management. *International Journal of Contemporary Hospitality Management*, *30*(1), 313–325. https://doi.org/10.1108/IJCHM-08-2016-0461

Majeed, S., Zhou, Z., Lu, C. & Ramkissoon, H. (2020). Online tourism information and tourist behavior: A structural equation modeling analysis based on a self-administered survey. *Frontiers in Psychology*, *11*, 599. https://doi.org/10.3389/fpsyg.2020.00599

Minkiewicz, J., Evans, J., & Bridson, K. (2014). How do consumers co-create their experiences? An exploration in the heritage sector. *Journal of Marketing Management*, *30*(1/2), 30–59.

Munar, A. M., & Jacobsen, J. K. S. (2014). Motivations for sharing tourism experiences through social media. *Tourism Management*, *43*, 46–54.

Murphy, L. (2001). Exploring social interactions of backpackers. *Annals of Tourism Research*, *28*(1), 50–67.

Park, S-Y. & Allen, J. P. (2013). Responding to online reviews: Problem solving and engagement in hotels. *Cornell Hospitality Quarterly*, *54*(1), 64–73.

Phillips, P., Zigan, K., Santos, M., & Schegg, R. (2015). The interactive effects of online reviews on the determinants of Swiss hotel performance: A neural network analysis. *Tourism Management*, *50*, 130–141.

Qu, H., & Lee, H. (2011). Travelers' social identification and membership behaviors in online travel community. *Tourism Management*, *32*(6), 1262–1270.

Randle, M., & Zainuddin, N. (2020). Value creation and destruction in the marketisation of human services. *Journal of Services Marketing*, *34*(3), 347–361.

Roozen, I., & Raedts, M. (2018). The effects of online customer reviews and managerial responses on travelers' decision-making processes. *Journal of Hospitality Marketing & Management*, *27*(8), 973–996.

Rubio-Escuderos, L., García-Andreu, H., Michopoulou, E., & Buhalis, D. (2021). Perspectives on experiences of tourists with disabilities: Implications for their daily lives and for the tourist industry. *Tourism Recreation Research*. https://doi.org/10.1080/02508281.2021.1981071

Sparks, B. A., So, K. K. F., & Bradley, G. I. (2016). Responding to negative online reviews: The effects of hotel responses on customer inferences of trust and concern. *Tourism Management*, *53*, 74–85.

Torres, E. N., & Orlowski, M. (2017). Let's 'Meetup' at the theme park. *Journal of Vacation Marketing*, *23*(2), 159–171.

Tsiotsou, R. H. (2021). Introducing relational dialectics on actor engagement in the social media ecosystem. *Journal of Services Marketing*, *35*(3), 349–366. https://doi.org/10.1108/JSM-01-2020-0027

Tsiotsou, R. H. (2022a). Identifying value-creating aspects in luxury hotel services via third-party online reviews: A cross-cultural study. *International Journal of Retail & Distribution Management*, *50*(2), 183–205. https://doi.org/10.1108/IJRDM-04-2021-0207

Tsiotsou, R. H. (2022b). Value creation in tourism through active tourist engagement: A framework for online reviews. In Correia, Antonia & Dolnicar, Sara (eds.), *Women's Voices in Tourism Research – Contributions to Knowledge and Letters to the Next Generation*. The University of Queensland, ISBN: 978-1-74272-357-0. https://uq.pressbooks.pub/tourismknowledge/chapter/value-creation-in-tourism-through-active-tourist-engagement-a-framework-for-online-reviews-contributions-by-rodoula-h-tsiotsou/

Tsiotsou, R. H. (2019). Rate my firm: Cultural differences in service evaluations. *Journal of Services Marketing*, *33*(7), 815–836. https://doi.org/10.1108/JSM-12-2018-0358

Tsiotsou, R. H. (2016). The social aspects of consumption as predictors of consumer loyalty: Online vs. offline services. *Journal of Service Management*, *27*(2), 91–116. https://doi.org/10.1108/JOSM-04-2015-0117

Tsiotsou, R. H. & Diehl, S. (2022). Delineating transformative value creation through service communications: An integrative framework. *Journal of Service Management*, *33*(4/5), 531–551. https://doi.org/10.1108/JOSM-11-2021-0420

Tsiotsou, R. H., & Wirtz, J. (2015). The three-stage model of service consumption. In Bryson, J. Jr & Daniels, P. W. (eds.), *Handbook of Service Business: Management, Marketing, Innovation and Internationalisation* (pp. 105–128). Cheltenham: Edward Elgar Publishing.

Um, T., & Chung, N. (2021). Does smart tourism technology matter? Lessons from three smart tourism cities in South Korea. *Asia Pacific Journal of Tourism Research*, *26*(4), 396–414.

Virabhakul, V., & Huang, C. H. (2018). Effects of service experience on behavioral intentions: Serial multiple mediation model. *Journal of Hospitality Marketing & Management*, *27*(8), 99–1016.

Wei, W., Lu, Y., Miao, L., Cai, L. A., & Wang, C.-Y. (2017). Customer-customer interactions (CCIs) at conferences: An identity approach. *Tourism Management*, *59*, 154–170.

Yachin, J. M. (2018). The 'customer journey': Learning from customers in tourism experience encounters. *Tourism Management Perspectives*, *28*, 201–210.

Yang, L., Henthorne, T. L., & George, B. (2020). Artificial intelligence and robotics technology in the hospitality industry: Current applications and future trends. In George, B. & Paul, J. (eds.), *Digital Transformation in Business and Society* (pp. 211–228). Basingstoke, UK: Palgrave Macmillan.

Ye, Q., Law, R., Gu, B., & Chen, W. (2011). The influence of user-generated content on traveler behavior: An empirical investigation on the effects of e-word-of-mouth to hotel online bookings. *Computers in Human Behavior*, *27*(2), 634–639.

Zhang, Z., Zhang, Z., & Yang, Y. (2016). The power of expert identity: How website-recognized expert reviews influence travelers' online rating behavior. *Tourism Management*, *55*, 15–24.

Zhou, Q., Zhang, J., Zhang, H., & Ma, J. (2015). A structural model of host authenticity. *Annals of Tourism Research*, *55*, 28–45.

# 30 Perspectives for communication in social media

## The case of thermal spas

*Vera Antunes, Gisela Gonçalves and Cristina Estevão*

### Introduction

This chapter analyses the comments on social networks and the opportunities that may arise when looking at the future of communication. The results come from a content analysis of the social networks of Termas de Chaves and Termas de São Pedro do Sul located in Portugal, where it is identified that people are becoming more motivated to experience thermal services and products in a fast-changing world. This is a situation that needs to be rethought, as we look to the future of thermalism and communication.

The relationship between communication and thermal tourism is still a very incipient area of knowledge (Antunes et al., 2022). Some studies are beginning to emerge. Social media, namely Facebook and Instagram, have been revolutionising the communication strategies used, through different mechanisms and dynamics of information dissemination in the digital context (Ladkin & Buhalis, 2016). We are facing a complete change in how we communicate on the Internet, and it is a priority to focus on future trends and what is happening to improve the future of communication. The contribution of this research is to understand in what way thermal spas use social media to communicate and build relationships with their guests and followers. The aim is to identify the best digital communication strategy to communicate thermalism efficiently and with empathy. Considering the orientations of this chapter the following research questions were raised. Q1: What is the contribution of social media in motivating people to enjoy experiences in thermalism? Q2: How can new technologies and sentiment analysis help to promote thermalism?

The netnographic research method was used, since the content generated in social media proved to be a daily activity for people to share feelings, ideas, beliefs, and experiences, in this case about thermalism, providing through this methodology relevant guidelines for peers (Kozinets, 2010; Kozinets et al., 2018). To identify the main themes, an in-depth analysis of the comments was carried out and, through an artificial intelligence algorithm, the polarity of feelings in the comments to the respective publications was identified. This research highlights strategic communication through valuable tools to obtain comprehensive results on the interests, needs, and habits of thermal-goers, and subsequently for targeted communication on social media.

The four sections of this chapter look at the academic perspective, the methodology, what people said on social media, and – considering the trends and opportunities for thermalism – looking to the future and the need to change the way digital communication works, namely social media, and consequently the need for the research that revealed the new conceptual model.

DOI: 10.4324/9781003291763-34

**Literature review**

To Kazakov and Oyner (2020) wellness tourism has been an abundant and proliferating research topic over the last 75 years, with an increase in emerging research expected that will contribute to the scientific knowledge of the sector. Wellness tourism, as a broad multidimensional concept, is composed of ten different offer components: medical tourism, hot springs, spas, care of the body and mind, sports, culture, enogastronomy, nature and environment, spirituality, and events (Dini & Pencarelli, 2021). Thermalism is for Jahić and Selimović (2015) a narrower concept than health tourism and implies a type of tourism that is carried out in a thermal spa to treat certain illnesses, improve psycho-physical health, or relax the body. Thermal treatments are an important therapeutic tool supported by centuries of experience and numerous scientific studies proving their effectiveness (Silva et al., 2020). For Mendonça et al. (2021), as the world becomes increasingly connected and people seek a healthier lifestyle, they report that the health and wellness tourism industry seems determined to continue its rapid growth, so it is an urgent need to identify opportunities to communicate thermalism (Navarrete & Shaw, 2020).

Digital destination communication and online sales have brought tourists closer together and allowed access to sharing the best experiences coupled with the best available offers (Leite et al., 2021). Online social media such as Twitter, Facebook, Tumblr, and YouTube have become the usual platforms for hundreds of millions of Internet users in order to facilitate the creation and maintenance of interpersonal relationships (Tanoli & Pais, 2020). Gaffar et al. (2022) found that social media marketing can effectively build tourism destination image through various measures, represented in five dimensions, namely online communities, content sharing, interaction, accessibility, and credibility. Digital interaction with other people, including colleagues, family, friends, service providers, and even strangers through social media, presupposes sharing opinions, suggestions, doubts, and memories related to trips. It is therefore crucial to explore online social contact behaviours and analyse experiences (Fan et al., 2019).

People will search for information on social networks and see the local community's feedback on the space they intend to visit, and it is fundamental to integrate the local community into the communication strategy and work on the image of the tourist who wants to visit the spa (Ghaffar et al., 2021).

We need to understand the reasons why people write positive reviews online. Assiouras et al. (2019) identified intrinsic motives, that is, they may share their experiences on social networks or face-to-face with friends and relatives because they feel good about the experience, or simply help someone who needs help (emotional reward). Helping others during the visit or for a future stay, experience can be another act of reciprocity that may have extrinsic motives (expecting a reward, offer of a stay). The evolution of digital media leads Buhalis et al. (2020) to refer to a new form of communication known as "electronic word of mouth" (eWOM). eWOM has gained a new dimension through the exploitation of modern technology, especially with the increased use of the Internet. It is necessary to evaluate the possibility of carrying out differentiated e-commerce actions for men and women to improve offline and online management and the quality of design and eWOM to better promote trust and satisfaction (Buhalis et al., 2020).

Humans are innate storytellers and we are also innately primed to be influenced by stories (Casillo et al., 2021) – storytelling being a preferred strategy for generating content on social media. Transmedia storytelling is the technique of telling a story or experience on multiple platforms and formats, using current digital technologies (Su et al., 2022). For Dionisio and Nisi (2021) these successful tools are able to engage, inspire, and bring together online and offline audiences as an increasingly important strategy for the tourism industry. Su et al. (2022) developed an integrated model that demonstrates how the tourist experience affects tourists' subjective wellbeing through memories and narratives, confirming the importance of storytelling. Storytelling helps tourists reflect on unforgettable experiences. To Thomaz et al. (2017) monitoring what is said online about the destination, product, service, or tourism organisation to consumers and tourists on social media offers enormous opportunities and benefits. Social media mining (SMM) is derived from Web content mining – extracting knowledge from multimedia data on the Web, such as images, videos, and audio, using associated textual data (Thomaz et al., 2017). Nilashi et al. (2019) have built a model that, based on online comments, allows us to define clusters: segments of tourists with very similar characteristics and tastes. An uncommon form of sentiment analysis is through the classification of extreme sentiments that represent the most negative and positive feelings about a given topic, object, or individual (Tanoli & Pais, 2020). The managers should focus on responses to positive and authentic comments and use them on their websites and social media. The tourist has to feel that the comment they are reading is authentic (Kim & Kim, 2020). When the content is truly authentic it can self-distribute virally through word of mouth and social media sharing (Mendon et al., 2021).

The world is increasingly connected and artificial intelligence and machine learning will play a key role in future scientific research in continuous interoperability (Chen et al., 2021). At the intersection of computer science and social sciences, artificial intelligence emerges as an essential tool for communication management by transforming unstructured information into interpretable patterns (Bharadwaj et al., 2020; Egger & Yu, 2021; Mariani et al., 2018). We are entering a new universe of communication in which companies, social networks, and people are aligned through virtual reality and the creation of the metaverse. This is a concept that unites augmented reality and virtual reality, with a tendency to be a future bet to live experiences only with real influences in this ecosystem (Dwivedi et al., 2022). Despite efforts to understand this new phenomenon in academia and industry, the metaverse is a vortex in constant motion. Gursoy et al. (2022) note that it will change the way the hospitality and tourism industry operates, classifying the future of the metaverse into three broad categories: staging experiences in the metaverse, understanding possible changes in consumer behaviour, and marketing strategies and operations in the metaverse. Buhalis and Karatay (2022) describe mixed reality as a very realistic real-world scenario for users. It is so realistic that users cannot distinguish virtual content from physical objects, providing a seamless experience between real and digitally constructed environments. As per Buhalis and Karatay (2022), tourist destinations require considerable modernisation if innovative and transformative experiences are to be realised. The same authors note that the metaverse in tourist and cultural heritage sites will undoubtedly co-create transformative experiences. Generation Z is the first social group to grow up with Internet connectivity and portable electronic devices from a young age. They are technologically savvy, they have grown up in a connected world, and therefore they are potential users, influencers, and transmitters of the messages for the future.

Although the future is an unknown, organisations should create digital spaces where people can share stories and also establish message boards through artificial intelligence that allow tourists to record the feelings of a trip and describe experiences.

## Methodology

In this research we considered the case study methodology. In this perspective, Termas de Chaves and Termas de São Pedro do Sul located in Portugal were the ones considered for the empirical study. The selection criteria were:

- Both have Facebook and Instagram accounts;
- Both have the highest number of followers in both platforms;
- Both have similar therapeutic indications;
- Both have the largest number of thermal-goers in Portugal.

Data collection took place from June to September 2021 and the netnographic analysis method was used as it is a research method based on online observation, which allows analysing comments and attitudes on social media (Kozinets, 2010). In netnographic research one should consider some different strategies for data collection. In this case, through the administrator accounts of the hot springs under study and using the Sudota comment tool the comments from the Facebook business page and Instagram business profile were downloaded.

Comments were analysed using Natural Language Processing (NLP) methodology that classifies comments based on a sentiment lexicon. Similarly to Tanoli and Pais (2020) who propose in their study an unsupervised and language-independent approach to detect the extreme sentiments of people in social media, in this research we consider the extreme sentiment algorithm as a personal feeling, extremely positive or negative, the neutral for unbiased comments and for identifying people in comments, and the positive or negative for like and dislike. This analysis was chosen because it is an industry in which emotions and sensations are a reflection of the thermal experience (Campon-Cerro et al., 2020) and consequently feelings are easily observed in the comments.

Furthermore, the NVivo software was used to analyse the most quoted words in the comments, a strategy that was used as support to identify the themes.

## Results and discussion

The thermal spas under analysis use social media, Facebook and Instagram, as a working tool in which they share events, products, and services in order to provoke relationships and social interactions. Mendon et al. (2021) note that the success factor lies in identifying the most occurring and relevant opinions among users related to the specific topic, in this case thermalism. Hu et al. (2019) and Li et al. (2020) advise that the analysis of comments is a way to identify competitive advantages and align marketing and promotion strategies. According to Chen et al. (2021) selecting the most appropriate information is key to generating valid results, considering that the text of social media posts and comments can also provide relevant information for further research and guidelines. Similar to Egger & Yu (2021) and considering the complexity of analysing short text data on social media, this study also extends the scope of state-of-the-art data science methods for knowledge extraction for the thermal industry.

An exploratory analysis of the comments on the social networks of the thermal spas under study was carried out, which allowed the following themes and users to be identified (Tables 30.1 and 30.2).

*Table 30.1* Analysis of comments: Themes

| Themes | Comments |
| --- | --- |
| **Animation** | "Good evening always good animation😊🧐"; "It was a good show, very cheerful"; "Beautiful! Continue always with these animations, that still make the thermal spa besides beautiful, more attractive"; "So many nights I spent watching these animations"; "Beautiful nights I spent there, it was dancing and singing with the river at our feet" |
| **Nature** | "Miss wonderful! beautiful City😊"; "Wonderful 😄 ♡ pure nature"; "With this landscape, each time they make us more envious and nostalgic"; "I really like these hot springs a very beautiful landscape all around" |
| **Emotions** | "My little corner, my refuge…I love this blessed place!"; "Oh how I miss ♡ everything Thermal, treatments, employees, food, environment and nature in the 15 years stay friends for life ♡ A wonder whose enjoyment helps us to overcome(one of these days I'll return)"; "My land many misses 🌷😊"; "My land is beautiful …"; "I already miss it, even drinking the hot water.🐚😵, my mother's town enchants me and I say it's my holiday town😲😅😶" |
| **Quality** (Products and services) | "I really like this Cream is very good"; "The moisturiser is very good, with a good price, I advise you to try it"; "There is no doubt that they are great, as I have been there for several years and I feel much better from my rheumatoid arthritis."; "Excellent products!! Love it!! 👆⚙️"; "I really like it there and the massage and pools with very good therapist!"; "The Immersion in Thermal Water Pool, with the duration of 30 minutes, has the cost of 10€." |
| **Employee competencies** | "It's beautiful! And the bathhouse staff are five stars I recommend"; "These professionals are very important! I like it a lot. I will be back soon. Exceptional service👆"; "I like it very much, they are all very nice, from the therapists not forgetting the doctor Francisco and the nurse Maria"; "We were surrendered to the power of your waters but, more than anything, to your hospitality and friendliness! Thank you very much for this relaxing and invigorating morning! A big hug" |
| **Benefits** (Treatments and therapeutic indications) | "Are they recommended for rhinitis and sinus problems?"; "They are the best in Europe"; "This is the best for the spine, being well massaged"; "I liked being in the bath, then I went as usual to have physiotherapy in the pool. And see you next year😊" |
| **Experience** | "It's worth trying, very good treatments, friendliness and professionalism, excellent water, whenever I go to Chaves I go there😊"; "What a great hot water I recommend"; "I'm going back to thermalism this year. Who doesn't know, doesn't know what he loses in physical and mental well-being!"; "I've been there many times in Chaves and never went to do any treatment but next time I will"; "Beta André Cris Palma we have to go there to wet our hands" |
| **Covid** | "What a big bummer, I happened to have been there a fortnight ago, it was close, let's have faith and hope, it's going to be ok, ♡ 😔." "How much I've missed you. And how I miss my thermal Spa I'm counting the days left, not long now;" "Access to treatments requires the presentation of only one of the following: Digital Certificate COVID of the European Union…"; " ♡ everything will be fine, I've been there doing thermal I didn't see any risky situations, on the contrary, everything with great care, and a lot of hygiene, let's wait for good news, courage and strength" |

*Table 30.2* Analysis of comments: User profiles

| User Profiles | Comments |
| --- | --- |
| **Residents** | "Dr.ᵃ Fátima show the claws of the Flavienses we need the thermal spa working for the good of all" ["Flavienses" is the person natural or resident of Chaves.] |
| **Emigrants** | "I miss of the thermal water, the beautiful landscapes, the gastronomy and… the people of S. Pedro do Sul". "You are very good at welcoming people and it felt so good to me these holidays. Until the year if GOD wills and long live my summer city" |
| **Employees** | "To my dear colleagues I hope you are all in good health. This year I unfortunately couldn't go to work because of this evil virus that affected me, I hope that soon everything will be back to normal" |
| **Thermal-goers** | "I feel like visiting. Congratulations!"; "I spent a weekend, I loved it! Very calm without stress😊 We did thermal circuit, it was fantastic…. On Monday when I returned to work, I had the feeling I was coming from a week's holiday😃" |
| **Potential thermal-goers** | "Patricia Sofia Ana Teresa Gonçalves if we are not going far away we can opt for something like this😃😃" |

Similarly to Paine (2011) when analysing the comments it was also possible to identify that most of the content was written ignoring the rules of spelling and grammar. The lexical and syntactical problems such as slang, abbreviations, word configurations, and use of emoticons may confuse the artificial intelligence algorithm which requires a thorough analysis by researchers.

The experiences with thermal water stimulate an increase of positive emotions and consequently of life satisfaction in the tourist (Campon-Cerro et al., 2020; Huang et al., 2019). According to Tanoli and Pais (2020) an extreme feeling is the worst or best view, judgement, or evaluation formed in someone's mind about a certain subject or person. This analysis makes it possible to understand users' feelings in real time. A total of 679 comments were ascertained. Despite the high number of neutral comments (47%), due to user interaction through small conversations, identification of friends, and other comments where no sentiment could be identified, they were nevertheless very useful for understanding the flow of information. Only 2% of the comments were classified as negative (some criticism of the lack of adequate means for reduced mobility), which we consider, however, to be aspects to improve in the future. The greatest evidence falls into positive feelings (34%) and extremely positive (17%) from the returns of the thermal experience that provoked wellbeing in the person. Similarly to Molinillo et al. (2019) who report that more emphasis needs to be placed on delivering emotional (affective) messages, in this research it is suggested that, in addition to affective messages, greater priority should be given to emigrant users and those visiting friends and family. The success factor of sentiment analysis lies in the identification of the most occurring and relevant opinions among users, related to thermalism. Depending on the strategy to be applied to each social network, the development of contents with strong and impactful ideas should be privileged in order to create a vision with authenticity of what thermalism is.

Disruption is linked to technology madness, evolution, automated society (Buhalis et al., 2019), and according to Rowan and Galanakis (2020) arise from a global drive to discover innovations that will lead to greater competitiveness, impact, and value for business and society. Considering that in the era of big data analysis, new predictive models of structural integrity and prescriptive models can be built (Yadav et al., 2021), it is time to

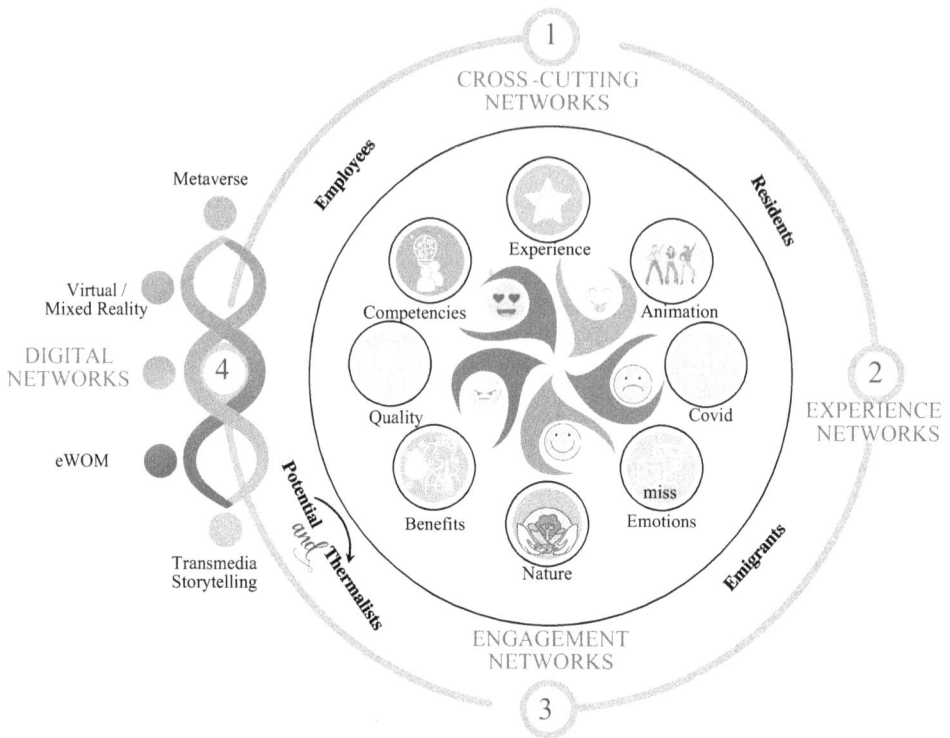

*Figure 30.1* Cross-cutting communication network model.

co-create and act accordingly with the analysed information. From this perspective we propose the following communication model (Figure 30.1).

The Internet is evolving at great speed and communication needs to keep up with it. This model is a reflection of people's comments with four future directions in the communication perspective:

1. Cross-cutting networks: Defining a strategy of transversal networks with the capacity to build links, involvement, and interaction, through a common denominator; in this particular case, through the identification of networks of production of health and wellbeing.
2. Experience networks: Develop a plan and create personalised communication strategies that appeal to feelings, capable of engaging people in what thermalism is all about and motivating new people to experience it. Positive thoughts and feelings create positive habits and generate movement. User types and comments should be analysed to consolidate emotions, through thoughts that become patterns of reasoning and behaviour.
3. Engagement networks: The strategy involves the creation of attractive narratives capable of engaging people. One should implement storytelling and design stories that emotionally touch the audience, create positive social change, and educate and connect the audience on a wide range of issues persuading people to act (Dionisio & Nisi, 2021). Gaffar et al. (2022) share the same opinion that interactivity needs to be improved by publishing posts with impact that encourage commenting and sharing, rather than merely informative texts. This strategy should be a more engaging portal between the

online community of local people, potential thermal-goers, management, and strangers. The emigrant community has huge potential to interact with other people and share experiences and thermal habits, evoking homesickness and localism.

4. Digital networks: This digital communication strategy involves different forms of inter-twined communication capable of influencing people's activities and experiences. There-fore, understanding the role that social contact plays in shaping the thermal experience is essential to explore its future impact on the sector. Digital communication tools will be privileged, with emphasis on virtual reality, metaverse, eWOM, transmedia storytell-ing, creation of segmented content, and an active digital presence, capable of attracting more qualified people to dialogue with users using communication about thermalism. For this purpose there is a need for special hardware, such as smart glasses, where the lenses are replaced by transparent screens and contain various sensors to track the user's environment, in the case of thermalism the experience should be connected with water (Buhalis & Karatay, 2022). In the future, thermal experiences, evoking emotions through sensory stimuli, may provide people with the future of the Internet, where new genera-tions create their new digital communication universe.

This model of online people engagement where participation and sharing are fundamental for most industries and sectors can be replicated and adapted to different themes. To do so, they should: identify cross-cutting networks for knowledge sharing; develop a plan with communication strategies that encourage experience; create attractive content using story-telling; and use advanced digital networks to provide experiences to people anywhere in the world.

Answering the first research question (Q1: What is the contribution of social media to motivate people to enjoy thermal experiences?), social media are fundamental to make thermal spas and their products known, to share the experiences, and to identify people, which motivates them to go. In fact, several questions were also identified with direct responses from the social media manager which could be a strategy to follow, because it is possible to align and refer the person immediately. When there are online comments, Liang and Li (2019) argue that one should respond to them. However, in their study they proved that tourists who receive a response have high expectations regarding the content of the response. Thus, managers should be aware of the importance of improving the quality of responses, and they should be informative and personalised, not using the standard responses that are usually the same for all customers.

Concerning Q2 (How can new technologies and sentiment analysis help to promote thermalism?), priority should be given to the specific wellbeing needs expressed by demand in comments (through sentiment analysis) to generate appealing narratives (through trans-media storytelling) in social networks. Avant-garde technologies should also be used to arouse curiosity to experiment and stimulate the demand for thermalism, provoking well-being in users and positively influencing people's lifestyles.

**Conclusions and perspectives for the future**

This chapter makes us reflect on the impact of digital communication strategies in thermal-ism. It opens a new frontier by using emotions as a portal to discover unconscious emo-tional patterns and instantly match them with their behaviour on social media. The main findings are the identification of the reasons for the comments, the emotions, and the unlimited potential that all people have to unconsciously generate new thermal experience habits with friends, family, and unknown people. The solution to the problem of absence and the manifest homesickness of the emigrant population is to make them feel present and this is possible through the production of attractive, creative, and sensorial contents using a strategy of content based on storytelling. Communication must be aligned, affective, and direct in order to remain in the foreseeable future.

We also observed that the vast majority of the information posted is harmless and that social media facilitate communication between people, visible in the feedback of comments, through the emergence of small conversations triggered by the post, and the identification of people in the post. With the aim of establishing notoriety indexes, use, opinion, and habits towards social media, it was concluded that artificial intelligence can be a valuable tool to obtain comprehensive results about the interests, needs, and habits of thermal users, and, subsequently, for a targeted communication on social media. Furthermore, aligned with the literature review, virtual reality and metaverse, through appropriate equipment, can influence people to have thermal experiences.

It is suggested that user profiles are analysed and characterised, as well as deepening the analysis of social media and sentiment to strengthen strategic communication in thermalism nationally and internationally. The results of this research have very relevant implications for the design of future experiences oriented towards the communication of the future, particularly in thermalism. Therefore, communicating innovation is based on scientific rigour, but also on the spirit of openness and creative intuition on which this chapter is based. The communication of science has conveyed theoretical representations of real life and physics that allow us to describe with high accuracy through artificial intelligence and the metaverse, predictions of behaviour that nature has in store for us in the near future. To this end, the following research questions are raised: Can the technology that drives social relationships displace engagement from real life to virtual reality? Will the technology be able to record the feelings of a trip and describe thermal experiences? Finally, a limitation was the fact that only two case studies were considered; it is suggested that the study is extended to another territory for greater consolidation of the results.

## References

Antunes, V., Gonçalves, G., & Estevão, C. (2022). A theoretical reflection on thermalism and communication: Future perspectives in times of crisis. *Journal of Hospitality and Tourism Insights*, ahead-of-print. https://doi.org/10.1108/JHTI-08-2021-0231

Assiouras, I., Skourtis, G., Giannopoulos, A., Buhalis, D., & Koniordos, M. (2019). Value co-creation and customer citizenship behavior. *Annals of Tourism Research*, *78*, 102742. https://doi.org/10.1016/j.annals.2019.102742

Bharadwaj, N., Ballings, M., & Naik, P. A. (2020). Cross-media consumption: Insights from super bowl advertising. *Journal of Interactive Marketing*, *50*, 17–31.

Buhalis, D., Harwood, T., Bogicevic, V., Viglia, G., Beldona, S., & Hofacker, C. (2019). Technological disruptions in services: Lessons from tourism and hospitality. *Journal of Service Management*, *30*(4), 484–506. https://doi.org/10.1108/josm-12-2018-0398

Buhalis, D., & Karatay, N. (2022). *Mixed Reality (MR) for Generation Z in Cultural Heritage Tourism Towards Metaverse* (pp. 16–27). Springer International Publishing. https://doi.org/10.1007/978-3-030-94751-4_2

Buhalis, D., Parra López, E., & Martinez-Gonzalez, J. A. (2020). Influence of young consumers' external and internal variables on their e-loyalty to tourism sites. *Journal of Destination Marketing & Management*, *15*, 100409. https://doi.org/10.1016/j.jdmm.2020.100409

Campon-Cerro, A. M., Di-Clemente, E., Hernandez-Mogollon, J. M., & Folgado-Fernandez, J. A. (2020). Healthy water-based tourism experiences: Their contribution to quality of life, satisfaction and loyalty. *International Journal of Environmental Research and Public Health*, *17*(6), Article 1961. https://doi.org/10.3390/ijerph17061961

Casillo, M., Santo, M. D., Lombardi, M., Mosca, R., Santaniello, D., & Valentino, C. (2021). Recommender Systems and Digital Storytelling to Enhance Tourism Experience in Cultural Heritage Sites. *Proceedings - 2021 IEEE International Conference on Smart Computing, SMARTCOMP 2021*.

Chen, J., Becken, S., & Stantic, B. (2021). Harnessing social media to understand tourist mobility: The role of information technology and big data. *Tourism Review*, ahead-of-print. https://doi.org/10.1108/TR-02-2021-0090

Dini, M., & Pencarelli, T. (2021). Wellness tourism and the components of its offer system: A holistic perspective. *Tourism Review*. https://doi.org/10.1108/Tr-08-2020-0373

Dionisio, M., & Nisi, V. (2021). Leveraging Transmedia storytelling to engage tourists in the understanding of the destination's local heritage. *Multimedia Tools and Applications*, *80*(26–27), 34813–34841. https://doi.org/10.1007/s11042-021-10949-2

Dwivedi, Y. K., Hughes, L., Baabdullah, A. M., Ribeiro-Navarrete, S., Giannakis, M., Al-Debei, M. M., Dennehy, D., Metri, B., Buhalis, D., Cheung, C. M. K., Conboy, K., Doyle, R., Dubey, R., Dutot, V., Felix, R., Goyal, D. P., Gustafsson, A., Hinsch, C., Jebabli, I., ... Wamba, S. F. (2022). Metaverse beyond the hype: Multidisciplinary perspectives on emerging challenges, opportunities, and agenda for research, practice and policy. *International Journal of Information Management*, *66*, 102542. https://doi.org/10.1016/j.ijinfomgt.2022.102542

Egger, R., & Yu, J. (2021). Identifying hidden semantic structures in Instagram data: A topic modelling comparison. *Tourism Review*. https://doi.org/10.1108/tr-05-2021-0244

Fan, D. X. F., Buhalis, D., & Lin, B. (2019). A tourist typology of online and face-to-face social contact: Destination immersion and tourism encapsulation/decapsulation. *Annals of Tourism Research*, *78*, 102757. https://doi.org/10.1016/j.annals.2019.102757

Gaffar, V., Tjahjono, B., Abdullah, T., & Sukmayadi, V. (2022). Like, tag and share: Bolstering social media marketing to improve intention to visit a nature-based tourism destination. *Tourism Review*, *77*(2), 451–470. https://doi.org/10.1108/tr-05-2020-0215

Ghaffar, S. F., Yee, C. P., Qing, H. M., & Zainuddin, M. R. B. (2021). Factors that are associated with wellness tourism in Malaysia. Hospitality, Tourism & Wellness Colloquium 4.0, Kota Bharu, Kelantan, Malaysia.

Gursoy, D., Malodia, S., & Dhir, A. (2022). The metaverse in the hospitality and tourism industry: An overview of current trends and future research directions. *Journal of Hospitality Marketing & Management*, *31*(5), 527–534. https://doi.org/10.1080/19368623.2022.2072504

Hu, F., Teichert, T., Liu, Y., Li, H., & Gundyreva, E. (2019). Evolving customer expectations of hospitality services: Differences in attribute effects on satisfaction and Re-Patronage. *Tourism Management*, *74*, 345–357. https://doi.org/10.1016/j.tourman.2019.04.010

Huang, Y.-C., Chen, C.-C. B., & Gao, M. J. (2019). Customer experience, well-being, and loyalty in the spa hotel context: Integrating the top-down & bottom-up theories of well-being. *Journal of Travel & Tourism Marketing*, *36*(5), 595–611. https://doi.org/10.1080/10548408.2019.1604293

Jahić, H., & Selimović, M. (2015). Balneological tourism in Fojnica–state and prospects. *Acta Geogr. Bosniae Herzeg*, *2*, 93–104.

Kazakov, S., & Oyner, O. (2020). Wellness tourism: A perspective article. *Tourism Review*. https://doi.org/10.1108/Tr-05-2019-0154

Kim, M., & Kim, J. (2020). The influence of authenticity of online reviews on trust formation among travelers. *Journal of Travel Research*, *59*(5), 763–776. https://doi.org/10.1177/0047287519868307

Kozinets, R. V. (2010). *Netnography Doing Ethnographic Research Online*. Thousand Oaks, CA: SAGE Publications Ltd.

Kozinets, R. V., Scaraboto, D., & Parmentier, M.-A. (2018). Evolving netnography: How brand auto-netnography, a netnographic sensibility, and more-than-human netnography can transform your research. *Journal of Marketing Management*, *34*(3–4), 231–242. https://doi.org/10.108 0/0267257x.2018.1446488

Ladkin, A., & Buhalis, D. (2016). Online and social media recruitment. *International Journal of Contemporary Hospitality Management*, *28*(2), 327–345. https://doi.org/10.1108/IJCHM-05-2014-0218

Leite, F., Correia, R. A. F., & Carvalho, A. (2021). *360° Integrated Model for the Management of Well-being Holistic Experiences in Tourist Destinations*. Iberian Conference on Information Systems and Technologies, CISTI.

Li, H., Hu, M., & Li, G. (2020). Forecasting tourism demand with multisource big data. *Annals of Tourism Research, 83*, 102912. https://doi.org/10.1016/j.annals.2020.102912

Liang, S., & Li, H. (2019). Respond more to good targets: An empirical study of managerial response strategy in online travel websites. *e-Review of Tourism Research, 16*(2/3), 215–223.

Mariani, M., Fatta, G., & Felice, M. (2018). Understanding customer satisfaction with services by leveraging big data: The role of services attributes and consumers' cultural background. *IEEE Access, 7*, 8195–8208. https://doi.org/10.1109/ACCESS.2018.2887300

Mendon, S., Dutta, P., Behl, A., & Lessmann, S. (2021). A hybrid approach of machine learning and lexicons to sentiment analysis: Enhanced insights from twitter data of natural disasters. *Information Systems Frontiers*. https://doi.org/10.1007/s10796-021-10107-x

Mendonça, V. J. D., Cunha, C. R., Correia, R. A. F., & Carvalho, A. M. O. (2021, 23–26 June 2021). Proposal for an Intelligent System to Stimulate the Demand for Thermal Tourism. *2021 16th Iberian Conference on Information Systems and Technologies (CISTI)*.

Molinillo, S., Anaya-Sánchez, R., Morrison, A. M., & Coca-Stefaniak, J. A. (2019). Smart city communication via social media: Analysing residents' and visitors' engagement. *Cities, 94*, 247–255. https://doi.org/10.1016/j.cities.2019.06.003

Navarrete, A. P., & Shaw, G. (2020). Spa tourism opportunities as strategic sector in aiding recovery from Covid-19: The Spanish model. *Tourism and Hospitality Research*. https://doi.org/Artn146735842097062610.1177/1467358420970626

Nilashi, M., Mardani, A., Liao, H., Ahmadi, H., Manaf, A. A., & Almukadi, W. (2019). A hybrid method with TOPSIS and machine learning techniques for sustainable development of green hotels considering online reviews. *Sustainability, 11*(21), 6013. https://doi.org/10.3390/su11216013

Paine, K. (2011). *Measure What Matters: Online Tools For Understanding Customers, Social Media, Engagement, and Key Relationships*. Hoboken, NJ: Wiley.

Rowan, N. J., & Galanakis, C. M. (2020). Unlocking challenges and opportunities presented by COVID-19 pandemic for cross-cutting disruption in agri-food and green deal innovations: Quo Vadis? *Science of the Total Environment, 748*, 141362. https://doi.org/10.1016/j.scitotenv.2020.141362

Silva, A., Oliveira, A. S., Vaz, C. V., Correia, S., Ferreira, R., Breitenfeld, L., Martinez-De-Oliveira, J., Palmeira-De-Oliveira, R., Pereira, C. M. F., Palmeira-De-Oliveira, A., & Cruz, M. T. (2020). Anti-inflammatory potential of Portuguese thermal waters. *Scientific Reports, 10*(1). https://doi.org/10.1038/s41598-020-79394-9

Su, L., Pan, L., Wen, J., & Phau, I. (2022). Effects of tourism experiences on tourists' subjective well-being through recollection and storytelling. *Journal of Vacation Marketing*. https://doi.org/10.1177/13567667221101414

Tanoli, I. K., & Pais, S. (2020). A Lexicon Based Approach to Detect Extreme Sentiments. *ICIMP 2020: The Fifteenth International Conference on Internet Monitoring and Protection*, Lisbon, Portugal.

Thomaz, G. M., Biz, A. A., Bettoni, E. M., Mendes-Filho, L., & Buhalis, D. (2017). Content mining framework in social media: A FIFA world cup 2014 case analysis. *Information & Management, 54*(6), 786–801. https://doi.org/10.1016/j.im.2016.11.005

Yadav, N., Verma, S., & Chikhalkar, R. D. (2021). eWOM, destination preference and consumer involvement–A stimulus-organism-response (SOR) lens. *Tourism Review, 77*(4), 1135–1152.

# 31 Urban mobility and mobility-as-a-service (MaaS) trends

*Xu Zhao, Claire Papaix and Yufang Zhou*

## Introduction

Travelling makes us happy, to the point where wellbeing has now become a focal point in tourism strategies (Pyke et al., 2016). Leisure travel has indeed an effect on tourists' overall satisfaction with life (sense of wellbeing) with a substantial overlap with information technology (Sirgy et al., 2011; Zhang et al., 2022). Recent advancements in information technology have contributed to the tourism revolution (Buhalis, 2020; Pedrana, 2014), giving tourists an improved experience when searching information, planning, and booking trips. However, current planning practices lack integration when it comes to information before/during/after trips, as the existence of multiple apps on the market illustrate for hotel booking, route planning, distance calculation, and price comparing. In addition, eco-friendly and green travel plans are most of the time not visible on such platforms, though they have become a priority. Therefore tourists rely mostly on car use, even when using Uber or self-driving options, and are not tempted by a shift towards public transport (PT), especially in unfamiliar contexts.

With the development of technology, mobility-as-a-service (MaaS) could be a way to solve such problems. This integrates existing and new mobility services into one single digital platform, providing customised door-to-door transport and offering personalised trip or package planning with payment options (Durand et al., 2018; Wong et al., 2018). It generates the search, with payment and booking functions of various transport services within one system (Kamargianni & Matyas, 2017; Polydoropoulou et al., 2020). The system also supports a shared transport system and improves visitors' comfort when travelling (Jittrapirom et al., 2017; Sochor et al., 2015). Thus, MaaS represents a novel concept to facilitate sustainable travel options and may well contribute to smarter, safer, and greener tourism overall (Alyavina et al., 2020; Fioreze et al., 2019).

By contrast with traditional transportation systems, mainly relying on extending transportation infrastructure and services, MaaS is based on the use of the existing transport network and combines various services to meet tourists' growing concern for safety, convenience, sustainability, and peacefulness while travelling. Research in tourism has transitioned from a heavy emphasis on computerised systems and efficiency to a focus on the attractiveness and ergonomy of travel management applications for a better adoption by stakeholders (Ali et al., 2020). The leisure travel experience itself is now the focus, as well as the level of wellbeing that can be derived from it. Such positive outcomes of MaaS also reach the employment sector as well as prompting new initiatives relating to the Sustainable Development Goals (SDGs) (Jones et al., 2017).

DOI: 10.4324/9781003291763-35

MaaS services are used in 41 cities across 17 countries. Among them, Whim is a mobility service provider in Finland, focusing on business innovations (Smith et al., 2018). In this case, the MaaS platform and bundled services and payments, including a-pay-per-ride option for users, are organised by the Finnish government (Aapaoja et al., 2017; Zhang & Zhang, 2021). Different in its design and governance, Beijing MaaS directly links the digital tool to the carbon market. Using open data, the Beijing MaaS platform launched the MaaS Mobility for Green City initiative in 2019, which allows travellers to book, pay, and use public transportation easily (Zhao et al., 2022). This initiative also rewards people for using PT which relies on the carbon market. While providing route planning and navigation services, the Beijing MaaS platform offers visitors who choose to travel by bus, subway, bicycle, or on foot carbon points based on the distance they travelled, which can be redeemed into vouchers or video platform membership (Mobility Transition in China, 2021). Furthermore, the platform allows users to check for new cases of COVID-19 during their trips so that they can make informed decisions, should infection levels increase.

The rest of this chapter focuses on the Beijing MaaS case study and discusses ways in which this digital platform contributes to Beijing's visitor economy leverage by comparing the results of the Beijing MaaS attitude surveys for 2020 and 2021 during the COVID-19 pandemic. The development of the MaaS platform in Beijing and its functioning are also discussed and compared to other schemes elsewhere in the world. Instead of developing more traditional travel infrastructure, digital technologies are herein promoted.

### Literature review

#### *MaaS and tourism sustainability*

Sustainability generally refers to a balance of economic, social, and environmental goals, including those that involve long-term, indirect, and non-marketed impacts (Figure 31.1) (Marletto & Mameli, 2012). With the emergence of the urban age, sustainability research

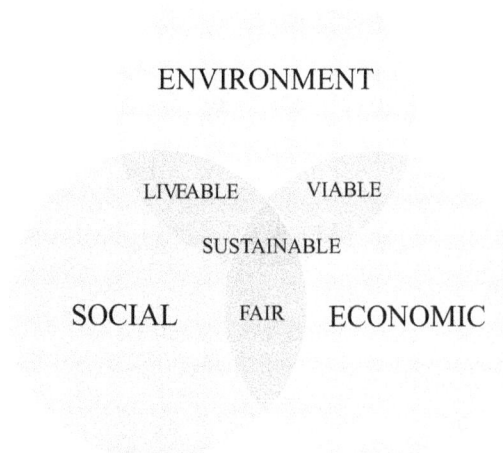

ENVIRONMENT

LIVEABLE   VIABLE

SUSTAINABLE

SOCIAL   FAIR   ECONOMIC

*Figure 31.1* Sustainability triptych.
Source: Based on Alhaddi (2015).

is facing a new frontier. Concern about sustainability is rooted in the growing awareness that human activities have significant environmental impacts that can impose economic, social, and ecological costs to current and future generations. Some common examples include global air pollution, rapid depletion of natural resources, such as fresh water and fisheries, and the cross-border nature of many environmental problems, highlighting the need to reconsider human impacts from a broad perspective. In fact, 2015 marked the adoption of the 2030 Agenda for Sustainable Development by governments, along with the SDGs. MaaS's application in tourism is consistent with the SDG framework, namely Goal 8 "to promote sustained, inclusive and sustainable economic growth, full and productive employment and decent work for all" as well as Goal 13 "to take urgent action to address climate change and its impacts" (Jones et al., 2017). Such a global policy framework targets the reduction of extreme poverty, flight inequality, and climate change for 2030.

Transport is a key contributing factor to the development and management of sustainable tourism destinations, particularly in an urban context (Coca-Stefaniak & Morrison, 2022; Pellegrino, 2021), where a systems-based approach to the management of destinations is of paramount importance (Morrison et al., 2018; Morrison & Maxim, 2022). The development of information and communication technologies (ICTs) has contributed to the reliability of public transport and improved the overall experience of travel for users with better prediction of travel flows and network availability (Rodrigue, 2020; Signorile et al., 2018). Despite this, tourists may remain discouraged from visiting cities, particularly when lengthy journeys and crammed conditions in transport services occur (Nikitas et al., 2017). As a result, visitors often opt for a single, less tiring transportation option, such as a private car (Karlsson et al., 2020; Weng et al., 2018), which contributes to traffic congestion problems in urban destinations.

By bringing together multiple modes of travel and combining different transport providers into a single service platform, that is, an integrated mobile application, MaaS alleviates this problem (Karlsson et al., 2020). However, MaaS platform efficiency often varies depending on the strategic priorities of local stakeholders (Kivimaa & Rogge, 2022). Table 31.1 outlines MaaS services as implemented in Europe.

### MaaS and tourist wellbeing

As a key enabler of sustainable mobility, MaaS can provide more convenient transport services for tourists. Travel time savings are found to improve tourists' satisfaction (Ram et al., 2014), and studies in eight European cities also reveal linkages between accessibility and tourists' wellbeing (Susilo & Cats, 2014). In comparison with our case study, the MaaS platform in China has had an impact on the wellbeing of transport system users (tourists and locals) through considering all stakeholders and addressing their concerns, needs, and values, particularly on the travel time savings dimension, as well as the improved convenience of journeys (Sarasini et al., 2017; Vij et al., 2020). Observations from the MaaS platform in China reported a struggle with the shared bike's barcode scanning system, being too time consuming. The interface was then upgraded with an automatic scanning functionality. Another example was the difficulty to find inner-city bus stations at stops due to their size and visibility. Indoor navigation itineraries were therefore designed and suggested to passengers, with an enhanced version for elderly and disabled groups.

As shown in Figure 31.2, investigating tourist activities has led to the identification of different definitions of wellbeing. Among those, objective wellbeing (OWB) relates to basic human needs, such as income, health, education, social and environmental surroundings,

*Table 31.1* Operating models of MaaS development in selected countries

| | Rejseplan | Whim\Tuup | UbiGo | Moovit | Wien Mobile | Moovel-Reach Now | NaviGoGo |
|---|---|---|---|---|---|---|---|
| Country City | Denmark Copenhagen | Finland Helsinki\Turku | Sweden Gothenburg, Stockholm | Europe Israel 100+ countries and regions | Austria Vienna | Germany Stuttgart Karlsruhe Aschaffenburg | UK Scotland |
| Starting time | 2018 | 2016 | 2013 | 2012 | 2016 | 2015 | 2018 |
| Functions | Booking, reservations | Users only need to pay a monthly fee to enjoy the packaged travel service | | The crowdsourced bus travel navigation application can help users in different countries and cities to easily query real-time bus conditions, and provide information such as the best travel mode, best route, and time | Planning, booking, and paying | Intelligent and seamless interconnection of different mobility products including booking and payment, providing solutions for urban personal transportation | Specifically for the daily urban or suburban travel needs of young people aged 16–25, opens up travel processes such as search, booking, and payment to create a one-stop travel service platform |
| Features | Shared electric vehicles and shared bicycles | OTP-based route planning | Pay on a monthly subscription basis | Crowdsourced data maintenance, 720,000 user community operation | One-stop travel service solution | Multiple travel options | Youth design concept |
| Effects | 3.7 million downloads | Demand response transportation services | 97% are willing to use, 50%+ satisfaction | 2017 Apple's Best Apps Official Transit Apps for Multiple Global Events | 500,000+ downloads on Google Play | 22 cities, 7.5 million+ users | 75% choose public transport and 80% pay for the whole travel chain |

*Source:* Li et al. (2020).

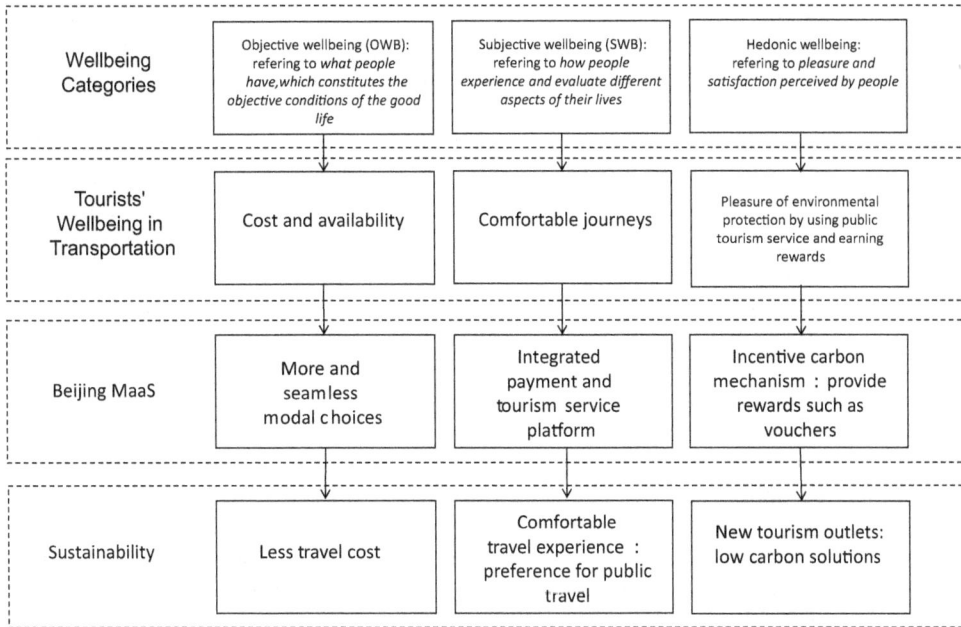

| Wellbeing Categories | Objective wellbeing (OWB): refering to *what people have, which constitutes the objective conditions of the good life* | Subjective wellbeing (SWB): refering to *how people experience and evaluate different aspects of their lives* | Hedonic wellbeing: refering to *pleasure and satisfaction perceived by people* |
|---|---|---|---|
| Tourists' Wellbeing in Transportation | Cost and availability | Comfortable journeys | Pleasure of environmental protection by using public tourism service and earning rewards |
| Beijing MaaS | More and seamless modal choices | Integrated payment and tourism service platform | Incentive carbon mechanism : provide rewards such as vouchers |
| Sustainability | Less travel cost | Comfortable travel experience : preference for public travel | New tourism outlets: low carbon solutions |

*Figure 31.2* Links between wellbeing, MaaS, and sustainability.

safety and security levels, and civil rights engagement (Reardon & Abdallah, 2013), for all of which transport is seen as a key daily experience. Subjective wellbeing (SWB) is associated with overall life evaluation and is usually found to be time invariant. It includes different domains, such as family sphere, job situation, sociality, and hobbies (Bergstad et al., 2010). Hedonic wellbeing refers to pleasure and dissatisfaction experienced by individuals while travelling for leisure (Kahneman et al., 1999). However, there is no general consensus on ways to define wellbeing, and it is therefore not an easy task to measure it accurately to prompt initiatives for sustainable tourism travel (Kahneman, 2000). From such initial definitions, a first pivotal element connecting tourism travel to wellbeing is the term "travel satisfaction" (De Vos et al., 2013). Travel satisfaction is found to result from the travel attributes (Fang et al., 2021), interests, and options of tourists, and to play a key role in explaining elements linked to the journey itself (e.g., mode selection, travel mindsets) (De Vos, 2018). Travel satisfaction is reflected by several dimensions, such as OWB (e.g., economic cost), SWB (e.g., comfort), and hedonic wellbeing (e.g., eco-friendliness of the mode). In addition, the notion of wellbeing in a transport context can take the form of the "actualisation of human potentials", according to the eudaimonic conception of the term, and/or "be conceptualised only as an internal subjective experience of each particular individual" (Alatartseva & Barysheva, 2015, p. 38).

MaaS could become the new transport paradigm as termed by Alyavina et al. (2022) with expected positive effects on the environment, traffic congestion, and accessibility, if such opportunities remain beyond the trendy Uberisation of the PT system, and if further empirical evidence is shown regarding health and wellbeing, real cost, employment effect, and service quality, as mentioned by Zhao et al. (2022). Cobbold et al. (2022) add to this list the enabled flexibility, attraction, and health benefits obtained through increased physical activity, stress relief, health satisfaction, emotional balance, and overall quality of life provided by the services if such benefits hold in multimodal contexts.

With societal and technological developments, holiday trip distances initially increased (e.g., in the case of the UK; Transport for London, 2020) with a resulting contribution to carbon emission of tourism travel, accelerating climate change (Wilson & Hannam, 2017). Specifically in Beijing, the transport sector contributed to 30% of all carbon dioxide emissions in 2020. With recent announcements of China's carbon neutrality target for 2060, encouraging the use of PT and notably MaaS solutions could be a viable option to cut carbon emissions while maintaining a profitable industry. In addition, a low-carbon tourism target has been set and promoted recently in China (Liu, Yang, & Huang, 2021), aiming to foster the tourism experience with the rollout of low carbon technologies and less energy intensive activities from transportation to entertainment (Bhaktikul et al., 2021). A great number of strategies have therefore been deployed to promote sustainable tourism travel behaviours. MaaS is a feasible illustration of such top-down solutions, providing a helicopter view on traffic conditions, and opportunities to balance transport flows (Audouin & Finger, 2018; Pangbourne et al., 2020). From a bottom-up perspective, MaaS can then be used to encourage citizens to use more sustainable transport modes, such as public and active transport, through economic incentives. For instance, travellers may accumulate reduced carbon emissions points or badges every time they complete a green trip. Hence, Beijing MaaS allows individuals to participate in the regional carbon trading scheme, where tourists can redeem gifts, for example, transport tickets, vouchers, or cash when using the system.

MaaS also helps to alleviate health concerns, particularly regarding tourists' travel intentions in a COVID-19 context (Li et al., 2021a). With almost half of the globe subjected to travel restrictions at the height of the pandemic (Raham et al., 2021) and now embarking on a post-COVID-19 route to recovery (Li et al., 2021b), one of the key challenges of destination management organisations around the world has been the reduction of tourists' perceptions of risk (Xie et al., 2021) and health concerns (Shin et al., 2022). At the same time, the pandemic has led to a surge in the use of digital technologies often linked to social distancing norms and nationwide lockdowns (Chadee et al., 2021). Although the crisis has had tragic consequences for many people, it has also provided a unique window of opportunity to shape the future of mobility, acting as a catalyst for the development of technological innovations and, more fundamentally, reset strategies for urban mobility towards more sustainable, resilient, and human-centric tourism systems (Papaix & Coca-Stefaniak, 2021).

## *MaaS stakeholder views*

MaaS development, as belonging to transport planning, requires the management of a bi-directional communication process and the coordination of specific programmes and skills from various stakeholders, and, in particular, conflicting interests and variables for anticipating problems (Le Pira et al., 2016). Stakeholder involvement should therefore be planned; Marletto and Mameli (2012) propose a consultation procedure with specialists, citizens, and stakeholders to engage in different ways with "top-down" stages (discussion on specialists' outcomes) and "bottom-up" stages (on MaaS users' experience).

Although the coordination between stakeholders is crucial for the development of MaaS, few researchers have discussed it from the full spectrum of actors (Moradi & Vagnoni, 2018). In fact, considering a large variety of stakeholders – for example, transport users, policy makers, public institutions, local communities, NGOs, public transport operators, service providers, specialists, retailers, the private sectors, and the third sector (Le Pira et al., 2016) – does not make planning costly and inconclusive but on the contrary enriches it and

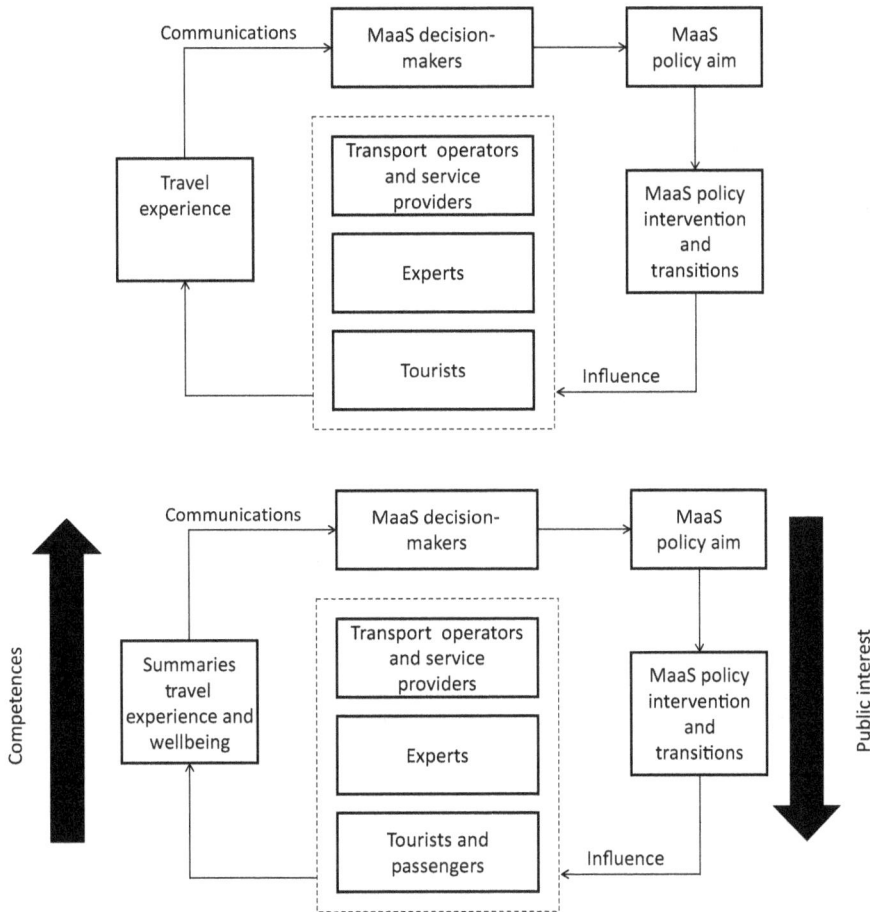

*Figure 31.3* Public stakeholder framework.

*Source*: Reardon & Abdallah (2013); Freeman & Medoff (1984); Le Pira et al. (2016); Cavoli et al. (2014); Mameli & Marletto (2009).

ensures its perennity. As shown in Figure 31.3, actors can be categorized into four classes: MaaS decision-makers, transport operators and service providers, experts, and MaaS users.

Stakeholders may indeed propose new problem definitions not previously considered by experts and suggest innovations to solve these and other problems. As a result, power is dispersed into networks of relationships (Knoke, 1990) after having been concentrated in the hands of a few (Smith, 1993). Thus, it can be argued that particularly powerful stakeholders may undermine participatory initiatives in transportation planning and as illustrated in the following case study.

## Case study: Beijing's MaaS for tourists

### Context of the Beijing MaaS platform development for tourists

It was in November 2019 that China's first MaaS platform, Beijing MaaS, emerged. The vision was to provide users with an integrated service platform which could be used for navigating and, most importantly, providing green incentives when doing so. Providing public data on PT services (i.e., bus, subway, railways linking cities and rural areas) as well as on active and shared mobilities (i.e., walking, cycling, and ride-hailing), information on departure/arrival times, carriage crowding, or reminders for transfer and disembarkation became available to users. Green travel carbon credits and trading mechanism were also made available. Beijing MaaS built on the existing mapping, navigation, and location-based service – Amap mobile app, owned by Alibaba. Figure 31.4 outlines some of its key features.

The following year in 2020, Beijing MaaS became operational and started to share such open data to summarise travellers' experiences and wellbeing. Special add-in functions were then built, such as that of checking for new COVID-19 cases. Carbon trade activity was launched in parallel, aiming at reducing emissions from citizen journeys using green travel records and rewarding them through ad hoc incentives. Relying on MaaS, the programme developed the Beijing Emission Reduction Methodology of Low Carbon Travel and a specific protocol measuring emissions dynamically, relying on big data technologies. Volunteers could enjoy all green travel modes and automatically gain rewards from the scheme. In 2021, Beijing MaaS signed its first carbon-incentive trading scheme that associated social benefits to citizens' willingness to reduce carbon emissions. At the start of 2022, more functionalities were explored, such as the display of bikes on the platform, and the increase of P+R (park and ride) facilities and payment services, as summarized in Figure 31.5.

In 2023, Beijing's population is 21 million, while the Beijing MaaS platform has already more than 30 million users, with over 300,000 of them registered to the carbon reduction activity (Beijing Municipal Commission of Transport, 2022). In what follows, we investigate Beijing MaaS's success story and the difficulties encountered to derive lessons for other places in the world currently experimenting with MaaS, such as Rome, Athens, Jakarta, and Rio de Janeiro (United Nations, 2019).

**Data collection and analysis**

A survey was administered to users of the Beijing MaaS platform in December 2020 and then again in 2021 to gain a better understanding of travellers' intentions, concerns, and needs during travelling, and to explore their perceptions and expectations with regards to the platform service. The online survey performed was divided into three sections: visitors' satisfaction and familiarity with the platform, travel intentions and actual behaviours related to green travel, and respondents' socio-demographic data. Figure 31.6 illustrates the geographical context of the areas of the city where the data were collected.

Overall, 260 questionnaires were received in 2020 and a further 1,580 were used in 2021 once cleaned, with the description of the randomized sample exhibited in Table 31.2.

*Rollout of the MaaS platform in 2020: Return on experience*

The survey carried out in 2020 revealed that 60% of Beijing visitors were aware of the existence of the Beijing MaaS platform, though 30% had no intention of using it. In 2020, visitors' satisfaction with the platform service was 76%, with its integrated door-to-door mobility service deemed to be the most valuable feature. Other services offered by the platform, such as its high degree of flexibility of modes and timing, were also praised by users.

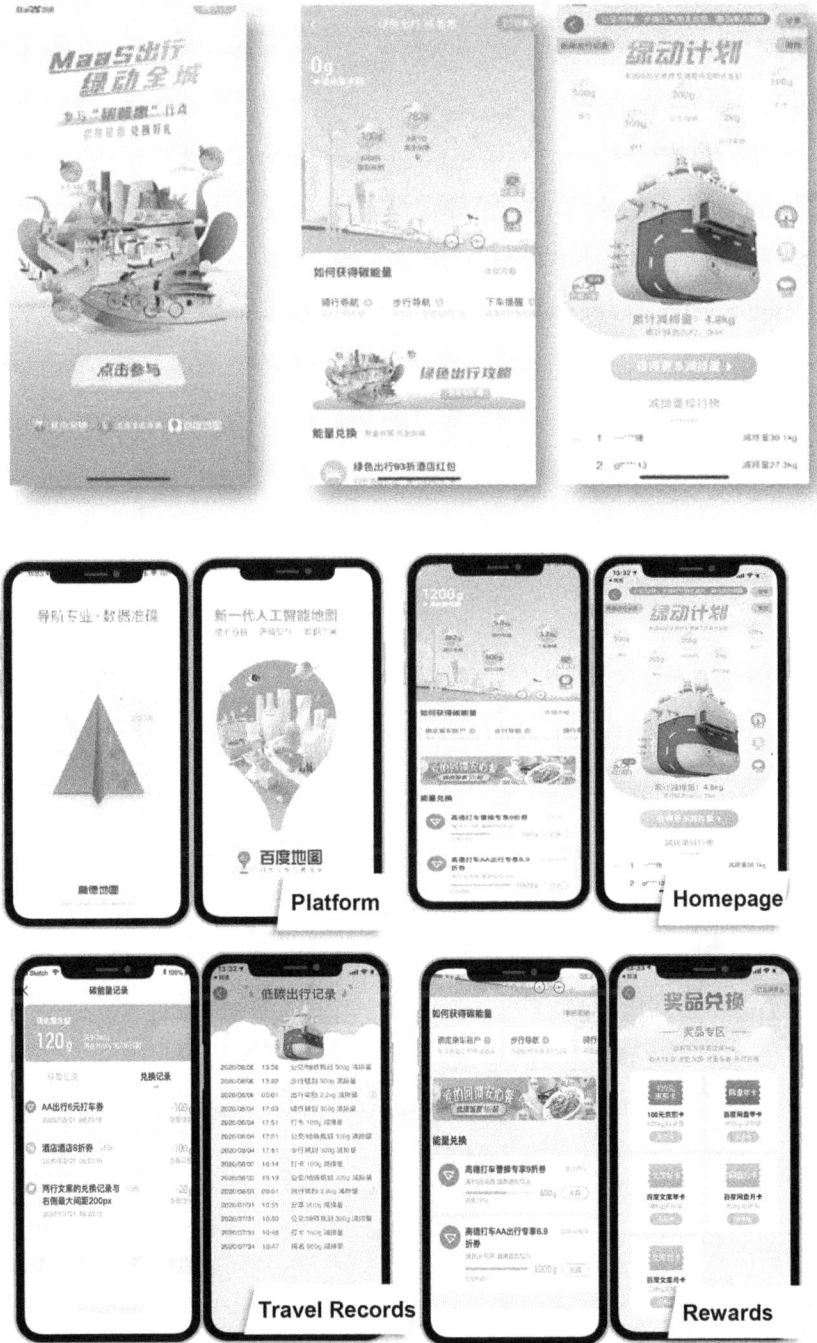

*Figure 31.4* App features of Beijing MaaS.

*Source*: Beijing Municipal Commission of Transport (2022).

**Government -enterprise cooperation:**
Beijing   MaaS service platform launched

**Green Incentive Mechanism:**
*MaaS Mobility for Green City*
launched

November 2019

April 2020

September 2021

November 2019

September 2020

**Data sharing mechanism:**
Published   *Beijing Traffic and
Travel Data Open Management
Measures*

**Travel service products:**
Real-time bus query, bus and
subway full load rate query
and other functions

**Green incentives :**
Carbon emission reduction
trading intention signed

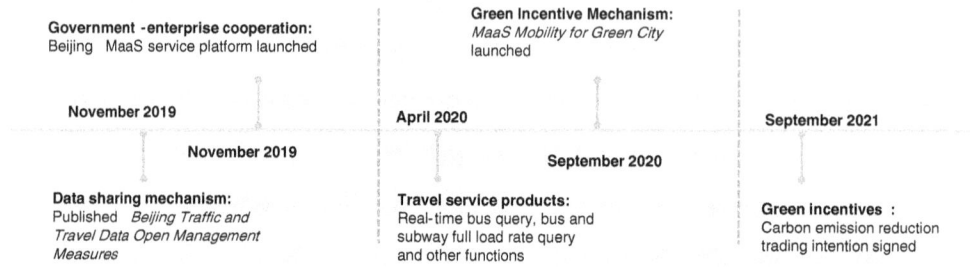

*Figure 31.5* Historical timeline of the development of Beijing MaaS.

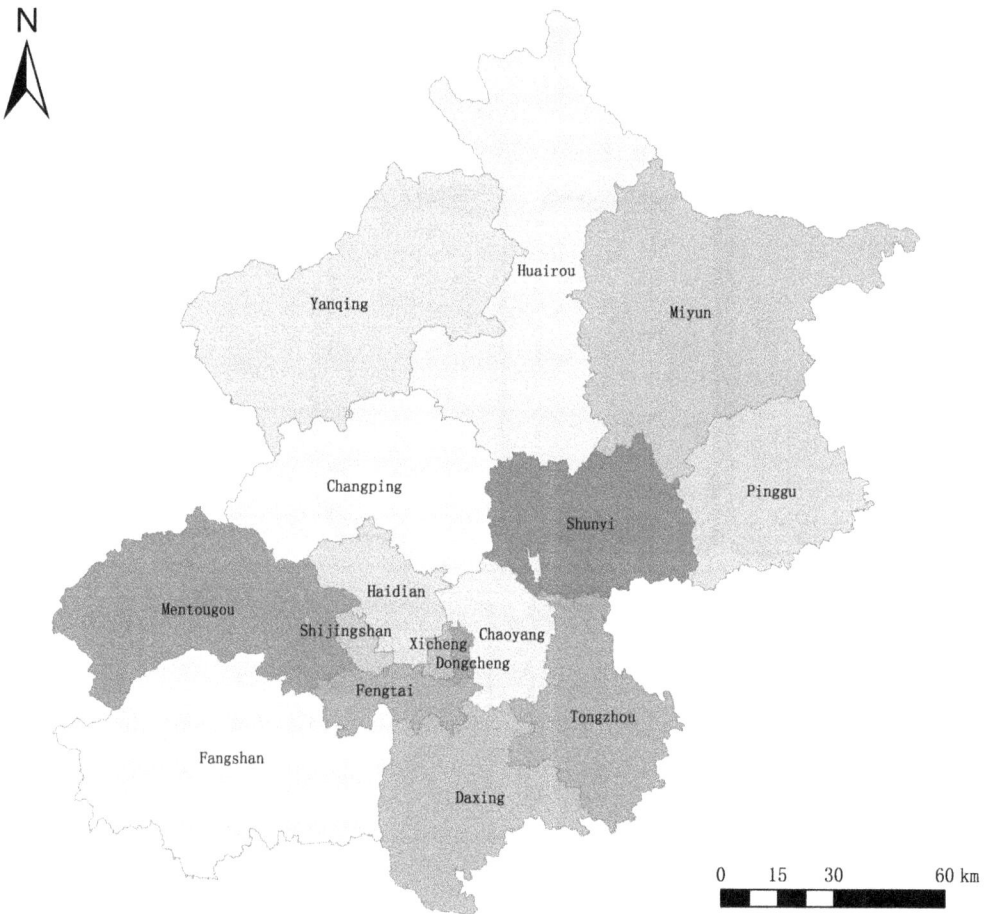

*Figure 31.6* Geographical context of the data collection of the study.

Most respondents were satisfied with Beijing's transportation system and its management, especially city buses, tubes, and bikes. Yet, specific mobility scenarios still made users feel anxious before, during, and/or after travel. For instance, whilst travelling, 70% of visitors expected Beijing MaaS to provide more accurate and inclusive travel plans, such as linking private vehicles to tubes.

*Table 31.2* Descriptive statistics of the sample of respondents among Beijing MaaS platform users

|  |  | 2020 | 2021 | 2020 (%) | 2021 (%) |
|---|---|---|---|---|---|
| Gender | Male | 738 | 128 | 46.6 | 49.2 |
|  | Female | 845 | 132 | 53.4 | 50.8 |
| Age | 18–25 | 321 | 53 | 20.3 | 20.4 |
|  | 26–31 | 392 | 46 | 24.8 | 17.7 |
|  | 31–40 | 377 | 101 | 23.8 | 38.8 |
|  | 41–60 | 424 | 58 | 26.8 | 22.3 |
|  | >60 | 69 | 2 | 4.4 | 0.8 |
| Education level | A level | 212 | 6 | 13.4 | 2.3 |
|  | High school | 508 | 8 | 32.1 | 3.1 |
|  | Undergraduate | 639 | 122 | 40.4 | 46.9 |
|  | Postgraduate | 224 | 124 | 14.2 | 47.7 |
| Occupation | Employee | 688 | 213 | 43.7 | 81.9 |
|  | Liberal professions | 728 | 13 | 46.0 | 5.0 |
|  | Army services | 75 | 2 | 4.7 | 0.8 |
|  | Students | 0 | 28 | 0.0 | 10.8 |
|  | Retired | 0 | 2 | 0.0 | 0.8 |
|  | Unemployed | 92 | 2 | 5.8 | 0.8 |
| Average income level p.a. (Yuan) | <50,000Yuan | 155 | 19 | 9.8 | 7.3 |
|  | 50,000–100,000 | 585 | 40 | 37.0 | 15.4 |
|  | 100,000–200,000 | 336 | 104 | 21.2 | 40.0 |
|  | 200,000–500,000 | 393 | 76 | 24.8 | 29.2 |
|  | >500,000 | 114 | 21 | 7.20 | 8.1 |
|  | **Total** | **1583** | **260** | **100.0** | **100.0** |

Contrary to previous research claiming that socially disadvantaged groups were more likely to experience transport-related exclusion during the COVID-19 pandemic (Yang et al., 2021), particularly in the tourism and hospitality sectors, most of our respondents found Beijing MaaS to be inclusive and equitable for all. That may be explained by the fact that, in the process of designing and launching the programme, all the stakeholders, including the public, government, traditional, and innovative transportation enterprises, were encouraged to participate and contribute to its development, with feedback.

The main reason for the inclusiveness of Beijing MaaS lies in its unique mechanism, which contains four levels. As shown in Figure 31.7, the first level concerns social performance management, as supported by government and policy makers. Such actors monitor transport emissions, transport users' experience, and resource allocation. At this level, the transportation carbon market should be highlighted, and users who use PT can get rewards from the carbon market mechanism. The second level is the core of the MaaS system, with its two sides, the supply side – integrated service providers – and the demand side – MaaS users. The third level is about the digital production tool itself, using digital technology to maintain Level 2. Level 4 comprises production factors, mixing traditional ones, for example, transportation infrastructure, and new ones, for example, big data developments.

Scholars have shown the importance of providing access to information on the payment data services of transport operators for third-party resale and use (Liimatainen & Mladenović, 2021). Yet, data onboarding brings potential shortcomings. Survey results indicate indeed that visitors' needs were difficult to predict. For instance, more than 50% of visitors showed an interest in such extra functions, for example, the monitoring of new COVID-19 infections. The survey also found that other functions, such as indoor

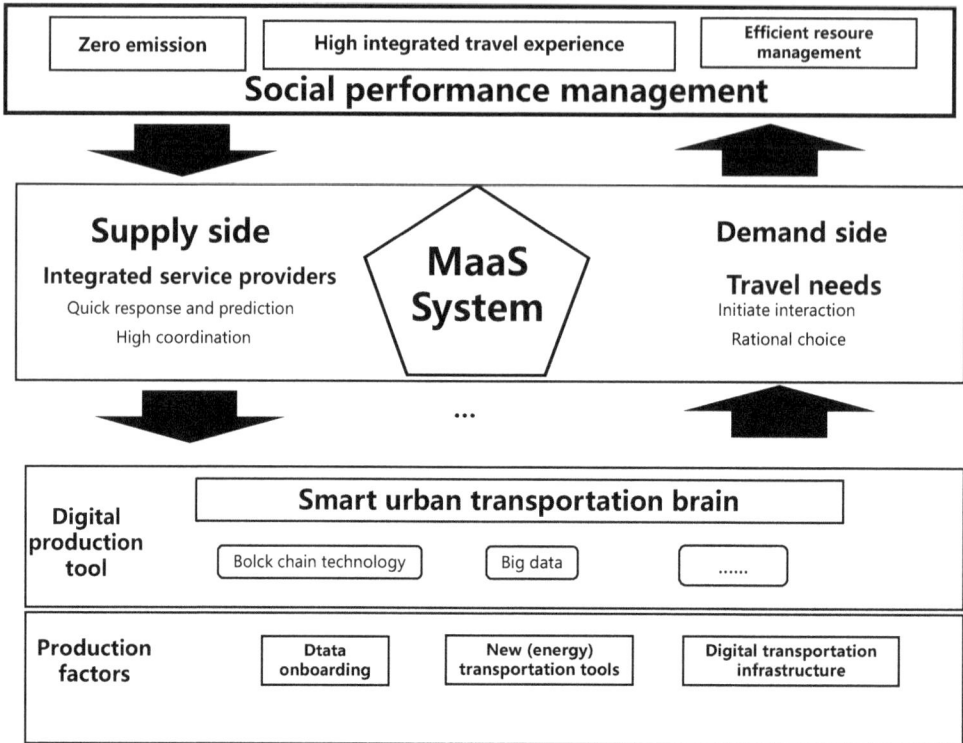

*Figure 31.7* MaaS development mechanism in Beijing.

navigation (e.g., finding routes at tourist attractions like museums or transport nodes such as airports) and accessible tips when taking the tube and bus, and cultural/language support functions (e.g., live tours, GPS guiding, travel tips at tourist attractions), were not that much used by visitors. In fact, more than 80% of survey respondents were not even aware of such functions.

When the survey was repeated in 2021, the data showed more positive results than in the first edition of the survey in 2020, which would suggest that the Beijing MaaS platform was becoming more effective at supporting tourism in Beijing. For instance, almost all sampled respondents in 2021 had used Beijing MaaS with a mean satisfaction level of 80%. More than 70% of people knew exactly what Beijing MaaS was. Among them, 81% of visitors had participated in the carbon incentive activity at least once. Relying on MaaS, Beijing MaaS developed the Beijing Emission Reduction Methodology of Low Carbon Travel and dynamic measuring technologies using big data on volunteer groups as shown in Figure 31.8.

All green travel modes were covered, and participants could gain rewards from using them partially or for the whole trip. This activity was matched with the Paris Agreement carbon emissions reduction goal (Rogelj et al., 2019). Urban traffic calls indeed for a zero-carbon transport system which is what Beijing innovatively attempted here, with its commitment to peak carbon emissions by 2030, and eventually achieving carbon neutrality by 2060. As one of the main sources of greenhouse gas emissions in China, the transport sector accounts for 9.7% of total carbon emissions, which can be seen as a barrier to sustainable development.

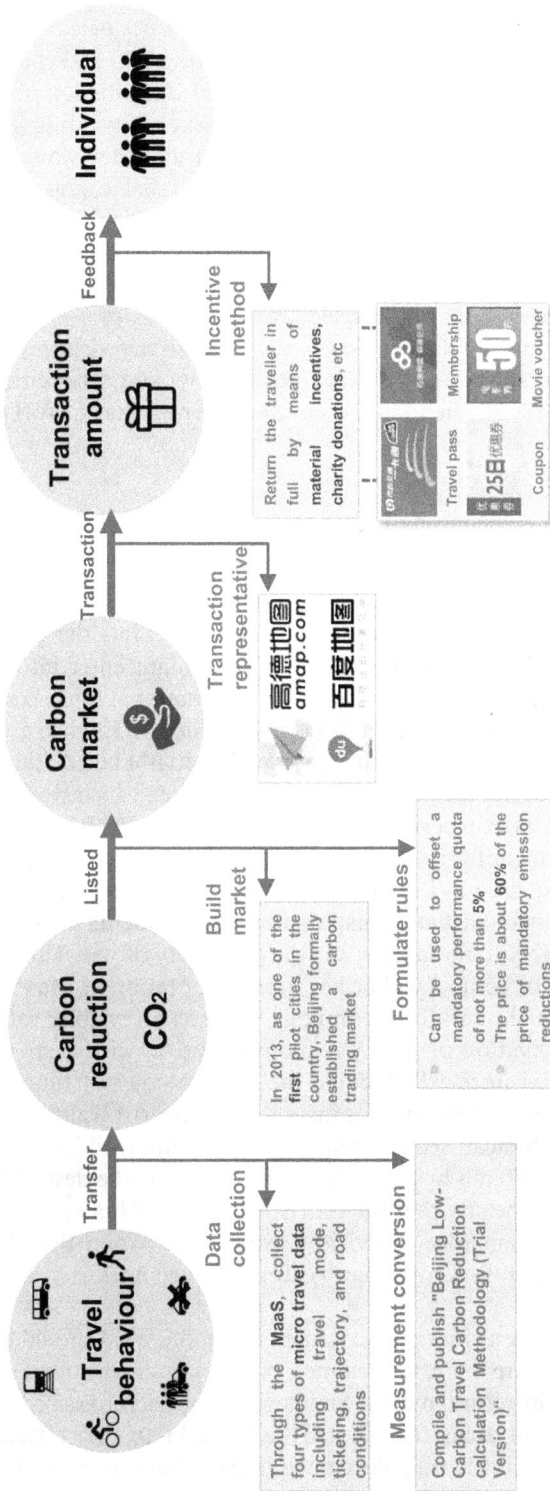

*Figure 31.8* Carbon incentive activity in Beijing MaaS.

According to the results of the study, Beijing MaaS has reached over three million people, saved 78,000 tons of carbon emissions, and switched 5% of car users to green travel modes, which forms a new win–win mode associated with multi-stakeholders including the public, enterprises, and government (Beijing Municipal Commission of Transport, 2022). Moreover, 89% of visitors expressed that they would like to switch to green travel if they could get an incentive for it; 55% of visitors expressed that they would also like to avoid using private vehicles if PT services were more punctual, reliable, and provided customised services; and 46% of visitors would use PT if it saved time through seamless services.

In the post-COVID-19 period, Beijing MaaS's new scenarios are not yet widely used. Beijing MaaS developed new scenarios in 2021, such as the integrated travel plan (e.g., the navigations of "bus + taxi + sharing bicycle" combined modes use), scanning the shared bicycle, and the riding hiring scheme. However, for inner-city visitors, only 11% of them have used these functionalities. For visitors across Beijing and other cities, just about 20% of travellers have used them. Such limited numbers proved that COVID-19 still creates a fear of travelling.

**Future trends and policy recommendations**

In the light of this brief international review of MaaS systems with a focus on the Beijing MaaS, it could be tempting for local decision makers in some global tourism cities to create specific transport lines/services for international tourists. Yet, the growing demand for authentic experiences, rather than "connected" ones, might enter into conflict with this (Nikitas et al., 2017). The tailoring of services for international tourists should be reduced to bare (but important) essentials, such as apps that include an English user option or bilingual signage that includes an English translation. It would not be advisable, for instance, to create specific transport lines only for international tourists which would detract from the authenticity of the transport experience.

Ease of use and payment for international visitors is also key in this respect. The Oyster card payment system used in London's transport system created a simple, easy to use, and affordable way of using urban transport for local residents as well as domestic and international tourists. Key to its success was the simplicity of use. This simplicity (as well as proved cost benefits, of course) will be key in the adaptation of MaaS for domestic and international visitors in the future. Technology remains an enabler rather than the sole focus, and that the navigation of MaaS remains intuitive, with clear recommendations given by the system to visitors, which could include a merger of functions with existing urban tourism apps in each location. For instance, if a tourist wants to embark on a thematic trail (digital or physical; see e.g., the digital walking trail in Suffolk in the UK in chroniclestories.co.uk), offered by a visitor app managed by the destination management organisation (DMO) of the city that the tourist is visiting, MaaS should integrate seamlessly into that so as to enhance the visitor experience. When the planning of urban transport connections to follow an urban visitor trail becomes an onerous task, visitors tend to lose interest.

Whilst tourists are using a city's public transport system, what should they watch out for aside from pickpockets? Are there, for instance, particular underground stations worth visiting briefly on their architectural merits? Similarly, as tourists cross a city, are there specific idiosyncrasies of different areas that should be brought to their attention? This could include, for instance, how Little Italy developed in New York, or why Golders Green is an interesting part of London from an ethnic perspective, or why the street book stalls by

Notre Dame cathedral in Paris merit a relaxed walk and a browse, with perhaps a coffee at a nearby cafe to rest one's legs and take in the atmosphere. These are experiential issues that visitors will appreciate as they contribute to the authenticity of their experience and encourage them to modify their plans slightly to discover parts of the city that maybe they had not planned to visit. MaaS should be instrumental in this.

In the future, will MaaS's user interface vary to suit the age and/or demographics of the user? This level of customisation of the user interface is worth exploring in the implementation of future schemes (Arias-Molinares & García-Palomares, 2020), adopting a future-based perspective, particularly given that the technology gap between younger and older generations is likely to increase in the future as the relentless pace of change of technology is likely to leave many of us behind, certainly the Baby Boomer Generation and Generation X.

Finally, how can the transport experience be used for edutainment? That is, sharing with international visitors insights about a city, whilst entertaining them during their travel around that city and still making sure they arrive safely where they want to get to? Does, for instance, virtual reality (Fan et al., 2022) have a role to play here? On a parallel front, we know that one of the main problems that urban environments will face in the future is people's sense of loneliness in spite of unprecedented levels of connectivity. How will MaaS help to alleviate this? These are sampled potential avenues of research as well as practical implications MaaS analysts as well as local decision makers may want to think of.

## Conclusion

Focusing on the implementation of Beijing MaaS in China, this chapter has reviewed and compared the status quo and governance of MaaS schemes aiming to achieve tourism transport carbon neutrality as well as leisure travel wellbeing in this post-COVID-19 era. Its attempt in Beijing seems to have been successful when addressing the inequity issue, for example, through considering vulnerable and often neglected traveller groups who mainly walk, cycle, and use public transportation, who benefit the most from the measure. Regardless of gender, age, ethnicity, education level, income, and disability, all visitors can also earn matched carbon trade benefits.

The Beijing MaaS example is unique in three ways: (i) it tightly links carbon trade and green travel relying on advanced big-data technologies turning the concept of carbon incentive into reality; (ii) it is inclusive of vulnerable groups; (iii) its scale of implementation in a world megacity, shifting the habits of three million people. The potential for the transformation and promotion of the scheme in other cities of the world have been demonstrated in this chapter.

## References

Aapaoja, A., Eckhardt, J., & Nykänen, L. (2017, November) Business models for MaaS. In *1st International Conference on Mobility as a Service.* pp. 28–29.

Alatartseva, E., & Barysheva, G. (2015) "Well-being: Subjective and objective aspects", *Procedia-Social and Behavioral Sciences*, 166, 36–42. https://doi.org/10.1016/j.sbspro.2014.12.479.

Alhaddi, H. (2015) "Triple bottom line and sustainability: A literature review", *Business and Management Studies*, 1(2), 6–10.

Ali, A., Rasoolimanesh, S. M., & Cobanoglu, C. (2020) "Technology in tourism and hospitality to achieve Sustainable Development Goals (SDGs)", *Journal of Hospitality and Tourism Technology*, *11*(2), 177–181.

Alyavina, E., Nikitas, A., & Njoya, E. T. (2020) "Mobility as a service and sustainable travel behaviour: A thematic analysis study", *Transportation Research Part F: Traffic Psychology and Behaviour*, 73, 362–381.

Arias-Molinares, D., & García-Palomares, J. C. (2020) "The Ws of MaaS: Understanding mobility as a service from a literature review", *IATSS Research*, 44, 253–263, ISSN 0386-1112.

Audouin, M., & Finger, M. (2018) "The development of Mobility-as-a-Service in the Helsinki metropolitan area: A multi-level governance analysis", *Research in Transportation Business & Management*, 27, 24–35.

Beijing Municipal Commission of Transport. (2022) *Beijing MaaS Platform and Carbon Reduction Mechanism*. Available at: http://jtw.beijing.gov.cn/xxgk/dtxx/202110/t20211029_2523797.html (In Chinese) [Accessed 29 May 2022].

Beijing Transport Institute. (2021) Beijing Transport development Annual Report. Available at: https://www.bjtrc.org.cn/Show/download/id/68/at/0.html [Accessed 29 September 2021].

Bergstad, C. J., Gamble, A., Gärling, T., Hagman, O., Polk, M., Ettema, D., & Olsson, L. E. (2010) "Subjective well-being related to satisfaction with Daily Travel", *Transportation*, *38*(1), 1–15. doi:10.1007/s11116-010-9283-z [Accessed 29 May 2022].

Bhaktikul, K., Aroonsrimorakot, S., Laiphrakpam, M. and Paisantanakij, W. (2021) "Toward a low-carbon tourism for sustainable development: A study based on a royal project for highland community development in Chiang Rai, Thailand", *Environment, Development and Sustainability*, 23(7), 10743–10762.

Buhalis, D. (2020) "Technology in tourism-from information communication technologies to eTourism and smart tourism towards ambient intelligence tourism: a perspective article", *Tourism Review*, 75(1), 267–272. https://doi.org/10.1108/TR-06-2019-0258

Cavallaro, F., Irranca Galati, O., & Nocera, S. (2021) "Climate change impacts and tourism mobility: A destination-based approach for coastal areas", *International Journal of Sustainable Transportation*, 15(6), 456–473.

Cavoli, C. M., & Jones, P. (2014) Future of the high street-Implications for transport policy and planning.

Chadee, D., Ren, S., & Tang, G. (2021) "Is digital technology the magic bullet for performing work at home? Lessons learned for post COVID-19 recovery in hospitality management", *International Journal of Hospitality Management*, 92, 102718.

Cobbold, A., Standen, C., Shepherd, L., Greaves, S., & Crane, M. (2022) "Multimodal trips, quality of life and wellbeing: An exploratory analysis", *Journal of Transport & Health*, 24, 101330.

Coca-Stefaniak, J. A., & Morrison A. M. (2022) "Cities", In: D. Buhalis (ed.), *Encyclopedia of Tourism Management and Marketing*. Cheltenham: Edward Elgar Publ.

De Vos, J. (2018) "Towards happy and healthy travellers: A research agenda", *Journal of Transport & Health*, 11, 80–85. https://doi.org/10.1016/j.jth.2018.10.009.

De Vos, J. (2020) "The effect of COVID-19 and subsequent social distancing on travel behavior", *Transportation Research Interdisciplinary Perspectives*, 5, 100121.

De Vos, J., Schwanen, T., van Acker, V., & Witlox, F. (2013) "Travel and subjective well-being: A focus on findings, methods and future research needs", *Transport Reviews*, 33, 421–442.

Donthu, N., & Gustafsson, A. (2020) "Effects of COVID-19 on business and research", *Journal of Business Research*, 117, 284–289.

Duan, J., Xie, C., & Morrison, A. M. (2021) "Tourism crises and impacts on destinations: A systematic review of the tourism and hospitality literature", *Journal of Hospitality & Tourism Research*. https://doi.org/10.1177/1096348021994194.

Durand, A., Harms, L., Hoogendoorn-Lanser, S., & Zijlstra, T. (2018) "Mobility-as-a-Service and changes in travel preferences and travel behaviour: A literature review", https://doi.org/10.13140/RG.2.2.32813.33760.

Fan, X., Jiang, X., & Deng, N. (2022) "Immersive technology: A meta-analysis of augmented/virtual reality applications and their impact on tourism experience", *Tourism Management*, 91, 104534, ISSN 0261-5177.

Fang, D., Xue, Y., Cao, J., & Sun, S. (2021) "Exploring satisfaction of choice and captive bus riders: An impact asymmetry analysis", *Transportation Research Part D: Transport and Environment*, 93, 102798.

Fioreze, T., De Gruijter, M., & Geurs, K. (2019) "On the likelihood of using Mobility-as-a-Service: A case study on innovative mobility services among residents in the Netherlands", *Case Studies on Transport Policy*, 7(4), 790–801.

Freeman, R. B., & Medoff, J. L. (1984) "What do unions do", *Industrial and Labor Relations Review*, 38, 244.

Guo, Y., Peeta, S., Agrawal, S., & Benedyk, I. (2022) "Impacts of Pokémon GO on route and mode choice decisions: Exploring the potential for integrating augmented reality, gamification, and social components in mobile apps to influence travel decisions", *Transportation*, 49(2), 395–444.

Jittrapirom, P., Caiati, V., Feneri, A. M., Ebrahimigharehbaghi, S., Alonso González, M. J., & Narayan, J. (2017) "Mobility as a service: A critical review of definitions, assessments of schemes, and key challenges", *Urban Planning*, 2(2), 13–25.

Jones, P., Wynn, M., Hillier, D., & Comfort, D. (2017) "The sustainable development goals and information and communication technologies", *Indonesian Journal of Sustainability Accounting and Management*, 1(1), 1–15.

Kádár, B., & Gede, M. (2021) "Tourism flows in large-scale destination systems", *Annals of Tourism Research*, 87, 103113.

Kahneman, D. (2000) Evaluation by moments: Past and future. *Choices, Values, and Frames*, pp. 693–708.

Kahneman, D., Diener, E., & Schwartz, N. (1999) *Well-Being: The Foundations of Hedonic Psychology*. New York: Russell Sage Foundation.

Kamargianni, M., & Matyas, M. (2017) "The business ecosystem of mobility-as-a-service", *Transportation Research Board*, 96. Available at: https://discovery.ucl.ac.uk/id/eprint/10037890/1/a2135d_445259f704474f0f8116ccb625bdf7f8.pdf [Accessed 29 September 2021].

Karlsson, I. C. M., Mukhtar-Landgren, D., Smith, G., Koglin, T., Kronsell, A., Lund, E., & Sochor, J. (2020) "Development and implementation of Mobility-as-a-Service–A qualitative study of barriers and enabling factors", *Transportation Research Part A: Policy and Practice*, 131, 283–295.

Kim, E. J., Kim, Y., Jang, S., & Kim, D. K. (2021) "Tourists' preference on the combination of travel modes under Mobility-as-a-Service environment", *Transportation Research Part A: Policy and Practice*, 150, 236–255.

Kimbu, A. N., Adam, I., Dayour, F., & de Jong, A. (2021) "COVID-19-induced redundancy and socio-psychological well-being of tourism employees: Implications for organizational recovery in a resource-scarce context", *Journal of Travel Research*. https://doi.org/10.1177/00472875211054571.

Kivimaa, P., & Rogge, K. S. (2022) "Interplay of policy experimentation and institutional change in sustainability transitions: The case of mobility as a service in Finland", *Research Policy*, 51(1), 104412.

Knoke, D. (1990) "Networks of political action: Toward theory construction", *Social Forces*, 68(4), 1041–1063.

Le Pira, M., Ignaccolo, M., Inturri, G., Pluchino, A., & Rapisarda, A. (2016) "Modelling stakeholder participation in transport planning", *Case Studies on Transport Policy*, 4(3), 230–238.

Li, J., Hallsworth, A. G., & Coca-Stefaniak, J. A. (2020) "Changing grocery shopping behaviours among Chinese consumers at the outset of the COVID-19 outbreak", *Tijdschrift voor economische en sociale geografie*, 111(3), 574–583.

Li, J., Nguyen, T. H. H., & Coca-Stefaniak, J. A. (2021a) "Coronavirus impacts on post-pandemic planned travel behaviours", *Annals of Tourism Research*. https://doi.org/10.1016/j.annals.2020.102964

Li, J., Nguyen, T. H. H., & Coca-Stefaniak, J. A. (2021b) "Understanding post-pandemic travel behaviours – China's Golden Week", *Journal of Hospitality and Tourism Management*, 49, 84–88. https://doi.org/10.1016/j.jhtm.2021.09.003

Liimatainen, H., & Mladenović, M. N. (2021) "Developing mobility as a service–user, operator and governance perspectives", *European Transport Research Review*, 13(1), 1–3.

Liu, D., Yang, D., & Huang, A. (2021) "Leap-based greenhouse gases emissions peak and low carbon pathways in China's tourist industry", *International Journal of Environmental Research and Public Health*, *18*(3), 1218.

Mameli, F., & Marletto, G. E. (2009) A participative procedure to select indicators of sustainable urban mobility policies.

Mao, Y., He, J., Morrison, A. M., & Coca-Stefaniak, J. A. (2020) "Effects of tourism CSR on employee psychological capital in the COVID-19 crisis: From the perspective of conservation of resources theory", *Current Issues in Tourism*, 24(19), 2716–2734.

Marletto, G., & Mameli, F. (2012) "A participative procedure to select indicators of policies for sustainable urban mobility. Outcomes of a national test", *European Transport Research Review*, 4, 79–89.

Mobility Transition in China. (2021) "Beijing Maas platform launches 'Maas Mobility for Green City' initiative", *Sustainable Transition China*. Available at: https://transition-china.org/mobilityposts/beijing-maas-platform-launches-maas-mobility-for-green-city-initiative/ [Accessed 29 September 2021].

Moradi, A., & Vagnoni, E. (2018) "A multi-level perspective analysis of urban mobility system dynamics: What are the future transition pathways"? *Technological Forecasting and Social Change*, 126, 231–243.

Morrison, A. M., Lehto, X., & Day, J. (2018) *The Tourism System*. New York: Kendall Hunt Publ.

Morrison, A. M., & Maxim, C. (2021) *World Tourism Cities: A Systematic Approach to Urban Tourism*. London: Routledge.

Morrison, A. M., & Maxim, C. (2022) *World Tourism Cities: A Systematic Approach to Urban Tourism*. London: Routledge.

Nikitas, A., Kougias, I., Alyavina, E., & Njoya Tchouamou, E. (2017) "How can autonomous and connected vehicles, electromobility, BRT, hyperloop, shared use mobility and mobility-as-a-service shape transport futures for the context of smart cities?" *Urban Science*, 1(4), 36.

Pangbourne, K., Mladenović, M. N., Stead, D., & Milakis, D. (2020) "Questioning mobility as a service: Unanticipated implications for society and governance", *Transportation Research Part A: Policy and Practice*, 131, 35–49.

Papaix, C., & Coca-Stefaniak, J. A. (2021) "Transport in tourism cities – Beyond the functional and towards an experiential approach", In: A. M. Morrison & J. A. Coca-Stefaniak (eds.), *Routledge Handbook of Tourism Cities*. London: Routledge. https://doi.org/10.4324/9780429244605-27

Pedrana, M. (2014) "Location-based services and tourism: Possible implications for destination", *Current issues in Tourism*, *17*(9), 753–762.

Pellegrino, F. (2021) "Transport and tourism relationship", In: F. Grasso & B. S. Sergi (eds.), *Tourism in the Mediterranean Sea*, pp. 241–256. Bingley: Emerald Publishing Limited.

Polydoropoulou, A., Pagoni, I., Tsirimpa, A., Roumboutsos, A., Kamargianni, M., & Tsouros, I. (2020) "Prototype business models for Mobility-as-a-Service", *Transportation Research Part A: Policy and Practice*, 131, 149–162.

Puchongkawarin, C., & Ransikarbum, K. (2021) "An integrative decision support system for improving tourism logistics and public transportation in Thailand", *Tourism Planning & Development*, 18(6), 614–629.

Pyke, S., Hartwell, H., Blake, A., & Hemingway, A. (2016) "Exploring well-being as a tourism product resource", *Tourism Management*, 55, 94–105, ISSN 0261-5177.

Raham, M. K., Gazi, M. A. I., Bhuiyan, M. A., & Rahaman, M. A. (2021) "Effect of Covid-19 pandemic on tourist travel risk and management perceptions", *Plos One*, 16(9), e0256486.

Reardon, L., & Abdallah, S. (2013) "Well-being and transport: Taking stock and looking forward", *Transport Reviews*. 33, 634–657.

Ritchie, B. W., & Jiang, Y. (2019) "A review of research on tourism risk, crisis and disaster management: Launching the annals of tourism research curated collection on tourism risk, crisis and disaster management", *Annals of Tourism Research*, 79, 102812. https://doi.org/10.1016/j.annals.2019.102812.

Rodrigue, J. P. (2020) *The Geography of Transport Systems*. London: Routledge.

Rogelj, J., Huppmann, D., Krey, V., Riahi, K., Clarke, L., Gidden, M., & Meinshausen, M. (2019) "A new scenario logic for the Paris Agreement long-term temperature goal", *Nature*, 573(7774), 357–363.

Santos, G., & Nikolaev, N. (2021) "Mobility as a service and public transport: A rapid literature review and the case of moovit", *Sustainability*, 13(7), 3666.

Sarasini, S., Sochor, J., & Arby, H. (2017, November) What characterises a sustainable MaaS business model. In *1st International Conference on Mobility as a Service (ICOMaaS)*, Tampere. pp. 28–29.

Shin, H., Nicolau, J. L., Kang, J., Sharma, A., & Lee, H. (2022) "Travel decision determinants during and after COVID-19: The role of tourist trust, travel constraints, and attitudinal factors", *Tourism Management*, 88, 104428.

Signorile, P., Larosa, V., & Spiru, A. (2018) "Mobility as a service: A new model for sustainable mobility in tourism", *Worldwide Hospitality and Tourism Themes*, 10(2), 185–200.

Sirgy, M. J., Kruger, P. S., Lee, D. J., & Yu, G. B. (2011) "How does a travel trip affect tourists' life satisfaction?" *Journal of Travel Research*, 50(3), 261–275.

Smith, G., Sochor, J., & Sarasini, S. (2018) "Mobility as a service: Comparing developments in Sweden and Finland", *Research in Transportation Business & Management*, 27, 36–45.

Smith, H. B. (1993) "Leisure travel market potential for a high speed civil transport", *Journal of Aviation/Aerospace Education & Research*, 5(1), 6.

Sochor, J., Strömberg, H., & Karlsson, I. M. (2015) "Implementing mobility as a service: Challenges in integrating user, commercial, and societal perspectives", *Transportation Research Record*, 2536(1), 1–9.

Susilo, Y. O., & Cats, O. (2014) "Exploring key determinants of travel satisfaction for multi-modal trips by different traveler groups", *Transportation Research Part A: Policy and Practice*, 67, 366–380.

Tirachini, A., & Cats, O. (2020) "COVID-19 and public transportation: Current assessment, prospects, and research needs", *Journal of Public Transportation*, 22(1), 1.

Transport for London. (2020) *Transport Supporting Paper*. Available at: https://www.london.gov.uk/sites/default/files/gla_migrate_files_destination/Transport%20Supporting%20Paper.pdf [Accessed 29 September 2021].

United Nations. (2019) Mobility as a Service Mobility as a Service Inland Transport Committee. Available at: https://unece.org/DAM/trans/main/wp5/publications/Mobility_as_a_Service_Transport_Trends_and_Economics_2018-2019.pdf.

Vij, A., Ryan, S., Sampson, S., & Harris, S. (2020) "Consumer preferences for Mobility-as-a-Service (MaaS) in Australia", *Transportation Research Part C: Emerging Technologies*, 117, 102699.

Wen, J., Kozak, M., Yang, S., & Liu, F. (2020) "COVID-19: Potential effects on Chinese citizens' lifestyle and travel", *Tourism Review*.

Weng, J., Di, X., Wang, C., Wang, J., & Mao, L. (2018) "A bus service evaluation method from passenger's perspective based on satisfaction surveys: A case study of Beijing, China", *Sustainability*, 10(8), 2723.

Wilson, S., & Hannam, K. (2017) The frictions of slow tourism mobilities: Conceptualising campervan travel. *Annals of Tourism Research*, 67, 25–36.

Wong, Y. Z., Hensher, D. A., & Mulley, C. (2018) "Emerging transport technologies and the modal efficiency framework: A case for mobility as a service (MaaS)", Available at: https://ses.library.usyd.edu.au/bitstream/handle/2123/19100/ITLS-WP-18-04.pdf?sequence=1&isAllowed=y [Accessed 29 September 2021].

Xie, C., Zhang, J., Morrison, A. M., & Coca-Stefaniak, J. A. (2021) "The effects of risk message frames on post-pandemic travel intentions: The moderation of empathy and perceived waiting time", *Current Issues in Tourism*, 24(23), 3387–3406.

Yang, Y., Cao, M., Cheng, L., Zhai, K., Zhao, X., & De Vos, J. (2021) "Exploring the relationship between the COVID-19 pandemic and changes in travel behaviour: A qualitative study", *Transportation Research Interdisciplinary Perspectives*, 11, 100450.

Zhan, L., Zheng, X, Morrison, A. M., Liang, H, & Coca-Stefaniak, J. A. (2020) "A risk perception scale for travel to a crisis epicenter: Visiting Wuhan after Covid-19", *Current Issues in Tourism*, 25(1), 150–167.

Zhang, S. N., Li, Y. Q., Ruan, W. Q., & Liu, C. H. (2022) "Would you enjoy virtual travel? The characteristics and causes of virtual tourists' sentiment under the influence of the COVID-19 pandemic", *Tourism Management*, 88, 104429.

Zhang, Z., & Zhang, N. (2021) "A novel development scheme of mobility as a service: Can it provide a sustainable environment for China?", *Sustainability*, 13(8), 4233.

Zhao, X., Zhang, Z., Guo, W., Zhou, Y., Papaix, C., & Sun, Q. (2022) "Evidence-based smart transition strategies for long-distance commuters in Beijing", *Frontiers in Future Transportation*, 8.

# 32 Digital nomads and destination characteristics

## A conceptual analysis

*Muhammet Necati Çelik*

## Introduction

In the last four centuries since the Industrial Revolution, several inventions and technologies, such as electricity, the computer, the Internet, and wireless networks, have been founded. These developments started a digital transformation process from which the global community has entered a digital age. This has also affected and changed organisational structures and human lifestyles. Expected worker skills and knowledge have been transformed digitally. New occupational groups such as online workers, Web and graphic designers, computer programmers, and freelancers in different areas have emerged. Many of these groups are remote knowledge workers who use the Internet (Wang, Schlagwein, Cecez-Kecmanovic, & Cahalane, 2020).

Since the Industrial Revolution, many crucial developments have occurred regarding employee rights. In the early 20th century, the principles of Taylorism were developed by Frederick Taylor, and they focus on productivity in a scientific manner (Ndaguba, Nzewi, Ijeoma, Sambumbu, & Sibanda, 2018). The principles include providing incentives to motivate workers for better performance (Britannica, 2022). Thereafter, the importance of worker motivation and their needs have been discovered by organisations. Gradually, legal working time periods (daily, weekly, and annually) have decreased. This is because, as Maslow mentioned in his hierarchy of needs theory, people have certain needs beyond physiological and safety-related needs, like forming relationships for social belonging, self-esteem, and self-actualisation (Yousaf, Amin, & Santos, 2018). Therefore, employees who work in 9 to 5 jobs have been in search of more flexible and yielding job opportunities. That social movement refers to a shift from the principles of Taylorism to Digital Taylorism or Neo-Taylorism (Günsel & Yamen, 2020; Wang et al., 2020).

Discourses on the Digital Taylorism paradigm concern two crucial dimensions of workplace transformation: job content and performance management (Gautié, Jaehrling, & Perez, 2020). Both dimensions are affected by the digital transformation process and human motivation. The job contents of organisations have changed because of technological developments and new technology-related jobs, such as business intelligence analysts, Web administrators, computer programmers, and Web and software developers, have emerged (Ha, Lee, Yun, & Coh, 2022). All of these jobs are performed by knowledge workers and can be carried out remotely. People desire to fulfil their self-esteem and self-actualisation needs. All this has resulted in the "digital nomadism flow" (Figure 32.1).

Digital nomadism is an emerging phenomenon that inspires people with a combination of remote work and continuing travel (Bozzi, 2020). In 1997, foresighted authors Makimoto and Manners threw out the term "digital nomad" to predict how technology would

DOI: 10.4324/9781003291763-36

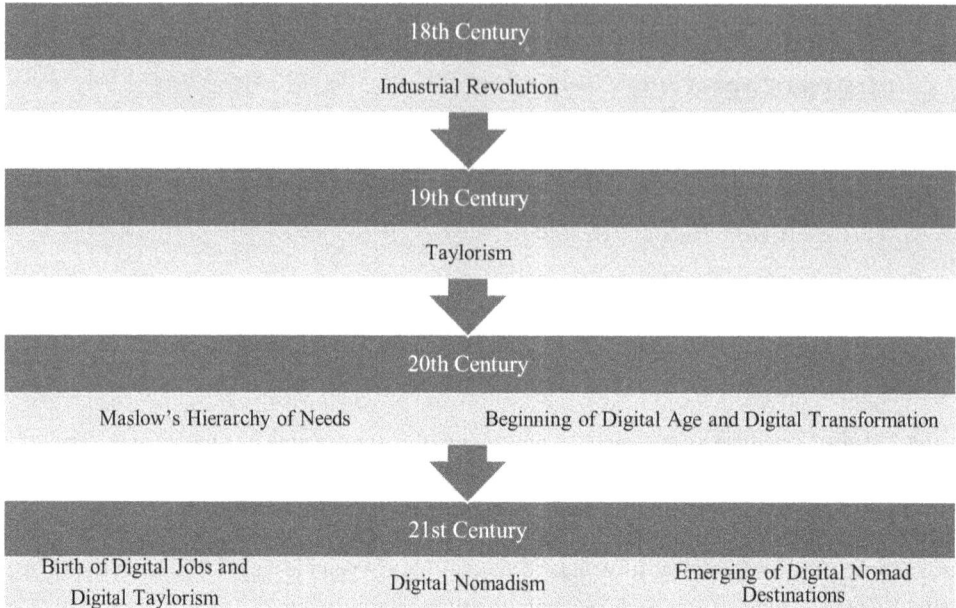

*Figure 32.1* Historical development of digital nomadism.

affect working conditions (Makimoto & Manners, 1997). In later years, many researchers supported their predictions and the importance of the term was understood (Czarniawska, 2014; Hannonen, 2020; Nash, Jarrahi, Sutherland, & Phillips, 2018; Reichenberger, 2018; Wang et al., 2020).

The term has had numerous definitions since its birth. In its early identification, authors emphasised that a digital nomad is someone who can talk with anyone through video calls and access any document from any place (Makimoto & Manners, 1997). With the triggering of advanced developments in information and communication technologies, new expressions have been added and the definition dramatically changed (Müller, 2016). For instance, the description of the term transformed to a knowledge worker that travels the world, aims to earn a high income in low-cost destinations, works in a digital environment, and busily uses the Internet and portable devices like laptops and smartphones (Wang et al., 2020).

In the last two decades, several books, reports, and articles have been written about this concept which is the dream of many people. Furthermore, more than 70% of publications about digital nomads or digital nomadism indexed by the Web of Science, Scopus, and Google Scholar databases were published between 2018 and 2022. Also, according to research conducted in 2021, 15.5 million workers in the USA defined themselves as digital nomads with an increase of 112% from 2019 (MBO Partners, 2021). All this evidence exhibits the importance of the emerging digital nomadism concept.

New destinations have been emerging to host and satisfy digital nomads. This growing concept creates an opportunity for destinations willing to accommodate such nomads. However, destinations need to provide some amenities and have unique characteristics for these people. If those required characteristics and amenities are identified, destinations could enhance their eligibility. Therefore, in this chapter I aim to determine the characteristics of a suitable destination that can be called "the utopian destination of digital nomads".

## Current research topics about digital nomads

*Definition of the concept*

Before 2000, nobody had any idea about the concepts "digital nomad" and "digital nomadism", except Makimoto and Manners. The potential change was foreseen by them, and the first definition of the concept emerged in 1997 (Makimoto & Manners, 1997; Mancinelli, 2022). In the long run, the concept has been transformed into a social phenomenon of work–life balance and a mobile lifestyle (Chevtaeva & Denizci-Guillet, 2021; Müller, 2016). Authors have been searching for a common understanding and a clear holistic definition of the growing phenomenon (Hannonen, 2020; Nash et al., 2018; Reichenberger, 2018; Shawkat, Rozan, Salim, & Shehzad, 2021). In line with this aim, it can be said that digital nomads are an ever-growing group of knowledge workers who use digital tools to earn while travelling.

The concept was divided into four levels, from zero to three, by Reichenberger in 2018. While in the first level, digital nomads are only location independent and work in an online environment, in level 1 they transfer the independence to mobility and refuse consistently to work in a particular individual office space. The possibility of work and travel at the same time exists in level 2. In the last level, it is important to not have a permanent residency (Reichenberger, 2018). Digital nomads are defined by one of the leading job platforms that provide independent professionals as "people who choose to embrace a location-independent, technology-enabled lifestyle that allows them to travel and work remotely, anywhere on the Internet-connected world" (MBO Partners, 2021). All digital nomads are remote workers, but not all remote workers are digital nomads, except regular travellers.

*Needs and motivations of digital nomads*

Digital nomads have some essential needs to sustain their lifestyle. As in the hierarchy of needs, there are some basic physiological, psychological, and self-fulfillment needs that are necessary for digital nomads in their lives. According to comprehensive research that analysed 336 posts and 7,134 comments of digital nomad communities, a workplace for a remote worker needs a decent and high-speed Internet connection, an accessible location, and an affordable price for the required services for basic physiological needs. The desire for socialising and belongingness are psychological needs demanded of the targeted destination. In the last of these needs, career development opportunities are vital to meet the need of self-fulfillment (Lee, Toombs, & Erickson, 2019a).

The motivations of digital nomads is a research topic. Schlagwein (2018) introduced three motivations intrinsic to a digital nomad: a low cost of living, being a part of a like-minded and interesting group of people, and a desire for cultural and personal experiences. Different from these, the escape from the a location-dependent, 9 to 5 working style was also suggested in the same year (Reichenberger, 2018). The desire for freedom and seeking a destination that has attractive leisure features, the opportunity to engage with friends or relatives, and hoped for markets or industries are placed among other motivating factors for digital nomads (Hall, Sigala, Rentschler, & Boyle, 2019). Similarly, while autonomy, flexible lifestyle, and life-changing decisions are presented, the role of wellbeing is also counted as one of the motivation factors (Von Zumbusch & Lalicic, 2020).

The needs and motivations of digital nomads are substantially interrelated. While the need is defined as "a condition of tension in an organism resulting from deprivation of something required for survival, wellbeing, or personal fulfillment", the concept of motivation is identified as "the variables, collectively, that alter the effectiveness of reinforcers" (American Psychological Association, 2022a; 2022b). The needs are vital conditions. People are also motivated to achieve their needs alongside their desires. After getting involved in the digital nomad community, such people are motivated by their basic physiological, psychological, and self-fulfillment needs and other desires.

### Co-living and co-working (co-spaces) for co-creation

Digital nomads tend to be part of like-minded social groups; and co-spaces are suitable places for co-creation (Chevtaeva, 2021). They are attracted to co-working and co-living spaces for several reasons, like cost-sharing, collaboration, socialisation, enjoyment, and co-creation (Aroles, Granter, & deVaujany, 2020; Kocaman, 2022; Spinuzzi, 2012). Put simply, co-living and co-working spaces are meant to be for sharing a work or living place with other people; and they are very common among digital nomads (Lee et al., 2019a, 2019b). According to a global study, the number of people working in co-working spaces was almost two million in 2020 and predicted to reach five million in 2024 (Coworking Resources and Coworker, 2021). The number of people living in co-living spaces across the world was around 3.5 million in 2020 (The Housemonk, 2022).

In co-working or co-living spaces, digital nomads are not only sharing the space, but also the costs, ideas, knowledge, and experience, which provides co-creation opportunities (Goermar, Barwinski, Bouncken, & Laudien, 2021; Putra & Agirachman, 2016). After the birth of these concepts, several businesses emerged to sell co-spaces (Hannonen, 2020) and they are actively used by digital nomads (Wang et al., 2020). As knowledge workers, it can be said that digital nomads have the ability to adequately consider crucial concepts such as collaboration, openness, community, accessibility, and sustainability. These concepts can be supported by co-working spaces (Nash, Jarrahi, & Sutherland, 2021), which have also been shown to be one of the localised dynamics of innovation (Capdevila, 2015).

Current research about digital nomads can be collected into three topics, as above. The digital nomadism movement is continuing to expand all over the world, resulting in emerging destinations called digital nomad destinations. Insight is needed to determine the characteristics of the most appropriate digital nomad destination with the purpose of contributing to the relevant literature. This determination can be explained using an analogy called the "Utopian Destination of Digital Nomads" (UDDN).

### The utopian destination of digital nomads

Although the characteristics of digital nomad destinations have been commonly evaluated by several websites and social media communities (nomadlist.com, digitalnomads.world), there is a limited amount of research about their characteristics. In this chapter I expect to fill this gap and contribute to the relevant literature with its theoretical and practical implications. Accordingly, I conducted a study using the content analysis method. A wide range of information resources, including 23 scientific publications (13 articles, 7 conference papers, 3 book chapters), 22 Web platforms (20 blogs and 2 community platforms), and 3 reports, were examined by me. First, the needs and motivations of digital nomads were determined. Then, qualifications were identified for preferring a destination by a digital nomad. Finally, the UDDN model was developed in order to present convenient characteristics for digital nomad destinations (see Figure 32.2).

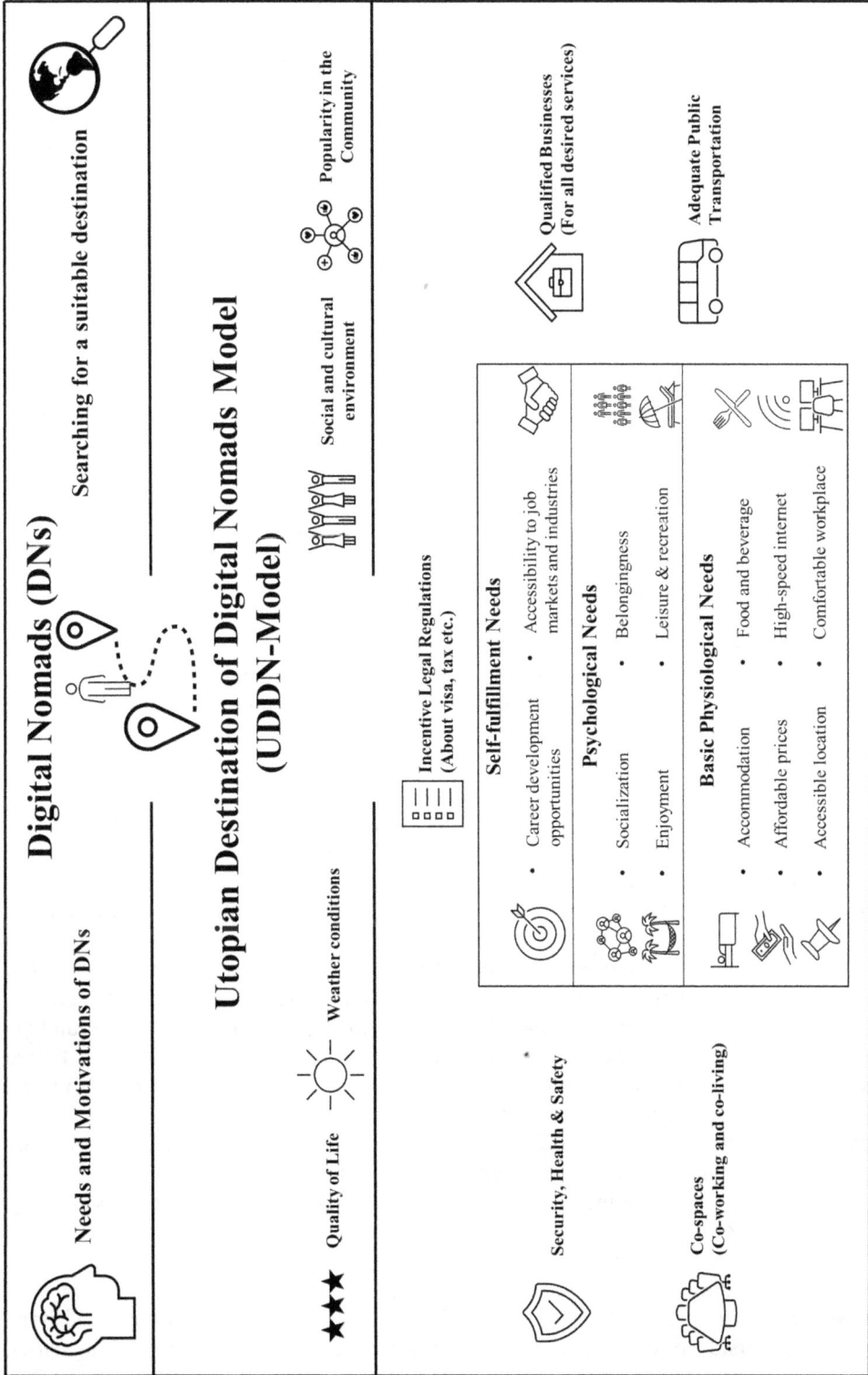

*Figure 32.2* The Utopian Destination of Digital Nomads model.

The UDDN model expresses all qualifications for digital nomads and provides the required knowledge for destinations. It is the first model for digital nomad destinations. It is expected to be a useful and effective guide for such destinations. All components of the UDDN model are explained in the following.

*Quality of life* is a concept related to the fulfillment of needs and the perception of needs (Costanza et al., 2007). Digital nomads have several needs to fulfill as mentioned in the UDDN model. Quality of life can be seen as a structure that includes the achievement level of the basic physiological, psychological, and self-fulfillment needs in a destination. This concept is an essential component for digital nomads in order to choose a destination. The level of quality of life can be compared by digital nomads with the level of quality life of their home destination or previously visited destination.

*Weather conditions* are another important component for digital nomads because they like to attend leisure and recreational activities during their stay. Appropriate weather conditions are demanded by them and, of course, whether they attend or not depends on their personal taste or preferred activities. A destination that has environmental issues like air pollution, flooding, or hurricane risks is not selected by digital nomads. Destinations with very hot and cold climates are also ignored by them.

*The social and cultural environment* of a destination is one of the leading fundamentals for digital nomads, because they like to live in a hospitable society and experience interesting cultural phenomena. They are stimulated in such an environment to be creative and productive. The prevalence of English-speaking residents is also another benchmark for the digital nomad community and is considered by them when making their decision to move to a destination.

*The popularity of the destination among digital nomads* is appealing to them and they are affected by that. Because they want to be together with their community members they are aware of established appropriate facilities presented by popular destinations. In spite of so many digital nomads living in trendy destinations, they tend to locate near job markets and industries. This brings along many job and networking opportunities.

*Incentivising legal regulations* such as facilitator visas or tax policies attract the attention of digital nomads and increase the popularity of the destination. Regarding digital nomads, several taxation policies are presented by different countries, such as lower taxation or exemptions, for certain levels of income and social security contributions (Tyutyuryukov & Guseva, 2021).

*Security, health, and safety* are interrelated issues. Digital nomads mention that they prefer to use their own mobile data rather than a public wi-fi connection (Nash et al., 2018). Also, they do not want to stay in a place that has a high crime rate or a war. Cyber and life security are significant for them. The importance of health infrastructure is also reflected on the Web platforms by digital nomads. If the destination provides a secure, healthful, and safe environment, demand will be increased.

*Relevant qualified businesses* are all service suppliers, such as accommodation enterprises, cafes, and restaurants. Businesses providing the required qualifications and satisfying digital nomads cause a positive impression about the destination and increase its popularity. Destinations are evaluated by digital nomads in terms of the quality of services on Web platforms. Accordingly, requirements to host digital nomads are needed to recognise and take action regarding requirements.

*Public transportation* facilities are vital for digital nomads as well as other visitors, because they tend to visit natural and cultural attractions. Especially in big cities, transportation is crucial to go shopping and to visit cafes, restaurants, and even co-working spaces.

*Co-spaces* include co-working spaces where digital nomads can collaborate, create knowledge, and produce innovative ideas, as well as co-living spaces where they can share costs and socialise with each other. The number of co-spaces is needed to enhance the destinations because of high demand generated by digital nomads. Co-working spaces can also be created by accommodation enterprises and cafes and encouraged by the authorities of the destination.

### The roles of stakeholders in digital nomad destinations (DNDs)

It is also important to determine the stakeholders of the DND and their roles in developing and managing it. Any group that is benefitted or affected by the actions of digital nomads, such as public authorities, destination management organisations (DMOs), businesses, non-governmental organisations (NGOs), local communities, and the media, can be counted as destination stakeholders (Çelik & Buhalis, 2022). All stakeholders have their own roles and interests which need to be unearthed. While some of them have strong relationships with digital nomads, others have weak relationships. The requirements of digital nomads at destinations are given in the UDDN model (Figure 32.2). Each destination stakeholder has some responsibility to provide a suitable environment for them.

Legal regulations such as visa or tax policies and the physical infrastructure like public transportation, healthcare, and security services should be prepared and strengthened by public authorities. Most of the time DMOs cooperate with public authorities in order to manage the destination. But, DMOs are responsible for several functions in the improvement of tourism activities, such as planning, management, and marketing. Designing a DND according to the order of the UDDN model is managed by DMOs. Preparing the statistics and reports about digital nomads can be counted as the other responsibilities of DMOs. Businesses (accommodation enterprises, cafes, restaurants, Airbnb) have strong commercial relations with the digital nomad community at the destination because they are actors who provide the required services for digital nomads, such as Internet access, accommodation, food and beverage, co-living and co-working spaces. Non-governmental organisations (NGOs), such as business associations related to specific business activities, are also placed at the destination. Sharing required knowledge about digital nomads and developing new products for them are accepted as the responsibilities of NGOs. Digital nomads also expect to find an English-speaking local community at the destination. The local community needs to warmly welcome digital nomads, and understand and respect their expectations. Regarding ensuring the expected environment, the media can play an important role in creating awareness about the digital nomad community among all stakeholders and informing them about the questions: Who are digital nomads? What do they do and need? Why do they come to the destination? How can stakeholders benefit? Increasing the popularity of the destination among digital nomads via using social media is another crucial role of the media.

### Further predictions about digital nomadism

The future of digital nomadism can be discussed under three headings: the development of technology, the philosophy of sustainable development, and expectations of the labour force. Each passing day, the use of technological solutions in processes in the workplace will continue to increase. The need for employees with technological knowledge will rise too. Considering that digital nomads are made up of people with this required knowledge, an increment in the number of digital nomads and a decrement in the number of regular 9 to 5 workers can be predicted.

The philosophy of sustainable development has been on the top of the global agenda since the last quarter of the 20th century. Most of the destinations will continue to consider sustainability issues and they will adapt their digital nomad facilities to the Sustainable Development Goals in the future. For instance, some incentive policies towards these goals could be implemented for businesses related to digital nomads. While waste management practices can be carried out by cafes and restaurants, energy-saving practices can be realised by accommodation enterprises. For sustainable destination development, tourist–host interaction needs to be considered by destination management authorities with a mutual gaze, which means understanding social interactions between locals and digital nomads (Lin & Fu, 2021). Some effort in order to improve the wellbeing of the locals and the digital nomad community could be made by the authorities.

It is necessary to know the expectations of the labour force for estimating the future of digital nomadism, because autonomy, life–work balance, and good economic conditions are demanded by these workers. People will choose to join the digital nomad community by increasing their knowledge according to the demand for digital jobs. If the increment continues at the 10% rate each year, the number of digital nomads in the USA alone will be more than 35 million in 2030. It seems that digital nomadism will continue to be a globally popular phenomenon in the future (MBO Partners, 2021). Many destinations, researchers, and politicians will look at digital nomadism as part of their activities in the near future.

## Opportunities and challenges

Digital nomadism brings several benefits for destinations and businesses as well as some challenges for local communities. With their long stays, knowledgeable consumption, and quality product demand, digital nomads are a profitable target market for destinations and businesses. Regular travel needs like accommodation, food, and beverage as well as sightseeing and co-working spaces are demanded by digital nomads. If an appropriate environment is organised by the destination, digital nomadism can convert a destination into a knowledge centre with its technology-skilled nomads. Creating a sustainable environment that contributes to the Sustainable Development Goals is another opportunity. For instance, if destinations can succeed in attracting knowledgeable digital nomads, a pool of labour that can produce added value and innovation is also hosted by them. A valuable and innovative labour resource is known to be an important factor that affects economic growth. Hence, decent work and economic growth (Goal 8), and industry, innovation, and infrastructure (Goal 9), can be contributed to by digital nomads at destinations. The contribution of digital nomads to economic growth has been demonstrated in empirical research (Demaj, Hasimja, & Rahimi, 2021). Digital nomads are a knowledgeable group who are aware of global issues. It will be easier to realise these practices in digital nomad destinations compared to many destinations whose target market is not similar. Therefore, it is also suggested to develop policies that contribute to the Sustainable Development Goals in digital nomad destinations and their businesses.

When a destination is chosen by many digital nomads, the local community may be affected negatively (O'Regan, Salazar, Choe, & Buhalis, 2022). For instance, because of increased demand for services like renting apartments and demanding co-working spaces in cafes and restaurants, prices will be expected to increase in the destination. But, if the local community can be included in the economic activities and the financial benefits are distributed to the locals, then locals will be affected positively. In conclusion, digital nomadism may become mainstream in the future and it is an opportunity that is not to be missed by destinations.

## Major theoretical and practical contributions of the chapter

The digital nomadism phenomenon has been explained in all its aspects in this chapter. First, the current situation and historical development of the phenomenon have been discussed with a model developed in order to clarify the concept. It is understood that digital nomadism has emerged because of cumulative knowledge from the Industrial Revolution to the 21st century. Subsequently, current research topics were presented and they came under three headings: the definition of the concept, needs and motivations of digital nomads, and co-living and co-working spaces. Currently, digital nomadism has become a social movement among digital workers and has encouraged a mobile working life. Also, it has been concluded that human needs have changed from the past to the present and the conditions of the work–life balance have been affected by this change. It can be said that digital nomadism has emerged as a result of this change. It is also concluded that co-spaces are vital for the creativity, collaboration, and productivity of digital nomads. Besides that, the created UDDN model is one of the most crucial contributions of this research and the components of the model have also been explained. All of this is a theoretical contribution of the chapter.

The UDDN model is also counted as a pathfinder for the destinations that are seeking guidance regarding digital nomads. If destination stakeholders can realise the following responsibilities, then they will easily attain success and become a digital nomad destination. First, businesses that serve digital nomads need to be ready and/or encouraged. DMOs should carry out the necessary marketing activities to attract digital nomads to the destination and conduct the strategic planning. The public authorities should prepare facilitator legal regulations and the infrastructure required for the nomads. Providing sufficient awareness among the local community and tourism-related NGOs is counted as another crucial success factor. The declaration of the relevant stakeholders and their roles, emphasising several implications, opportunities, and challenges, and future predictions for destinations, businesses, and other stakeholders can be shown to be the practical contributions of this chapter. Digital nomadism is an opportunity for destinations that want to attract knowledge workers and which need to rejuvenate themselves (Figure 32.3). It could also be an opportunity to realise practices towards the Sustainable Development Goals. If seasonal destinations can meet the expectations of digital nomads and succeed at keeping them there in their low season, digital nomadism could also be a solution for seasonality. This can be seen as a major opportunity for seasonal destinations in the low season to fill their empty resources. The number of digital nomads is estimated to rise and it is not wrong to say that most of the destinations and their stakeholders are expected to use the opportunity soon.

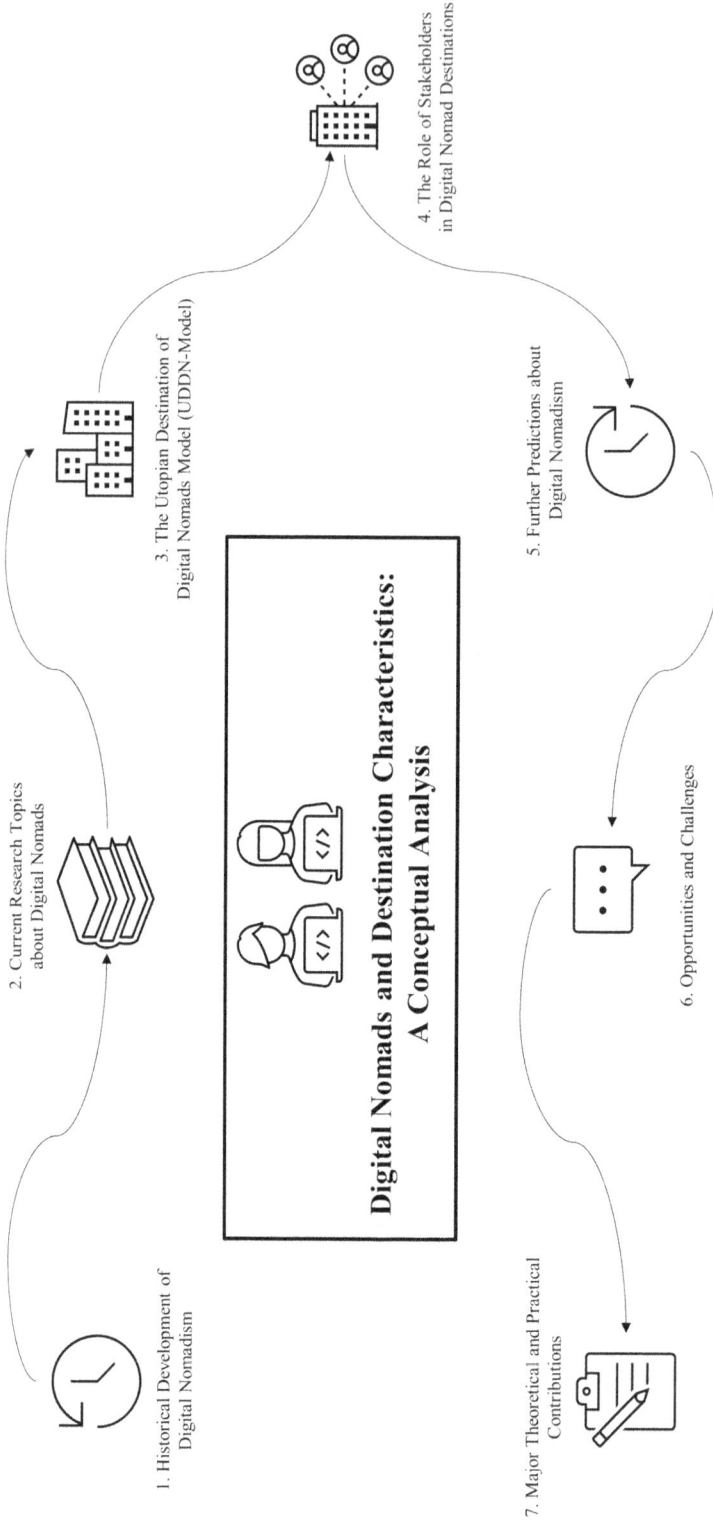

4. The Role of Stakeholders in Digital Nomad Destinations

3. The Utopian Destination of Digital Nomads Model (UDDN-Model)

2. Current Research Topics about Digital Nomads

5. Further Predictions about Digital Nomadism

**Digital Nomads and Destination Characteristics: A Conceptual Analysis**

6. Opportunities and Challenges

1. Historical Development of Digital Nomadism

7. Major Theoretical and Practical Contributions

*Figure 32.3* Digital nomads and destination characteristics.

# References

American Psychological Association. (2022a, February 15). *need*. Retrieved from dictionary.apa.org: https://dictionary.apa.org/need

American Psychological Association. (2022b, February 15). *motivation*. Retrieved from dictionary.apa.org: https://dictionary.apa.org/motivation

Aroles, J., Granter, E., & deVaujany, F.-X. (2020). 'Becoming mainstream': The professionalisation and corporatisation of digital nomadism. *New Technology, Work and Employment*, 35(1), 114–129. doi:10.1111/ntwe.12158

Bozzi, N. (2020). #digitalnomads, #solotravellers, #remoteworkers: A Cultural Critique of the Traveling Entrepreneur on Instagram. *Social Media + Society*, April–June 2020, 1–15. doi:10.1177/2056305120926644

Britannica, T. Editors of Encyclopaedia. (2022, February 10). *Taylorism*. Retrieved from Encyclopedia Britannica: https://www.britannica.com/science/Taylorism

Capdevila, I. (2015). Co-working spaces and the localised dynamics of innovation in Barcelona. *International Journal of Innovation Management*, 19(3), 1–28. doi:10.1142/S1363919615400046

Çelik, M. N., & Buhalis, D. (2022). Sustainable Destination Development. In D. Buhalis (ed.), *Encyclopedia of Tourism Management and Marketing*. Cheltenham, UK: Edward Elgar Publishing.

Chevtaeva, E. (2021). Coworking and Coliving: The Attraction for Digital Nomad Tourists. In W. Wörndl, C. Koo, & J. L. Stienmetz (eds.), *Information and Communication Technologies in Tourism 2021* (pp. 202–209). Cham: Springer. doi:10.1007/978-3-030-65785-7_17

Chevtaeva, E., & Denizci-Guillet, B. (2021). Digital nomads' lifestyles and coworkation. *Journal of Destination Marketing & Management*, 21, 1–11. doi:10.1016/j.jdmm.2021.100633

Costanza, R., Fisher, B., Ali, S., Beer, C., Bond, L., Boumans, R., ... Snapp, R. (2007). Quality of life: An approach integrating opportunities, human needs, and subjective well-being. *Ecological Economics*, 61, 267–276. doi:10.1016/j.ecolecon.2006.02.023

Coworking Resources and Coworker. (2021, February 20). *Global Coworking Growth Study 2020*. Retrieved from www.coworkingresources.org: https://en.coworkingresources.org/hubfs/Coworking/Global-Coworking-Study-2020.pdf

Czarniawska, B. (2014). Nomadic work as life-story plot. *Computer Supported Cooperative Work*, 23, 205–221. doi:10.1007/s10606-013-9189-3

Demaj, E., Hasimja, A., & Rahimi, A. (2021). Digital Nomadism as a New Flexible Working Approach: Making Tirana the Next European Hotspot for Digital Nomads. In M. Orel, O. Dvouletý, & V. Ratten (eds.), *The Flexible Workplace Coworking and Other Modern Workplace Transformations* (pp. 231–257). Cham, Switzerland: Springer Nature Switzerland AG. doi:10.1007/978-3-030-62167-4_13

Gautié, J., Jaehrling, K., & Perez, C. (2020). Neo-taylorism in the digital age: Workplace transformations in French and German Retail Warehouses. *Relations industrielles/Industrial Relations*, 75(4), 774–795. doi:10.7202/1074564ar

Goermar, L., Barwinski, R. W., Bouncken, R. B., & Laudien, S. M. (2021). Co-creation in coworking-spaces: Boundary conditions of diversity. *Knowledge Management Research & Practice*, 19(1), 53–64. doi:10.1080/14778238.2020.1740627

Günsel, A., & Yamen, M. (2020). Digital Taylorism as an Answer to the Requirements of the New Era. In B. Akkaya (ed.), *Agile Business Leadership Methods for Industry 4.0* (pp. 103–119). Bingley: Emerald Publishing Limited.

Ha, T., Lee, M., Yun, B., & Coh, B.-Y. (2022). Job forecasting based on the patent information: A word embedding-based approach. *IEEE Access*, 10, 7223–7233. doi:10.1109/ACCESS.2022.3141910

Hall, G., Sigala, M., Rentschler, R., & Boyle, S. (2019). Motivations, Mobility and Work Practices; The Conceptual Realities of Digital Nomads. In J. Pesonen, & J. Neidhardt (eds.), *Information and Communication Technologies in Tourism 2019* (pp. 437–449). Cham: Springer. doi:10.1007/978-3-030-05940-8_34

Hannonen, O. (2020). In search of a digital nomad: Defining the phenomenon. *Information Technology & Tourism*, 22, 335–353. doi:10.1007/s40558-020-00177-z

Kocaman, S. (2022). Co-working Spaces. In D. Buhalis (ed.), *Encyclopedia of Tourism Management and Marketing*. Cheltenham, UK: Edward Elgar Publishing. doi:10.4337/9781800377486.co-working. spaces

Lee, A., Toombs, A. L., & Erickson, I. (2019a). Infrastructure vs. Community: Co-spaces Confront Digital Nomads' Paradoxical Needs. *CHI EA '19: Extended Abstracts of the 2019 CHI Conference on Human Factors in Computing Systems* (pp. 1–6). New York: Association for Computing Machinery. doi:10.1145/3290607.3313064

Lee, A., Toombs, A. L., Erickson, I., Nemer, D., Ho, Y.-S., Jo, E., & Guo, Z. (2019b). The social infrastructure of co-spaces: Home,work, and sociable places for digital nomads. *Proceedings of the ACM on Human-Computer Interaction*, 3(CSCW), 1–23. doi:10.1145/3359244

Lin, B., & Fu, X. (2021). Gaze and tourist-host relationship – State of the art. *Tourism Review*, 76(1), 138–149. doi:10.1108/TR-11-2019-0459

Makimoto, T., & Manners, D. (1997). *Digital Nomad*. Chichester: Wiley.

Mancinelli, F. (2022). Digital Nomads. In D. Buhalis (ed.), *Encyclopedia of Tourism Management and Marketing*. Cheltenham, UK: Edward Elgar Publishing. doi:10.4337/9781800377486.digital. nomads

MBO Partners. (2021). *The Digital Nomad Search Continues*. Ashburn, VA: MBO Partners.

Müller, A. (2016). The digital nomad: Buzzword or research category? *Transnational Social Review*, 6(3), 344–348. doi:10.1080/21931674.2016.1229930

Nash, C., Jarrahi, M. H., Sutherland, W., & Phillips, G. (2018). Digital Nomads Beyond the Buzzword: Defining Digital Nomadic Work and Use of Digital Technologies. In G. Chowdhury, J. McLeod, V. Gillet, & P. Willett (eds.), *Transforming Digital Worlds. iConference 2018*. Lecture Notes in Computer Science, vol 10766 (pp. 207–217). Cham: Springer. doi:10.1007/978-3-319-78105-1_25

Nash, E. C., Jarrahi, M. H., & Sutherland, W. (2021). Nomadic work and location independence: The role of space in shaping the work of digital nomads. *Human Behavior and Emerging Technologies*, 3(2), 271–282. doi:10.1002/hbe2.234

Ndaguba, E. A., Nzewi, O. I., Ijeoma, E. C., Sambumbu, M., & Sibanda, M. M. (2018). Using Taylorism to make work easier: A work procedure perspective. *South African Journal of Economic and Management Sciences*, 21(1), 1–10. doi:10.4102/sajems.v21i1.2120

O'Regan, M., Salazar, N. B., Choe, J., & Buhalis, D. (2022). Unpacking overtourism as a discursive formation through interdiscursivity. *Tourism Review*, 77(1), 54–71. doi:10.1108/TR-12-2020-0594

Putra, G. B., & Agirachman, F. A. (2016). Urban Coworking Space: Creative Tourism in Digital Nomads Perspective. *Proceedings of the 6th International Conference of Arte-Polis* (pp. 169–178). Bandung. Retrieved from https://www.researchgate.net/publication/316472768_Urban_Coworking_Space_Creative_Tourism_in_Digital_Nomads_Perspective

Reichenberger, I. (2018). Digital nomads – A quest for holistic freedom in work and leisure. *Annals of Tourism Research*, 21(3), 364–380. doi:10.1080/11745398.2017.1358098

Schlagwein, D. (2018). 'Escaping the Rat Race': Justifications in Digital Nomadism. *Twenty-Sixth European Conference on Information Systems (ECIS2018)* (pp. 1–7). Portsmouth. Retrieved from https://aisel.aisnet.org/ecis2018_rip/31/

Shawkat, S., Rozan, M. Z., Salim, N. B., & Shehzad, H. M. (2021). Digital Nomads: A Systematic Literature Review. *2021 7th International Conference on Research and Innovation in Information Systems (ICRIIS)* (pp. 1–6). Bahru: IEEE. doi:10.1109/ICRIIS53035.2021.9617008

Spinuzzi, C. (2012). Working alone, together: Coworking as emergent collaborative activity. *Journal of Business and Technical Communication*, 26(4), 399–441. doi:10.1177/1050651912444070

The Housemonk. (2022, February 20). *Global Coliving Report 2020*. Retrieved from www.thehousemonk.com: https://thehousemonk.com/wp-content/uploads/2020/12/Global-Coliving-Report-2020.pdf

Tyutyuryukov, V., & Guseva, N. (2021). From remote work to digital nomads: Tax issues and tax opportunities of digital lifestyle. *IFAC-PapersOnLine*, 54(13), 188–193. doi:10.1016/j.ifacol.2021.10.443

Von Zumbusch, J. S., & Lalicic, L. (2020). The role of co living spaces in digital nomads' well being. *Information Technology & Tourism*, 22, 439–453. doi:10.1007/s40558-020-00182-2

Wang, B., Schlagwein, D., Cecez-Kecmanovic, D., & Cahalane, M. C. (2020). Beyond the factory paradigm: Digital nomadism and the digital future(s) of knowledge work post-COVID-19. *Journal of the Association for Information Systems*, 21(6), 1379–1401. doi:10.17705/1jais.00641

Yousaf, A., Amin, I., & Santos, J. A. (2018). Tourists' motivations to travel: A theoretical perspective on the existing literature. *Tourism and Hospitality Management*, 24(1), 197–211. doi:10.20867/thm.24.1.8

# 33 Gamification in museum tourism

*Ige Pırnar, Duygu Çelebi and Muruvvet Deniz Sezer*

## Introduction

The concept of gamification has been considered one of the fundamental tools for increasing the quality of interaction with participants (Xu and Buhalis, 2021). From actual experience, gamification is accepted as an effective, innovative tool for improving the involvement and motivation of participants. The potential significance of gamification, which enhances tourist engagement and knowledge of the destination, has recently been widely recognised in the tourism industry. As predicted in the report published in 2011 by the World Traffic Market (WTM), gamification was expected to be the primary trend. In alignment with these predictions, gamification is a major trend, and its applications in tourism have gradually increased in recent years (Xu and Buhalis, 2021; Xu et al., 2013). Gamification is also increasingly meaningful in promoting cultural sites and museum collections. The gamification technique is useful for museum tourism to increase customer loyalty by enhancing motivation with the interactive game technique or by making the visit enjoyable by using the visitors' engagement. Through rethinking the traditional role of museums, it offers to participants the possibility of finding different options that open up ways to develop their potential experience (Madsen, 2020). The application of gamification contributes to the increased intention of potential visitors and the expected number of visitors (Bieszk-Stolorz et al., 2021).

In light of recent technological developments, the blend of gamification and museum tourism needs to be addressed together, but there are limited studies discussing this subject. There is a need to inquire into the gamification applications in the tourism industry. Thus, we aim to present insight into the different types of gamification techniques applied in museums.

## Literature review

### Gamification in general

Gamification comes from the mutation of a widespread word, which is as old as humanity itself – "game". The idea of gamification is the incorporation of game design tools into a non-game environment as a motivating action for competition and cooperation (Xu and Buhalis, 2021). It aims to increase the motivation of participants and encourage them to solve problems by applying the characteristics of game elements. This term was coined in 2002 by Nick Palin, but the concept was widely recognised by late 2010 (Jayawardena et al., 2021). Since 2011, the topic of gamification has attracted the attention of many researchers and practitioners (Leclercq et al., 2020), and it has been identified as the use of game-based

DOI: 10.4324/9781003291763-37

elements to motivate, promote, and engage people by providing training, interaction, and dealing with problems.

Gamification has been applied in a wide range of fields such as commerce, health, education, marketing, and advertising (Xu et al., 2013; Xi & Hamari, 2019; Pasca et al., 2021), aiming to increase the user experience through the interactive and full engagement of the participants. Volchek (2022) has described gamification as using game tools in a non-game environment as a strategic initiative that aims to strengthen systems, organisations, and activities by motivating and engaging participants with similar experiences while playing a game. Buhalis and Sinarta (2019) described gamification as a technology that will increase customer satisfaction and help increase visitor participation in the future. This concept is based on the three fundamental roles of game components – educating, engaging, and entertaining – by ensuring the involvement of users in a real experience (Anastasiadis et al., 2018). The gamification application enables self-sufficient participants to increase their gamification-based performance. Pasca et al. (2021) noted that gamification has several roles, such as edutainment, enhancement of sustainable behaviour, and providing engagement factors.

According to Wünderlich et al. (2020), the implementation of gamification dynamics contributes to augment and enhance the loyalty of customers, as well as enhancing stakeholders' (e.g., visitors, volunteers, donors, museum staff, educators, museum institutions, collectors, curators, artists, universities, local government, associations, civil society, and industry) engagement. Businesses gain benefit from gamification, which is a novel tool to create cooperation with participants and improve the behaviour of their employees, investors, and customers (Baptista & Oliveira, 2017). Gamification is an effective way to form interactions and connect people to create added value for business. It is seen that global companies use the concept of gamification as a strategic tool in their marketing activities (Zhang et al., 2017) and employee training and development (Armstrong & Landers, 2018). In this sense, businesses can apply game-based learning to teach and reinforce their workers' skills to develop strong and long-term relationships between partners, which helps business to gain strategic advantages.

### Gamification in the tourism industry

The development of the technologies has led to transforming traditional experiences into a novel type of experience, such as gamification in the tourism industry (Trigo-De la Cuadra et al., 2020). The successful outcomes of the gamification trend have already started to apply in different tourism contexts (Xu et al., 2013, 2016; Zica et al., 2018; Cilingir UK & Gultekin, 2021; Sigala & Nilsson, 2021), such as the destination (e.g., Foursquare), hospitality industry (e.g., Marriott Hotels, Starwood Hotels), the airline industry (e.g., American Airlines, British Airways, Lufthansa, Turkish Airlines), and food and beverage industry (e.g., Starbucks, Pizza Hut, Domino's Pizza). Stamp books, loyalty programmes, storytelling, competitions, and reward membership refers to gaming elements of tourism organisations which enable participants to collect points, badges, or bonuses for real presents or gifts (Pasca et al., 2021).

The increasing number of these similar adoptions proves that gamification applications are increasingly important for tourism. According to Xu et al. (2013) the usage of gamification applications provides not only internal (e.g., human resources, employee training, crowd-sourcing, productivity enhancement) but also external (e.g., customer engagement, sales, and marketing) advantages for the supply side of the tourism industry. As confirmed by Serravalle et al. (2019), gamification is one of the new instruments in the tourism sector, which is based on the entertainment services to increase the attraction of visitors and gain a competitive advantage eventually. Through the effective usage of tourism-based gamification strategies, tourists can get detailed tourism-relevant information about the destination or any type of specific touristic product before they visit the area, creating interest, curiosity, and desire to travel to the destination (Trigo-De la Cuadra et al., 2020). Through these new and simultaneous developments tourists can engage with tourism not only pre-visit but also during their real-time trip. The significant existence of gamification in the tourism context has been divided into two in practice as "social games" and "location-based games" (Xu et al., 2016). Within the context of social games, participants can play tourism games on many social platforms without physical engagement with the relevant destination. Location-based games require the real-time participation of players, and the related games start while the participant is physically at the destination. These two types of games are designed for different purposes. While social games aim to engage potential customers in the firm's branding and marketing practices, the local-based games are designed to provide more engagement for tourists when they are on-site.

When compared with other fields, the use and integration of gamification strategies in the tourism industry are still in the infancy stages. Although relatively rare in practice and the existing literature, the blend of gamification and tourism has started to be debated thoroughly by many scholars (Xu et al., 2013, 2016, 2017; Abou-Shouk & Soliman, 2021). According to these scholars, gamification strategies and the integration of the tourism content with these strategies are strikingly crucial and include various benefits for the future of the tourism industry and tourism marketing. These benefits can be summarised as:

- Increase the attraction of potential visitors;
- Increase the motivation of tourists;
- Encourage tourist engagement;
- Improve the customer experience;
- Increase the interaction of visitors;
- Enhance customer loyalty;
- Increase customer satisfaction (pre-visit, during visit, post-visit);
- Increase employee satisfaction;
- Increase brand awareness;
- Improve awareness of the touristic destination or product.

The benefits of gamification in the tourism field are varied and concerned with the demand and supply side of the industry. Although mostly applied in the general tourism context already, gamification practices have started to be applied in different tourism forms such as museum tourism to differentiate, enhance the experience of visitors, and gain a competitive advantage eventually.

### Museum tourism and gamification in museum tourism

Museums are an important tourist attraction of leading global destinations. There has been a rapid change in museums' passive role since their existence – reasons, coverage, and target segment plans have changed due to altering trends and lifestyle shifts of local and international visitors. As museum tourism products change and develop, the customers of this product and their needs also change, making the museum tourism experience more dynamic and exciting (Liu & Idris, 2018). Hence, museums that have proper and active communication with their visitors provide a more comfortable and entertaining environment, making their experiences more meaningful with interesting learning and entertainment opportunities (Madsen, 2020).

> With the development of social economy, the increasing demand for the culture life of public, further study of museum and cognition development, museums are transforming from collection-based to visitor-oriented. To provide better exhibitions and education for visitors, museums have to open up their materials and invite people to participate, create and link to each other.
>
> (He et al., 2019, p. 196)

Gamification applications in museum tourism have been gaining importance to enhance visitor expectations and increase satisfaction and involvement levels (Nofal et al., 2020). The trends that encourage gamification applications in museums may be summarised as (Pırnar & Sarı, 2013; Liu & Idris, 2018):

- Edutainment – which stands for entertainment through film, videos, and other media – leading to education and providing knowledge. Internet and mobile usage in information gathering, searching, social interaction, and any other use is encouraged in museums.
- Museums focus on social integration, and they are more people-oriented rather than object-oriented as they were in the past. They are more alive and play an active role in involving local people in museums by various appeals and motivation. Hence, this participation leads to increasing quality of life levels of local people in a social sense.
- They are focusing on dialogues with visitors by concentrating on audience development and visitor satisfaction research issues. Gamification applications have become important since museums make an effort to encourage visitors to participate in museum functions. They also play a new public service role by gathering the whole of society to cultural, artistic, historical, natural, and scientific understandings.
- Museums try to specialise more in special events and exhibitions to attract visitors. Instead of classic steady and static displays, they also prefer motivational and dynamic ones. They also promote themselves as a location to be visited frequently rather than once.

Due to all these trends, gamification applications are becoming more common and widely adapted by many museums since they fully appeal to educational, marketing, motivational, and personal visitor development aims.

Museum gamification applications also support sustainable museum development (Bieszk-Stolorza et al., 2021). The museum gamification processes appeal to all pillars, namely, cultural, environmental, social, and economic. Another advantage of the gamification applications is measurable results in behaviour change in museums. Since people often like playing games, the gamification applications help museum visitors to develop entertainment, excitement, and motivational experiences while increasing their visitor engagement levels (Madsen, 2020). Integration of gamification methods and involvement of smart objects have become very common in many museums like the National Museum of Scotland, the Rijksmuseum in Amsterdam, and The British Museum. One approach for integrating smart objects into museums is by means of mobile phones acting as nexuses between exhibits and visitors. Combined they can provide added value to museums through advanced visualisation of exhibitions or gamification techniques (López-Martínez et al., 2020).

Hence, for successful museum gamification applications, the games should be designed in accordance with the expectations of various age segments, providing the necessary intelligence and emotional experiences.

## Methodology

As a specific methodology, the usefulness of the case study has been proven numerous times as a benchmark for policymakers and scholars. It lays the foundation for proper procedures and shows the correct approach for doing things. As a qualitative method, the use of case studies allows researchers to examine and scrutinise a variety of aspects, including a single phenomenon, concepts, individuals, events, and places in their natural environment. The primary goal of the case study method is to generalise the plethora of units and create a detailed insight into the all-encompassing nature of a complex issue (Heale & Twycross, 2018). The primary justification for employing the case study approach in this chapter lies in the essence of explaining the significant details of a successful implementation. Despite the apparent importance of the blend of gamification and museum tourism, there is a lack of a general view in the existing literature. The Louvre Museum and Jamtli Museum were chosen as case studies which represent two different crucial gamification techniques – which are utterly different in terms of how they disseminate the tools (digital or non-digital) of gamification to their visitors. Data were obtained with secondary data collection techniques.

## Case 1: The Louvre Museum (Paris, France)

One of the contemporary examples among museums and modern entertainment firms was launched by a partnership between the tech giant Nintendo and the world-known Louvre Museum in Paris, France. The Nintendo 3DS Louvre Guide was designed in 2012 to enhance the visitor's museum engagement and cultural enjoyment through the replication of reality by using GPS and 3D imaging. These rentable 3D units allow users to visit the world's largest museum easily by providing a definite layout that prevents them from getting lost and enables them to go where they want (Gerval & Le Ru, 2015; Osterman, 2018; Othman et al., 2021). The official Nintendo DS Louvre Guide is comprised of high-resolution photographs of artworks, stories, fun facts, audio and video commentaries, 3D models, virtual tours, and interactive maps (Lütfi & Yüksel, 2020; Corona, 2021).

"Language selection" is the first option that displays on this virtual guide console. After this, the menu options vary according to the user's interests and preferences. In this respect, the guide tries to respond to the demands of different types of tourists or visitors. So, visitors who prefer to take independent tours can choose to visit the museum freely in general by using the virtual maps offered by this interactive console. "Masterpieces Tour" is another option that directs a tour for visitors with the help of clear audio and visual directions and commentaries. "Visit a must-see work" is the third alternative tour that includes a standard package for visitors who do not want to miss must-see artworks (e.g., Aphrodite, Venus de Milo, Mona Lisa, Rebellious Slave, and Liberty Leading the People) that are highly recommended by the Louvre Museum. "Temporary Exhibition" tour is the last option and introduces the current exhibition to visitors to catch their attention and extend their visit. Whichever option is chosen by visitors, virtual integrated maps reinforce this process through the following easy-to-understand directions. By playing audio commentaries, users can listen to a detailed explanation (e.g., common name, artist(s), date, and size) and presentation of historical insights about the artefacts, artworks, paintings, or collections they are interested in. They can also focus on hidden details and view the piece in high definition by using the console's zoom, rotate, or spin functions. They can identify and view related artefacts in 3D by touring the museum to get more points. So, users can examine the statues from all different angles. With the help of this interactive guide, users can explore the works easily and make some comparisons with other related ones.

It must be underlined that the Nintendo 3DS Louvre Guide targets not only existing visitors but also potential ones. A home version of this audiovisual guide was launched in 2013 and released at the Louvre's gift shop and on its e-shop. Through the additional offerings, users can visit the Louvre Museum virtually anytime and anywhere in the world. Similarly, they can select an artwork that they wish to discover by using audio and visual commentaries and virtual maps while they are touring the museum virtually. This educational and informative feature of the application makes the art more accessible for any segment and increases the potential visitors' museum experience, engagement, and attention to transform their virtual visit into a real one. This shows how gamification acts as one beneficial way for museums to provide visitors with more immersive and engaging experiences (Tayara & Yılmaz, 2020). Nintendo 3D XLs Louvre Edition is the new version of the 3DS Louvre Guide (Amitrano et al., 2021) that was released to present updates to all users; it does not matter whether it exists or is potential.

**Case 2: Jamtli Museum (Östersund, Sweden)**

Jamtli Museum is a renowned indoor and open-air museum located in Östersund, Sweden (Malm, 2021). This museum consists of not only permanent but also temporary exhibitions. So, visitors can attend the museum for two different purposes: to learn about the region's past they choose permanent exhibitions, and to examine arts and handicrafts they choose the temporary exhibitions. Jamtli Museum supports joyful and lifelong alternative learning techniques to enhance participants' practical skills, engagement, and experience by storytelling and role playing instead of using advanced technology (Jamtli, n.d.). Based on these standpoints, the museum offers various activities and programmes to attract all segments ranging from children to older people.

Time travelling is the primary attraction tool offered by Jamtli Museum, which allows museum visitors to discover Swedish history by presenting different time eras, items from those periods, and medieval actors. Through this, participants can be part of history and meet actors who wear traditional costumes and tell local historical stories of the time. This is the illusion that Jamtli Museum creates with an aim to visualise "what daily life was like", "what people did", "how they looked", and "how they worked" between the years 1785 to 1975 (Hansen, 2016).

Within the context of gamification, the museum offers a specific game for its guests. The aforementioned game consists of a variety of tasks which should be solved in a meaningful manner. Visitors should engage in five different medieval time periods (1975, 1956, 1942, 1895, and 1785) via their time-travelling passport given at the reception of the museum. To complete and accomplish this game, visitors need to collect stamps from places where they visit by partaking in activities (e.g., planting, harvesting, farming, cooking, or making cheese) and re-enact historical events. At the end of this game, the passport used during the game transforms into a souvenir. Being one of the most critical elements of gamification, providing learning activities, creating a real experience, and interaction among visitors are impeccably reflected by Jamtli Museum.

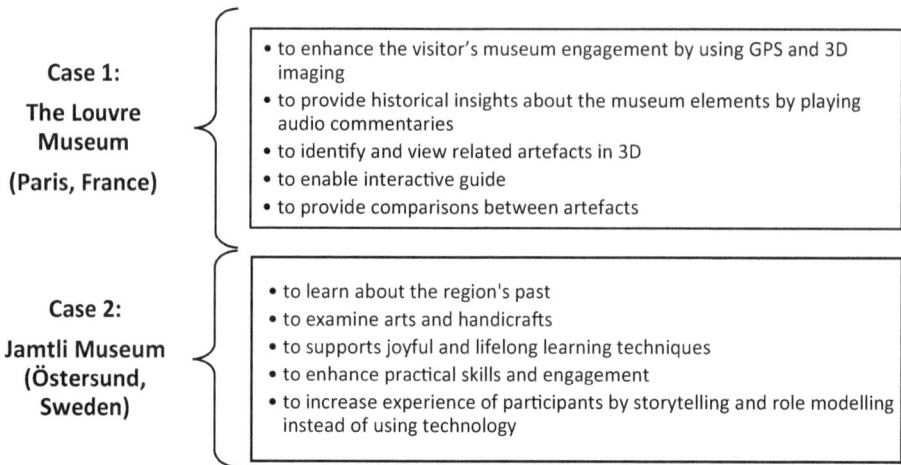

Case 1:
The Louvre
Museum
(Paris, France)

- to enhance the visitor's museum engagement by using GPS and 3D imaging
- to provide historical insights about the museum elements by playing audio commentaries
- to identify and view related artefacts in 3D
- to enable interactive guide
- to provide comparisons between artefacts

Case 2:
Jamtli Museum
(Östersund,
Sweden)

- to learn about the region's past
- to examine arts and handicrafts
- to supports joyful and lifelong learning techniques
- to enhance practical skills and engagement
- to increase experience of participants by storytelling and role modelling instead of using technology

*Figure 33.1* Differences between the Louvre Museum and Jamtli Museum.

The two different critical gamification techniques of The Louvre Museum and Jamtli Museum are shown in Figure 33.1.

**Conclusion**

The aim of gamification in tourism is the engagement of visitors by providing different tasks to enhance their touristic experience by the utilisation of gaming elements. This chapter has focused on studies related to gamification applications in the tourism industry and museum gamification applications and presents two well-known gamification practices in museum tourism. Gamification practices in museums are quite diverse, such as the active participation of visitors directly engaged in the process or the passive participation of visitors in the story. The main theoretical contribution of this study has been to investigate gamification applications in museum tourism from different perspectives. As a practical contribution, its potential usage for the tourism industry is still significant. Thus, the museum industry should integrate novel gamification practices into their processes to remain sustainable in the long term.

The Louvre Museum and Jamtli Museum were selected as case studies that provide two crucial gamification techniques that can be applied in museum tourism. The main result of this chapter is to indicate that gamification not only focuses on the technology-based methods but also on the real-time experience of visitors, increasing their motivations and involvement.

As a practical implication, destination marketers should keep pace with technology developments for integrating game elements with the new intelligent tools. Gamification plays an active role in achieving Sustainable Development Goal 9: the Industry, Innovation, and Infrastructure target. Effective marketing strategies should be developed by tourism organisations considering changing customer needs. The game designer is a critical part of the gamification concept, and thus strong collaboration between destination marketers and game designers should be established by tourism organisations. Gamification strategies can be applied to enhance citizens' motivations and relationships. In line with the Sustainable Development Goal 11: Sustainable Cities and Communities, the concept of gamification can be achieved by creating a sustainable partnership in the long term. Gamification takes on educational roles with mutual interaction and an entertaining museum experience offered to visitors. Gamification in museums encourages participants by creating a gateway to a better learning experience and providing interaction of the processes. As Bonacini and Giaccone (2022) pointed out, gamification practices can be considered as a strategic tool to encourage cultural heritage and tourism apart from their "learning by playing" role. "Learning by interacting" has led to increased visitor engagement and provides innovative approaches for improved marketing purposes in the tourism industry (Jayawardena et al., 2021). In addition, stakeholders such as research institutions, educators, collectors, and artists should act in coordination with the development of educational content.

Thus, Sustainable Development Goal 4: Quality Education is Essential is achieved. As a policy-maker implication, local government should provide incentives to encourage participants, and they should make the latest technological investments in this field. Local government should also enhance awareness of this novel technique through education and training.

Future predictions for 2030, 2040, and 2050 are varied. In light of the continuity of museums in the near future, gamification represents a promising integration tool. Through the adaptation of new technologies, gamification applications will continue to evolve within the context of museum tourism. This situation will create an opportunity for museums which have already applied gamification techniques. In contrast, it provides a challenge to traditional museums, which object to integrating new technologies in the near future. As a prediction, it can be said that more museums will adopt gamification technologies into their services to catch the requirements of an era and ensure their continued existence. Another crucial prediction stems from the changing nature of tourists. As a new entrant into the tourism industry, Generation Z will be leading segment in the future and dominate through their changing travel-based needs, motivations, expectations, experiences, and preferences. They are more familiar with games, gaming elements, and digitalisation when compared with previous generations. So, it is anticipated that new tourist characteristics will challenge the current travel structure and require new adaptations. The blend of gamification and other tourism types will create new tourism markets for new generations in future years.

## References

Abou-Shouk, M., & Soliman, M. (2021). The impact of gamification adoption intention on brand awareness and loyalty in tourism: The mediating effect of customer engagement. *Journal of Destination Marketing & Management, 20*, 1–10. https://doi.org/10.1016/j.jdmm.2021.100559.

Amitrano, C. C., Russo Spena, T., & Bifulco, F. (2021). Digital Engagement and Customer Experience. In *Digital Transformation in the Cultural Heritage Sector* (pp. 119–136). Cham, United States: Springer. https://doi.org/10.1007/978-3-030-63376-9_6.

Anastasiadis, T., Lampropoulos, G., & Siakas, K. (2018). Digital game-based learning and serious games in education. *International Journal of Advances in Scientific Research and Engineering, 4*(12), 139–144. http://doi.org/10.31695/IJASRE.2018.33016.

Armstrong, M. B., & Landers, R. N. (2018). Gamification of employee training and development. *International Journal of Training and Development, 22*(2), 162–169. http://doi.org/10.1111/ijtd.12124.

Baptista, G., & Oliveira, T. (2017). Why so serious? Gamification impact in the acceptance of mobile banking services. *Internet Research, 27*(1), 118–139. https://doi.org/10.1108/IntR-10-2015-0295.

Bieszk-Stolorz, B., Dmytrów, K., Eglinskiene, J., Marx, S., Miluniec, A., Muszyńska, K., Miluniec, A., Muszyńska, K., Niedoszytko, G., Podlesińska, W., Rostoványi, A., Swacha, J., Vilsholm, R., & Vurzer, S. (2021). Impact of the availability of gamified e-guides on museum visit intention. *Procedia Computer Science*, *192*, 4358–4366. https://doi.org/10.1016/j.procs.2021.09.212.

Bonacini, E., & Giaccone, S. C. (2022). Gamification and cultural institutions in cultural heritage promotion: A successful example from Italy. *Cultural Trends*, *31*(1), 3–22. https://doi.org/10.1080/09548963.2021.1910490.

Buhalis, D., & Sinarta, Y. (2019). Real-time co-creation and nowness service: Lessons from tourism and hospitality. *Journal of Travel & Tourism Marketing*, *36*(5), 563–582. https://doi.org/10.1080/10548408.2019.1592059.

Cilingir Uk, Z. & Gultekin, Y. (2021). Gamification Applications in Hospitality and Airline Industries: A Unified Gamification Model. In F. Xu & D. Buhalis (Ed.), *Gamification for Tourism* (pp. 83–99). Bristol, Blue Ridge Summit: Channel View Publications. https://doi.org/10.21832/9781845418236-007

Corona, L. (2021). Museums and communication: The case of the louvre museum at the covid-19 age. *Humanities and Social Science Research*, *4*(1), 15–26. https://doi.org/10.30560/hssr.v4n1p15.

Gerval, J. P., & Le Ru, Y. (2015). Fusion of Multimedia and Mobile Technology in Audio-Guides for Museums and Exhibitions. In: D. Sharma, M. Favorskaya, L. Jain, & R. Howlett (Eds.), *Fusion of Smart, Multimedia and Computer Gaming Technologies*. Intelligent Systems Reference Library, vol. 84. Cham, Switzerland: Springer.

Hansen, A. (2016). Learning to feel well at Jamtli Museum: A case study. *Journal of Adult and Continuing Education*, *22*(2), 168–183. https://doi.org/10.1177/1477971416672327.

He, H., Li, Z., Cheng, X., & Wu, J. (2019, July). Gamified participatory museum experience for future museums. In *International Conference on Human-Computer Interaction* (pp. 195–208). Cham: Springer. https://doi.org/10.1007/978-3-030-23538-3_15.

Heale, R., & Twycross, A. (2018). What is a case study? *Evidence-Based Nursing*, *21*(1), 1–2. http://doi.org/10.1136/eb-2017-102845.

Jamtli. (n.d.). *About Jamtli*. Retrieved March 15, 2022, from https://www.jamtli.com/en/about-jamtli/

Jayawardena, N. S., Ross, M., Quach, S., Behl, A., & Gupta, M. (2021). Effective online engagement strategies through gamification: A systematic literature review and a future research agenda. *Journal of Global Information Management*, *30*(5), 1–25. 10.4018/JGIM.290370.

Leclercq, T., Poncin, I., & Hammedi, W. (2020). Opening the black box of gameful experience: Implications for gamification process design. *Journal of Retailing and Consumer Services*, 52, 1–9. https://doi.org/10.1016/j.jretconser.2019.07.007.

Liu, S., & Idris, M. Z. (2018). Constructing a framework of user experience for museum based on gamification and service design. In *MATEC Web of Conferences* (Vol. 176, p. 04007). EDP Sciences. https://doi.org/10.1051/matecconf/201817604007.

López-Martínez, A., Carrera, Á., & Iglesias, C. A. (2020). Empowering museum experiences applying gamification techniques based on linked data and smart objects. *Applied Sciences*, 10(16), 5419. doi:10.3390/app10165419.

Lütfi, A. T. A. Y., & Yüksel, T. (2020). Use of technology in museums, sample applications. *Journal of Global Tourism and Technology Research*, *1*(2), 91–105.

Madsen, K. M. (2020). The Gamified Museum: A Critical Literature Review and Discussion of Gamification in Museums. In T. Jensen, O. Ertløv Hansen, & C. A. Foss Rosenstand (Eds.), *Gamescope: The Potential for Gamification in Digital and Analogue Places Aalborg Universitetsforlag*. Aalbord, Denmark: Aalborg Universitetsforlag.

Malm, C. J. (2021). Social Innovations in Museum and Heritage Management. In *New Approach to Cultural Heritage* (pp. 201–219). Singapore: Springer. https://doi.org/10.1007/978-981-16-5225-7.

Nofal, E., Panagiotidou, G., Reffat, R. M., Hameeuw, H., Boschloos, V., & Moere, A. V. (2020). Situated tangible gamification of heritage for supporting collaborative learning of young museum visitors. *Journal on Computing and Cultural Heritage (JOCCH)*, *13*(1), 1–24.

Osterman, M. D. (2018). Museums of the future: Embracing digital strategies, technology and accessibility. *Museological Review*, *22*, 10–17.

Othman, M. K., Aman, S., Anuar, N. N., & Ahmad, I. (2021). Improving children's cultural heritage experience using game-based learning at a living museum. *Journal on Computing and Cultural Heritage (JOCCH)*, *14*(3), 1–24. https://doi.org/10.1145/3453073.

Pasca, M. G., Renzi, M. F., Di Pietro, L., & Guglielmetti Mugion, R. (2021). Gamification in tourism and hospitality research in the era of digital platforms: A systematic literature review, *Journal of Service Theory and Practice*, 31(5), 691–737. https://doi.org/10.1108/JSTP-05-2020-0094.

Pırnar, I., & Sarı, F. Ö. (2013). The changing role of museums: For tourists or local people? In *Conference Paper: International Conference on Sustainable Cultural Heritage Management*. Rome.

Serravalle, F., Ferraris, A., Vrontis, D., Thrassou, A., & Christofi, M. (2019). Augmented reality in the tourism industry: A multi-stakeholder analysis of museums. *Tourism Management Perspectives*, *32*, 1–11. https://doi.org/10.1016/j.tmp.2019.07.002.

Sigala, M. & Nilsson, E. (2021). Innovating the Restaurant Industry: The Gamification of Business Models and Customer Experiences. In F. Xu & D. Buhalis (Ed.), *Gamification for Tourism* (pp. 100–117). Bristol, Blue Ridge Summit: Channel View Publications. https://doi.org/10.21832/9781 845418236-008.

Tayara, M., & Yilmaz, H. (2020). The Gamification of Museum Attractions: The Perspective of Visitors. In *Heritage Tourism Beyond Borders and Civilizations*, 31–43. Springer, Singapore. https://doi.org/10.1007/978-981-15-5370-7.

Trigo-De la Cuadra, M., Vila-Lopez, N., & Hernandez-Fernández, A. (2020). Could gamification improve visitors' engagement?. *International Journal of Tourism Cities*, *6*(2), 317–334. https://doi.org/10.1108/IJTC-07-2019-0100.

Volchek, K. (2022). Book review "gamification for tourism". *Information Technology & Tourism*, *24*, 157–159. https://doi.org/10.1007/s40558-021-00209-2.

Wünderlich, N. V., Gustafsson, A., Hamari, J., Parvinen, P., & Haff, A. (2020). The great game of business: Advancing knowledge on gamification in business contexts. *Journal of Business Research*, *106*, 273–276. https://doi.org/10.1016/j.jbusres.2019.10.062.

Xi, N., & Hamari, J. (2019). "Does gamification satisfy needs? A study on the relationship between gamification features and intrinsic need satisfaction", *International Journal of Information Management*, *46*, 210–221. https://doi.org/10.1016/j.ijinfomgt.2018.12.002.

Xu, F., and Buhalis, D. (2021). *Gamification in Tourism*, Channel View Publications, Bristol ISBN:9781845418229

Xu, F., Weber, J., & Buhalis, D. (2013). Gamification in Tourism. In Zheng Xiang and Iis Tussyadiah (Eds.), *Information and Communication Technologies in Tourism 2014* (pp. 525–537). Springer, Cham.

Xu, F., Tian, F., Buhalis, D., Weber, J., & Zhang, H. (2016). Tourists as mobile gamers: Gamification for tourism marketing. *Journal of Travel & Tourism Marketing*, *33*(8), 1124–1142.

Xu, F., Buhalis, D., & Weber, J. (2017). Serious games and the gamification of tourism. *Tourism Management*, *60*, 244–256.

Zhang, C., Phang, C. W., Wu, Q., & Luo, X. (2017). Nonlinear effects of social connections and interactions on individual goal attainment and spending: Evidence from online gaming markets. *Journal of Marketing*, *81*(6), 132–155. https://doi.org/10.1509/jm.16.0038.

Zica, M. R., Ionica, A. C., & Leba, M. (2018). Gamification in the context of smart cities. In *IOP Conference Series: Materials Science and Engineering* (Vol. 294, No. 1, p. 012045). IOP Publishing.

# 34 User Generated Content Contribution to Museum Experience Design

*Jonatan Gómez Punzón and Nuria Recuero Virto*

## Introduction

Museums provide exceptional value and have been required to adapt to new conditions, moving from their typical presentation activity to a non-explored endeavour (Verde & Valero, 2021). During the last decade, and more in the last two years, museums have increased their online content, as well as the importance given to their positioning in relation to their online engagement (Agostino, Arnaboldi, & Lampis, 2020). During the COVID-19 pandemic, their cultural presentations rose on online cultural activities, taking place through social media and multimedia channels. This online proof has inspired further suggestions on the future trend of digitally empowered styles of museology and its amusement.

Many social networks have been used to make these interactions between online visitors and museums, as there has been a substantial increase of museums on Twitter, delivering cultural content, rather than sharing basic information or just marketing content. Social media has opened new ways to interact, becoming a strong channel to encourage messages on the importance of heritage, its preservation, and appreciation (Rivero, Navarro-Neri, & García-Ceballos, 2020).

Some museums have already begun to research with the big data generated during these lockdown, virtual, cultural experiences, as for example keyword-based Web search and analysis, and linked data as a tool to connect their databases (which might comprise millions of records), performing multi-layered and multimedia analysis, focusing on the data generated by these online visitors.

Much of this effort is still in the early stages of technological progress in the long term, it will likely become a framework for creating strong knowledge on how to use most of these virtual tools. This would allow museums to become much more interactive in the real and digital world, interacting through 360° strategies with their audiences, no matter what future disasters arise, such as COVID-19 (Kahn, 2020).

There have been many museum visitors who began to gain interest in finding which technologies were available to enjoy the museum experience during the lockdowns and the pandemic. As we can see in Figure 34.1, that there were some spikes of interest to get the most and to learn as much as possible about the possibilities of enjoying the museum's experiences through their technology.

Museums have to be aware of how important it is to maintain insightful conversations with visitors. No matter the channel used in social media, no matter the level of analyses that museum research has used to learn more about visitors and potential visitors, it is especially important to analyse which are the consequences of this interaction when

DOI: 10.4324/9781003291763-38

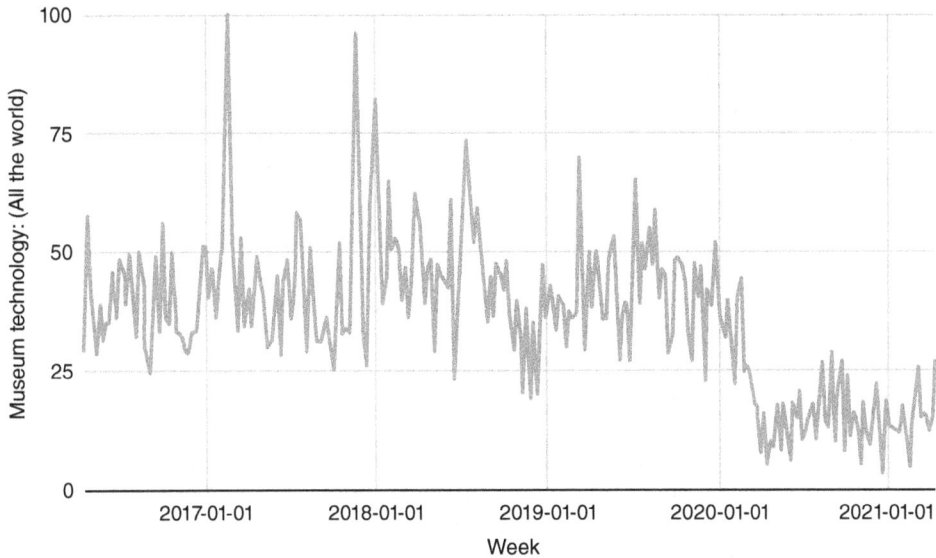

*Figure 34.1* Search trends on "museum technology" in Google Trends report (2017–2021).

researching the user generated content (UGC) that refers to the museum on online platforms reviewing the quality of the experience provided by the museum.

The role of Tripadvisor and many other online reviewing platforms on learning, agreeing, or discussing whether to attend a museum, no matter digital or face to face, show that there are much more to learn on UGC and on the presence of museums in these online platforms.

UGC has become a common practice in tourism during the last decade, but not at the museum level yet. The online reputation of a museum and the height of visitor engagement affect museums visitor decision-making (Fernández-Hernández, Vacas-Guerrero, & García-Muiña, 2020).

Due to the COVID-19 crisis, this study aims to answer the recent call for improving visitors' physical experience in museums. Furthermore, it offers insights regarding the analysis of sentiments, which will present useful information supporting collaborative learning.

## Literature review

### *Importance of social networks in the museum sector*

When museums realise the importance of internally showcasing the positive impact of social media strategies to be implemented, their online curators should keep in mind crowdsourcing museums on social media.

These contributions became an active role for museums to build opportunities that reply to these online interaction requests from content shared on social media networks, such as Snapchat or Instagram (Villaespesa & Wowkowych, 2020). Museums have not seized the opportunity offered, especially regarding Instagram and its importance, which seems to be underestimated (Amanatidis, Mylona, Mamalis, & Kamenidou, 2020).

With museums being asked to embrace the participatory potential of social media, no matter the channel, there are still significant differences in the readiness to change museum openness on social media (Booth, Ogundipe, & Røyseng, 2020). But more and more museums use Twitter to attract visitors, engage effectively with them, and facilitate the construction and strengthening of communities, where visitors create their own authentic, trustworthy content (Kydros & Vrana, 2021).

As an example of this content, there are numerous activities as the Twitter museum account takeover by any world-famous artist spreads the voice and showcases the museum's collection (Martín, & Aguirre, 2020). Twitter has been one of the main social networks used for museums to interact with fans and online visitors, no matter the type of museum (Rivero, Navarro-Neri, García-Ceballos, & Aso, 2020).

### Social media and the museum experience design

During the lockdowns, museum fans appreciated the addition of a digital flavour to exhibitions and art galleries. Improved experiences, better experience, and overall higher satisfaction result from the introduction of virtualisation in museums (Zollo, Rialti, Marrucci, & Ciappei, 2021). One of the causes of increasing use of social media is the engagement through current onsite experiences, since, despite the level of online experience of museums, the online experience design is increasing (Romolini, Fissi, & Gori, 2020). This museum experience design is expected to change, rearranging how experiences will be offered, and inevitably led to more digital-real (virtual reality and augmented reality) (Kargas, Karitsioti, & Loumos, 2020).

Regarding the content used to design these experiences, a participatory environment of social media can help museums improve comprehension by providing new ways of seeing, as a particular potential in that digital context may relate to a humorous discourse showcasing real content (Najda-Janoszka, & Sawczuk, 2020).

Prior to the pandemic, while most museums maintained active Twitter accounts, few used the social media platform in a productive manner, increasing their own experience design, since museums just used Twitter to receive suggestions and changes these museums should make to improve their experience, rather than engage and communicate as it happens now (Baker, 2017). Twitter represents an opportunity for communication and experience management in museums (Caerols-Mateo, 2017). Twitter is used by museums in order to maintain contact with existing customers and gain new clients. Increasing the online experience production has been adopted only recently (Bertoldi, Giachino, Stupino, & Mosca, 2018).

### User generated content and sentiment analysis concerning museum experience design

But we should wonder how museums were able to generate viewers who saw their UGC posts during lockdowns. Sentiment and UGC analysis are useful to engage new audiences, by repurposing UGC for traditional media and their metrics (Fernández-Hernández, Vacas-Guerrero, & García-Muiña, 2020). It is clear that museum visitors want to be educated and inspired, and UGC metrics reveal how to do this more effectively. Using visitors' experiences to drive the content strategy reflects the institution's "visitor-first" standard and extends it beyond the museum's UGC experience (Naiditch, Gertz, & Chamorro, 2017). UGC results affect museums in many ways, as for example user-generated reviews pose challenges to museums, and museum visitors pay attention to the usefulness of UGC as a source of data on museum visitors (Alexander, Blank, & Hale, 2018). Museum UGC

*Table 34.1* Main previous studies on UGC and museums

| Authors | Description |
| --- | --- |
| Colladon, Grippa, and Innarella (2020) | Analysis of online conversations using Semantic Brand Score to measure the brand importance of European museums. |
| Bailey-Ross et al. (2017) | QRator project. Exploring how mobile devices and interactive digital labels can create new models of UGC for visitors. |
| Baldoni, Baroglio, Patti, & Schifanella (2013) | ArsEmotica, an application software that combines an ontology of emotional UGC sentiment lexicons. |
| Hellin-Hobbs (2010) | Constructivist Museum. Using UGC research to find out how UGC is being incorporated into museums. |

should contribute to sustaining value co-creation and service innovation, encouraging the participation of audiences by enhancing experiences and interactions, and driving participants to create knowledge on cultural heritage (Romanelli, 2020). We have to consider that museums are pieces of the destination experience as a whole, and how visitors would perceive destination UGC as a global atmosphere. Leisure and recreation, social environment, culture, history, and art, are also affected by the image of the museum experience design and their infrastructure (Qi, & Chen, 2019).

In Table 34.1, we can check one of the most recent studies which used sematic analysis to track the importance of European museum brands. Variations in brand importance aligned with changes in the perceptions of experience design. In this case the sentiment perceived was not relevant in order to be useful for predicting potential changes on the visitor side.

The second piece of research is a project to explore how the technology provided by mobile devices showcases new types of models to research UGC in museums. Here an interpretation was attempted as to how culture could become innovative to provide good quality-based content from and for museums, to enable engagement in a deeper way.

Another interesting study was done in 2013 by designing new software to gather emotions provided by the UGC analysed using sentiment lexicons. Visitors from the most important museums were beginning to provide valuable content as reviews, and it is important to check which activity they reflect and what the value is of what they provide. The sentiment analysis shows that the community of visitors and fans of the museums are usually moved once the experience has been lived.

The fourth study which has been included as one of the main previous studies on UGC refers to an earlier stage of ten years ago, when UGC was beginning to be considered interesting and be incorporated in the research of museums to try to understand as best as possible the behaviour of visitors. In this research the aim was to discover how UGC could be incorporated considering that there are official interpretations that, at the time, were difficult to reshape.

We can see a clear evolution from the moment when UGC was not really considered as a useful tool when worthy UGC is not seem as sufficient and there is also analysis of semantics and brand sentiment.

### Research questions

This study followed an approach based on Research Questions (RQs), which is particularly interesting for motivating learning processes (Schumaker et al., 2016) and stimulating the generation of new ideas.

Table 34.1 details studies that have confirmed that the most trendy and latest topics can be recognised by investigating UGC on social media channels. Thus, these RQs are proposed:

- **RQ1:** *Is it feasible to try to recognise the most popular museum technology topics conducting an analysis of UGC on Twitter?*

    Sentiment analysis has been widely used to distinguish the feelings expressed and categorised into positive, neutral, and negative sentiments (Saura et al., 2020). Accordingly, the following question is:
- **RQ2:** *Will the recognised museum technology topics in UGC on Twitter be related to different feelings?*

    It has been affirmed that UGC from Twitter advances collective learning (Stephansen & Couldry, 2014; Tang & Hew, 2017). Therefore, all the conclusions from this study offer pertinent information, for instance, to advance museum experience design.
- **RQ3:** *Is it feasible to motivate collective learning of how to improve museum experience design by grouping the identified museum technology topics in relation to positive, neutral, and negative feelings?*

**Research method**

This study implements the three-stage methodology for text-data mining employed in Saura and Bennett's study (2019). In the first stage the Latent Dirichlet allocation (LDA) was employed. Python was used to evaluate the large dataset and identify the associated datasets. First, 617 tweets were extracted and cleaned within the hashtags #edchat and #edtech. LDA was employed to categorise these tweets to classify the museum topics most discussed on Twitter based on the above-mentioned dataset extracted (i.e., 617).

In the second stage, the Support Vector Machine (SVM) type algorithm was implemented in order to operate sentiment analysis. This phase permitted the detection of the feelings for each topic by labelling these sentiments as positive, negative, or neutral. Finally, Atlas.ti was implemented for text-data mining within the results to order them regarding the weight of repetition of words and the system of measurement acknowledged as weighted percentage (WP), which distinguishes the words that are repeated the most.

*Data collection*

The data extraction was accomplished using Python software 3.9.2 for Mac OS X linked to the public Twitter Application Programming Interface (API). The collected UGC was in English and Spanish and contained as keywords any of the following hashtags: #musetech and #musesocial (Figure 34.2 and Table 34.2).

The dataset was cleaned by eliminating repeated tweets, retweets, and not readable tweets, which left a final sample of $n = 617$ tweets. The data collection was realised from 15 December 2020 to 1 February 2021 and comprised the Christmas holiday period, and the pre- and post-vacation phases. The museum tags #musesocial and #musetech in Twitter were chosen as these are the most used hashtags for museum related tweets.

*Topic detection*

The LDA model employed in this study was based on Jia's (2018) study and takes into account a two-step mathematical and probabilistic approach. The first phase classifies the

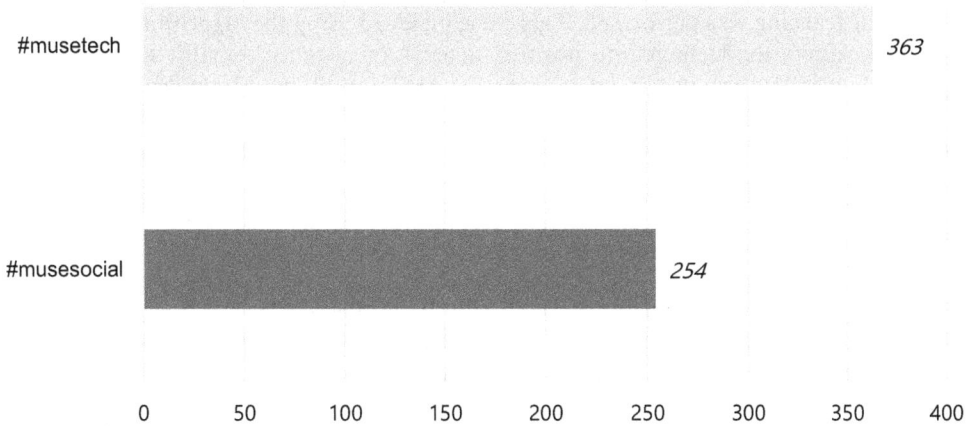

*Figure 34.2* Tweets collected using the popular hashtags #musetech and #musesocial used in Twitter.

*Table 34.2* Tweet collection concerning the hashtags employed

| Tag | Number of tweets |
| --- | --- |
| #musesocial | 254 |
| #musetech | 363 |

keywords within a database, where each word is encoded in an independent file. During the second stage, the topics are randomly acknowledged, and the themes are identified (Equation 34.1).

$$p\left(\beta_{1:k},\theta_{1:D},Z_{1:D},\omega_{1:D}\right)=\Pi_{i=1}^{K}\left(\beta_i\right)\left(\beta_1\right)\times\Pi_{d=1}^{D}\,p\left(\theta_d\right)\times\sum_{n=1}^{N}p(Z_{d,n}\mid\theta_d)p\left(W_{d,n}\mid\beta_{1:K},Z_{d,n}\right)\ (34.1)$$

where:

- $\beta_i$ is the distribution of words in topic $i$, $K$ topics altogether;
- $\theta d$ is the proportion of topics in document $d$, in all $D$ documents;
- $Z_d$ is topic designation in document $d$;
- $Z_{d,n}$ is the topic designation for the $n$th word in document $d$, in all $N$ words;
- $W_d$ is spotted words for document $d$;
- $W_{d,n}$ is the $n$th word for document $d$.

Then, the identification of the topics and words is prearranged following Equation 34.2, using Gibbs sampling (Jia, 2018):

$$\rho\left(\beta_{1:k},\theta_{1:D},Z_{1:D}\mid\omega_{1:D}\right)=\frac{\rho\left(\beta_{1:k},\theta_{1:D},Z_{1:D}\omega_{1:D}\right)}{\rho\left(\omega 1:D\right)}\qquad(34.2)$$

*Sentiment analysis*

Once the most popular museum topics in Twitter were identified, a Python algorithm based on machine learning was performed. This stage included using the algorithm for text-data mining to classify the feelings into positive, neutral, or negative. Finally, Krippendorff's Alpha Value (KAV) was used to determine the precision of the sentiment analysis. The reliability of the results was established in the three sentiments, as it concludes that when α ≥ 0.800 the reliability of the results is high, when α < 0.667 the results are low; the limit for tentative results is α ≥ 0.667 (Krippendorff, 2004).

*Textual data analysis*

This stage was implemented using Atlas.ti software. Text-data mining was accomplished and labelled into the three sentiments (i.e., positive, neutral, negative). The dataset was structured following three processes: (1) recognising the frequency of repetition of the words; (2) establishing the keywords' total weight measured as a WP; and (3) filtering the words that were not appropriate for the research objectives (Newton-John, 2018). WP labels the weight of the indicators gathered into nodes, which concurs with the number of times they are repeated (Newton-John, 2018). Figure 34.3 clarifies the three stages fulfilled in this research.

## Results

The results of LDA estimation revealed 13 museum-related topics, which are shown in Table 34.3. Throughout this procedure, LDA categorises the words into topics that were cautiously controlled and named after examining the group of words (Büschken & Allenby, 2016; Jia, 2018; Miller et al., 2017; Saura & Bennett, 2019). The name of the topic was given by considering the 10 to 20 most repeated words; meticulous descriptions of the topics were established bearing in mind the content of the topics.

Sentiment analysis was fulfilled by following Saura et al.'s (2020) recommendations, where the tweets of each topic were individually studied. The sentiment analysis algorithm was trained to achieve the endorsed probability of success (Saura et al., 2020).

Consequently, a sample of 160 posts was managed with data-mining techniques to train the algorithm. Hence, KAVs were estimated; the results are above the thresholds. Table 34.4 details the reliability of the sentiment analysis.

The results of the textual analysis phase are presented in Figure 34.4, where the WP of each topic is identified in relation to the feeling expressed (Krippendorff, 2004).

## Discussion

After analysing all the results collected in this chapter, we can discuss the research question results compared with the insights achieved. Regarding RQ1, the main technology-based comments are those which are perceived as greedy, and also some comments related to the online dimension of this innovation. The analysis reveals differences between UGC attributes and UGC feelings according to the UGC types. In fact, UGC attributes have positive effects on eWOM, brand attitude, and visit intention (Yu, & Ko, 2021).

Hence, we cannot perceive, and we cannot recognise, popular museum topics using these technology-based concept analyses on Twitter. A positive concept that we can perceive

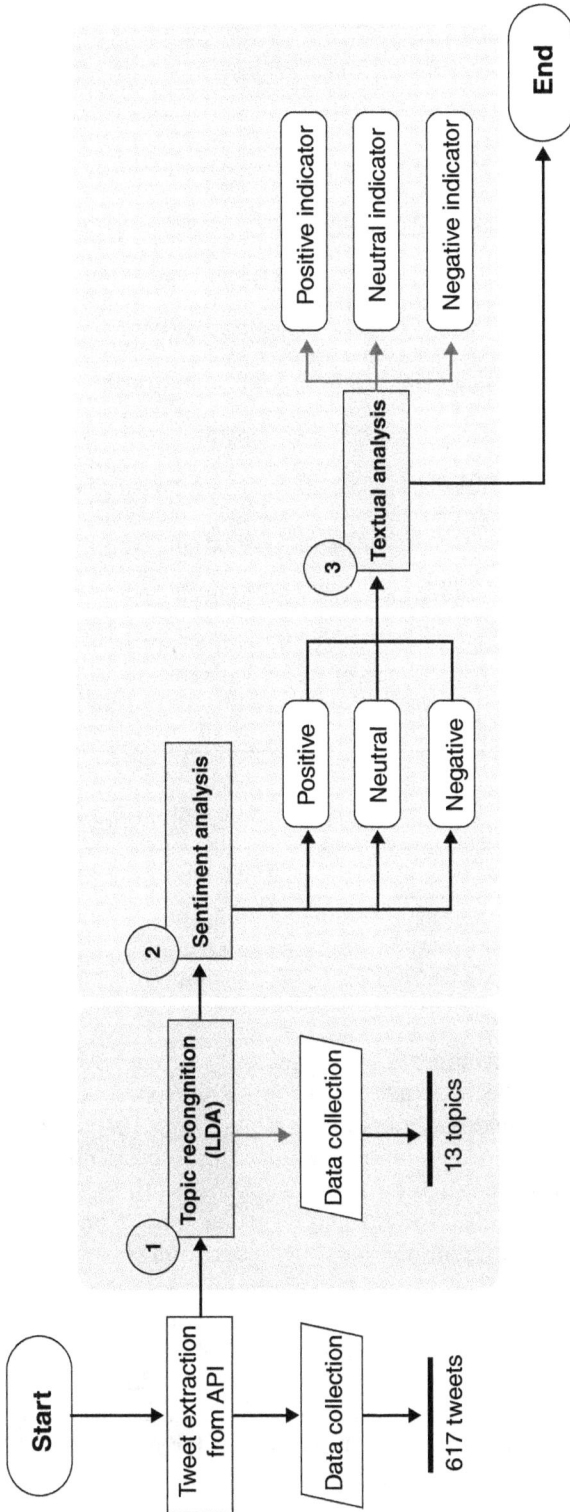

*Figure 34.3* Stages of the research methodology.

*Table 34.3* Recognised topics related to #musetech and #musesocial in UGC

| Topic name | Topic description | WP | Sentiment |
|---|---|---|---|
| Avarice | Comments in relation to museum technology where these are perceived as greedy. | 0.76 | negative |
| Wired | Observations concerning the online dimension of these innovations. | 0.66 | positive |
| Experiential marketing | Mentions regarding the significance of creative experiences for the engagement. | 0.49 | positive |
| HR updating | References of the employee requirements to update, look for innovations, and learn their use. | 0.45 | negative |
| Great | Mentions the impacts of these technologies to obtain immersive experiences that are perceived as cool and relaxed. | 0.41 | positive |
| Missed opportunities | Annotations regarding the lack of adaptation to many of these advancements. | 0.32 | negative |
| Management supervision | Remarks relative to the team supervision of these innovations. | 0.32 | neutral |
| Digital | Recognition of the lack of adaptation concerning digital improvements that can be implemented in museums. | 0.31 | negative |
| Time | References to all the technological advances that have been historically adopted. | 0.22 | neutral |
| Free | Recognition of the economic benefits from all these technological devices. | 0.18 | positive |
| Jobs | Discussion of the advantages for the creation of employment. | 0.16 | positive |
| Recent | Recognition of the fastness of these advancements. | 0.16 | positive |
| Demand | Comments and opinions regarding visitors who are asking for technological innovations. | 0.15 | negative |

*Table 34.4* Reliability of the sentiment analysis conclusions (Krippendorff's alpha)

| Reliability | KAV | Sentiment | Average KAV |
|---|---|---|---|
| High | $\alpha \geq 0.800$ | Positive | 0.875 |
| Tentative | $\alpha \geq 0.667$ | Negative | 0.775 |
| Low | $\alpha < 0.667$ | Neutral | 0.685 |

regards how the technological innovations are created to engage with the visitors at distance. In relation to RQ2, it has been affirmed that the UGC from Twitter advances collective learning (Stephansen & Couldry, 2014; Tang & Hew, 2017). Therefore, all the conclusions from this study may offer pertinent information, for instance, to advance museum experience design. In this regard we can discuss what happened with RQ3.

Is it feasible to motivate collective learning regarding how to improve museum technology experience design by grouping the identified education topics in relation to positive, neutral, and negative feelings?

We can find positive, negative, and neutral feelings which allow us to reach topics on museum technology experience design through motivating collective learning. UGC is a valid source for the study of tourism destination images (including museums), confirming the need to adopt a holistic and attribute-based approach to this concept (Alarcón-Urbistondo, Rojas-de-Gracia, & Casado-Molina, 2021). We cannot forget that not all attributes, either positive or negative, influence the overall idea of a museum and its technologies.

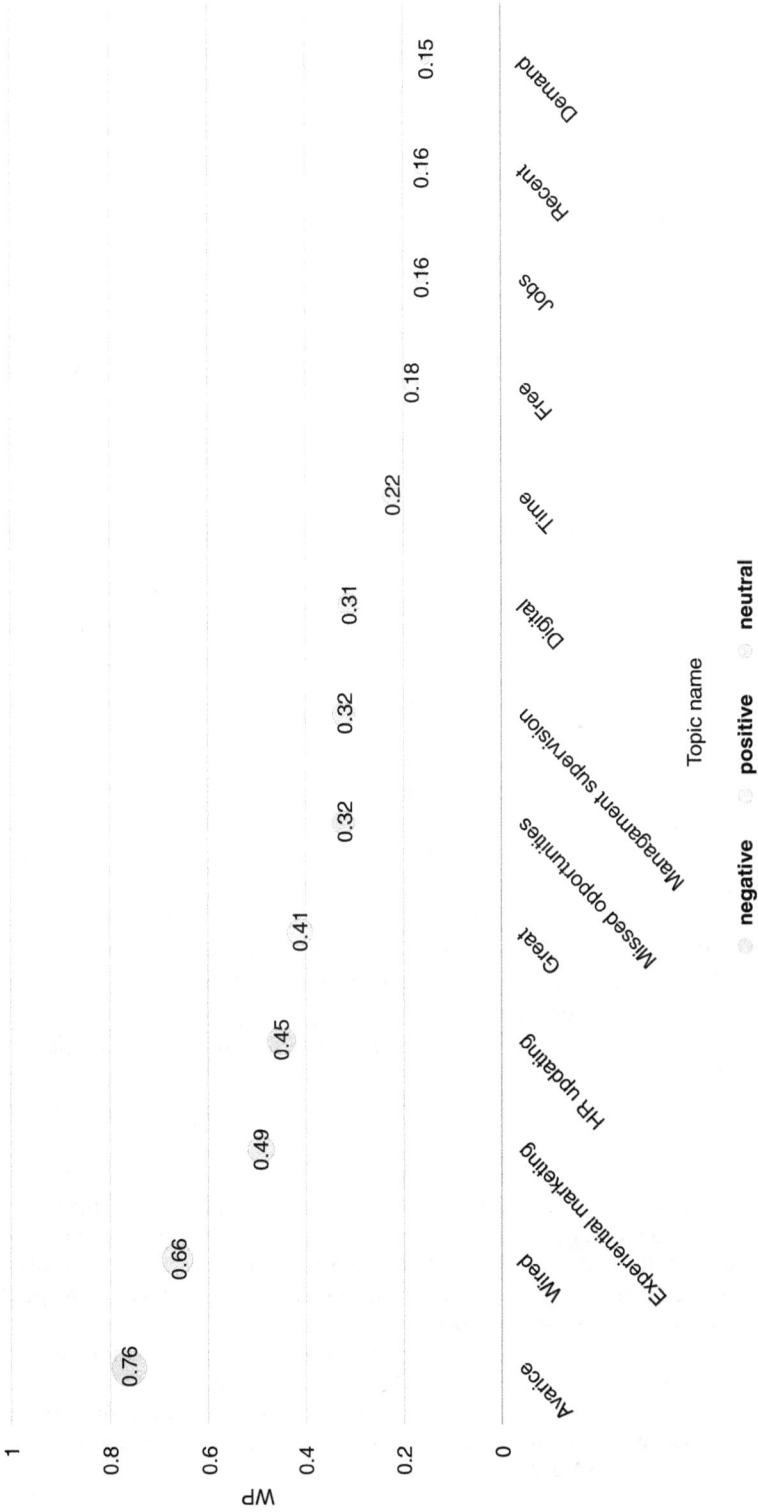

*Figure 34.4* Recognised museum topics in relation to the sentiment expressed and WP.

## Conclusions

The latest analysis and visualisation methods on UGC are required for gathering useful insights from the massive volumes of data generated by new technologies and data sharing stages. We have tried to find a groundwork for such practices so that big data may also be involved in the stage of experience, vision, and understanding. Museums and technology have built a deep relationship in the present time.

Visitors and museums share a plentiful supply of UGC on social networks and digital platforms. We have checked how useful technological interactions can be obtained and envisaged from examples of easily accessible UGC, in this case, the topics analysed in tweets from the social network Twitter. As a conclusion, we have to consider, based on the results obtained, that insights related to innovative encounters between museums and visitors can be fruitful to improve strategies and technological developments and involvement in museums in the near future.

### Theoretical implications

This research has provided three main contributions. Firstly, it provides knowledge regarding the influence of UGC sentiment analysis on museum technology management. Secondly, it offers 13 topics that can be employed to improve collective learning in Twitter. Thirdly, it identifies the feelings that are aroused from each of the topics. This study could be improved by using other social media channels and raising the time lapse of the data collection (Tang & Hew, 2017).

### Managerial contributions

This research could assist decision-making processes as it improves the knowledge of some of the trendiest issues regarding museum technology implementations discussed on Twitter. The three most popular topics were avarice, wired, and experiential marketing, where avarice is related to negative emotions, and wired and experiential marketing to positive feelings. This finding is useful for museum managers as they come to know that much museum technology is perceived as greedy, which denotes the importance of promoting positive experience among users, visitors, and employees regarding these devices. Besides, online and creative experiences are well perceived, which is significant as these initiatives can be increased.

## References

Agostino, D., Arnaboldi, M., & Lampis, A. (2020). Italian state museums during the Covid-19 crisis: From onsite closure to online openness. *Museum Management and Curatorship*, *35*(4), 362–372.

Alarcón-Urbistondo, P., Rojas-de-Gracia, M. M., & Casado-Molina, A. (2021). Proposal for employing user-generated content as a data source for measuring tourism destination image. *Journal of Hospitality & Tourism Research*, *47*(4), 10963480211012756.

Alexander, V. D., Blank, G., & Hale, S. A. (2018). TripAdvisor reviews of London museums: A new approach to understanding visitors. *Museum International*, *70*(1–2), 154–165.

Amanatidis, D., Mylona, I., Mamalis, S., & Kamenidou, I. E. (2020). Social media for cultural communication: A critical investigation of museums' Instagram practices. *Journal of Tourism, Heritage & Services Marketing (JTHSM)*, *6*(2), 38–44.

Bailey-Ross, C., Gray, S., Ashby, J., Terras, M., Hudson-Smith, A., & Warwick, C. (2017). Engaging the museum space: Mobilizing visitor engagement with digital content creation. *Digital Scholarship in the Humanities*, *32*(4), 689–708.

Baker, S. (2017). Identifying behaviors that generate positive interactions between science museums and people on Twitter. *Museum Management and Curatorship*, *32*(2), 144–159.

Baldoni, M., Baroglio, C., Patti, V., & Schifanella, C. (2013). Sentiment Analysis in the Planet Art: A Case Study in the Social Semantic Web. In: C. Lai, G. Semeraro, & E. Vargiu (eds.), *New Challenges in Distributed Information Filtering and Retrieval*. Studies in Computational Intelligence, vol. 439. Berlin, Heidelberg: Springer.

Booth, P., Ogundipe, A., & Røyseng, S. (2020). Museum leaders' perspectives on social media. *Museum Management and Curatorship*, *35*(4), 373–391.

Büschken, J., & Allenby, G. M. (2016). Sentence-based text analysis for customer reviews. *Marketing Science*, *35*(6), 953–975. https://doi.org/10.1287/mksc.2016.0993

Caerols-Mateo, R. (2017). Social networking sites and museums: Analysis of the Twitter campaigns for International Museum Day and Night of Museums. *Revista Latina de Comunicación Social*, *72*, 220–234.

Colladon, A. F., Grippa, F., & Innarella, R. (2020). Studying the association of online brand importance with museum visitors: An application of the semantic brand score. *Tourism Management Perspectives*, *33*, 100588.

Fernández-Hernández, R., Vacas-Guerrero, T., & García-Muiña, F. E. (2020). Online reputation and user engagement as strategic resources of museums. *Museum Management and Curatorship*, *33*(6), 1–16.

Hellin-Hobbs, Y. (2010). Constructivist Museum. Using UGC research how user-generated content is being incorporated into museums. *EVA'10, Proceedings of the 2010 International Conference on Electronic Visualisation of the Arts*, July, 72–78.

Jia, S. (2018). Leisure motivation and satisfaction: A text mining of yoga centres, yoga consumers, and their interactions. *Sustainability*, *10*, 44–58. https://doi.org/10.3390/su10124458

Kahn, R. (2020). *Locked Down Not Locked Out–Assessing the Digital Response of Museums to Covid-19*. Impact of Social Sciences Blog.

Kargas, A., Karitsioti, N., & Loumos, G. (2020). Reinventing Museums in 21st Century: Implementing Augmented Reality and Virtual Reality Technologies alongside Social Media's Logics. In *Virtual and Augmented Reality in Education, Art, and Museums* (pp. 117–138). Hershey, PA: IGI Global.

Krippendorff, K. (2004). Measuring the reliability of qualitative text analysis data. *Quality and Quantity*, *38*(6), 787–800. https://doi.org/10.1007/s11135-004-8107-7

Kydros, D., & Vrana, V. (2021). A Twitter network analysis of European museums. *Museum Management and Curatorship*, *36*, 569–589.

Martín, E. L., & Aguirre, B. M. (2020). Residencias artísticas virtuales en twitter: Retos, motivaciones y experiencias. *Kepes*, *17*(21), 195–223.

Miller, M., Banerjee, T., Muppalla, R., Romine, W. &, Sheth, A. (2017). What are people tweeting about Zika? An exploratory study concerning its symptoms, treatment, transmission, and prevention. *JMIR Public Health Surveill*, *3*(2): e38. https://doi.org/10.2196/publichealth.7157

Mosca, Fabrizio, Bertoldi, Bernardo, Giachino, Chiara, & Stupino, Margherita (2018). Facebook and Twitter, social networks for culture. An investigation on museums. *Mercati & Competitivit? FrancoAngeli Editore*, *2018*(2), 39–59.

Naiditch, M., Gertz, R., & Chamorro, E. (2017). How do you museum? Marketing user-generated content to engage audiences. *MW17: Museums and the Web 2017*.

Najda-Janoszka, M., & Sawczuk, M. (2020). Cultural authority with a light touch: Museums using humor in social media communication. *Romanian Journal of Communication and Public Relations*, *22*(2 (50))).

Newton-John, T. R. (2018). Qualitative data analysis in health psychology: Testing theoretical models using qualitative data. *Sage Research Methods*. doi: https://doi.org/10.4135/9781526427700

Qi, S., & Chen, N. (2019). Understanding Macao's destination image through user-generated content. *Journal of China Tourism Research*, *15*(4), 503–519.

Rivero, P., Navarro-Neri, I., García-Ceballos, S., & Aso, B. (2020). Spanish Archaeological Museums during Covid-19 (2020): An Edu-Communicative Analysis of Their Activity on Twitter through the Sustainable Development Goals. *Sustainability*, *12*(19), 8224.

Romanelli, M. (2020). Museums and Technology for Value Creation. In *Technology and Creativity* (pp. 181–210). Cham: Palgrave Macmillan.

Romolini, A., Fissi, S., & Gori, E. (2020). Visitors' engagement and social media in museums: Evidence from Italy. *International Journal of Digital Culture and Electronic Tourism*, *3*(1), 36–53.

Saura, J. R., & Bennett, D. (2019). A three-stage methodological process of data text mining: A UGC business intelligence analysis. *Symmetry 11*(4), 519. https://doi.org/10.3390/sym11040519

Saura, J. R., Reyes-Menendez, A., & Thomas, S. (2020). Gaining a deeper understanding of nutrition using social networks and user-generated content. *Internet Interventions*, *20*, 100–312. https://doi.org/10.1016/j.invent.2020.100312

Schumaker, R. P. A., Jarmoszko, T., & Labedz, C. (2016). Predicting wins and spread in the Premier League using a sentiment analysis of Twitter. *Decision Support Systems*, *88*, 76–84. https://doi.org/10.1016/j.dss.2016.05.010

Stephansen, Hilde C., & Couldry, N. (2014). Understanding micro-processes of community building and mutual learning on Twitter: A 'small data' approach. *Information, Communication & Society*, *17*(10), 1212–1227. https://doi.org/10.1080/1369118X.2014.902984

Tang, Y., & Hew, K. F. (2017). Using Twitter for education: Beneficial or simply a waste of time? *Computers & Education*, *106*, 97–118. https://doi.org/10.1016/j.compedu.2016.12.004

Verde, A., & Valero, J. M. (2021). Virtual museums and Google arts & culture: Alternatives to the face-to-face visit to experience art. *International Journal of Education and Research*, *9*(2), 43–54.

Villaespesa, E., & Wowkowych, S. (2020). Ephemeral storytelling with social media: Snapchat and Instagram stories at the Brooklyn museum. *Social Media+Society*, *6*(1), 2056305119898776.

Yu, J., & Ko, E. (2021). UGC attributes and effects: Implication for luxury brand advertising. *International Journal of Advertising*, *40*(6), 1–23.

Zollo, L., Rialti, R., Marrucci, A., & Ciappei, C. (2021). How do museums foster loyalty in tech-savvy visitors? The role of social media and digital experience. *Current Issues in Tourism*, *25*(18), 2991–3008.

## 35 Presence in virtual hotel experience and purchase intention

### The mediating role of decision comfort

*Sima Rahimizhian, Farzad Safaeimanesh, Mobina Beheshti and Olayinka Afolabi*

### Introduction

The leisure and travel industry emphasises customer experience as one of the single most efficient competitive discriminators that marketers encounter. Recently, the progression of technologies such as augmented reality (AR), virtual reality (VR), and mixed Reality (MR) has revolutionised the way people experience their environments (Buhalis & Karatay, 2022; Jung et al., 2015). Businesses running within the travel industry have been particularly receptive to implementing VR technology for justifiable reasons. For marketers, VR acts as an efficient tool to give consumers a little taste of the experience before the actual purchase (Huang et al., 2016; Rahimizhian et al., 2020). VR is becoming a more convenient source of information for customers than traditional visual media (Fransen et al., 2015). It changes the processes in which tourism goods and services are handled, distributed, and promoted on the market, which means a similar evolution in the way tourists are motivated, book, plan, and experience travel (An et al., 2021; Buhalis & Law, 2008; Neuhofer et al., 2014; Tussyadiah et al., 2018).

VR technologies and headsets provide a magnificent opportunity for users to transform their surroundings virtually into a distinct environment. The creative use of VR as a source of travel information can significantly help travellers to reduce the information search process. Previous studies in tourism have discussed the potential applications of VR in tourism marketing (Williams & Hobson, 1995; Guttentag, 2010; Huang et al., 2016; Rahimizhian et al., 2020). VR has the potential to become a powerful marketing tool as it helps customers to visualise, interact with, and experience a product before the actual purchase which also enhances their future intentions (Subawa et al., 2021). Suh & Lee (2005) proved that VR experience as a marketing tool intensifies customers' knowledge of the product, and leads to more positive attitudes and purchase intentions.

The fact that most tourists do not begin their research to book with a particular brand leaves a large area of untouched opportunity for hotels. Some goods and services need to be experienced to be known and truly appreciated. Customers expect VR to provide immersive advantages while reducing decision-making uncertainty (Dacko, 2017). Moreover, the intangible essence of tourism products with the inability to provide an exploration of conditions is one of the principal deficiencies of, and a challenge for, hospitality businesses. Special consumer requirements, such as the sensation of presence in the physical environment as well as the comforting impact of an immersive VR experience for decision-making, have received limited consideration in the study of the VR experience and tourists' behavioural intentions. Therefore, tourism marketers need a more comprehensive knowledge of

DOI: 10.4324/9781003291763-39

advanced technology as a way to enhance customer experience, interconnectivity, the opinion of a brand, and the impact of VR on decision-making (Buhalis & Leung, 2018).

As stated by Guttentag (2010), many tourism providers offer various VR-type technologies as high-resolution and low-cost portable VR headsets, also known as head-mounted devices (HMDs), with built-in head tracking capability to enhance tourists' experience. This immersive user experience might also be utilised by hotel management and travel companies to enhance bookings and assist customers in making decisions. Despite VR's enormous potential, for the hotel industry, just a few brands (e.g., Thomas Cook, Hilton, Choice Hotels, Best Western, and Marriott International) have utilised this innovative technology (Israel et al., 2019). As a result, the purpose of this chapter is to investigate the role of presence in VR experiences provided by HMDs on travellers' decision comfort and purchase intentions based on the stimulus-organism-response (SOR) theoretical framework. This emphasises the potential of VR presence in clarifying the process through which VR-based experience translates into favourable customer evaluations of online service experiences, in terms of decision comfort.

## Literature review

### *Technology-based marketing of tourism services*

The advancement of technology has had a significant impact on the innovation of management in the tourism industry (Law et al., 2014), it has created a more demanding competitive market among the different tourism services (Buhalis et al., 2019), and it has increased the expectations of the touristic experience in tourism destinations (Chathoth et al., 2016; Huang et al., 2016). New innovative technology has been able to connect tourists, the destination, and tourism service marketers in different ways. First, in the pre-visit phase which is where the marketers can promote the destination to improve the number of tourists (Neuhofer et al., 2015; Jung et al., 2017); second, the use of technology during the trip brings competitiveness among the different tourism services, the tourist having to choose between different hotels and different restaurants (Baker, 2016) – the advancement of VR technology has become a significant component of destination marketing (Van Kerrebroeck et al., 2017); third, in the post-visit stage with electronic word of mouth (Litvin et al., 2008).

Williams and Hobson (1995) were among the first researchers to address the importance that VR would come to have on the marketing and promotion of tourism products. Since then, VR has grown and been discussed in different research on its implications in influencing the promotion of tourism destinations, services, and products (Chathoth et al., 2016; Disztinger et al., 2017; Rahimizhian et al., 2020). VR has become a tool tourism providers use to help tourists in choosing their brand or their service (Disztinger et al., 2017; Buhalis et al., 2019). In employing VR as a commercial tool to promote the hotel, Leung et al. (2020) found that VR commercials had higher immediate effects than traditional commercials, particularly among participants with greater cognitive ability. Additionally, VR can boost hotel brand awareness, brand attitudes, and purchase intentions (Leung et al., 2020). Furthermore, Lyu et al. (2021) discovered that VR commercials increase customer vividness and interactivity, which positively influences attitudes toward the commercial, brand attitudes, and hotel booking intentions. Tourists are requiring more from destinations, and also from the tourism service providers; they want to have specifications for their hotel or accommodation room, and they want to have a feel of the scenery and aesthetics of the place, without having to stay in the hotel yet or visit the physical hotel's location (Bogicevic et al., 2019). The technology provides interactivity, visual communication, and engagement that aids the user's decision-making process and helps them make a final decision about a choice of destination or tourism service (Jung & tom Dieck, 2017).

**VR and tourist purchase intentions**

The tourism industry is a combination of different service industries (Otto & Ritchie, 1996); the tourist while having a touristic experience makes use of different services in the destination. Transportation, accommodation, attractions, restaurants, and events jointly make up the experience tourists have at a destination (Lugosi & Walls, 2013). Tourism has continued to grow and develop and, along with it, the service sector is growing; hotels have also recorded increased revenue as a result of the rise in the market (Chen, 2016). This has encouraged more hotels to be built, and existing hotels to expand their brand and portfolio across different tourism services (Wang & Chung, 2015). At the initial stage when a tourist is deciding to travel, the destination is the focus, location, experience, and so on; but further into the trip, the type of hotel and accommodation to use, and the touristic services to choose from, are very important as they can make the experience more enjoyable or otherwise (Han et al., 2011). This makes different hotel brands seek strategic marketing plans to communicate the best touristic experience that their hotel can provide at different points when the tourist has to decide on a service (Baker, 2016). VR is an important aspect of tourist response and attitude to different tourism services and different brand options in a destination (Van Kerrebroeck et al., 2017).

VR is different from artificial reality because it provides "physical immersion and psychological presence" to the users in the virtual world. VR is a 3D environment that creates a real-time simulation of a place where the users can interact and simulate their five senses as if they were in the physical location itself (Guttentag, 2010). The virtual world is shown to users by using an HMD output device, which can be a Google helmet, and a sophisticated technology device that can be used as a visual output device placed directly in the user's view like wearing glasses. One of the major features, and even considered the most important feature of VR, is the visuals it provides, so the quality of the 3D images must be very good to create a virtual experience that resonates positively with the user (Guttentag, 2010). The performance and development of VR devices have a positive effect on the use of VR, for promoting the tourist's experience from pre-visit when they are still making decisions about the destination to visit (Neuhofer et al., 2015), to during the visit when they are at the tourism destination and they need to choose between the different tourism services available without first experiencing it (Bogicevic et al., 2019).

With the growth and continuous use of VR in travel, tourism, and among consumers (Gartner, 2015; Marasco et al., 2018) the research trend on it has also progressed, since the first tourist acceptance of VR was investigated (Disztinger et al., 2017). Huang et al. (2016) studied VR using the technology acceptance model to predict how tourists use VR. The results of the study showed that VR can attract tourists that are online searching for a destination, or tourists that are already at the destination and trying out different tourism services. Several studies investigated VR's influence on tourists' behavioural intentions, in different tourism contexts (Jung & tom Dieck, 2017), and VR has an influence on destination visit and revisit intentions (Marasco et al., 2018). Tussyadiah et al. (2018) studied tourists' attitudes and visitation intentions with the use of tourism content that is presented or advertised to the consumer with VR. The study found VR positively influences the attitude of tourists. In the study by Zhang et al. (2019) virtual technology was shown to influence online consumer purchase intention, and further studies in the domain of consumers considered VR as an influence on real estate purchase intention (Talukdmix & Yu, 2018).

**Theoretical underpinning**

The theory used in this chapter is the SOR framework, first introduced by Mehrabian and Russell (1974), and after more than a decade, adjusted by Jacoby (2002). This framework argues that several factors in the environment influence the emotional and cognitive mindset of a person, and that emotion or cognition influences behavioural outcomes (Donovan & Rossiter, 1982). This theory has been used in different domains to explain the responses to computer experience (Eroglu et al., 2003). The theory has been used to explain website experience (Mollen & Wilson, 2010) and customer engagement (Islam & Rahman, 2017), and it has been used in the domain of tourism to explore consumer behaviour in VR tourism (Kim et al., 2020). The application of SOR in this study is towards tourists' behaviour and response.

The SOR theory observes three elements: stimulus, organism, and response. The first element, stimulus, is defined as "the influence that arouses the individual" (Eroglu et al., 2001, p. 179). VR influences behavioural reactions to virtual stimuli (Schuemie et al., 2001), a virtual presence which is the feeling a person can have while experiencing VR that can make them feel like they are in the atmosphere, based on the virtual presence (Keng et al., 2018). In this chapter, the stimulant is considered as the presence in the virtual environment being shown by the tourism service brand; the second element which is the organism is the middle stage of the customer's response as it is between the input which is the stimulus, and the response which is the output (Loureiro & Ribeiro, 2011). The organism is the affective and cognitive state of the customer, which reflects their emotions; the stimuli influence their decisions (Loureiro & Ribeiro, 2011). The organism state is where the consumers consider everything about the acquisition (Eroglu et al., 2001); the organism that would make the user give a response is the variable "decision comfort", which is a soft positive emotion (Parker et al., 2016). The response, which is the third element in the SOR framework, is the customer's output in terms of avoidance, behaviour, or approach (Donovan & Rossiter, 1982) – positive behaviours like visit intentions and purchase have been studied with the SOR framework (Kim et al., 2020). This study considers the purchase intention to be the output of the tourists towards the sense of presence of a tourism service VR video and the organism of decision comfort.

**Presence and purchase intention**

When considering the use of information technology systems in tourism experiences, the quality of the technology is an important factor (Guttentag, 2010; Jung et al., 2015). Virtual presence is the use of the context of presence in the effectiveness of VR. Presence is the feeling of being engaged in the virtual environment, and the extent to which the user feels physically "present" in the mediated environment (Schuemie et al., 2001). Presence is a complex experience, with different perceptives which can be determined by cognitive cues and multi-sensory information (Diemer et al., 2015). In tourism, VR has been studied as a means to improve the touristic experience (Baker, 2016). Presence in VR can elevate enjoyment, preference, and the like regarding a tourist destination (Tussyadiah et al., 2018).

The quality of the imaging and aesthetics in a VR enhances pre-consumption expectations among consumers of tourism and bolsters a sense of presence in the brand (Rodríguez-Ardura & Martínez-López, 2014). VR preview interactivity of Web tours of hotels, that is "the extent to which the user can modify the environment," improves tourists' brand experience of the hotel (Bogicevic et al., 2019). In the context of online shopping, presence has been studied to stimulate attitude and purchase intentions (Peng & Kim, 2014). The sense of virtual presence in VR videos of theme parks has been studied to influence the experience and behavioural intentions of the tourists and visitors to the park (Wei et al., 2019). Appealing AR improves the perceived ease of use and usefulness of AR and has a significant influence on destination visit intention (Chung et al., 2015). Virtual technology influences the purchase decision process and finally the purchase intentions of consumers (Zhang et al., 2019). AR services offered by organisations can achieve behavioural intentions that are favourable, such as purchase behaviour, and positive word of mouth (Dacko, 2017). Using VR videos that make the user feel like they are experiencing and present in the virtual environment by tourism marketers can influence the tourist purchase intention of the product, like watching a VR video of a hotel before or during the visit can influence the tourist to purchase the hotel service.

## Decision comfort and purchase intention

Virtual technology allows users to be immersed in the virtual objects as if it is real life. Previous studies have shown that users expect that decision uncertainty would reduce as a result of the influence of virtual technology (Dacko, 2017). Decision comfort was first conceptualised by Parker et al. (2016) as the extent to which a customer feels comfortable and contented about a decision. Furthermore, decision comfort is a positive emotion (Parker et al., 2016) that is different from decision confidence; while decision confidence reflects the level of certainty or uncertainty about making the best choice, decision comfort is an affect-based sense of reflection on the ease of making a choice (Hilken et al., 2017). Virtual technology is deployed as a tool to enhance and help customers in their decision-making process (Hilken et al., 2017). Retailers have stated the need for customers to be comfortable, and that this can influence having an easier purchase process (Heller et al., 2019). Parker et al. (2016) called for further research to extend the decision comfort on the purchase and repurchase intention of brands, and comfort has been studied on its affect on behavioural intentions and experience (Sweeney et al., 2000). Additionally, decision comfort has been studied on its influence on behavioural intentions and word of mouth (Hilken et al., 2017); it has also been studied for its influence on choice and word of mouth (Heller et al., 2019).

## VR presence and decision comfort

Studies have shown that customers seek additional value from virtual technology (Dacko, 2017); customers' decisions are influenced by stimuli (Petrova & Cialdini, 2018). Decision comfort is very important for VR frontline technology (Heller et al., 2019). The real-life quality that VR technology provides to customers makes them feel more comfortable about making a decision (Schubert & Koole, 2009). The theoretical underpinning for AR mental imagery was studied to positively influence decision comfort (Heller et al., 2019). Spatial presence was studied by Hilken et al. (2017) and shown to influence decision comfort. When a virtual simulation of an environment is real-life-like, it provides spatial presence. Customers' conviction that they are experiencing an authentic virtual simulation influences

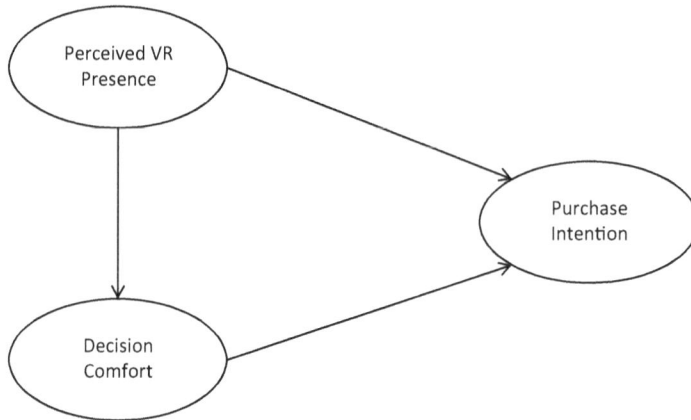

*Figure 35.1* Proposed research model.

their decision comfort (Wirth et al., 2007). Visual simulation has the most influence on behaviour and emotions in VR (Wu et al., 2016; Marasco et al., 2018). Decision comfort is a positive emotion (Parker et al., 2016) and VR influences positive emotions (Vishwakarma et al., 2020). With the presence of virtual technology as the input, this stimulus can influence decision comfort in the tourist experiencing the brand through VR.

Using the SOR framework, VR presence would positively influence decision comfort in the post-decision stage and that would in turn influence purchase intention (Figure 35.1).

## Methodology

Using judgemental sampling (Judd et al., 1991), data were collected from potential tourists who wanted to travel in the coming future, who came to three travel agencies in Famagusta, a city in Northern Cyprus, in order to buy a touristic package or getting a consultant for their next holiday trip. Using the back-translation method (Mcgorry, 2000), the English questionnaire was prepared in Turkish. A pilot study of ten tourists was conducted to check the face validity of the data. Consequently, a total of 136 questionnaires were collected through Google Forms from volunteer tourists whose anonymity and confidentiality were assured.

The perceived VR presence (PVP) was adapted and assessed based on subjective measures of spatial presence as conceptualised and operationalised in Wirth et al. (2007) and Vorderer et al. (2004). The decision comfort (DC) was assigned with a five-item scale from Parker et al. (2016). The purchase intention (PI) was assigned with a three-item scale from Chiang and Jang (2007). All the measurement items were assessed on the five-point scale of 1 (strongly disagree) to 5 (strongly agree).

We conducted case and variable screening prior to doing estimates. In terms of skewness and kurtosis, we found reasonably normal distributions for our latent factor indicators, which is consistent with the criteria of Sposito et al. (1983) who recommend ±3.3 as the upper threshold for normality.

A confirmatory factor analysis was performed using AMOS 24.0 to address issues of convergent validity (Straub, 1989). The reliability of the constructs was measured using Cronbach's alpha coefficient, a composite reliability statistic (Bagozzi & Yi, 1988). Several fit statistics for the assessment of the measurement model in confirmatory factor analysis (CFA) were used.

In order to analyse the mediation effect, the macro PROCESS model 4, V.3.4 for SPSS V.25 using a bootstrapped 5,000 sample size via the 95% confidence interval (Hayes & Rockwood, 2020) was utilised. Moreover, the mediating relationships were tested and reported with the Aroian version of the Sobel test, as suggested in Baron and Kenny (1986).

The results revealed that all measurements were reliable where the composite reliability scores (PVR = 0.795, DC = 0.854, PI = 0.907) and the coefficient of Cronbach's alpha (PVR = 0.806, DC = 0.849, PI = 0.913) were greater than 0.60 and 0.70 respectively (Bagozzi & Yi, 1988). All the factor loadings of each construct were greater than 0.50 with significant t-values. The varieties of model fit indices were analysed and affirm that the three-factor measurement model fits the validity of the proposed model (Hurley et al., 1997). The result of assessing the HTMT ratio shows that discriminant validity does not appear to be a concern since all these ratios are less than 0.85 (Henseler et al., 2015; Voorhees et al., 2016).

The findings highlighted that PVP was positively linked to DC (B = 0.485, p < 0.001), while DC depicted a positive association with PI (B = 0.746, p < 0.001). Therefore, hypotheses 2 and 3 were supported. Moreover, we found support for the positive linkage between PVP and PI (B = 0.198, p < 0.05) resulting in support for hypothesis 1. The indirect impact of PVP on PI through DC was 0.362 (lower level confidence interval, LLCI = 0.235, upperlevel confidence interval, ULCI = 0.508). The confidence intervals did not contain zero. These results revealed that DC mediated the linkage between PVP and PI, providing support to hypothesis 4, which indicates the partial mediation role of DC.

## Discussion and conclusions

With the rapid growth of technology and its influence on the tourism industry, where the tourists choose their flight and make reservations for hotels and restaurants even before they arrive at the destination, tourism destination service providers continuously need to find ways to present, market, and advertise their services to tourists online. VR has created a way for marketers to present their services to the tourist in ways that can make them feel like they are in the physical location (Marasco et al., 2018). This chapter has concentrated on how VR presence can influence DC and make tourists who view the VR purchase the tourism service being presented. Specifically, the purpose was to investigate the effect of presence in VR videos on the users' DC and the impact of DC on the tourist purchase intention.

The results showed that the perceived VR presence was positively related to tourists' purchase intentions. Specifically, the results showed that the Atlantis hotel of Dubai viewed through VR aided by HMD made the users feel like they were present in the hotel, sightseeing the scenery and aesthetics of the hotel, such that they were motivated to book based on the VR video they had seen. This result is consistent with previous research that determines the value of the immersive sense of being there that can contribute to customer benefits and the customer's booking intention (Israel et al., 2019); the study by Peng and Kim (2014) also indicates that presence can influence attitude and intentions.

The results further showed the interesting influence of DC on the tourists purchase intention towards the Atlantis hotel. The result supports the hypothesis and shows that when the user has high DC, that is they are comfortable about the decision they make, they would make a purchase. The ability of the VR system to make a user feel present and immersed in the virtual environment is sufficient to make them purchase, and it is also important if users have positive DC as it in turn affects their purchase intention. This result is consistent with previous studies on DC significantly related to positive word of mouth and behavioural intentions by Heller et al. (2019).

The outcomes also demonstrated that perceived VR presence has a positive effect on DC. This means that if users feel immersed in the VR experience and they feel present, as though they were in that physical location, they would be comfortable with the decision they make. This result is supported by previous research on the influence of virtual presence on users' decisions (Hilken et al., 2017). When users feel like they have had a first-hand experience with VR as if it was a real world experience, they feel at ease about their decision. Finally, the study analysed the mediating role of DC. The result supported the partial mediation effect of DC on perceived VR presence and purchase intentions, and the results show that DC has a major influence on intention, which is supported by SOR theory.

### Theoretical implications

This chapter contributes to the literature on service marketing and offers theoretical implications as well. First, the empirical study performed shows the importance of VR presence and experience that is made possible using HMD devices as effective tools that tourism destination service providers' marketers can use. The tourism industry is having a serious revolution with the influence of advanced technology devices and VR in tourism experiences (Marasco et al., 2018). As also stated by Leung et al. (2020), based on the projection that future VR headsets would look like regular pairs of glasses, consumers agreed that VR technology would eventually become as popular as smartphones and revolutionise the way people buy. Second, research in the tourism industry has not yet sufficiently covered the effect of DC on tourist purchase intentions. Although previous studies have considered the influence of presence in VR, DC is a new construct that has not been fully explored in the tourism domain, and the mediating role of DC provides an interesting result for further research. From a theoretical perspective, this study has introduced perceived VR presence and DC as two major factors that influence purchase intentions. The results further prove the effect of SOR theory. Furthermore, the effect of virtual presence and DC was studied concerning the virtual experience of a real hotel. The actual existing environments in VR studies can help to adequately conceptualise the effect and role of VR in shaping actions and attitudes.

### Managerial implications

This chapter has provided the managerial implications which are especially useful for tourism service providers' marketers, that is the hotels, restaurants, and extra services that different tourism destinations provide that tourists now choose in the pre-visit stage. Managers should include marketing strategies that start targeting the tourist from the pre-visit stage, as the level of presence the user has while experiencing the virtual environment of the service can determine their intention to purchase and book before even visiting; then they can take steps by improving the quality of their VR environment and including VR in their service communication, marketing, and promotion tools. The results have shown that the decision a user feels while experiencing VR can influence their purchase intention. The findings of this study have shown the mediating role of DC from virtual presence to purchase intention. Hence, a further implication for managers and organisations is to max-imise the likelihood of users feeling DC, as this is a positive emotion that encourages pos-itive intentions (Rafaeli et al., 2017). Additionally, the findings also provide implications for virtual technology designers as they can implement some guidelines in the design of VR technology that would support positive attitude; they can also implement advanced VR to create a sense of presence that would further cover the gap between online and phys-ical experience.

### Limitations and future research

This study has some limitations that should be considered in future research. First, the study uses subjective measurements of VR presence and DC, which are experienced during VR, based on participants' most memorable hotel experiences. It is recommended that further research could be undertaken by sensors or on-site, to obtain more reliable emo-tional responses while adopting VR systems and reduce potential inclination. Second, a sample of potential travellers in only one city was used as the target population in this research, which might limit the generalisability of the findings. It would be valuable to rep-licate this study using broader samples, particularly international travellers with diverse cultural backgrounds. For example, future research may focus on the cross-validation tech-nique to test several populations and apply longitudinal studies in various cultures. Finally, this study has solely focused on the outcomes of VR presence, while it would also be help-ful to expand the study's model based on the combined antecedents and consequents of VR presence in the same context. For example, user characteristics and media characteris-tics are the two most important variables that influence the sense of presence in VR. More-over, the prior experience might be added as a moderating variable.

The VR-enabled experience employed in this study was realised with the help of high-resolution 360-degree videos. In this context, it would be interesting to add more advanced features like animations to demonstrate virtual service staff members to investigate the impact of virtual staff on the perception of telepresence. A virtual service staff person, for example, may join the consumer on the tour and answer questions regarding specific pur-chase decisions.

### References

An, S., Choi, Y., & Lee, C. K. (2021). Virtual travel experience and destination marketing: Effects of sense and information quality on flow and visit intention. *Journal of Destination Marketing & Management*, 19, 100492.

Bagozzi, R. P., & Yi, Y. (1988). On the evaluation of structural equation models. *Journal of the Acad-emy of Marketing Science*, 16(1), 74–94.

Baker, M. A. (2016). Managing customer experiences in hotel chains. In Maya Ivanova, Stanislav Ivanov, & Vincent P. Magnini (Eds.), *The Routledge handbook of hotel chain management*. Routledge: London.

Baron, R. M., & Kenny, D. A. (1986). The moderator-mediator variable distinction in social psycho-logical research: Conceptual, strategic, and statistical considerations. *Journal of Personality and Social Psychology*, 51(6), 1173.

Bogicevic, V., Seo, S., Kandampully, J. A., Liu, S. Q., & Rudd, N. A. (2019). Virtual reality presence as a preamble of tourism experience: The role of mental imagery. *Tourism Management*, 74, 55–64.

Buhalis, D., & Karatay, N. (2022, January). Mixed Reality (MR) for Generation Z in Cultural Herit-age Tourism Towards Metaverse. In *ENTER22 e-Tourism Conference* (pp. 16–27). Cham: Springer.

Buhalis, D., & Law, R. (2008). Progress in information technology and tourism management: 20 years on and 10 years after the Internet—The state of eTourism research. *Tourism Management*, 29(4), 609–623.

Buhalis, D., & Leung, R. (2018). Smart hospitality—Interconnectivity and interoperability towards an ecosystem. *International Journal of Hospitality Management*, 71, 41–50.

Buhalis, D., Harwood, T., Bogicevic, V., Viglia, G., Beldona, S., & Hofacker, C. (2019). Technological disruptions in services: Lessons from tourism and hospitality. *Journal of Service Management*, 30(4), 484–506.

Chathoth, P. K., Ungson, G. R., Harrington, R. J., & Chan, E. S. (2016). Co-creation and higher order customer engagement in hospitality and tourism services: A critical review. *International Journal of Contemporary Hospitality Management*, 28, 222–245.

Chen, M. H. (2016). A quantile regression analysis of tourism market growth effect on the hotel industry. *International Journal of Hospitality Management*, 52, 117–120.

Chiang, C. F., & Jang, S. S. (2007). The effects of perceived price and brand image on value and purchase intention: Leisure travelers' attitudes toward online hotel booking. *Journal of Hospitality & Leisure Marketing*, 15(3), 49–69.

Chung, N., Han, H., & Joun, Y. (2015). Tourists' intention to visit a destination: The role of augmented reality (AR) application for a heritage site. *Computers in Human Behavior*, 50, 588–599.

Dacko, S. G. (2017). Enabling smart retail settings via mobile augmented reality shopping apps. *Technological Forecasting and Social Change*, 124, 243–256.

Diemer, J., Alpers, G. W., Peperkorn, H. M., Shiban, Y., & Mühlberger, A. (2015). The impact of perception and presence on emotional reactions: A review of research in virtual reality. *Frontiers in Psychology*, 6, 26.

Disztinger, P., Schlögl, S., & Groth, A. (2017). Technology Acceptance of Virtual Reality for Travel Planning. In *Information and Communication Technologies in Tourism 2017* (pp. 255–268). Cham: Springer.

Donovan, R., & Rossiter, J. (1982). Store atmosphere: An environmental "psychology dad" de Ulrich Beck. *Economia, Sociedad y Territorio*, 10(32), 275–281.

Eroglu, S. A., Machleit, K. A., & Davis, L. M. (2001). Atmospheric qualities of online retailing: A conceptual model and implications. *Journal of Business Research*, 54(2), 177–184.

Eroglu, S. A., Machleit, K. A., & Davis, L. M. (2003). Empirical testing of a model of online store atmospherics and shopper responses. *Psychology & Marketing*, 20(2), 139–150.

Fransen, M. L., Verlegh, P. W., Kirmani, A., & Smit, E. G. (2015). A typology of consumer strategies for resisting advertising, and a review of mechanisms for countering them. *International Journal of Advertising*, 34(1), 6–16.

Gartner, I. (2015). Gartner's 2015 hype cycle for emerging technologies identifies the computing innovations that organizations should monitor. *Gartner Web Site*. Available at: https://www.gartner.com/en/newsroom/press-releases/2015-08-18-gartners-2015-hype-cycle-for-emerging-technologies-identifies-the-computing-innovations-that-organizations-should-monitor (accessed 05 January 2021).

Guttentag, D. A. (2010). Virtual reality: Applications and implications for tourism. *Tourism Management*, 31(5), 637–651.

Han, H., Kim, W., & Hyun, S. S. (2011). Switching intention model development: Role of service performances, customer satisfaction, and switching barriers in the hotel industry. *International Journal of Hospitality Management*, 30(3), 619–629.

Hayes, A. F., & Rockwood, N. J. (2020). Conditional process analysis: Concepts, computation, and advances in the modeling of the contingencies of mechanisms. *American Behavioral Scientist*, 64(1), 19–54.

Heller, J., Chylinski, M., de Ruyter, K., Mahr, D., & Keeling, D. I. (2019). Let me imagine that for you: Transforming the retail frontline through augmenting customer mental imagery ability. *Journal of Retailing*, 95(2), 94–114.

Henseler, J., Ringle, C. M., & Sarstedt, M. (2015). A new criterion for assessing discriminant validity in variance-based structural equation modeling. *Journal of the Academy of Marketing Science*, 43(1), 115–135.

Hilken, T., de Ruyter, K., Chylinski, M., Mahr, D., & Keeling, D. I. (2017). Augmenting the eye of the beholder: Exploring the strategic potential of augmented reality to enhance online service experiences. *Journal of the Academy of Marketing Science*, 45(6), 884–905.

Huang, Y. C., Backman, K. F., Backman, S. J., & Chang, L. L. (2016). Exploring the implications of virtual reality technology in tourism marketing: An integrated research framework. *International Journal of Tourism Research*, 18(2), 116–128.

Hurley, A. E., Scandura, T. A., Schriesheim, C. A., Brannick, M. T., Seers, A., Vandenberg, R. J., & Williams, L. J. (1997). Exploratory and confirmatory factor analysis: Guidelines, issues, and alternatives. *Journal of Organizational Behavior*, 18(6), 667–683.

Islam, J. U., & Rahman, Z. (2017). The impact of online brand community characteristics on customer engagement: An application of Stimulus-Organism-Response paradigm. *Telematics and Informatics*, 34(4), 96–109.

Israel, K., Zerres, C., & Tscheulin, D. K. (2019). Presenting hotels in virtual reality: Does it influence the booking intention? *Journal of Hospitality and Tourism Technology*, 10(3), 443–463.

Jacoby, J. (2002). Stimulus-organism-response reconsidered: An evolutionary step in modeling (consumer) behavior. *Journal of Consumer Psychology*, 12(1), 51–57.

Judd, C. M., Smith, E. R., & Kidder, L. L. H. (1991) *Research methods in social relations*. New York: Holt, Rinehart, and Winston.

Jung, T. H., & tom Dieck, M. C. (2017). Augmented reality, virtual reality and 3D printing for the co-creation of value for the visitor experience at cultural heritage places. *Journal of Place Management and Development*, 10(2), 140–151.

Jung, T., tom Dieck, M. C., Lee, H., & Chung, N. (2016). Effects of virtual reality and augmented reality on visitor experiences in museum. In: A. Inversini & R. Schegg (eds.), *Information and communication technologies in tourism 2016*. Cham: Springer. https://doi.org/10.1007/978-3-319-28231-2_45

Jung, T., Chung, N., & Leue, M. C. (2015). The determinants of recommendations to use augmented reality technologies: The case of a Korean theme park. *Tourism Management*, 49, 75–86.

Keng, C. J., Chen, Y. H., & Huang, Y. H. (2018). The influence of mere virtual presence with product experience and social virtual product experience on brand attitude and purchase intention: Conformity and social ties as moderators. 交大管理學報, 38(2), 57–94.

Kim, M. J., Lee, C. K., & Jung, T. (2020). Exploring consumer behavior in virtual reality tourism using an extended stimulus-organism-response model. *Journal of Travel Research*, 59(1), 69–89.

Law, R., Buhalis, D., & Cobanoglu, C. (2014). Progress on information and communication technologies in hospitality and tourism. *International Journal of Contemporary Hospitality Management*, 25(5), 727–750.

Leung, X. Y., Lyu, J., & Bai, B. (2020). A fad or the future? Examining the effectiveness of virtual reality advertising in the hotel industry. *International Journal of Hospitality Management*, 88, 102391.

Litvin, S. W., Goldsmith, R. E., & Pan, B. (2008). Electronic word-of-mouth in hospitality and tourism management. *Tourism Management*, 29(3), 458–468.

Loureiro, S., & Ribeiro, L. (2011). The Effect of Atmosphere on Emotions and Online Shopping Intention: Age Differentiation. In *Australian and New Zealand Marketing Academy Conference*, Perth, Australia.

Lugosi, P., & Walls, A. R. (2013). Researching destination experiences: Themes, perspectives and challenges. *Journal of Destination Marketing and Management*, 2(2), 51–58.

Lyu, P., & Ma, Q. (2021). Research on the new development direction of visual communication based on VR virtual reality. *Journal of Physics: Conference Series*, 1992, 022020, https://doi.org/10.1088/1742-6596/1992/2/022020

Marasco, A., Buonincontri, P., van Niekerk, M., Orlowski, M., & Okumus, F. (2018). Exploring the role of next-generation virtual technologies in destination marketing. *Journal of Destination Marketing & Management*, 9, 138–148.

McGorry, S. Y. (2000). Measurement in a cross-cultural environment: Survey translation issues. *Qualitative Market Research: An International Journal*, 3, 74–81.

Mehrabian, A., & Russell, J. A. (1974). *An approach to environmental psychology*. Cambridge, MA: The MIT Press.

Mollen, A., & Wilson, H. (2010). Engagement, telepresence and interactivity in online consumer experience: Reconciling scholastic and managerial perspectives. *Journal of Business Research*, 63(9–10), 919–925.

Neuhofer, B., Buhalis, D., & Ladkin, A. (2014). A typology of technology-enhanced tourism experiences. *International Journal of Tourism Research*, 16(4), 340–350.

Neuhofer, B., Buhalis, D., & Ladkin, A. (2015). Smart technologies for personalized experiences: A case study in the hospitality domain. *Electronic Markets*, 25(3), 243–254.

Otto, J. E., & Ritchie, J. B. (1996). The service experience in tourism. *Tourism Management*, 17(3), 165–174.

Parker, J. R., Lehmann, D. R., & Xie, Y. (2016). Decision comfort. *Journal of Consumer Research*, 43(1), 113–133.

Peng, C., & Kim, Y. G. (2014). Application of the stimuli-organism-response (SOR) framework to online shopping behavior. *Journal of Internet Commerce*, 13(3–4), 159–176.

Petrova, P. K., & Cialdini, R. B. (2018). Evoking the imagination as a strategy of influence. In *Handbook of consumer psychology* (pp. 510–528). London: Routledge.

Rafaeli, A., Altman, D., Gremler, D. D., Huang, M. H., Grewal, D., Iyer, B., … & de Ruyter, K. (2017). The future of frontline research: Invited commentaries. *Journal of Service Research*, 20(1), 91–99.

Rahimizhian, S., Ozturen, A., & Ilkan, M. (2020). Emerging realm of 360-degree technology to promote tourism destination. *Technology in Society*, 63, 101411.

Rodríguez-Ardura, I., & Martínez-López, F. J. (2014). Another look at 'being there' experiences in digital media: Exploring connections of telepresence with mental imagery. *Computers in Human Behavior*, 30, 508–518.

Schubert, T. W., & Koole, S. L. (2009). The embodied self: Making a fist enhances men's power-related self-conceptions. *Journal of Experimental Social Psychology*, 45(4), 828–834.

Schuemie, M. J., Van Der Straaten, P., Krijn, M., & Van Der Mast, C. A. (2001). Research on presence in virtual reality: A survey. *CyberPsychology & Behavior*, 4(2), 183–201.

Sposito, V. A., Hand, M. L., & Skarpness, B. (1983). On the efficiency of using the sample kurtosis in selecting optimal lpestimators. *Communications in Statistics-simulation and Computation*, 12(3), 265–272.

Straub, D. W. (1989). Validating instruments in MIS research. *MIS Quarterly*, 13(2), 147–169.

Subawa, N. S., Widhiasthini, N. W., Astawa, I. P., Dwiatmadja, C., & Permatasari, N. P. I. (2021). The practices of virtual reality marketing in the tourism sector, a case study of Bali, Indonesia. *Current Issues in Tourism*, 24, 3284–3295.

Suh, K. S., & Lee, Y. E. (2005). The effects of virtual reality on consumer learning: An empirical investigation. *MIS Quarterly*, 29(4), 673–697.

Sweeney, J. C., Hausknecht, D., & Soutar, G. N. (2000). Cognitive dissonance after purchase: A multidimensional scale. *Psychology & Marketing*, 17(5), 369–385.

Talukdar, N., & Yu, S. (2018). A serial mediation effect of immersive virtual reality on purchase intention in real estate and the moderating role of psychological distance. *ACR Asia-Pacific Advances*, 12, 15–16.

Tussyadiah, I. P., Wang, D., Jung, T. H., & tom Dieck, M. C. (2018). Virtual reality, presence, and attitude change: Empirical evidence from tourism. *Tourism Management*, 66, 140–154.

Van Kerrebroeck, H., Brengman, M., & Willems, K. (2017). When brands come to life: Experimental research on the vividness effect of Virtual Reality in transformational marketing communications. *Virtual Reality*, 21(4), 177–191.

Vishwakarma, P., Mukherjee, S., & Datta, B. (2020). Travelers' intention to adopt virtual reality: A consumer value perspective. *Journal of Destination Marketing & Management*, 17, 100456.

Voorhees, C. M., Brady, M. K., Calantone, R., & Ramirez, E. (2016). Discriminant validity testing in marketing: An analysis, causes for concern, and proposed remedies. *Journal of the Academy of Marketing Science*, 44(1), 119–134.

Vorderer, P., Wirth, W., Gouveia, F. R., Biocca, F., Saari, T., Jäncke, L., ... & Klimmt, C. (2004). Mec spatial presence questionnaire. *Retrieved Sept*, 18, 2015.

Wang, Y. C., & Chung, Y. (2015). Hotel brand portfolio strategy. *International Journal of Contemporary Hospitality Management*, 27(4), 561–584.

Wei, W., Qi, R., & Zhang, L. (2019). Effects of virtual reality on theme park visitors' experience and behaviors: A presence perspective. *Tourism Management*, 71, 282–293.

Williams, P., & Hobson, J. P. (1995). Virtual reality and tourism: Fact or fantasy? *Tourism Management*, 16(6), 423–427.

Wirth, W., Hartmann, T., Böcking, S., Vorderer, P., Klimmt, C., Schramm, H., ... & Jäncke, P. (2007). A process model of the formation of spatial presence experiences. *Media Psychology*, 9(3), 493–525.

Wu, D., Weng, D., & Xue, S. (2016). Virtual reality system as an affective medium to induce specific emotion: A validation study. *Electronic Imaging*, 2016(4), 1–6.

Zhang, T., Wang, W. Y. C., Cao, L., & Wang, Y. (2019). The role of virtual try-on technology in online purchase decision from consumers' aspect. *Internet Research*, 29(3), 529–551.

# 36 Co-designing the smart tourism experience for all-inclusive hotels as a new trend in staycation experience

*Nabila Norizan and Norhazliza Halim*

## Introduction

Technology in tourism is increasingly associated with connectedness, interactivity, and smartness. Smart tourism has received attention from the research community and governments as technological progress and the convergence of information and communication technology (ICT) has produced new potentials (Buhalis & Amaranggana, 2014; Gretzel, Sigala, Xiang, & Koo, 2015a). Accordingly, ICT not only enables change in the way businesses and consumers interact in tourism, but also how and by whom tourism services, products, knowledge, and experiences are designed, created, and consumed (Errichiello & Marasco, 2017; Liburd, 2012; Neuhofer, 2016; Tribe & Liburd, 2016).

Big data contributes to a better understanding of the consumer market and propels collaboration, value co-creation, and open innovation in the tourism industry (Xiang, Stienmetz, & Fesenmaier, 2021). In response to the COVID-19 crisis, the importance of big data is to understand, control, predict, and modify the behaviour of tourists as a means to generate revenue and market control. Big data functions as a new toolbox for smart tourism development and enables the design and implementation of smart destinations (Kachniewska, 2021). However, data can only be transformative if one knows how to use it. Finding meaning in the relationships that emerge and to transform big data into insights and plans cannot be solved without a level of human engagement and focus on design processes. Design processes depend on the human aspects of continuously monitoring, rethinking, and fine-tuning current ways of doing things.

Extant theories of smart tourism focus on digital opportunities and data-driven developments to enhance destination competitiveness, tourist experiences, and support for new forms of collaboration and value creation (Boes et al., 2015; Gretzel et al., 2015a, 2015b). An "output view" embedded in data structures and technology is extensive in smart tourism research (Buhalis & Amaranggana, 2013; Wang et al., 2013; Zacarias et al., 2015; Zhu et al., 2014). However, an overt focus on formal outcomes such as services, products, and technology can divert attention from how things and operations are actually achieved (Sproedt & Heape, 2014). Therefore, an attention to the nuance of practice is needed to provide valuable insights (Heape et al., 2015). Accordingly, this chapter explores how the concept of smart tourism can be developed to reach beyond a data-driven, technical, and end-user outcome logic.

Previous literature provided various examples of frameworks, evaluation criteria, and technologies involved in smart tourism (Buhalis & Amaranggana, 2013; Huang, Goo, Nam, & Yoo, 2017; Mehraliyev, Chan, Choi, Koseoglu, & Law, 2020; Wang et al., 2013; Zacarias et al., 2015; Zhu et al., 2014). However, there is a lack of studies that address the

DOI: 10.4324/9781003291763-40

socio-technical, situated, and creative innovation processes, whereby different stakeholders wish to engage with smart tourism development (Liburd & Becken, 2017; Nielsen, 2019). Therefore, the chapter addresses the gap by using co-designing as a learning and experiment-driven process that is socially oriented. It extends co-designing approaches in the field of smart tourism to leverage the communicative interaction between stakeholders through both – rather than either/or – approaches.

The current assumptions and preliminary discussions of smart tourism unfold in a technical, data-driven, and system-oriented way, as described by service-dominant logic theory. A paradigm shift towards this theory occurred within the field of services marketing and management. This has resulted in major implications for the design and creation of experiences, which orients towards a more consumer-oriented experience co-design from economic and firm-centric principles (Neuhofer, 2014). Experiences and value are co-created through conjoint resource integration by consumers (tourists) and businesses (Edvardsson, Tronvoll, & Gruber, 2011; Ramaswamy, 2011; Vargo & Akaka, 2012). The experiences and value co-created result in generating tourist-generated big data information. However, big data implementation and its transformation into meaningful outcomes that generate business value remain unexplored (Mikalef, Pappas, Krogstie, & Pavlou, 2020).

**Co-designing smart tourism experiences**

Current notions of smart tourism, despite its evolution, focus on complex systems, environments, networks, and technology infrastructures supported by ICT (Buhalis & Amaranggana, 2013; Gretzel et al., 2015a; Guo, Liu, & Chai, 2014; Wang et al., 2016). Smart tourism emphasises the integration of information and the intensive use of technology to increase the quality of experience consumption, optimise service delivery, and improve business and destination management (Del Chiappa & Baggio, 2015; Errichiello & Micera, 2015; Lamsfus et al., 2015; Xiang et al., 2015a; Zhu et al., 2014). In order to understand the context, necessity, and future directions of smart tourism development, Wang et al. (2013) propose service-dominant (SD) logic (Lusch & Vargo, 2014; Vargo & Lusch, 2004, 2008). The SD logic in service ecosystems is described as "relatively self-adjusting, self-contained systems of resource-integrating actors that are connected by shared institutional logics and mutual value creation through service exchange" (Vargo & Lusch, 2004, p. 242).

The SD logic underpinning smart tourism depends on generic actors co-creating value through the integration of resources and exchange of services, coordinated through actor-engendered institutions in overlapping service ecosystems (Liburd & Becken, 2017). Afsarmanesh and Camarinha-Matos (2000) describe SD logic as the new paradigm of smart organisations: "for tourism is then a temporary consortium of different organisations representing service providers such as travel agencies, organisers of leisure programmes, accommodation providers, or public tourism organizations" (p. 456). These organisations join their resources and skills to either offer an aggregated or integrated service, or to respond to a better business opportunity, and provide cooperation supported by computer networks. Therefore, this seems to rest on an attitude of mind borne on by the SD principle of "we deliver what we think you need", where collaboration, innovation, and value creation are technology driven (Liburd & Becken, 2017).

Tourism co-design advocates the participatory nature of people's communicative interaction (Stacey, 2001). In particular, it is essential for those involved, including all stakeholders, to develop a sensitivity to the emergent and dynamic nature of the holistic interrelationship (Liburd & Becken, 2017). Smarter tourism occurs when practices change in the emergent processes of finding new opportunities, insights, thinking, meaning, and doing of smart tourism (Buhalis, 2020). Hence, it can be inferred that smarter tourism happens when practices change in the emergent processes of negotiating new meaning, new opportunities, new insights, new thinking, and new doing of smart tourism (Liburd & Becken, 2017). Hence, one can consider tourism co-design as a continuous process of learning and becoming, where many people change their practices, rather than solely considering tourism co-design regarding the development or resolution of a specific solution, product, or technology (Tamura, 2012). All in all, tourism co-design is described as a process that is unfolding instead of a foreclosure, whereby smarter concepts, services, or products emerge from the relational positioning of those involved.

Experience design is a practice of designing processes, services, products, and environments with a focus on the quality of the user's total experiences. Therefore, the multifaceted nature of smart tourism is grasped by a more holistic experience design. Experience design is not a matter of simply staging a theme park or developing a tour package. It implies designing tourist experiences to have at the park or on the tour (Breiby et al., 2020). Furthermore, experience design leverages participation at each stage of the design process and includes active engagement between users and stakeholders. As a result, opportunity arises for design practices to address wider societal values and needs and to function as a resource for smart tourism development (Sanders & Stappers, 2008).

## Big data value creation

The emergence of the big data paradigm provides a framework for smart tourism destination into three main layers: (1) a smart information layer (for data collection); (2) a smart exchange layer (for interconnectivity); and (3) a smart processing layer (for data analysis). Accordingly, smart tourism capitalises on utilising advanced technologies to transform data through social connections and physical infrastructures into value creation (Gretzel et al., 2015b). The value creation includes new business models for efficiency and on-site experiences. Different researchers highlight the importance of big data for tourism firms to gain insights on the experiences, preferences, behaviours, interests, and opinions of tourists (Marine-Roig & Clavé, 2015; Raguseo, Neirotti & Paolucci, 2017; Xiang et al., 2015). However, it is argued that such data are not widely used by authorities and policymakers to create value (Miah, Vu, Gammack, & McGrath, 2016).

Within the emergence of the big data paradigm, there is a framework for describing a smart tourism destination according to three main layers: a smart information layer (for data collecting), a smart exchange layer (for inter-connectivity), and a smart processing layer (for data analysis). Smart tourism is largely focused on the usage of advanced technologies to transform data collected and aggregated through social connections and physical infrastructures into on-site experiences and new business models for efficiency and value creation (Gretzel et al., 2015b). Smart tourism destinations reflect the requirements of a large community of stakeholders to make available for them tourism products, services, spaces, and experiences.

Based on this exhaustive literature review, this chapter integrates two theories: SD logic and resource-based view (RBV). To address the research gap of the overtly technical orientations of smart tourism, socio-technical orientation through co-designing approach is introduced. Smart tourism unfolds in service logic as indicated, and thus RBV theory is advocated to bridge the gap as demonstrated in the conceptual framework of the research.

Extant SD logic theory does not explain how its defining characteristics lead to the achievement of competitive advantage (Evans, 2016). The theory has yet to address the importance of recognising value creation dimensions resulting from co-designing tourism. Therefore, the gap is addressed by integrating RBV theory to the SD logic framework. RBV theory highlights the importance of recognising resources as dynamic capabilities for a firm to achieve sustainable competitive advantage. RBV also emphasises the firm's ability to integrate, recreate, and reconfigure its resources and capabilities to attain and maintain its competitive advantage (Wang & Ahmed, 2007). Based on the context of the research, the resources refer to the big data information generated from the co-designing of smart tourism experiences. The value co-created from the co-designing processes and activities leading to innovation will, thus, be addressed by RBV theory. Therefore, integrating RBV and SD logic reveals potential values and resources of value creation processes which function as key drivers of organisation performance.

### Smart tourism experience in all-inclusive hotels as a new trend in staycation experience

The hotel industry is one of the leading sectors in the tourism industry. In the context of smart tourism, technology is used extensively in hotels as a key success strategy in creating memorable guest experiences (Del Vecchio et al., 2018). Hotel guest experience and satisfaction have been widely studied in hospitality management research (Line et al., 2020; Xiang et al., 2015). Researchers can obtain very different results of factors constituting guest experience and the reasons leading to guest satisfaction, subject to research methods and design (Crotts, Mason, & Davis, 2009). Conventional methods usually rely on a set of predefined hypotheses and justified using a previous and existing body of knowledge. Attempts are then made in the direction of either accepting or rejecting such hypotheses. However, the case does not apply for big data analytics (Xiang et al., 2015). Through an analytical process, researchers are able to utilise data to reveal patterns reflective of guests' evaluation of their actual experiences with products and services (hotels in this case).

Application of big data and smart technology in the hospitality industry include an adaptation of the tourism service consumption model by Berzina, Grizane, and Jurgelane (2015). The study employs massive datasets of visitor consumption at the destination level to be applicable to harnessing big data generated by smart technology in the hospitality context. A further aspect of smart technology deployment in hotels is the usage of proximity sensors for hotel management applications to provide micro-locationing inside the hotel for context-sensitive information to guests (Hemchand, 2016). A specific example of such application is by Marriott hotels, which provide location-based notifications for guests depending on their proximity to related offers. Consequently, the usage of micro-locationing allows hotel staff to respond to automated service request allocations to the nearest available staff and to track the turnaround of guest services so as to evaluate productivity. To conclude, this trend provides further opportunity for the industry to optimise its business to a previously unprecedented level.

### Methodology

Using netnography as a tool to understand online guests' opinions, experiences, and behaviours (demand side), which in turn generate valuable big data information for all-inclusive hotel staycation businesses (supply side), this process was conducted to explore social interactions among stakeholders in co-designing processes. The process enrols different materials, processes, and people (Alvarez et al., 2019; Vaajakallio & Mattelmäki, 2014) into a creative innovation process with the purpose of activating big data in a meaningful way (Nielsen, 2019).

Many studies argue that big data information from the netnography process can only generate value when multiple stakeholders conduct the co-designing process to uncover and unfold the data (Alvarez et al., 2019; Vaajakallio & Mattelmäki, 2014). By collaboratively identifying latent potentials in socio-technical relations, the main objective of netnography is to work with elements of innovation towards developing smart tourism experiences in the hotel. A roadmap will present the analyses and entry points to the case examples, which will be constructed by vignettes.

Netnography presents a systematic approach to obtaining and working with the data collected for this research. The methodology process begins first by identifying and choosing the most relevant and significant online communities or user-generated content sites. In relation to Kozinets's study (2015), the online communities or user-generated content sites we have chosen are based on the criteria of: (1) a more focused and research question-relevant segment, group, or topic; (2) more descriptively or detailed rich data; (3) postings with higher traffic; (4) more interactions between members; and (5) larger numbers of discrete comments. Therefore, for this research, the online-generated content site served as the big data platform for the hotel supply side that was selected for the case study across the selected hotels' social media platform. As an example, M Live enables Marriott International's guests to engage with its users across multiple social media platforms such as Instagram, Facebook, and Twitter (Buhalis and Sinarta, 2019).

**Pilot results**

Koens et al. (2019) argue that, although the Smart City Hospitality framework outlines a framework for stakeholders to jointly discuss smart tourism development, it is not evident that this will happen. Instead, stakeholders "talk about each other instead of *with* each other" (Melissen et al., 2016, p. 149). The challenge includes implementation of collaborative principles in practice. Through conducting the case study, the overall pilot research findings include the understanding of co-designing as a learning and experiment driven process. Thus, the co-designing for the smart tourism experience will ensure that smart tourism is socially oriented.

The early findings shows that hotel companies use the research findings on emergent types of experiences and apply the right practices in their marketing and management strategies to improve their sales. The research finding also has determined the understanding of tourists' perceived risk towards COVID-19 and its impacts on tourists' decision-making processes, segmentation profiles, and future intentions. Consequently, the main contribution of the smart tourism experience is for the supply side – for hoteliers need to understand how their guests perceive socio-technical dimensions and possibly how this affects their future hotel stays. The findings help to optimise the hotel's decision-making and investments in socio-technical strategic orientations to evoke the smart tourism experience.

**Conclusions**

To conclude, in smart tourism, key concepts for enhancing tourism experiences are described as "value-adding experiences" and "value co-creation" mediated by technology, smart devices, and real-time data. This research has provided a central argument and contribution in theorising that the potential value of big data to evoke innovation is only as great as the successful interplay of sensemaking and co-designs underpinning big data practices. The overall aim was to test the SD-RBV model in describing collaborative design (co-design) processes on how big data can be utilised so as to be transformed into value creation to evoke smart tourism experiences. By employing the netnography methodology, expected key findings include that smart tourism is collaborative and that its successful generation requires the creating and sustaining of new relations through working with and managing changes on multiple levels.

## References

Afsarmanesh, H., & Camarinha-Matos, L. M. (2000). Future smart-organizations: A virtual tourism enterprise. In: *Web Information Systems Engineering, 2000* (Vol. 1, pp. 456–461). IEEE.

Alvarez, J., Irrmann, O., Djaouti, D., Taly, A., Rampnoux, O., & Sauvé, L. (2019). Design games and game design: Relations between design, codesign and serious games in adult education. In: *From UXD to LivXD: Living eXperience Design* (pp. 229–253).

An, W., & Alarcon, S. (2021). From netnography to segmentation for the description of the rural tourism market based on tourist experiences in Spain. *Journal of Destination Marketing & Management*, *19*, 100549.

Ardito, L., Cerchione, R., Del Vecchio, P., & Raguseo, E. (2019). Big data in smart tourism: Challenges, issues and opportunities. *Current Issues in Tourism*, *22*(15), 1805–1809.

Bas, B., & Stuart, C. (2007). Sustainable tourism infrastructure planning: A GIS-supported approach. *Tourism Geographies*, *9*, 1–21.

Benslama, T., & Jallouli, R. (2020). Clustering of Social Media Data and Marketing Decisions. In: *International Conference on Digital Economy* (pp. 53–65). Cham: Springer.

Berzina, I., Grizane, T., & Jurgelane, I. (2015). The tourism service consumption model for the sustainability of the special protection areas. *Procedia Computer Science*, *43*, 62–68.

Boes, K., Buhalis, D., & Inversini, A. (2015). Conceptualising Smart Tourism Destination Dimensions. In: *Information and Communication Technologies in Tourism 2015* (pp. 391–403). Cham: Springer.

Breiby, M. A., Duedahl, E., Øian, H., & Ericsson, B. (2020). Exploring sustainable experiences in tourism. *Scandinavian Journal of Hospitality and Tourism*, *20*(4), 335–351.

Brown, B., Chui, M., & Manyika, J. (2011). *Are You Ready for the Era of 'Big Data'?* McKinsey, & Company (Eds.). McKinsey Global Institute.

Browning, V., So, K. K. F., & Sparks, B. (2013). The influence of online reviews on consumers' attributions of service quality and control for service standards in hotels. *Journal of Travel & Tourism Marketing*, *30*(1–2), 23–40.

Buhalis, D., & Amaranggana, A. (2013). Smart Tourism Destinations. In: Z. Xiang & I. Tussyadiah (Eds.), *Information and Communication Technologies in Tourism 2014* (pp. 553–564). Cham: Springer International Publishing.

Buhalis, D., & Amaranggana, A. (2014). Smart Tourism Destinations. In: Z. Xiang, & I. Tussyadiah (Eds.), *Information and Communication Technologies in Tourism* (pp. 553–564). Dublin: Springer.

Buhalis, D., & Amaranggana, A. (2015). Smart Tourism Destinations Enhancing Tourism Experience through Personalisation of Services. In: *Information and Communication Technologies in Tourism 2015* (pp. 377–389). Cham: Springer.

Buhalis, D., Amaranggana, A., Tussyadiah, I., & Inversini, A. (2015). Smart tourism destinations enhancing tourism experience through personalisation of services. In: *ENTER 2015 Proceedings*. Lugano: Springer-Verlag, Wien.

Buhalis, D., & Sinarta, Y. (2019). Real-time co-creation and nowness service: lessons from tourism and hospitality. *Journal of Travel & Tourism Marketing*, *36*(5), 563–582. https://doi.org/10.1080/10548408.2019.1592059

Buhalis, D. (2020). Technology in tourism- From Information Communication Technologies to eTourism and Smart Tourism towards Ambient Intelligence Tourism: A perspective article, *Tourism Review*, 75(1), 267–272. https://doi.org/10.1108/TR-06-2019-0258

Buonincontri, P., & Micera, R. (2016). The experience co-creation in smart tourism destinations: A multiple case analysis of European destinations. *Information Technology & Tourism*, 16(3), 285–315.

CAREC. (2021). Impact of COVID-19 on CAREC aviation and tourism. *Asian Development Bank*. http://hdl.handle.net/11540/13146

Campos, A. C., Mendes, J., Valle, P. O. D., & Scott, N. (2018). Co-creation of tourist experiences: A literature review. *Current Issues in Tourism*, 21(4), 369–400.

Cheewinsiriwat, P. (2009). GIS application for the maps of tourist attractions and ethnic groups of Nan Province, Thailand. *Manusya: Journal of Humanities*, 12(2), 19–30.

Cho, Y. S., & Linderman, K. (2020). Resource-based product and process innovation model: Theory development and empirical validation. *Sustainability*, 12(3), 913.

Choi, H. C., & Sirakaya, E. (2006). Sustainability indicators for managing community tourism. *Tourism Management*, 27(6), 1274–1289.

Colombo, M. G., Piva, E., Quas, A., & Rossi-Lamastra, C. (2016). How high-tech entrepreneurial ventures cope with the global crisis: Changes in product innovation and internationalization strategies. *Industry and Innovation*, 23(7), 647–671.

Crotts, J. C., Mason, P. R., & Davis, B. (2009). Measuring guest satisfaction and competitive position in the hospitality and tourism industry: An application of stance-shift analysis to travel blog narratives. *Journal of Travel Research*, 48(2), 139–151.

Cuomo, M. T., Tortora, D., Foroudi, P., Giordano, A., Festa, G., & Metallo, G. (2020). Digital transformation and tourist experience co-design: Big social data for planning cultural tourism. *Technological Forecasting and Social Change*, 162, 120345.

Dallle, J. (2015, December 16). TraknProtext used to find missing things in your house; now it handles hotels. *Chicagoinno*. Retrieved from: https://www.bizjournals.com/chicago/inno/stories/news/2015/12/16/traknprotect-used-to-find-missing-things-in-your.html

De Mauro, A., Greco, M., & Grimaldi, M. (2016). A formal definition of big data based on its essential features. *Library Review*, 65(3), 122–135.

Del Chiappa, G., & Baggio, R. (2015). Knowledge transfer in smart tourism destinations: Analyzing the effects of a network structure. *Journal of Destination Marketing & Management*, 4(3), 145–150.

Del Vecchio, P., Mele, G., Ndou, V., & Secundo, G. (2018). Creating value from social big data: Implications for smart tourism destinations. *Information Processing & Management*, 54(5), 847–860.

Del Vecchio, P., Mele, G., Passiante, G., Vrontis, D., & Fanuli, C. (2020). Detecting customers knowledge from social media big data: Toward an integrated methodological framework based on netnography and business analytics. *Journal of Knowledge Management*, 24, 799–821.

Dumay, J. (2016). A critical reflection on the future of intellectual capital: From reporting to disclosure. *Journal of Intellectual Capital*, 17(1), 168–184.

Eboy, O. V., & Chan Kim Lian, J. (2021). Application of GIS in identifying potential site for river tourism activities along the Petagas River. *İlköğretim Online*, 20(4), 743–752.

Edvardsson, B., Tronvoll, B., & Gruber, T. (2011). Expanding understanding of service exchange and value co-creation: A social construction approach. *Journal of the Academy of Marketing Science*, 39(2), 327–339.

Elia, G., Polimeno, G., Solazzo, G., & Passiante, G. (2020). A multi-dimension framework for value creation through big data. *Industrial Marketing Management*, 90, 617–632.

Erickson, S., & Rothberg, H. (2014). Big Data and knowledge management: Establishing a conceptual foundation. *Electronic Journal of Knowledge Management*, 12(2), 108–116.

Errichiello L. & Micera R. (2015). Smart tourism destination governance. In: J. S. Spender, G. Shiuma, & V. Albino (Eds.), Culture, Innovation and Entrepreneurship: Connecting the knowledge dots (pp. 2179–2191), *Proceedings of IFKAD 2015 - International Forum on Knowledge Asset Dynamics*, 10–12 June, Bari.

Errichiello, L. & Marasco, A. (2017). Tourism Innovation-Oriented Public-Private Partnerships for Smart Destination Development. In: N. Scott, M. De Martino, & M. Van Niekerk (Eds.), *Knowledge Transfer to and within Tourism* (pp. 147–166). Bingley, UK: Emerald Publishing Limited.

EsadeEcPol. (2020). European (geo) politics in times of COVID-19. *Global Agenda*, *1*, 21, https://dobetter.esade.edu/en/covid-19-questions-geopolitics

Evans, N. G. (2016). Sustainable competitive advantage in tourism organizations: A strategic model applying service dominant logic and tourism's defining characteristics. *Tourism Management Perspectives*, *18*, 14–25.

Fadahunsi, J. T. (2011). Application of geographical information system (GIS) technology to tourism management in Ile-Ife, Osun state, Nigeria. *Pacific Journal of Science and Technology*, *12*, 274–283.

Gallego, I., & Font, X. (2020). Changes in air passenger demand as a result of the COVID-19 crisis: Using Big Data to inform tourism policy. *Journal of Sustainable Tourism*, *29*, 1470–1489.

Giuliano, G., & Golob, J. M. (1995). Los Angeles smart traveler information kiosks: A preliminary report. *Transportation Research Record*, *1516*, 11–19.

Gossling, S., Scott, D., & Hall, C. M. (2020). Pandemics, tourism and global change: A rapid assessment of COVID-19. *Journal of Sustainable Tourism*, *29*(1), 1–20.

Gretzel, U., Fuchs, M., Baggio, R., Hoepken, W., Law, R., Neidhardt, J., ... & Xiang, Z. (2020). e-Tourism beyond COVID-19: A call for transformative research. *Information Technology & Tourism*, *22*, 187–203.

Gretzel, U., Sigala, M., Xiang, Z., & Koo, C. (2015a). Smart tourism: Foundations and developments. *Electronic Markets*, *25*(3), 179–188.

Gretzel, U., Werthner, H., Koo, C., & Lamsfus, C. (2015b). Conceptual foundations for understanding smart tourism ecosystems. *Computers in Human Behavior*, *50*, 558–563.

Guerrero, J. V. R., Gomes, A. A. T., de Lollo, J. A., & Moschini, L. E. (2020). Mapping potential zones for ecotourism ecosystem services as a tool to promote landscape resilience and development in a Brazilian Municipality. *Sustainability*, *12*(24), 10345. https://doi.org/10.3390/su122410345

Guo, Y., Liu, H., & Chai, Y. (2014). The embedding convergence of smart cities nd tourism internet of things in China: An advance perspective. *Advances in Hospitality and Tourism Research*, *2*(1), 54–69.

Gursoy, D., & Chi, C. G. (2020). Effects of COVID-19 pandemic on hospitality industry: Review of the current situations and a research agenda. *Journal of Hospitality Marketing & Management*, *29*(5), 527–529.

Hall, C. M., Prayag, G., & Amore, A. (2017). *Tourism and Resilience: Individual, Organisational and Destination Perspectives*. Bristol, UK: Channel View Publications.

Heape, C. R. A. (2007). *The Design Spaces - the design process as the construction, exploration and expansion of a conceptual space* (PhD dissertation). University of Southern Denmark, Denmark.

Heape, C., Larsen, H., & Revsbæk, L. (2015). Participation as taking part in an improvised temporal unfolding. In: *5th Decennial Aarhus Conference*, Aarhus, Danmark.

Hemchand, S. (2016). Adoption of sensor based communication for mobile marketing in India. *Journal of Indian Business Research*, *8*(1), 65–76.

Hofstede, G., Hofstede, G. J., & Minkov, M. (2005). *Cultures and Organizations: Software of the Mind* (Vol. 2). New York: McGraw-Hill.

Höjer, M., & Wangel, J. (2015). Smart Sustainable Cities: Definition and Challenges. In: L. Hilty & B. Aebischer (Eds.), *ICT Innovations for Sustainability. Advances in Intelligent Systems and Computing* vol. 310. Cham: Springer.

Hollands, R. G. (2008). Will the real smart city please stand up? *City*, *12*(3), 303–320.

Huang, C. D., Goo, J., Nam, K., & Yoo, C. W. (2017). Smart tourism technologies in travel planning: The role of exploration and exploitation. *Information & Management*, *54*(6), 757–770.

Ingold, T. (2000). *The Perception of The Environment: Essays in Livelihood, Dwelling and Skill*. London and New York: Routledge.

Jennings, G. (2010). *Tourism Research* (2nd edn.). Australia: John Wiley & Sons Australia Ltd.

Kachniewska, M. (2021). Smart Tourism: Towards the Concept of a Data-Based Travel Experience. In: A. Lubowiecki-Vikuk, B. M. B. de Sousa, B. M. Đerčan, & W. Leal Filho (Eds.), *Handbook of Sustainable Development and Leisure Services*. World Sustainability Series. Cham: Springer.

Karim, W., Haque, A., Anis, Z., & Ulfy, M. A. (2020). The movement control order (mco) for covid-19 crisis and its impact on tourism and hospitality sector in Malaysia. *International Tourism and Hospitality Journal*, *3*(2), 1–7.

Kim, J. Y., & Canina, L. (2015). An analysis of smart tourism system satisfaction scores: The role of priced versus average quality. *Computers in Human Behavior*, *50*, 610–617.

Koens, K., Melissen, F., Mayer, I., & Aall, C. (2019). The smart city hospitality framework: Creating a foundation for collaborative reflections on overtourism that support destination design. *Journal of Destination Marketing & Management*, *19*, 100376.

Komninos, P. (2011). Intelligent cities: Variable geometries of spatial intelligence. *Journal of Intelligent Buildings International*, *3*, 1–17.

Koubaa, H., & Jallouli, R. (2019, April). Social networks and societal strategic orientation in the hotel sector: Netnographic study. In: *International Conference on Digital Economy* (pp. 87–109). Cham: Springer.

Kozinets, R. V. (2015). *Netnography: Redefined*. London: Sage.

Lamsfus, C., Martín, D., Alzua-Sorzabal, A., & Torres-Manzanera, E. (2015). Smart Tourism Destinations: An Extended Conception of Smart Cities Focusing on Human Mobility. In: I. Tussyadiah & A. Inversini (Eds.), *Information and Communication Technologies in Tourism 2015* (pp. 363–375). Cham: Springer International Publishing.

Larsen, H., & Sproedt, H. (2013). Researching and teaching innovation practice. In: *Proceedings from 14th International CINet Conference, "Business Development and Co-creation"* 8–11 September 2013, Nijmegen, Netherlands.

Lave, J. & Wenger, E. (1991) *Situated Learning*. Cambridge: Cambridge University Press.

Li, J., Xu, L., Tang, L., Wang, S., & Li, L. (2018). Big data in tourism research: A literature review. *Tourism Management*, *68*, 301–323.

Liburd, J. J. (2012) Tourism Research 2.0. *Annals of Tourism Research*, *39*(2), 883–907.

Liburd, J. J. (2013). *The Collaborative University. Lessons from Tourism Education and Research*. Professorial Dissertation. University of Southern Denmark. Odense: Print & Sign.

Liburd, J. J., & Becken, S. (2017). Values in nature conservation, tourism and UNESCO World Heritage Site stewardship. *Journal of Sustainable Tourism*, *25*(12), 1719–1735.

Liburd, J. J., Carlsen, J. & Edwards, D. (Eds.) (2013). *Networks for Sustainable Tourism Innovation: Case studies and Cross-Case Analysis*. Melbourne: Tilde University Press.

Line, N. D., Dogru, T., El-Manstrly, D., Buoye, A., Malthouse, E., & Kandampully, J. (2020). Control, use and ownership of big data: A reciprocal view of customer big data value in the hospitality and tourism industry. *Tourism Management*, *80*, 104106.

Lusch, R. F., & Vargo, S. L. (2014). *The Service-Dominant Logic of Marketing: Dialog, Debate, and Directions*. London: Routledge.

Malaysian Association of Hotel (MAH). (2020). *Tourism Comes to a Standstill*. Retrieved form: https://www.hotels.org.my/press/22578-tourism-comes-to-a-standstill

Marine-Roig, E., & Clavé, S. A. (2015). Tourism analytics with massive user-generated content: A case study of Barcelona. *Journal of Destination Marketing & Management*, *4*(3), 162–172.

Marriott International. (2020). Marriott bonvoy benefits. *Marriott*. Retrieved from: https://www.marriott.com/loyalty.mi

Mattelmäki, T., & Visser, F. S. (2011). Lost in Co-X: Interpretations of co-design and co- creation. In: *4th World Conference on Design Research (IASDR 2011)*, Delft, The Netherlands.

Mehraliyev, F., Chan, I. C. C., Choi, Y., Koseoglu, M. A., & Law, R. (2020). A state-of-the-art review of smart tourism research. *Journal of Travel & Tourism Marketing*, *37*(1), 78–91.

Mehraliyev, F., Choi, Y., & Köseoglu, M. (2019). Progress on smart tourism research. *Journal of Hospitality and Tourism Technology*, *10*(4), 522–538.

Melissen, F., Koens, K., Brinkman, M., & Smit, B. (2016). Sustainable development in the accommodation sector: A social dilemma perspective. *Tourism Management Perspectives*, *20*, 141–150.

Miah, S. J., Vu, H. Q., Gammack, J., & McGrath, M. (2016). A Big Data analytics method for tourist behaviour analysis. *Information & Management, 54*(6), 771–785.

Mikalef, Patrick, Pappas, Ilias O., Krogstie, John, & Pavlou, Paul A. (2020). Big data and business analytics: A research agenda for realizing business value. *Information & Management, 57*(1): 103237. ISSN 0378-7206. https://doi.org/10.1016/j.im.2019.103237

Mirzekhanova, Z. G., & Debelaia, I. D. (2015). Tourism zoning of area for sustainable development of the example of Khabarovsky Krai. *Mediterranean Journal of Social Sciences, 6*(3 S5), 283.

MMA. (2014). M live platform: M live geofencing surprise and delight platform. *MMA Global* Retrieved from: https://www.mmaglobal.com/case-study-hub/case_studies/view/46372

Morabito, V. (2015). *Big Data and Analytics: Strategic and Organizational Impacts.* Cham, Switzerland: Springer.

Nadkarni, S., Kriechbaumer, F., Rothenberger, M., & Christodoulidou, N. (2019). The path to the hotel of things: Internet of things and big data converging in hospitality. *Journal of Hospitality and Tourism Technology, 11*(1), 93–107.

Nam, K., Dutt, C. S., Chathoth, P., & Khan, M. S. (2019). Blockchain technology for smart city and smart tourism: Latest trends and challenges. *Asia Pacific Journal of Tourism Research, 26,* 454–468.

Nasir, M. A., Nasir, N. F., Nasir, M. N. F., & Faiz, M. The effective marketing strategies for the tourism and hospitality industry in Malaysia: Reviewing of marketing mix from the hotel perspective. (2020). *International Journal of Interdisciplinary Innovative Research and Development, 5*(1), 65–68.

Ndou, V., & Beqiri, M. (2014). Introduction to the special issue "Unlocking the value of Big Data". *EJASA: Decision Support Systems and Services Evaluation, 5*(1).

Neuhofer, B. (2016). Innovation through Co- creation: Towards an Understanding of Technology-Facilitated Co-creation Processes in Tourism. In: *Open Tourism - Open Innovation, Crowdsourcing and Co- creation Challenging the Tourism Industry.* New York Dordrecht London: Springer-Verlag.

Neuhofer, B. E. (2014). *An exploration of the technology enhanced tourist experience.* Doctoral dissertation, Bournemouth University.

Neuhofer, B., Buhalis, D., & Ladkin, A. (2012). Conceptualising technology enhanced destination experiences. *Journal of Destination Marketing & Management, 1*(1), 36–46.

Neuhofer, B., Buhalis, D., & Ladkin, A. (2015). Smart technologies for personalized experiences: A case study in the hospitality domain. *Electronic Markets, 25*(3), 243–254.

Nicolaides, C., Avraam, D., Cueto-Felgueroso, L., González, M. C., & Juanes, R. (2020). Hand-hygiene mitigation strategies against global disease spreading through the air transportation network. *Risk Analysis, 40*(4), 723–740.

Nielsen, T. K. (2019). *Co-designing smart tourism – evoking possible futures through speculation and experimentation.* PhD dissertation, Department of Communication and Design, University of Southern Denmark.

Nyangwe, S., & Buhalis, D. (2018). Branding Transformation through Social Media and Co-creation: Lessons from Marriott International. In: B. Stangl & J. Pesonen (Eds.), *Information and Communication Technologies in Tourism 2018.* Cham: Springer.

Raguseo, E., Neirotti, P., & Paolucci, E. (2017). How small hotels can drive value their way in info-mediation. The case of "Italian hotels vs. OTAs and TripAdvisor". *Information & Management, 54*(6), 745–756.

Rageh, A., Melewar, T. C., & Woodside, A. (2013). Using netnography research method to reveal the underlying dimensions of the customer/tourist experience. *Qualitative Market Research: An International Journal, 16*(2), 126–149.

Rahman, S. A. A., Yusof, M. A., Nakamura, H., & Nong, R. A. (2020). Challenges of smart tourism in Malaysia eco-tourism destinations. *Planning Malaysia, 18*(14), 442–451.

Ramaswamy, V. (2011). It's about human experiences… and beyond, to co-creation. *Industrial Marketing Management, 40*(2), 195–196.

Rodriguez-Anton, J. M., & Alonso-Almeida, M. D. M. (2020). COVID-19 impacts and recovery strategies: The case of the hospitality industry in Spain. *Sustainability, 12*(20), 8599.

Rowley, J. (2002). Using case studies in research. *Management Research News, 25*(1), 16–27.

Roy, S. K., Balaji, M. S., Soutar, G., & Jiang, Y. (2020). The antecedents and consequences of value co-creation behaviors in a hotel setting: A two-country study. *Cornell Hospitality Quarterly*, *61*(3), 353–368.

Saha, A., Dutta, A., & Sifat, R. I. (2021). The mental impact of digital divide due to COVID-19 pandemic induced emergency online learning at undergraduate level: Evidence from undergraduate students from Dhaka City. *Journal of Affective Disorders*, *294*, 170–179.

Sanders, E. B.-N., & Stappers, P. J. (2008). Co-creation and the new landscapes of design. *CoDesign*, *4*(1), 5–18.

Secundo, G., Del Vecchio, P., Dumay, J., & Passiante, G. (2017). Intellectual capital in the age of big data: Establishing a research agenda. *Journal of Intellectual Capital*, *18*(2), 242–261.

Shapiro, J. M. (2006). Smart cities: Quality of life, productivity, and the growth effects of human capital. *The Review of Economics and Statistics*, *88*(2), 324–335.

Shaw, A. (2020). Netnography and a summative content analysis approach to market research. *Journal of Emerging Trends in Marketing and Management*, *1*(1), 12–22.

Sigala, M. (2020). Tourism and COVID-19: Impacts and implications for advancing and resetting industry and research. *Journal of Business Research*, *117*, 312–321.

Smith Travel Research. (2020, November 6). *STR: Asia Pacific Hotel Performance for October 2020*. Hendersonville, TN: STR.

Sproedt, H., & Heape, C. (2014). Cultivating Imagination across Boundaries: Innovation practice as learning through participatory inquiry. In: *Paper presented In 15th International CINet Conference Continuous Innovation Network*. Continuous Innovation Network (CINet).

Stacey R. (2001). *Complex Responsive Processes in Organisations*. London and New York: Routledge.

Stacey, R. (2003). Learning as an activity of interdependent people. *The Learning Organization*, *10*(6), 325–331.

Suchman, L. A. (1987). *Plans and Situated Actions: The Problem of Human-Machine Communication*. Cambridge: Cambridge University Press.

Tamura, H. (2012). Keynote presentation. In: E. Tunstall & J. Buur (Eds.), *Proceedings Participatory Innovation Conference 2012: PIN-C 2012*. University of Southern Denmark.

Tarantola, S., Giglioli, N., Jesinghaus, J., & Saltelli, A. (2002). Can global sensitivity analysis steer the implementation of models for environmental assessments and decision-making? *Stochastic Environmental Research and Risk Assessment*, *16*, 63–76.

Tribe, J., & Liburd, J. J. (2016). The tourism knowledge system. *Annals of Tourism Research*, *57*, 44–61.

Tsaur, S. H., Lin, Y. C., & Lin, J. H. (2006). Evaluating ecotourism sustainability from the integrated perspective of resource, community and tourism. *Tourism Management*, *27*(4), 640–653.

United Nations. (2021). *Goal 11: Make Cities and Human Settlements Inclusive, Safe, Resilient and Sustainable*. Retrieved February 8, 2021 from: https://unstats.un.org/sdgs/report/2016/goal-11

Vaajakallio, K. (2012). *Design games as a tool, a mindset and a structure*. PhD Thesis defended at the Aalto University, Department of Design.

Vaajakallio, K., & Mattelmäki, T. (2014). Design games in codesign: As a tool, a mindset and a structure. *CoDesign*, *10*(1), 63–77.

Vargo, S. L. and Akaka, M. A., 2012. Value co-creation and service systems (re)formation: A service ecosystems view. *Service Science*, *4*(3), 207–217.

Vargo, S. L., & Lusch, R. F. (2004). Evolving to a new dominant logic for marketing. *Journal of Marketing*, *68*(1), 1–17.

Vargo, S. L., & Lusch, R. F. (2008). Service-dominant logic: Continuing the evolution. *Journal of the Academy of Marketing Science*, *36*(1), 1–10.

Verka, J., & Angelina, N. (2008). The application of GIS and its components in tourism. *Yugoslav Journal of Operations Research*, *2*(18), 261–272.

Walsh, L., & Kahn, P. (Eds.) (2010). *Collaborative Working in Higher Education. The Social Academy*. New York and London: Routledge.

Wamba, S. F., Akter, S., Edwards, A., Chopin, G., & Gnanzou, D. (2015). How 'big data' can make big impact: Findings from a systematic review and a longitudinal case study. *International Journal of Production Economics*, *165*, 234–246.

Wang, C., Hu, R., & Zhang, T. C. (2020). Corporate social responsibility in international hotel chains and its effects on local employees: Scale development and empirical testing in China. *International Journal of Hospitality Management*, *90*, 102598.

Wang, C. L., & Ahmed, P. K. (2007). Dynamic capabilities: A review and research agenda. *International Journal of Management Reviews*, *9*(1), 31–51.

Wang, D., & Xiang, Z. (2012). The new landscape of travel: A comprehensive analysis of smartphone apps. In: *Information and Communication Technologies in Tourism*, 308–319.

Wang, D., Li, X. (Robert), & Li, Y. (2013). China's 'smart tourism destination' initiative: A taste of the service-dominant logic. *Journal of Destination Marketing & Management*, *2*(2), 59–61.

Wang, X., Li, X. (Robert), Zhen, F., & Zhang, J. (2016). How smart is your tourist attraction? Measuring tourist preferences of smart tourism attractions via a FCEM-AHP and IPA approach. *Tourism Management*, *54*, 309–320.

World Bank. (2020). World Bank Group COVID-19 crisis response approach paper: Saving lives, scaling-up impact and getting back on track. Washington, DC: World Bank Group. http://documents.worldbank.org/curated/en/136631594937150795/World-Bank-Group-COVID-19-Crisis-Response-Approach-Paper-Saving-Lives-Scaling-up-Impact-and-Getting-Back-on-Track

Xiang, Z., Schwartz, Z., Gerdes, J. H., & Uysal, M. (2015a). What can big data and text analytics tell us about hotel guest experience and satisfaction? *International Journal of Hospitality Management*, *44*, 120–130.

Xiang, Z., Stienmetz, J., & Fesenmaier, D. R. (2021). Smart tourism design: Launching the annals of tourism research curated collection on designing tourism places. *Annals of Tourism Research*, *86*, 103154.

Xiang, Z., Tussyadiah, I., & Buhalis, D. (2015b). Smart destinations: Foundations, analytics, and applications. *Journal of Destination Marketing & Management*, *4*(3), 143–144.

Xiang, Z., Wang, D., O'Leary, J. T., & Fesenmaier, D. R. (2015c). Adapting to the Internet: Trends in travellers' use of the Web for trip planning. *Journal of Travel Research*, *54*(4), 511–527.

Ye, B. H., Ye, H., & Law, R. (2020). Systematic review of smart tourism research. *Sustainability*, *12*(8), 3401.

Zacarias, F., Cuapa, R., De Ita, G., & Torres, D. (2015). Smart tourism in 1-Click. *Procedia Computer Science*, 56, 447–452.

Zhu, W., Zhang, L., & Li, N. (2014). Challenges, function changing of government and enterprises in Chinese smart tourism. *Information and Communication Technologies in Tourism*, *10*, 553–564.

# 37 A digital safe-zone tourism network

## Are we ready to travel again?

*Norhazliza Halim, Nabila Norizan and*
*Thinaranjeney Thirumoorthi*

## Introduction

In the last 40 years, the entire world has seen a number of severe epidemics/pandemics, but none have had the same worldwide economic impact as the COVID-19 pandemic. COVID-19 is not as contagious as measles, and it is less likely to kill an infected person than Ebola, but it can spread for many days before symptoms appear (Bai et al., 2020; Rothe et al., 2020). The novel coronavirus (COVID-19) which originated in the city of Wuhan, China, has challenged the world (Gossling et al., 2020) and it is widening social and economic gaps, especially in developing countries (Jamal & Higham, 2021). This includes the tourism industry which plays important roles in economies and was largely disrupted in March 2020, with international travel bans affecting over 90% of the world's population and widespread restrictions on public gatherings and social mobility (Gossling et al., 2020).

Tourism globally has reported the increasing tightening of restrictions and bans on foreign and domestic travel, as well as on daily life, gatherings, shopping, and recreation (Angguni, & Lenggogeni, 2021). This is supported by Gossling et al. (2020) and Ocheni et al. (2020) who stated that unprecedented worldwide travel restrictions and stay-at-home orders are wreaking havoc on the global economy in ways that have not been seen since World War II. COVID-19 has slowed global tourism and provided a time to halt, reposition, and reassess (Lew et al., 2020). The tourism industry encompasses a wide range of businesses, including aviation, accommodation, meetings, incentives, conferencing & exhibitions (MICE), sporting events, restaurants, and cruises (Gossling et al., 2020). The pandemic had an impact on all of these industries in some manner and all tourism-dependent communities have now turned into "crisis communities" (Nepal, 2020).

One of the most important findings of pandemic research is that travel is crucial to epidemiology and disease surveillance (Hon, 2013; Khan et al., 2019). This also entails acknowledging that travel and tourism are contributors to illness propagation and economic implications, as well as being significantly impacted by non-pharmaceutical interventions (NPIs) (Nicolaides et al., 2019). Numerous industry associations have already issued estimations of COVID-19's impact on global tourism in 2020 and these forecasts should be viewed with extreme caution because it is still unclear how the pandemic will evolve and how travel restrictions and severe job losses will affect visitor demand during the critical Northern season. It is critical to recognise the socio-economic implications on vulnerable and marginalised populations as well as the ecological impacts on varied ecologies while developing the recovery programme (Rastegar et al., 2021) and ensuring tourism resilience. According to Menon (2020), measures put in place to tackle the pandemic are

DOI: 10.4324/9781003291763-41

having far-reaching economic consequences, harming tourism and travel, supply chains, and labour supply, and contributing to decreased economic growth.

The global economy and tourism industry has experienced similar significant global crisis events including economic recession and pandemics in the past (Gossling et al., 2020). Nonetheless, history has shown that the tourism industry is used to and has become resilient in bouncing back from these crises (Sigala, 2020). However, the COVID-19 crisis is profound and resulting in long-term changes towards the socio-economic and structural transformation of the tourism industry (Sigala, 2020). Countries that predominantly rely on tourism are faced with major economic consequences as it contributes as their main economic power. Therefore, an urgent need to understand the tourism supply chain is required to identify necessary interventions and to manage through the COVID-19 crisis recovery in order to build tourism resilience.

**Integrative framework of a digital safe-zone tourism network**

In the aftermath of the COVID-19 epidemic, a fresh approach to the situation is required for the tourism industry to establish a new normal (Sigala, 2020). Although COVID-19 cases are continuing to rise, some countries have been successful in containing the pandemic through the adoption of several public health and social measures (Bank, 2020). Such measures, however, have negatively affected economic growth. In an attempt to revamp the economy and reactivate the tourism industry, countries are re-establishing connections and partnering to develop virus-free travel bubbles, which offer a safe environment for the development of tourist activities, protecting tourists, and the local population.

There are three phases in reopening businesses in travel bubbles, which are: (i) greater harmonisation of standards; (ii) greater leadership in building the infrastructure and institutions in order to lay the foundations for a South East Asian "safe travel zone"; and (iii) a formal fortnightly review process.

In the short term, the COVID-19 outbreak is changing travellers' preferences to closer and safer areas. Therefore, these three phases are needed in building tourist and business confidence for domestic and international tourism. Initiatives during the first phase will enforce and establish common rules. Provisions must also be made for unforeseen changes in circumstances and unexpected scenarios given the volatile nature of the pandemic. A way must be found: to resolve disputes, such as over border controls, should the existing guidelines prove to be inadequate; to focus on skills development, including through the use of technology; to adopt and implement common quality standards for tourism services, including health and safety protocols (CAREC, 2021).

During this phase, the main aim will be increasing market share in lucrative international markets by improving connectivity, infrastructure, and regulatory procedures in the priority tourism clusters. Initiatives during this phase could include, among others, improvement of the aviation sector and air connectivity at low fares; development of hubs with stopover features in the priority clusters; improvement of last-mile access and border crossing points.

The main goal of phase three should be to strengthen the development plans of phases one and two within the priority cluster groups, and to explore expanding toward the creation of secondary destinations (outside the priority clusters) if there is sufficient demand and ownership from member countries. During this period, initiatives could involve, among other things, tourism promotion targeting for specific areas within the priority clusters;

further improvement of infrastructure and accessibility; and development of new tourism products to further diversify the offer of the different clusters.

According to new research (Polyzos et al., 2021; Cave & Dredge, 2020) it can be expected that a new era is coming for the tourism industry. One of the ways is by implementing tourism zoning which comprises traveller decision making, tourism destination/tourism-related business, and tourism in response to COVID-19.

**Methodology**

Zoning is a scientific systematic methodology that uses an imaginary segmentation of a specific territory into parts based on concrete elements, defined criteria, and the understanding that if the selected criteria changed, the zoning pattern would definitely change. Modern territory zoning principles allow for the inclusion of a variety of criteria, which can result in a variety of overlapping zoning systems (Mirzekhanova & Debelaia, 2015). This also applies to tourism zoning which is an excellent approach of systematising information about various ranked territories in order to justify the direction of sustainable growth. This is supported by Kim and Kim (2021) who stated that tourism zoning served as a comprehensive and systematic strategy to increase the number of visitors. According to Roman et al. (2007), zoning and the application of Limits of Acceptable Change (LAC) are two viable tourism management solutions that involve the gathering and combining of ecological and socioeconomic data. These data can be used in developing a geo-environmental model based on geographic information systems (GISs) to spatially identify places with increased ability for tourism promotion (Roque Guerrero et al., 2020).

Luo (2018) found, in setting tourism zoning, that it may include a few variables such as the economy, efficiency, effectiveness, and environmental quality. Figure 37.1 shows the proposed framework in identifying digital tourism zoning mapping which in this chapter we narrow down to developing the tourism zoning parameter.

These variables need to be ranked using weighing indicators, also known as objective weight. Its importance derives from the fact that weights can have a significant effect on the rankings of analysed regions and subsequent policymaking. This is because relative indicator weights may significantly differ depending on the chosen weighting procedure in which larger weights are consequently assigned to indications that are more essential than others, which is intended when they are aggregated. Basically, there are two types of approaches to obtain weights that reflect an indicator's importance: opinion-based (subjective) and data-centric (objective) approaches. To eliminate the differences between evaluation unit and dimension, and to improve data comparability, indicator data were normalised with the range standardised method. In opinion-based approaches the information about importance is extracted from subjective judgements. Well-known approaches comprise the Delphi method or expert panel surveys where importance is attached to indicators by means of rating-, ranking-, or constant-sum scales. A variant of the constant-sum scale, the budget allocation technique, has been used in the European Internal Market Index (Tarantola et al., 2002). In tourism-related studies, the Delphi method has been used for the selection of indicators (e.g., Choi and Sirakaya, 2006) and their weighting (e.g., Tsaur et al., 2006).

Opinion-based approaches may, however, also involve techniques like conjoint analysis or the analytical hierarchy process (AHP). In Saha et al. (2021), the AHP was used on five indicators that were created to help with priority setting and decision-making. Several threats have been considered and weighted in order to determine the multi-hazard risk zone

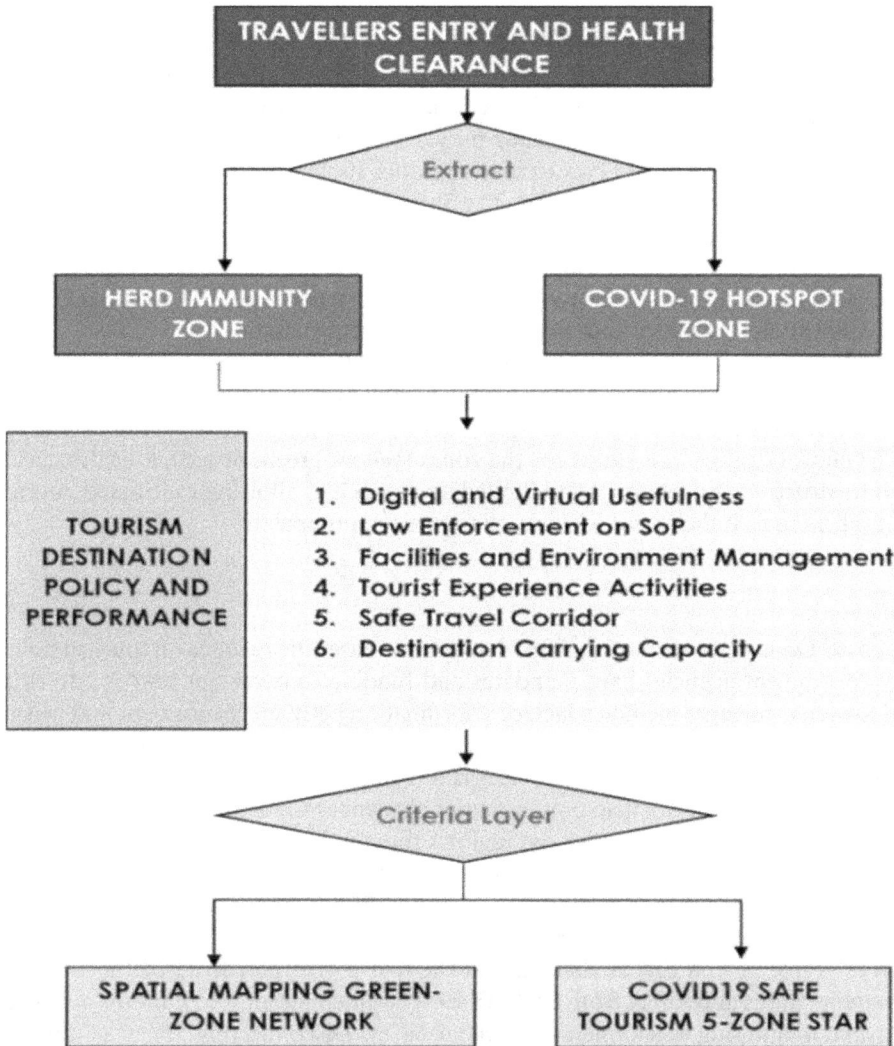

*Figure 37.1* Evaluation framework of digital tourism zoning.

based on researchers' experiences and local people's viewpoints. AHP approaches further provide the ability to determine the nature of the consistency of preferences provided by the report by employing a consistency ratio. The information entropy weight (IEW) approach is also one of two weight computational methods that are based on statistical characteristics and measurement data (Luo, 2018). Luo (2018) recommends dividing the performance into a percentage index, where the greater the importance of the parameter, the higher the index percentage that may be proposed.

The tourism zoning concept identifies tourism hotspot zoning labelling with five different colours: blue, green, yellow, red, and black. The parameters of the five zoning colours refer to: 5 (blue) score of more than 80%, 4 (green) score of 70–79%, 3 (yellow) score of 50–69%, 2 (red) score of 20–49%, and 1 (black) score of less than 20%. Following the sequence of the coloured zoning parameter, tourists are well informed about the COVID-19 record of the country they intend to visit (e.g., blue: zero COVID in 15 days versus black: 100/10,000 of the population with COVID in 15 days). The law enforcement of the blue zone is 100%, with excellent Health, Safety, Security, and Cleanliness (HSSC) facilities, whereas there is no enforcement in the black zone and insufficient HSSC facilities. As an example, through digital zoning, Langkawi provides information about its yellow health check: 13 COVID-19 cases with 42% vaccination status. Its access covers business travellers with a less than three day working permit. The Safe-Zone Travelling Network is crucial as it helps to build confidence for visitors and the market. The new norm that aims to implement tourism zoning for countries includes ensuring the safety of the traveller's journey, creating a diversity of tourism offer, providing trusted information on the destinations, and fostering innovation to scale up responsiveness to market demand.

The methodology used measures Safe-Travel Zoning Parameters and Boundaries. The network will also provide an interactive map for Digital Tourism Zoning using GIS applications in tourism. The chapter also tests pilot destinations with the reopening of tourism spots. The network provides a status for the zones (yellow, green, or red), a health check, quality infrastructure and facilities, and limited access such as allowing vaccinated international travellers to visit the island without quarantine requirements.

### Findings

The proposed Digital Safe-Zone Tourism Network increases the reliance on tourism policy as many government agencies have mandates and funding to carry out tourism development. However, challenges include a lack of coordination between government and private sectors as well as different levels of stakeholders. The lack of core tourism skills, overlapping jurisdiction and harmonisation of tourism-related policies, and public acceptance towards the proposed network also present major challenges towards the Network's implementation. This would help to revive the region's tourism industry with the focus on key recovery strategies to build a sustainable and resilient future.

Based on the parameters and zoning classification, this research established an ASEAN Digital Tourism Zoning Guideline (see Table 37.1).

Based on Table 37.1, it can be defined that the higher the total score of the zone, the better the zone status and safety from COVID-19. The guideline aims to limit the spread of a virus while minimising economic and societal harm. Disconnecting zones as much as possible disrupts transmission chains, reducing the possibility of virus reintroductions. In addition, based on the colour code, this research also established the safe tourism zone star for each zone. The star colour shows how safe the tourism zone is from COVID-19. The star was classified as shown in Figure 37.2.

The expected results from these two parameters cover the benefits and risks associated with the Digital Safe-Zone Tourism Network. The tourism zoning presents financial risks and benefits. Due to the extensive involvement of technology such as the utilisation of digital and virtual platforms to implement the zoning, the cost could be high. Some governments may lack support, funding, grants, and incentives to implement the initiative. However, the returns of investment are two-fold. Second, the tourism zoning initiative provides confidence for tourists and tourism businesses in that there are strict health travel certificates and standard operating procedures in place. However, risks could include the lack of real-time determination of tourism safe areas. Third, environmental benefits focus on destination carrying capacity. Due to the social distancing restrictions in place, the tourism zoning initiative would be beneficial in limiting the number of tourists to a destination for

*Table 37.1* ASEAN digital tourism zoning guideline

| ASEAN zoning parameter | 5 (Blue) | 4 (Green) | 3 (Yellow) | 2 (Red) | 1 (Black) |
|---|---|---|---|---|---|
| **COVID19 Record (weightage: 30/100)** | 0 COVID in 15 days | < 10/10,000 population COVID in 15 days | 11–30/10,000 population COVID in 15 days | 31–100/10,000 population COVID in 15 days | >100/10,000 population COVID in 15 days |
| **Vaccination Status (weightage: 30/100)** | >90% (herd community) | 70–90% (herd community) | 50–69% | 20–49% | <20% |
| **Digital and Virtual Platform/ Mechanism (weightage: 5/100)** | VR/AR, Realtime game, big data | Dynamic website and mobile apps | Static website | No website, but sharing info | No info |
| **Law Enforcement (weightage: 5/100)** | 100% enforcement | 80% enforcement | 50–80% enforcement | <50% enforcement | No enforcement |
| **HSSC Facility (weightage: 5/100)** | Excellent | Very good | Complete but not quality | Not complete and bad condition | Not enough |
| **HSSC Environment (weightage: 5/100)** | Excellent (low carbon concept) | Good (eco-friendly concept) | Medium (minimalist) | Environmental awareness only | Do not care |
| **Tourism Buffer Zone (Weightage: 5/100)** | 10 km from main point | 15 km from main point | 20 km from main point | >20 km from main point | No buffer |
| **Safe Travel Corridor (weightage: 10/100)** | <3 days with no transit | 3–5 days with less 5 short haul transit areas | 5–7 days with less than 10 medium haul transit areas | >7 days with more than 10 medium/long haul transit areas | >14 days with more than 10 long haul staging areas |
| **Destination Carrying Capacity (weightage: 5/100)** | <30 tourist/1ha/ hour | 3–50 tourist/1ha/ hour | 50–100 tourist/1ha/ hour | 100–200 tourist/1ha/ hour | No carrying capacity (mass) |
| **TOTAL SCORE Score/Total X 100** | >80% | 70–79% | 50–69% | 20–49% | <20% |

*Source*: Findings from ASEAN ITTP-COVID19, 2021.

*Note*: Basic assumption of zoning parameter, which depends on stakeholders acceptance and detail evaluation.

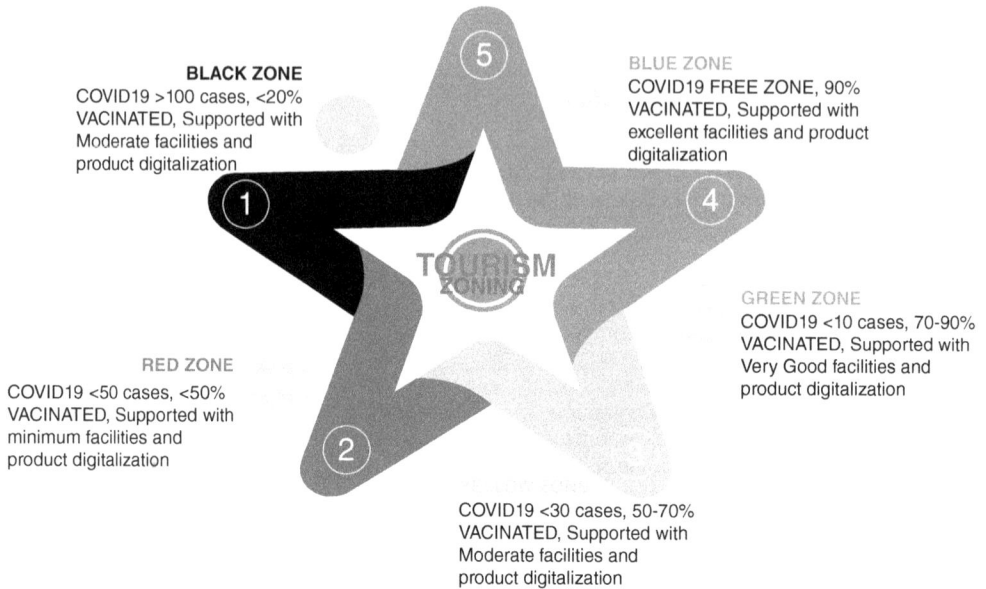

*Figure 37.2* ASEAN tourism zoning star.

health, safety, and environmental reasons. All in all, a responsibility centre or a focal point should be established to manage the coordination of risks and benefits from all related countries in working towards shared benefits to achieve successful collaboration.

Using opinion-based statistical evaluation, a public acceptance opinion has been gathered using an online survey. From the findings, half of the total respondents are under 31 years old and 20.2% are under 41 years old. Those under 51 years old represent 18.4% of the total respondents. Students represent 48.6% of total respondents, the highest composition, and academics at 34.2% are the second highest. The rest are government agencies (12.9%) and tourism industry stakeholders (4.3%). These percentages represent young and working-age people, including singles, couples, and small families. These groups are essential for mass tourism and domestic demands (Figure 37.3).

The first dimension in the tourism zoning framework is the pre-visitation criteria that are important for the public. Pre-visitation concentrates on the destination selected during the booking phase and criteria that influence public decisions. Figure 37.4 shows that the COVID-19 pandemic significantly increases public concern on health and safety procedures that reduce mortality rates as the 70.2% of respondents selected. Two criteria with a

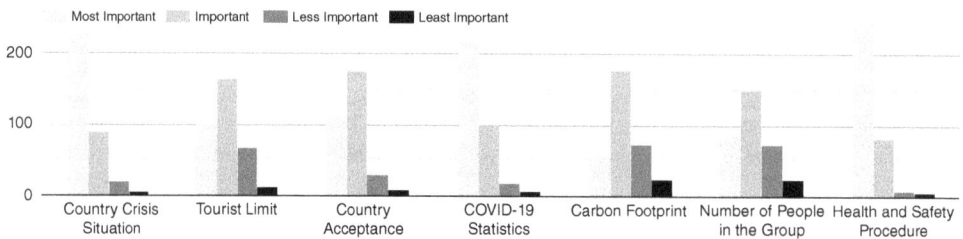

*Figure 37.3* Ranks of importance in tourism destination selection.

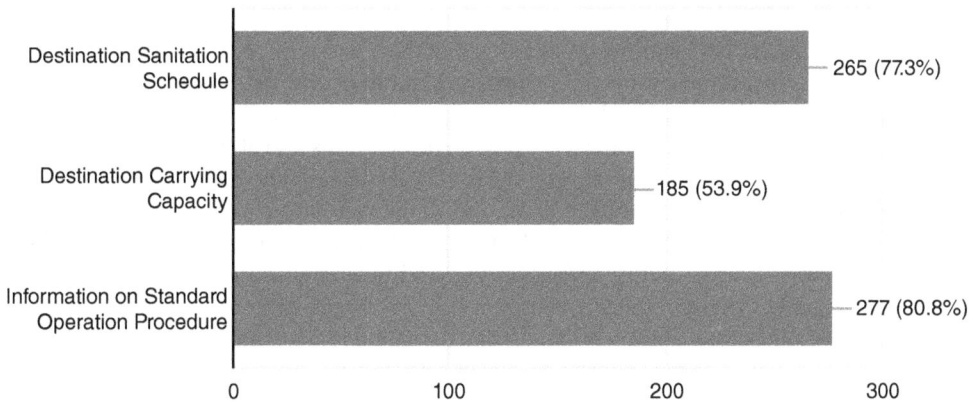

*Figure 37.4* Important factors for tourist cleanliness, healthiness, safety, and environment.

slight difference are the COVID-19 statistic (62.5%) and country case situation (65.3%). Country acceptance is also considered significant to the public, with 32% of respondents selecting it as the highest priority. Hence, the situation in the country resembles the image of the preparedness of a country to receive tourists. However, the tourist limit (28.4%), carbon footprint (15.8%), and the number of people (24.4%) in the group are the least priority for the public during tourism destination selection. Interestingly, these reasons also lead to alternative tourism destinations if the first destination fails to meet the general requirement for travel.

Respondents agreed to receive information on standard operational procedures (80.8%) as the highest importance in tourism destination selection, as shown in Figure 37.4. Slightly behind is the destination sanitation schedule, which 77.3% of respondents chose. The least important factor is the destination carrying capacity (53.9%). However, the overall public is aware that cleanliness, healthiness, safety, and environment information at the tourism destination is necessary to the pre-visitation phase. Interestingly, the public is willing to pay extra costs for the cleanliness, healthiness, safety, and environmental equipment offered at the destination due to long term confinement. Tourism destination or tourism-related business is the third dimension in the survey which focuses on the pandemic's impact on suppliers and their preparedness to receive future tourists. The public perceived that safety procedure (37%) is the most important for the short recovery phase to ensure business reputation. The highest priority for the second phase is product enhancement, chosen by 49.9% of the respondents. Also included are private financial aid (47.9%), upgrading infrastructure (47%), and government assistance (46.1%). These data show that the medium recovery phase should focus on rebuilding, repairing, and upgrading infrastructure before tourist arrival by private and government funding. Interestingly, safety procedures also need to be implemented to avoid the pandemic hitting again in the long recovery phase.

The Co-Designing of Safe-Zone Tourism Network using the Geographic Information System (GIS) provides the potential for tourism development through the use of attractions/destination maps, digital maps, and digital files for the Internet and mobile phones (Verka & Angelina, 2008). Tourism planning refers to the integrated planning of attractions, services, and transportation facilities. As a result, GIS can geometrically, conceptually, and topologically define and identify tourism infrastructure aspects (Bas & Stuart, 2007).

GIS also allows resource managers to do searches and analysis on large volumes of spatial and non-spatial data. Furthermore, it has better accuracy and is faster than manual analysis. GIS can also aid in the promotion of tourist destinations and the administration of the tourism industry (Fadahunsi, 2011). According to Cheewinsiriwat's (2009) research, GIS applications are utilised to create maps of tourism destinations and ethnic groupings. A GIS database was created by merging map layers from various sources, such as administrative boundaries, highways, rivers, contour lines, and village sites. It could then be utilised to assist tourists (Valentine & Chan, 2021). The network provides a status for the zones: yellow, green or red, health check, quality infrastructure and facilities, and limited access, such as allowing vaccinated international travellers to visit the island without quarantine requirements. For better travelling, ASEAN green zone travelling is also being adapted from green zone travelling: A Pan-European strategy to save tourism (EsadeEcPol, 2020). In this green zone travelling, the tourist can travel to another zone if the zone turns or changes into a green zone three days prior to travelling. The pilot destinations will also be classified based on the zoning stars and parameters. The Safe-Zone Tourism Network Analysis also tested a few pilot destinations with the reopening of tourism spots. The survey was conducted and distributed among the participants in the one-day webinar. Google Form was used, and feedback was received from the 349 participants. All the items consist of close-ended questions and a ranking scale. The survey aimed to measure the priorities of the Tourism Zoning criteria based on public perception. The proposed Cross-Border Digital Tourism Zoning Guideline increases the reliance on tourism policy as many government agencies have mandates and funding to carry out tourism development. Based on the guideline, tourists can identify green zones and this helps to ensure safe mobility and allows them to travel from one green zone to another green zone (see Figure 37.5).

Based on the star colour, the tourist can identify immediately the COVID-19 status at each place. Other than that, the tourist can make a choice to travel to a certain zone which is nearer to a higher probability zone that will turn into a green zone. This helps a lot to

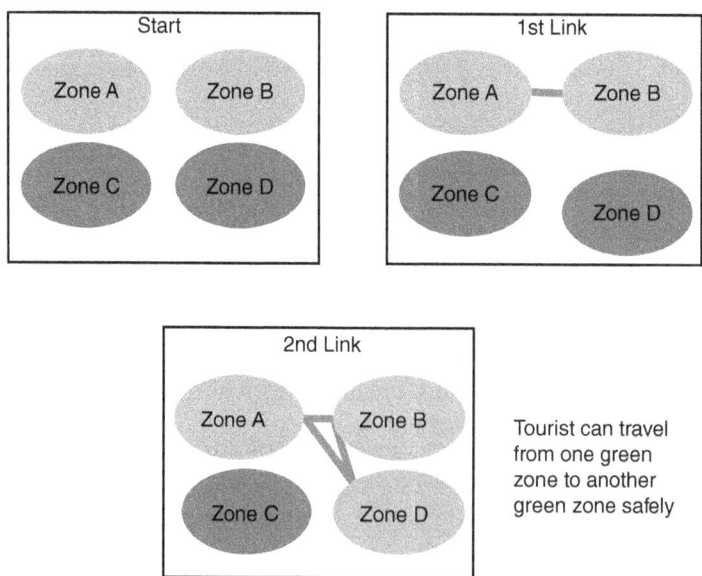

*Figure 37.5* Illustration of Green Zone Network.

avoid the red zone. With the aid of Safe-Zone Tourism Network big data sharing from neighbouring countries, and the Green Zone Travelling Network, this helps the public to travel in a safer manner and be comfortable during travelling. As a result, "green zoning" refers to a policy that depends entirely on mobility restrictions and public health measures based on the epidemiological status of well-defined zones. Green zoning tries to reduce the spread of an infectious disease while also allowing zones where the virus has been contained to impose limitations and resume regular economic and social activity.

## Conclusions

Strengthening the resilience of the tourism industry during the COVID-19 era requires enhancing its destination capacity. The ultimate aim is to reduce destination vulnerability and increase preparedness in managing tourism shocks. Therefore, the establishment of the Digital Safe-Zone Tourism Network is crucial. The Network would also grow as the spread of the virus is progressively contained. As a result, this ensures safe mobility for the industry through allowing travelling into green zones (EsadeEcPol, 2020). Therefore, this finding provides a new insight into tourism zoning guidelines for providing assurance to travellers in mobility mapping in COVID-19 friendly travelling. This supports the Green-Zone Travelling Network in gaining confidence for visitors and the markets, including ensuring the safety of the traveller, creating diversity of tourism offers, providing trusted information on the destinations, and fostering innovation to scale up responsiveness for market demand. A risk-informed tourism planning is the agent of change and one of the transformation tools for revitalising our susceptible tourism areas into sustainable and resilient regions.

## References

Aliperti, G., & Cruz, A. M. (2019). Investigating tourists' risk information processing. *Annals of Tourism Research*, *79*, 102803.

Angguni, F., & Lenggogeni, S. (2021). The impact of travel risk perception in COVID 19 and travel anxiety toward travel intention on domestic tourist in Indonesia. *Jurnal Ilmiah MEA (Manajemen, Ekonomi, & Akuntansi)*, *5*(2), 241–259.

Araña, J. E., & León, C. J. (2008). The impact of terrorism on tourism demand. *Annals of Tourism Research*, *35*(2), 299–315.

Bai, Y., Yao, L., Wei, T., Tian, F., Jin, D. Y., Chen, L., & Wang, M. (2020). Presumed asymptomatic carrier transmission of COVID-19. *JAMA*, *323*(14), 1406–1407.

Barney, J. (1991). Firm resources and sustained competitive advantage. *Journal of Management*, *17*(1), 99–120.

Barney, J. B. (2001). Resource-based theories of competitive advantage: A ten-year retrospective on the resource-based view. *Journal of Management*, *27*(6), 643–650.

Cave, J., & Dredge, D. (2020). Regenerative tourism needs diverse economic practices. *Tourism Geographies*, *22*(3), 503–513.

Dallke, J. (2015). TraknProtect used to find missing things in your house; now it handles hotels. Chicago Inno, available at: https://goo.gl/ohKx9y

Dolnicar, S. (2005). Understanding barriers to leisure travel: Tourist fears as a marketing basis. *Journal of Vacation Marketing*, *11*(3), 197–208.

Gossling, S., Scott, D., & Hall, C. M. (2020). Pandemics, tourism and global change: A rapid assessment of COVID-19. *Journal of Sustainable Tourism*, *29*(1), 1–20.

Hon, K. L. (2013). Severe respiratory syndromes: Travel history matters. *Travel Medicine and Infectious Disease*, *11*(5), 285–287.

Jamal, T., & Higham, J. (2021). Justice and ethics: Towards a new platform for tourism and sustainability. *Journal of Sustainable Tourism, 29*(2–3), 143–157.

Jernsand, E. M., Kraff, H., & Mossberg, L. (2015). Tourism experience innovation through design. *Scandinavian Journal of Hospitality and Tourism, 15*(supl), 98–119.

Karpen, I. O., Bove, L. L., & Lukas, B. A. (2012). Linking service-dominant logic and strategic business practice: A conceptual model of a service-dominant orientation. *Journal of Service Research, 15*(1), 21–38.

Khan, M. J., Chelliah, S., Khan, F., & Amin, S. (2019). Perceived risks, travel constraints and visit intention of young women travelers: The moderating role of travel motivation. *Tourism Review, 74*(3), 721–738.

Lew, A. A., Cheer, J. M., Haywood, M., Brouder, P., & Salazar, N. B. (2020). Visions of travel and tourism after the global COVID-19 transformation of 2020. *Tourism Geographies, 22*(3), 455–466.

Lockett, A., Thompson, S., & Morgenstern, U. (2009). The development of the resource-based view of the firm: A critical appraisal. International *Journal of Management Reviews, 11*(1), 9–28.

Menon, J. (2020). COVID-19 in East Asia: Impacts and Response. *Thailand and the World Economy, 38*(2), 119–127.

Nepal, S. K. (2020). Adventure travel and tourism after COVID-19–business as usual or opportunity to reset?. *Tourism Geographies, 22*(3), 646–650.

Ocheni, S. I., Agba, A. O., Agba, M. S., & Eteng, F. O. (2020). COVID-19 and the tourism industry: Critical overview, lessons and policy options. *Academic Journal of Interdisciplinary Studies, 9*(6), 114.

Polyzos, S., Samitas, A., & Spyridou, A. E. (2021). Tourism demand and the COVID-19 pandemic: An LSTM approach. *Tourism Recreation Research, 46*(2), 175–187.

Rastegar, R., Higgins-Desbiolles, F., & Ruhanen, L. (2021). COVID-19 and a justice framework to guide tourism recovery. *Annals of Tourism Research, 91*, 103161.

Rothe, C., Schunk, M., Sothmann, P., Bretzel, G., Froeschl, G., Wallrauch, C., … Hoelscher, M. (2020). Transmission of 2019-nCoV infection from an asymptomatic contact in Germany. *New England Journal of Medicine, 382*(10), 970–971.

Shedroff, N. (2001). *Experience design 1*. Indianapolis, Indiana: New Riders Publishing.

Tu, Q. & Liu, A. (2014). Framework of smart tourism research and related progress in China. In *International Conference on Management and Engineering (CME 2014)*, pp. 140–146. Lancaster, Pennsylvania: DEStech Publications.

Tussyadiah, I. P. (2014). Toward a theoretical foundation for experience design in tourism. *Journal of Travel Research, 53*(5), 543–564.

Wetter-Edman, K., Sangiorgi, D., Edvardsson, B., Holmlid, S., Grönroos, C., & Mattelmäki, T. (2014). Design for value co-creation: Exploring synergies between design for service and service logic. *Service Science, 6*(2), 106–121.

Xiang, Z., Schwartz, Z., & Uysal, M. (2015). What types of hotels make their guests (un) happy? Text analytics of customer experiences in online reviews. In I. Tussyadiah & A. Inversini (Eds.), *Information and Communication Technologies in Tourism 2015: Proceedings of the International Conference* in Lugano, Switzerland, February 3–6, 2015, pp. 33–45. Cham, Switzerland: Springer.

Zuboff, S. (2015). Big other: Surveillance capitalism and the prospects of an information civilization. *Journal of Information Technology, 30*(1), 75–89.

# 38 Reconstructing tourism development in China

## The role of the internet industry

*Lina Zhong and Mengyao Zhu*

## Introduction

China's Internet industry has made a world-renowned leap in a short period and has taken on unique Chinese characteristics. China's Internet industry features a large scale, high growth rate, and great potential. According to the China Internet Network Information Center (CNNIC, 2021), by December 2020, Internet users in China had reached 989 million, with an Internet penetration rate of 70.4%. The number of mobile phone users stood at 989 million. Nearly one billion Internet users constitute the world's most extensive Internet user group. The growth rate in Internet applications such as online shopping, live streaming, and payment is remarkable. China has seen the development of Internet Platform Companies (IPCs) such as Alibaba, Tencent, Baidu, and NetEase. These companies have competed fiercely around various segments, extending their ecology and enhancing their core competencies. They have gradually become dominant in the market over the past decade (Zeng & Mackay, 2019).

In this context, Chinese IPCs are actively transforming and upgrading through platform business patterns, shifting from focusing on the consumer Internet to developing the industrial Internet. As the behavioural habits of Chinese consumers have become highly digital, this transformation has helped tourism companies accumulate data from tourists for product design and decision-making. There is no doubt that platform businesses help small and medium-sized enterprises gain more profits (Li et al., 2019). Moreover, the platform business pattern application facilitated communication between different stakeholders in the tourism industry. It enhanced the relationship between consumers and suppliers, providing favourable digital opportunities for Chinese companies (Candelon et al., 2019).

The "Internet + Tourism" model has been one of the driving forces behind China's regional economic development (Ding & Huang, 2021). As ICTs integrated with tourism and hospitality (Buhalis & Law, 2008), tourist experiences and the industry were affected.

Internet applications from the consumer side to the supply side have become a trend. The industrial Internet provided essential support for industries' intelligent upgrading and local economies (Chinese Academy of Engineering, 2020). The consumer Internet makes people's travel experiences more convenient by applying mobile terminals to provide services. The industrial Internet is about digitalising the supply side of tourism services with the help of lower-cost sensors, data storage, and faster data analysis capabilities. This improved the quality and efficiency of tourism service supply, thus meeting the deep-seated personalised needs of travellers. Moreover, it also comprehensively increased the efficiency of the tourism industry and reshaped the tourism experience.

DOI: 10.4324/9781003291763-42

A typical example that could help us understand this impact and trend is the GO-Yunnan smart travel platform. This tourism project was led by Yunnan province's Department of Culture and Tourism. As a Chinese IPC, Tencent was responsible for the development and operation of the platform.

The generation of this smart travel platform was attributed to government demand because the local administration in Yunnan needed a tool to solve the unhealthy problems in the tourist market. The local government needs to govern the tourism market and create better experiences for travellers and improve public services. As an IPC, Tencent also faced transformation from the consumer Internet to the industrial Internet. Based on the needs of both parties, they started to have a connection. Yunnan province's Department of Culture and Tourism turned to Tencent to help digitally transform the local tourism industry. Then, the GO-Yunnan smart travel platform was born.

This smart travel platform provided the government with a better response to consumer demands and public service supply. It connected stakeholders such as the local government, hotels, travel agencies, and transportation companies in the tourism industry. In practice, the platform effectively helped build a new ecology of industry connectivity, directly contributing to the development of the local digital economy. It reflected the trend of Chinese Internet companies moving from the consumer Internet to the industrial Internet, affecting the development of the Chinese tourism industry.

How has the Internet industry reconstructed China's tourism industry? The answer is in the trends. The transformation and reconstruction were equal to using a platform pattern in tourism, which continued to explore digitalisation from the consumer Internet to the industrial Internet. This chapter will discuss the role of the Internet industry in restructuring China's tourism industry from the aspects of trends, issues, stakeholders, future opportunities, challenges, and catalysts. In the end, we propose some implications (Buhalis, D, 2020).

**Major trends and issues**

*Trends*

*Digital transformation of the supply side in the tourism industry*

The new generation of information technology has enabled linking people, things, and services in the tourism industry (Buhalis, D, 2020). The data on consumer behaviour generated by the development of the consumer Internet has become a new factor of production, changing the allocation of resources on the supply side. Technology is a crucial driver of digitalisation and data analysis. It facilitates the digital transformation of the tourism industry and improves its efficiency (Mourtzis et al., 2021). The integration of technology into the information management of the tourism industry will significantly enhance its management and make the information and data management platforms more intelligent (Li, 2019). The power of the industrial Internet has developed from value transformation to value creation, which improves the efficiency of the tourism industry. Tourism products become diverse, and the cycle of design and planning is effectively shortening.

Smart travel platforms, like Go-Yunnan, used Internet technology to connect tourism-related stakeholders such as the government, scenic areas, travel agencies, hotels, or other related services. Furthermore, this also improves efficiency and organisational performance (Enz, 2012). The digital ecology consisting of tourists, enterprises, and regulators was built. The platform pattern regrouped and reallocated destinations' resources and created an information-driven multi-supply system (Siwei & Fang, 2014). Thus, the tourism industry realised the supply side's digital and intelligent transformation and promoted the service supply standard.

### Reshaping the travel consumer experience

In terms of reshaping the tourism experience, information technology provides a personalised experience by establishing a value network based on consumer demand. Miniprogramme and tourism mobile applications with high-tech attributes provide convenience for travellers (Berry et al., 2002), saving time and reducing travel costs (Davis, 1989; Wu & Wang, 2005). The development of the mobile Internet has influenced travellers' willingness to continue using mobile tourism applications (Xu et al., 2019). Industry and technology are working to create a new sense of experience on the consumer side of tourism. Tourism companies, technology companies, and related stakeholders joined to create new and innovative products to satisfy tourism consumption.

New tourism types emerged, such as digital scenic areas (Zhu & Huang, 2013) and digital museums (Guo et al., 2021). Virtual tourism enabled visiting from offline to online and triggered visual enjoyment, thus further stimulating tourists' consumption behaviour for online tourism products and services (Melo et al., 2022). Using AR and VR technologies and systems, tourism destinations' culture and cultural heritage truly come to life (Loureiro et al., 2020). Integrating various techniques with the profound meaning of culture created many immersive scenes, shaping travellers' experiences. The appearance of immersive restaurants and games in tourism destinations greatly enriched the tourism product system.

From 2017, the Go-Yunnan smart travel platform was helping tourists buy tickets, identify local flowers and plants, and do many other things. Using this one-stop app or miniprogramme, tourists in Yunnan can enjoy the scenic spots and access route maps on their mobile terminals. Using the platform, they can easily find what they want. The services start before they enter the province and provides customised travel products.

### Empowering government to improve public services and governance

In China, Chinese IPCs were helping the government to provide highly sophisticated public and travel services by a modern approach such as building smart travel platforms. It was a path for the government to attract new visitors and give travellers a better experience. From another point of view, the form of platform business assisted the government to utilise data-driven approaches and provide better public services (Jiang et al., 2019). Moreover, it also helped governments improve governance capacity and solve urban issues such as scenic area management and traffic problems (Große-Bley & Kostka, 2021). The further development of smart travel platforms improves urban public administration. It strengthens the administrative and political accountability of the government (Liou, 2007).

With forward-thinking, the Chinese government continued to invest in the digitisation of the tourism industry, making the Internet an essential infrastructure for sustainable tourism development. This pattern laid an excellent foundation for digital economic growth. It promoted cooperation between local government and IPCs, efficiently supporting the government's digitalisation process.

For the past few years, the persistence of the pandemic has made tourism digitalisation a new normal in China. As is well-known, the Chinese government set an excellent example of epidemic prevention and adopted active policies. They provided subsidies to support related tourism enterprises using robots and intelligent technologies to supply touchless services for creating a safe tourism environment. In China, tourists need to use QR codes to enter destinations and other public places such as hotels, shopping malls, subways, trains, and airports. With cloud computing, big data, and the Internet of Things (Ye et al., 2020), tourism officials can quickly trace and contact potentially infected tourists, thus improving their ability. Therefore, it enabled governments to provide better public services and governance.

## Issues

The tourism industry is connected with many other industries. The enterprises are also equally heterogeneous (Mare & Graham, 2013). Different tourism enterprises are at various stages of development in the industry chain. They had different degrees of digitalisation and operational approaches. So it was not easy for the government to combine all stakeholders in one tourism industry chain. If one refused or had difficulties with technical transformation, the whole chain may not have a complete economic cycle. The small and medium-sized tourism enterprises may face difficulties in affording the cost of purchasing technology devices to reconfigure the organisation. In addition, some enterprises clung to traditional models in terms of innovative thinking and are unwilling to respond to digital transformation or upgrading guidance. Some of them may lack an understanding of the value of data sharing, thus affecting the ecological restructuring of the industry.

## Stakeholders and catalysts

As the Internet industry reshapes China's tourism industry, it is also the path of the consumer Internet moving towards the industrial Internet through the accumulated momentum of the past. The process of digitisation and intelligence in China's tourism industry results from all parties in society (Bakos, 1998). The digital marketplace connects individuals or organisations on the demand and supply sides through platforms. With the development of smart tourism Internet platforms, a wide range of travel services created a larger market, higher industry efficiency, and closer connectivity between stakeholders through the Internet. The stakeholders were subjects and driving factors. Technology, policy support, and consumer upgrades were mainly critical catalysts.

In China, the government always played an important role in reform and local development. They actively made policies and adopted different approaches to providing better public services for local people and tourists. In terms of technological factors, technology enhances the capabilities and satisfies the needs of each subject in the industry chain. Governments and companies gather consumer consumption and behavioural data to empower the industry's supply-side and enhance integrating resources. The supply side of tourism resources had been integrated through digitisation. Moreover, the government's public services and digital governance capabilities were enhanced. Also, the transformation of travel-related platform enterprises from the consumer Internet to the industrial Internet had gone one step further.

Due to the upgrading of consumption, tourists' needs have shifted from functional consumption to emotional consumption. Therefore, the traditional way cannot offer a good tourism experience. The development of the Internet industry has led to the digitisation of traditional and emerging industries. It has provided more convenient, diverse, and personalised tourism products. Tourists will gain detailed services that meet their consumption needs.

**Opportunities and challenges**

In the future, China's internet industry will continue to drive technological innovation and industrial change based on the development of new infrastructures. It has further promoted digitalisation and intelligence on the supply and consumption sides of the tourism industry. The government, enterprises, and tourists will continue to form a strong triangular value creation network. The tourism industry will benefit from intelligence and digitalisation (Filieri et al., 2021). Digital governance will facilitate the transformation of government functions from regulation to service, and data sharing between merchants will drive the process of standardisation. New technologies such as 5G will be thoroughly combined with new energy and blockchain to create unique consumption scenes in the future. For cultural tourism, digitalisation will also impact the supply and demand side of the industry (Ammirato et al., 2021).

Opportunities and challenges will always coexist. Firstly, the development of new technologies will challenge ethical issues. When artificial intelligence and big data empowered the construction of smart tourism destinations, it also brought many legal and ethical challenges (Cain et al., 2019) on the Internet. For example, cyber security issues (Braun et al., 2018) and data privacy issues (Kabadayi et al., 2019) must be prevented. Furthermore, various cities in China have been actively building smart cities in recent years. Smart tourism has become an essential part of smart city construction (Khan et al., 2017). Therefore, how to connect the development of the local industrial Internet with the destination's original smart tourism has also become one of the issues that local tourism officials should consider. It will be the key to realising the integration of local tourism resources and the digital transformation of local platform enterprises.

**Implications**

The rapid development of China's Internet industry has been made possible by promoting policies and the strong support of local governments (Yang et al., 2017). Therefore, local governments should actively make policies to promote the digitisation of the supply side, such as scenic areas, hotels, restaurants, and transportation. Furthermore, local governments should enhance the digitisation of the infrastructure in tourism destinations and provide policy support for region-wide connectivity.

The government should actively create an environment to develop new technology and promote platform patterns in the tourism industry when facing increased competition. They can help industries and enterprises accelerate digital upgrades by helping them strengthen their technical foundation to capture the next wave of business growth. Tourism officials need to cooperate with enterprises to enhance the destination's image and brand. To provide a better public and travel service quality for tourists, constructing infrastructures such as 5G networks and big data centres should also be considered.

Tourism companies need to clarify the position of enterprises in the tourism industry to transform from the consumer Internet to the industrial Internet. Actively promoting the digital upgrading of the supply side of the enterprise and giving full play to the advantages of the platform business model is essential. Travel-related IPCs should learn to integrate tourism resources and develop more profound cultures in tourism destinations. It is favourable for stakeholders to share data with subjects in the tourism industry chain and uphold the value of co-creation, which will optimise their product, innovation, and design. In this way, the digital process in the tourism industry may be accelerated from the consumer side to the supply side.

**Major contributions**

The unique context and speed of Internet development in China have received sustained global attention. This chapter has briefly introduced the GO-Yunnan smart travel platform. This is a representative case of the Internet's reshaping of China's tourism industry, which draws out the current trends. We have provided further insights into the new trends, issues, opportunities, and challenges. Furthermore, we explained how the Internet industry impacts tourism in China. At the same time, we outlined the driving factors in the process, clarifying the relationships between stakeholders and catalysts that drive change. Finally, by interpreting the impact of China's Internet industry and digital development in the tourism industry, we offered insights and suggestions for government and enterprises to achieve digital transformation.

**References**

Ammirato, S., Felicetti, A. M., Linzalone, R., & Carlucci, D. (2021). Digital business models in cultural tourism. *International Journal of Entrepreneurial Behavior and Research*. https://doi.org/10.1108/ijebr-01-2021-0070

Bakos, Y. (1998). The emerging role of electronic marketplaces on the Internet. *Communications of the ACM, 41*(8), 35–42. https://doi.org/10.1145/280324.280330

Berry, L. L., Seiders, K., & Grewal, D. (2002). Understanding service convenience. *Journal of Marketing, 66*(3), 1–17. https://doi.org/10.1509/jmkg.66.3.1.18505

Braun, T., Fung, B. C. M., Iqbal, F., & Shah, B. (2018). Security and privacy challenges in smart cities. *Sustainable Cities and Society, 39*, 499–507. https://doi.org/10.1016/j.scs.2018.02.039

Buhalis, D., & Law, R. (2008). Progress in information technology and tourism management: 20 years on and 10 years after the Internet – The state of eTourism research. *Tourism Management, 29*(4), 609–623. https://doi.org/10.1016/j.tourman.2008.01.005

Buhalis, D. (2020). Technology in tourism- From Information Communication Technologies to eTourism and Smart Tourism towards Ambient Intelligence Tourism: A perspective article. *Tourism Review, 75*(1), 267–272. https://doi.org/10.1108/TR-06-2019-0258

Cain, L. N., Thomas, J. H., & Alonso, M. (2019). From sci-fi to SCI-fact: The state of robotics and AI in the hospitality industry. *Journal of Hospitality and Tourism Technology, 10*(4), 624–650. https://doi.org/10.1108/jhtt-07-2018-0066

Candelon, F., Yu, C., & Wang, J. (2019). *Get Ready for the Chinese Internet's Next Chapter*. https://www.bcg.com/en-cn/publications/2019/get-ready-for-chinese-internet-next-chapter

Chinese Academy of Engineering. (2020). *Industrial Internet*. Singapore: Springer. https://doi.org/10.1007/978-981-15-7490-0

CNNIC. (2021). *The 47th China Statistical Report on Internet Development*. Beijing: CNNIC.

Davis, F. D. (1989). Perceived usefulness, perceived ease of use, and user acceptance of information technology. *Mis Quarterly, 13*(3), 319–340. https://doi.org/10.2307/249008

Ding, R. J., & Huang, M. (2021). The spatial difference of "internet plus tourism" in promoting economic growth. *Sustainability, 13*(21), Article 11788. https://doi.org/10.3390/su132111788

Enz, C. A. (2012). Strategies for the implementation of service innovations. *Cornell Hospitality Quarterly, 53*(3), 187–195. https://doi.org/10.1177/1938965512448176

Filieri, R., D'Amico, E., Destefanis, A., Paolucci, E., & Raguseo, E. (2021). Artificial intelligence (AI) for tourism: An European-based study on successful AI tourism start-ups. *International Journal of Contemporary Hospitality Management, 33*(11), 4099–4125. https://doi.org/10.1108/ijchm-02-2021-0220

Große-Bley, J., & Kostka, G. (2021). Big data dreams and reality in Shenzhen: An investigation of smart city implementation in China. *Big Data & Society, 8*(2). https://doi.org/10.1177/20539517211045171

Guo, K. X., Fan, A. L., Lehto, X., & Day, J. (2021). Immersive digital tourism: The role of multisensory cues in digital museum experiences. *Journal of Hospitality & Tourism Research*, Article 1096 3480211030319. https://doi.org/10.1177/10963480211030319

Jiang, J. Y., Meng, T. G., & Zhang, Q. (2019). From Internet to social safety net: The policy consequences of online participation in China. *Governance-an International Journal of Policy Administration and Institutions*, *32*(3), 531–546. https://doi.org/10.1111/gove.12391

Kabadayi, S., Ali, F., Choi, H., Joosten, H., & Lu, C. (2019). Smart service experience in hospitality and tourism services A conceptualization and future research agenda. *Journal of Service Management*, *30*(3), 326–348. https://doi.org/10.1108/josm-11-2018-0377

Khan, M. S., Woo, M., Nam, K., & Chathoth, P. K. (2017). Smart city and smart tourism: A case of Dubai. *Sustainability*, *9*(12), Article 2279. https://doi.org/10.3390/su9122279

Li, S., Yu, C., Wang, J., Zhu, Y., Xiao, P., Cheng, X., ... Wang, Q. (2019). *Chinese Internet Economy White Paper 2.0: Decoding the Chinese Internet 2.0.*

Li, X. (2019). Research on tourism industrial cluster and information platform based on Internet of things technology. *International Journal of Distributed Sensor Networks*, *15*(7). https://doi.org/10.1177/1550147719858840

Liou, K. T. (2007). E-government development and China's administrative reform. *International Journal of Public Administration*, *31*(1), 76–95. https://doi.org/10.1080/01900690601052597

Loureiro, S. M. C., Guerreiro, J., & Ali, F. (2020). 20 years of research on virtual reality and augmented reality in tourism context: A text-mining approach [Article]. *Tourism Management*, *77*, 21. https://doi.org/10.1016/j.tourman.2019.104028

Mare, D. C., & Graham, D. J. (2013). Agglomeration elasticities and firm heterogeneity. *Journal of Urban Economics*, *75*, 44–56. https://doi.org/10.1016/j.jue.2012.12.002

Melo, M., Coelho, H., Goncalves, G., Losada, N., Jorge, F., Teixeira, M. S., & Bessa, M. (2022). Immersive multisensory virtual reality technologies for virtual tourism A study of the user's sense of presence, satisfaction, emotions, and attitudes. *Multimedia Systems*. https://doi.org/10.1007/s00530-022-00898-7

Mourtzis, D., Angelopoulos, J., & Panopoulos, N. (2021). Smart manufacturing and tactile internet based on 5G in industry 4.0: Challenges, applications and new trends. *Electronics*, *10*(24). https://doi.org/10.3390/electronics10243175

Siwei, Z., & Fang, F. (2014). Development of a model for a cluster-based virtual tourism supply chain. *Tourism Tribune*, *29*(2), 46–54.

Wu, J.-H., & Wang, S.-C. (2005). What drives mobile commerce? An empirical evaluation of the revised technology acceptance model. *Information & Management*, *42*(5), 719–729. https://doi.org/10.1016/j.im.2004.07.001

Xu, F., Huang, S., & Li, S. (2019). Time, money, or convenience: What determines Chinese consumers' continuance usage intention and behavior of using tourism mobile apps? *International Journal of Culture Tourism and Hospitality Research*, *13*(3), 288–302. https://doi.org/10.1108/ijcthr-04-2018-0052

Yang, K. H., Yuan, C. H., & Guo, J. J. (2017, Jul 09–13). B2B platform development in electronics manufacturing supply chain of China. In *Portland International Conference on Management of Engineering and Technology [2017 Portland International Conference on Management of Engineering and Technology (PICMET)]. Portland International Conference on Management of Engineering and Technology (PICMET)*, Portland, OR.

Ye, Q., Zhou, J., & Wu, H. (2020). Using information technology to manage the COVID-19 pandemic: Development of a technical framework based on practical experience in china. *JMIR Medical Informatics*, *8*(6), Article e19515. https://doi.org/10.2196/19515

Zeng, J., & Mackay, D. (2019). The influence of managerial attention on the deployment of dynamic capability: A case study of Internet platform firms in China. *Industrial and Corporate Change*, *28*(5), 1173–1192. https://doi.org/10.1093/icc/dty057

Zhu, Y., & Huang, Z. F. (2013). Application of virtual reality technology on digital scenic spot. In *Applied Mechanics and Materials 2nd International Conference on Measurement, Instrumentation and Automation (ICMIA 2013)*, Guilin, China.

# Index

Pages in *italics* refer to figures and pages in **bold** refer to tables.

For Product Safety Concerns and Information please contact our EU
representative GPSR@taylorandfrancis.com
Taylor & Francis Verlag GmbH, Kaufingerstraße 24, 80331 München, Germany

www.ingramcontent.com/pod-product-compliance
Lightning Source LLC
Chambersburg PA
CBHW080122220326
41598CB00032B/4919